MITOCHONDRIAL GENOMICS

MITOCHONDRIAL GENOMICS

Insights Into Dysfunction and Disease

ARUNA PAL

West Bengal University of Animal & Fishery Sciences, Kolkata, West Bengal, India

KESHAV K. SINGH

The University of Alabama at Birmingham, Birmingham, AL, United States

ELSEVIER

Elsevier
Radarweg 29, PO Box 211, 1000 AE Amsterdam, Netherlands
125 London Wall, London EC2Y 5AS, United Kingdom
50 Hampshire Street, 5th Floor, Cambridge, MA 02139, United States

ISBN: 978-0-443-26655-3

For Information on all Elsevier publications
visit our website at https://www.elsevier.com/books-and-journals

Publisher: Andre Wolff
Acquisitions Editor: Jessica Mack
Editorial Project Manager: Kritika Sharma
Production Project Manager: Jenela Priscy J
Cover Designer: Mark Rogers

Typeset by MPS Limited, Chennai, India

Contents

6. Mitogenomics and viral disease

7. Mitogenomics and bacterial disease

8. Mitogenomics and parasitic disease

9. Mitogenomics and reproduction

10. Mitogenomics and neurodegenerative disorder

15. Whole mitochondrial genome sequencing and its implication in diseases in animal models

16. Implication of gene editing in mitochondrial repair

CHAPTER

1

Overview of mitochondria

Introduction

The major basic factor for life is bioenergetics and nonmendelian inheritance. Mitochondria is not merely a cell organelle, it has some independent existence of its own, unlike other cell organelles. It has its own set of DNA with 37 genes coding for 13 polypeptides coding for essential functions, tRNA, rRNA and one connecting D-loop region.

It is quite interesting to note how diseases are caused by mitochondrial involvement:

1. Mutations in the mitochondrial gene
2. Copy number variation for mitochondrial gene
3. Heteroplasmy for mitochondria
4. Interaction with nuclear genes
   a. Involved with mitochondrial function
   b. Involved with oxidative phosphorylation, forming a complex with mitochondrial proteins
5. Through microRNA produced from mitochondria
6. Epigenetic regulations through methylation and acetylation

Mitochondria—an important cellular organelle

Mitochondria are rod-shaped, double membrane-bound cellular organelles which act as the power generators of the cell, where they convert oxygen and nutrients into adenosine triphosphate (ATP). ATP is the chemical energy regarded as the 'currency' of the cell, facilitating the cell's metabolic operations. This is processed through aerobic respiration mean respiration in the presence of oxygen. Animals breathe oxygen for the conduction of aerobic respiration. Aerobic respiration is the prime requirement for higher animals. In the absence of mitochondria (singular, mitochondrion), the existence of higher animals would not survive, as in that case, their cells would have acquired energy from anaerobic respiration (in the absence of oxygen), a much less effective process than aerobic respiration. Mitochondria allow cells to produce 15 times more ATP compared to other processes. It is noteworthy that complex animals, including humans, need large amounts of energy in order to exist.

Mitochondrial Genomics
DOI: https://doi.org/10.1016/B978-0-443-26655-3.00011-1

Architecture of mitochondria

What is a mitochondria?

Mitochondria word is composed of two Greek words: 'mitos' means 'thread' and 'chondros' means 'granule'. It is regarded as the 'powerhouse' of the cell since most of the energy is produced inside the mitochondria. Mitochondria was first discovered by Albert von Kolliker in 1857, who was a physiologist. In 1886, Richard Altman named it 'bioblasts' (life germs), which was later termed as 'mitochondria' by Carl Benda.

Structure of mitochondria

The mitochondria consist of the outer membrane, inner membrane, cristae, matrix, mitochondrial DNA (mtDNA), and ribosomes, schematically presented in Fig. 1.1.

Outer mitochondrial membrane

The outer mitochondrial membrane contains porin proteins and is involved in the movement of ions in and out of mitochondria. It also possesses enzymes responsible for fatty acid elongation and adrenaline oxidation.

Intermembrane space

The space present between the external and inner membranes of mitochondria is called the intermembrane space. It plays an important part in the regulation of respiration.

Inner mitochondrial membrane

The inner mitochondrial membrane consists of several enzymes. It comprises ATP synthase, which generates ATP, and transport proteins that regulate the passage of molecules into and out of the matrix. Matrix is the area within the inner mitochondrial membrane is referred to as the matrix.

It's the main point of aerobic respiration, the Krebs cycle or tricarboxylic acid and adipose acid cycles. It contains enzymes, RNA, DNA, and ribosomes.

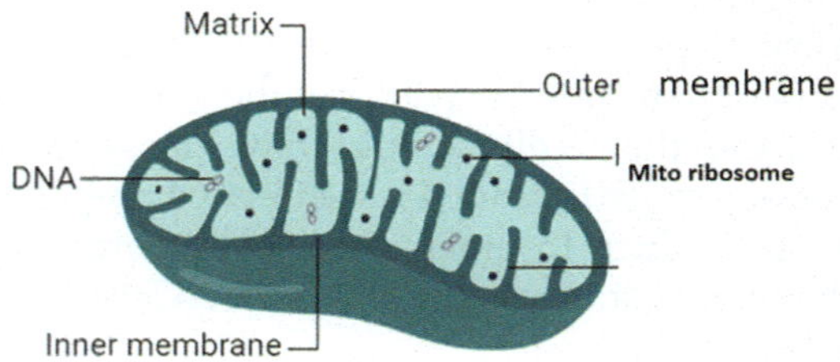

FIGURE 1.1 A diagram of a mitochondrion.

Cristae: In contrast to the outer membrane, the inner membrane has many folds referred to as cristae. It imparts a wrinkled appearance to the inner membrane. It increases the surface area for the chemical responses to take place, similar as oxidative phosphorylation.

mtDNA: Mitochondria has its own DNA known as mtDNA, hence it's a tone replicating organelle. mtDNA is small, indirect DNA present transferred from the parent cell to offspring cells. It contains genes for the conflation of enzymes involved in energy-producing pathways.

Ribosomes: Ribosomes in mitochondria are involved in the conflation of proteins. The number of mitochondria present in a cell depends upon the metabolic conditions of that cell, and may range from a single large mitochondrion to thousands of the organelles. Mitochondria, which are set up in nearly all eukaryotes, including plants, animals, fungi, and protists, are large enough to be observed with a light microscope and were first discovered in the 1800s. The names of the organelles was chased to reflect the way they looked to the first scientists to observe them, stemming from the Greek words for 'thread' and 'scrap'. For numerous times after their discovery, mitochondria were generally believed to transmit heritable information. It was not until the mid-1950s when a system for segregating the organelles completely was developed that the ultramodern understanding of mitochondrial function was worked out. The elaborate structure of a mitochondrion is veritably important to the functioning of the organelle (see Fig. 1.1).

Each mitochondrion in a cell is encased by two distinct membranes, which separate the organelle into a narrow intermembrane space and a significantly larger internal matrix, both of which predominantly contain specific proteins. The external membrane of a mitochondrion contains numerous channels formed by the protein porin and acts like a sieve, filtering out molecules that are too big. Also, the inner membrane, which is largely sophisticated so that a large number of infoldings called cristae are formed, also allows only certain materials to pass through it and is much more selective than the external membrane. To ensure that only elements vital to the matrix are permitted entry, the inner membrane employs a set of transport proteins that exclusively facilitate transport. Together, the various compartments of a mitochondrion are suited to working in harmony to produce ATP in a complex multistep process.

Mitochondria are generally oblong organelles, which range in size between 1 and 10 μM in length and vary in number according to the metabolic activity. The organelles are relatively flexible, still and time-consuming studies of living cells have demonstrated that mitochondria change shape fleetly and move about in the cell nearly constantly. Movements of the organelles appear to be linked in some way to the microtubules present in the cell, and are presumably transported along the network with motor proteins. Accordingly, mitochondria may be organised into long travelling chains, packed tightly into fairly stable groups, or appear in numerous other conformations depending on the particular requirements of the cell and the characteristics of its microtubular network.

Electron microscopic structure of mitochondria (animal cell micrograph)

Fig. 1.2 displays a digitised image of the mitochondrial network in the ovarian tissue, as observed under a fluorescence optical microscope. The intricate interconnected

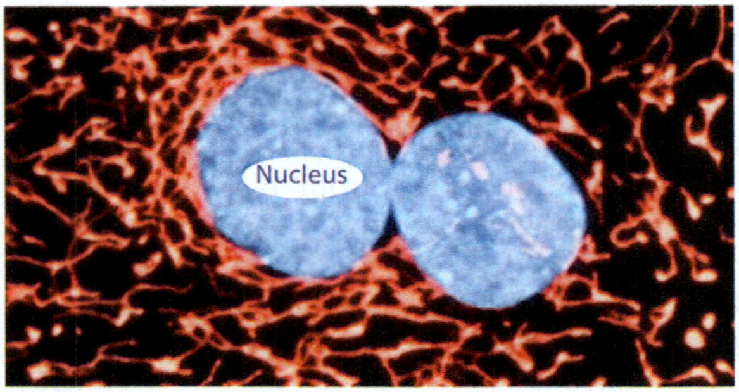

FIGURE 1.2 A fluorescence microscopy image shows a cell containing two prominent nuclei surrounded by an extensive mitochondrial network.

network is marked with a synthetic dye called MitoTracker Red (red fluorescence), which localises in the respiring mitochondria of cultured live cells. The uncommon twin nuclei in this cell were counterstained with a blue dye (cyan fluorescence) to indicate their central position relative to the mitochondrial network. Fluorescence microscopy is a crucial instrument employed by scientists to investigate the architecture and functionality of intra cellular organelles.

Staining of mitochondria

In the above Fig. 1.2, the mitochondria are stained red through MitoTracker red dye. The detailed protocol for staining is described elsewhere later in this chapter for assessment of their various biological functions.

Isolation of mitochondria

Most of the methods for isolation of mitochondria are based on differential centrifugation, a two-step centrifugation conducted at low speed for the removal of intact cells, cell and tissue debris, and nuclei from whole cell extracts. The next step is high-speed centrifugation to concentrate mitochondria, which is used for the separation from other organelles. However, the methodology for the disruption of cells and tissues varies. Moreover, density gradient centrifugation or affinity purification of the organelle is employed for further purification of mitochondria or separation of different populations of the organelle.

The protocols for the isolation of mitochondria from different cells and tissues, as well as approaches for the assessment of the purity and integrity of isolated organelles, are described below.

Methods

I. Isolation of mitochondria from rat brain—a case study

There existed a high level of heterogeneity among mitochondria from even a single region of the brain, as per their morphological, histochemical, and enzymatic characteristics.

Most methods are designed for the isolation of three distinct populations of mitochondria from rat brain:

1. nonsynaptic mitochondria, also known as 'free mitochondria' (FM);
2. synaptosomal mitochondria (synaptic), that be further subdivided into two fractions based on sedimentation properties, heavy (HM) and light (LM).

Synaptosomal mitochondria are responsible for the regulation of neurotransmitter release and synaptic vesicle formation. On the contrary, nonsynaptic mitochondria are derived from multiple cell types and from neuronal soma and are responsible for microRNA regulation and energy production.

Rationale

The initial material for this preparation is usually a dissected rat or mouse brain, sometimes pooled. Crude mitochondria are isolated from brain extracts through the process of differential centrifugation. Different populations of mitochondria are isolated from crude mitochondria with the help of discontinuous gradients (sucrose, Ficoll, or metrizamide).

Materials, equipment, and reagents

Reagents: The pH of all solutions is checked daily.
Solution A: Sucrose 0.32 M, EDTA-K^+ 1.0 mM, Tris-HCl 10 mM, pH 7.4
Solution B: Sucrose 0.32 M, EDTA-K^+ 50 µM, Tris-HCl 10 mM, pH 7.4
Solution C: Tris-HCl 6 mM, pH 8.1
Solution D: Mannitol 0.24 M, Sucrose 60 mM, EDTA-K^+ 50 µM, Tris-HCl 10 mM, pH 7.4
Solution E: Ficoll 3% (w/w), Mannitol 0.12 M, Sucrose 30 mM, EDTA-K^+ 25 µM, Tris-HCl 5 mM, pH 7.4
Solution F: Mannitol 0.22 M, Sucrose 0.07 M, Tris-HCl 50 mM, EDTA 1 mM, pH 7.2
Reagents for gradients

- Gradient I:
- Ficoll 7.5% (w/w) in solution B
- Ficoll 12% (w/w) in solution B
- Gradient II:
- Ficoll 4.5% (w/w) in solution D
- Ficoll 6% (w/w) in solution D

Equipment

- For tissue homogenisation
- Potter-Elvehjem homogeniser consisting of a large clearance (0.1–0.15 mm, volume 10 mL) teflon pestle driven into a glass vessel mechanically (Braun S homogeniser)
- For differential centrifugation
- Centrifuge: Beckman J2–21 or Sorvall RC-5B
- Fixed Angle Rotor: Beckman JA-20 or Sorvall SS-34
- Tubes: Polyallomer thick-wall, 10 mL
- For gradients

- Ultracentrifuge: Sorvall OTD65B or Beckman L5–50
- Swinging bucket rotor: AH-650 or SW 50.1
- Tubes: Polyallomer thin-wall, 5 mL

Protocols

Preparation of two-step discontinuous gradients. Preparation of the solutions for gradients is done just prior to use and kept on ice. Similarly, gradients are also prepared prior to use and kept in the refrigerator.

Gradient I:

Put 1.9 mL of the 12% Ficoll solution in a 5 mL ultracentrifuge tube.

Gently place 1.9 mL of the Ficoll 7.5% step on top of the 12% Ficoll step, using caution to avoid mixing of the steps.

Store at 4°C.

Gradient II:

Place 1.4 mL of the 6% Ficoll solution in a 5 mL ultracentrifuge tube.

Gently place 2.6 mL of the 4.5% Ficoll solution on top of the 6% Ficoll step, using caution to avoid mixing of the steps.

Store at 4°C.

Preparation of crude mitochondria from rat brain. All solutions used are at 0°C–4°C. Each step of homogenisation is conducted at 0°C–4°C.

The procedure described below is based on the preparation from one rat brain (ca. 120 mg). Rats are sacrificed as per the guidelines of ethical procedures.

Place rat brain in a prepared refrigerated box (0°C–4°C) within 15 seconds of euthanization and separate hemicortexes from hippocampus and striatum.

Rinse and place the hemicortexes (freed from blood and debris) in 2.5 mL of solution A.

Homogenise the tissue in a teflon-glass homogeniser by five up-and-down passes of the pestle at 800 rpm for a maximum of 1 minute.

Transfer the homogenate to a prechilled centrifuge tube, resuspend residual cell extract in the homogeniser in 2.5 mL of solution A, and transfer this material to the centrifuge tube.

Centrifuge the homogenate by gradually increasing the centrifugation rate to 1000 × g over a period of 4 minutes, with centrifugation at 1000 × g for an additional 11 seconds. Decant the supernatant into a new centrifuge tube and maintain on ice.

Resuspend the pellet with 2 mL of solution A and homogenise the resuspended pellet at 300 rpm for 60 seconds and three passages of the pestle, and transfer the homogenate to a fresh centrifuge tube.

Resuspend residual cell extract in the homogeniser in 2.0 mL of solution A and transfer this material to the centrifuge tube containing the homogenate from step 7.

Centrifuge the homogenate as in steps 6–7 and pool the supernatant with the supernatant from the first round of homogenisation.

Carry out one more round of homogenisation of the pellet as in steps 8–9.

Centrifuge the three pooled supernatants at 15,000 × g for 20 minutes.

The pellet obtained is crude mitochondria.

Isolation of synaptonemal light and heavy fractions and free mitochondria from crude mitochondrial preparations

Discard the supernatant and resuspend the 'crude' mitochondrial pellet from step 12 above (which contains nonsynaptic and synaptic mitochondria) to a final volume of 0.7 mL in solution A, and followed by homogenisation at 150 rpm and three passages of the pestle.

Gently apply the suspension to gradient I using caution to avoid mixing the gradient steps.

Centrifuge in a swinging bucket rotor at 73,000 × g for 24 minutes. Myelin and synaptosomal mitochondria are isolated into two distinct bands. FM are isolated in the pellet.

To isolate synaptosomal mitochondria, eliminate the myelin band, and retrieve the synaptosomal band at the interface of the 7.5%–12% gradients using a Pasteur pipette, then transfer it to a centrifuge tube. Dilute the synaptonemal band with solution A to a final volume of 5 mL and centrifuge at 15,000 × g for 20 minutes.

Dilute the synaptonemal band with solution A to a final volume of 5 mL and centrifuge at 15,000 × g for 20 minutes.

To isolate FM, resuspend the pellet obtained from gradient I in solution F to a final volume of 0.6–0.8 mL. This is the final FM fraction.

To separate HM and LM fractions, reconstitute the pellet from step 5 in 5 mL of solution C and homogenise at 150 rpm for 2 minutes with four pestle strokes.

Transfer the homogenate to a prechilled centrifuge tube and centrifuge the suspension at 14,000 × g for 30 minutes.

Resuspend the pellet in 5 mL of solution C and centrifuge the suspension at 14,000 × g for 30 minutes.

Resuspend the pellet in solution E to a final volume 0.6 mL.

Gently apply the suspension to gradient II, using caution to avoid mixing.

Centrifuge at 10,000 × g for 30 minutes. The HM fraction is in the pellet, and the LM fraction is at the interface of the 4.5%–6% steps.

To obtain the LM fraction, discard the band at the interface of the 3%–4.5% steps. Transfer the LM fraction at the 4.5%–6% interface to a new centrifuge tube using a Pasteur pipette and dilute it with an equivalent volume of solution A.

Centrifuge the suspension at 15,000 × g for 30 minutes.

Discard the supernatant and resuspend the pellet in solution F to a final volume of 0.6–0.8 mL. This is the final LM fraction.

To obtain the HM fraction, resuspend the pellet obtained from gradient II in solution F to a final volume of 0.6–0.8 mL. This is the final HM fraction.

Store all fractions at −80°C.

Analysis and statistics

The protein content is quantified using the Lowry method (Lowry, Rosebrough, Farr, & Randall, 1951). The anticipated yields of mitochondrial protein from various fractions are around 5 mg/mL for FM, 1.50 mg/mL for LM, and 1.60 mg/mL for HM, which represent 10.8% (FM), 2.0% (LM), and 2.4% (HM) of the homogenate protein, respectively.

This approach facilitates the synthesis of highly pure mitochondrial fractions from the rat brain's cerebral cortex, appropriate for assessing respiratory chain enzymatic activity (Pallotti & Lenaz, 2007). Mitochondrial fractions are typically pulse sonicated five times for 10 seconds per minute at 150 W in an ice bath under nitrogen gas to guarantee nicotinamide adenine dinucleotide availability before measuring enzymatic activity (Pallotti et al., 1998). The cytochrome content can be assessed by the methodologies established by Vanneste (1966) and Nicholls (1976). Mitochondrial preparations can be evaluated for coenzyme Q10 content using reversed-phase HPLC analysis following extraction with methanol and light petroleum (Battino et al., 1995).

Important considerations for mitochondria

The mitochondrion is different from most other organelles owing to its own circular DNA (similar to the DNA of prokaryotes) and its independent mode of reproduction in which it is found; an apparent case of **endosymbiosis**. Scientists speculate that millions of years ago, small, free-living prokaryotes were engulfed, but not consumed, by larger prokaryotes, perhaps because they were able to resist the digestive enzymes of the host organism. Accordingly, a symbiotic relationship was evolved among the two organisms over time, the larger organism supplies the smaller with ample nutrients, and the smaller organism, in turn, gives ATP molecules to the larger one. With time, the larger organism developed into the eukaryotic cell and the smaller organism into the mitochondrion.

mtDNA is present in the matrix, which also houses several enzymes and ribosomes for protein synthesis. Numerous essential metabolic processes of cellular respiration are facilitated by enzymes that can diffuse across the mitochondrial matrix. The other proteins involved in respiration, including the enzyme that generates ATP, are embedded within the mitochondrial inner membrane. Infolding of the cristae dramatically increases the surface area available for hosting the enzymes responsible for cellular respiration.

Mitochondria resemble plant chloroplasts in that both organelles can generate energy and metabolites essential for the host cell. Mitochondria are the loci of respiration, producing chemical energy in the form of ATP by metabolising carbohydrates, lipids, and other substrates with the aid of molecular oxygen. Chloroplasts are exclusively present in plants and algae, serving as the principal locations for photosynthesis. These organelles function distinctly to transform solar energy into the production of essential organic nutrients, utilising carbon dioxide and water. Similar to mitochondria, chloroplasts possess their own DNA and can grow and replicate autonomously within the cell.

In the majority of animal species, mitochondria are predominantly inherited through the maternal line; however, emerging data indicate that, in exceptional cases, paternal inheritance of mitochondria may occur. A sperm typically has mitochondria in its tail, serving as an energy source for its extensive travel to the egg. Upon fertilisation, when the sperm adheres to the egg, the flagellum detaches. Therefore the sole mitochondria typically acquired by the young organism are derived from the egg supplied by its mother. Consequently, in contrast to nuclear DNA, mtDNA is not recombined each generation, leading to the assumption that it evolves at a more gradual pace, which is advantageous for the examination of human evolution. mtDNA serves as a tool in forensic research for the identification of corpses or body parts and has been associated with several hereditary disorders, including Alzheimer's disease and diabetes.

Roles of mitochondria: The primary roles of mitochondria are:

- Mitochondria serve as the cell's powerhouse and are the locus for ATP synthesis (see Fig. 1.3 and Fig. 1.4).
- It catalyses the conversion of ADP into ATP. ATP is referred to as the 'energy currency' and supplies energy for chemical reactions. An increased quantity of mitochondria within a cell correlates with elevated ATP production.

• It participates in thermogenesis inside brown adipose tissue via the electron transport chain.

• It is crucial in the process of apoptosis.

• It sequesters and stores calcium ions.

Thus we can summarise that mitochondria play central roles in energy metabolism, calcium transporters, cell proliferation and differentiation, free radical production, apoptosis and autophagy, ageing, signalling, immunity, growth, reproduction and others (see Fig. 1.5 and Fig. 1.6).

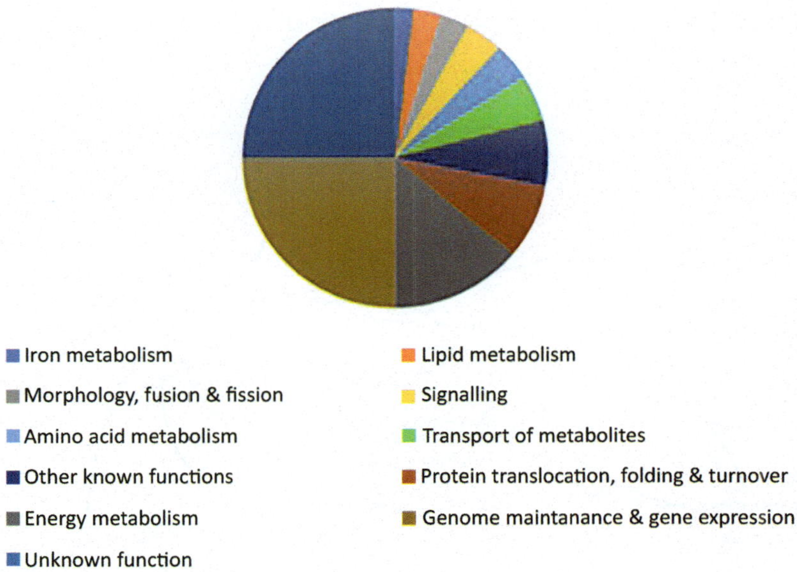

FIGURE 1.3 Multifaceted role of mitochondria.

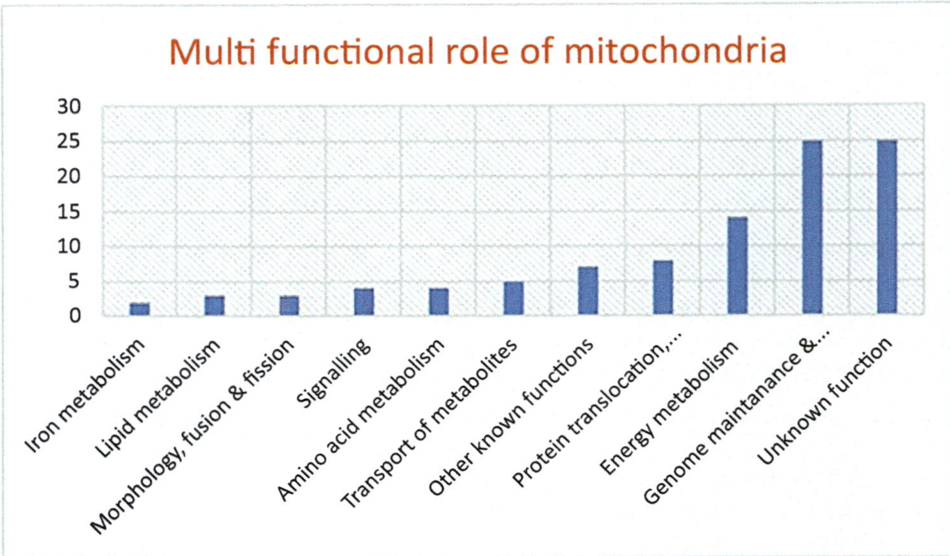

FIGURE 1.4 Functional role of mitochondria.

- ❑***Mitochondria is the power house of the cell providing energy.***
-

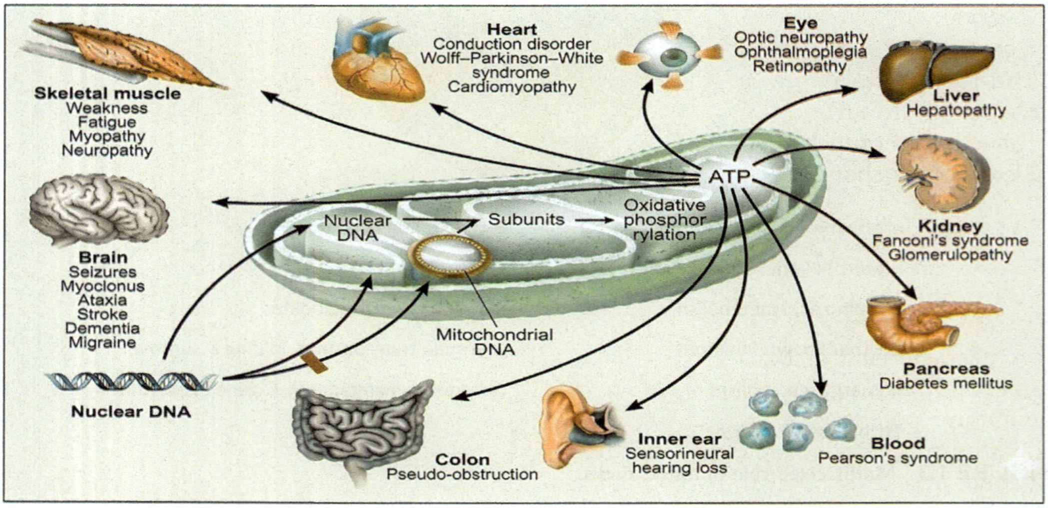

FIGURE 1.5 The tissues demanding high level of energy.

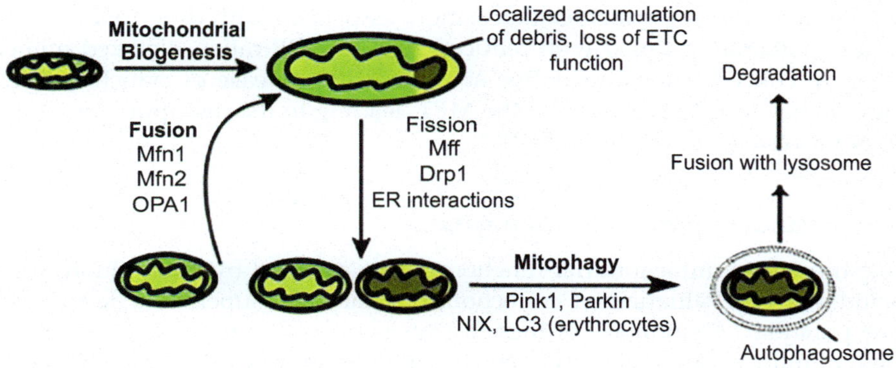

FIGURE 1.6 A schematic representation of the mitochondrial life cycle, indicating mitochondrial fusion, fission and mitophagy.

Mitochondrial functional assay

The impairment of mitochondrial function can be identified by many fluorescence-based tests, including:

1. Measurements of mitochondrial calcium,
2. Superoxide,
3. Mitochondrial permeability transition (MPT), and
4. Membrane potential.

We describe the following points:

1 Mitochondrial function changes in apoptosis
2 Overview of mitochondria function assays
3 Detection of dynamic changes in mitochondrial membrane potential and metabolism
4 End-point assays in mitochondrial membrane potential
5 Detection of mitochondrial superoxide production
6 Detection of mitochondrial calcium
7 Detection of changes in the mitochondrial permeability transition pore (MPTP)

Mitochondrial function changes in apoptosis

A characteristic of the initial phases of programmed cell death is the impairment of mitochondrial function. This mitochondrial disruption encompasses modifications in membrane potential, a critical aspect of mitochondrial integrity, and variations in the oxidation-reduction potential of the mitochondria. The inner mitochondrial membrane potential is essential in Ca^{2+} uptake and storage, reactive oxygen species (ROS) generation, detoxification, and most importantly, the synthesis of ATP by oxidative phosphorylation. Consequently, the depolarisation of the membrane serves as a reliable signal of mitochondrial malfunction, which is increasingly associated with drug toxicity. Alterations in membrane potential, coupled with reduced ATP to ADP ratios, elevated mitochondrial matrix calcium concentrations, oxidative stress, and the release of cytochrome c into the cytosol, are all believed to be linked to the MPT, leading to the disruption of ion and small molecule homoeostasis through the MPTP.

Summary of mitochondrial function assays

The impairment of mitochondrial function can be assessed by many fluorescence-based tests, including the evaluation of mitochondrial calcium, superoxide levels, MPT, and membrane potential. A summary of these assays is presented in Table 1 below. While the impairment of mitochondrial activity correlates with apoptosis, the aforementioned features are not exclusive to apoptotic cell death. Due to the nonspecificity of the previously described apoptosis-related mitochondrial function parameters and their variability among cell types, it is frequently beneficial to employ multiparametric assays that incorporate specific markers for the activation of distinct caspase proteases (see Fig. 1.7).

Visualisation of mitochondrial superoxide in live cells with specificity and ease of use, MitoSOX green possesses certain advantages as:

Highly selective for mitochondrial superoxide, not other reactive nitrogen species (RNS)

Can be imaged easily with a standard fluorescein isothiocyanate (FITC)/green fluorescent protein (GFP) fluorescence filter on a fluorescent microscope

Can be multiplexed easily with other popular live cell dyes

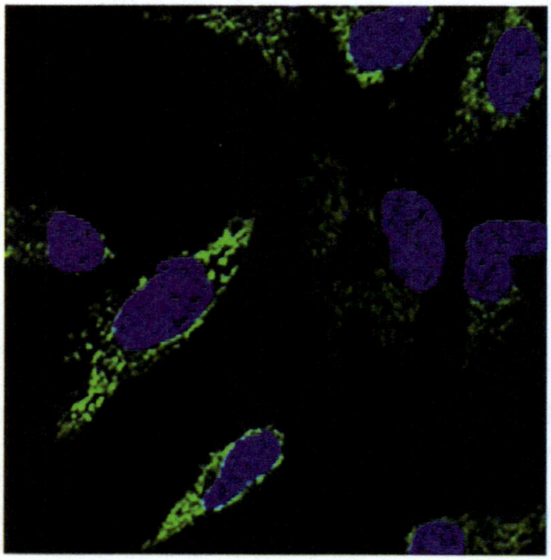

FIGURE 1.7 Microscopic view of mitochondria—with MitoSOX green for mitochondrial superoxide detection reagent.

Summary of mitochondria function assays.

	Mitochondria function assays			
	Membrane potential and metabolism	**Superoxide production**	**Calcium**	**Mitochondrial permeability transition (MPT)**
To be identified	Cells with healthy, metabolically active mitochondria. Two ways: 1. ***Detection of dynamic changes*** in relative mitochondrial membrane potential (reversible detection). 2. ***End-point assays*** for detecting relative mitochondrial membrane potential at a specific time point (irreversible detection).	Irreversible detection of mitochondrial superoxide production in live cells.	Irreversible detection of mitochondrial calcium in live cells.	Irreversible detection of the MPT subsequent to dysregulation of the MPT pore.
Basis of assay	Measure ***fluorescence*** of reagents that accumulate in active (healthy) mitochondria with intact membrane potentials.	Measure the increasing ***fluorescence of a superoxide detection*** reagent that accumulates in mitochondria in a membrane potential-dependent fashion, where it is selectively oxidised and subsequently binds nucleic acids.	Measure the increasing ***fluorescence of a calcium indicator*** with increasing calcium concentration in the mitochondria.	***Loss of mitochondrial calcein fluorescence*** due to MPT.
Fixability of the assay	Only a small selection of the endpoint assays for detecting mitochondrial membrane potential are fixable. None of the dynamic assays is fixable.	Not compatible with fixation.	Not compatible with fixation.	Not compatible with fixation.

Detection of dynamic changes in mitochondrial membrane potential.

	JC-1	MitoProbe JC-1 assay kit	MitoProbe $DiOC_2(3)$ assay kit
Principle	Active mitochondria exhibit a brighter red fluorescence signal compared to mitochondria with lower membrane potential, which fluoresce green. Changes in the red/green fluorescence signal ratio can be used to determine healthy versus depolarised mitochondria.		

Ex/Em (nm)	514/529 (monomer, green) 514/590 (aggregate, red)		485/497 (monomer, green) 485/650 (aggregate, far-red)
Suggested filters	Tetramethylrhodamine isothiocyanate (TRITC)	FITC and PE	FITC and Allophycocyanin
Instrument platform	Imaging microscopy	Flow cytometry	Flow cytometry
Sample type	Live cells	Live cells	Live cells
Compatibility with fixation	No	No	No
Format	5 mg	Kit contents: JC-1, DMSO carbonyl cyanide 3-chlorophenylhydrazone (CCCP) (a mitochondria membrane potential disrupter in DMSO), 10x PBS	Kit contents: $DiOC_2(3)$ in DMSO CCCP, (a mitochondrial membrane potential disrupter in DMSO)

We use dual emission, ratiometric reagents, and single-emission reagents. *PBS*, phosphate-buffered saline.

End-point assays in mitochondrial membrane potential.

	MitoTracker probes—imaging				
MitoTracker green free mitochondria (FM)	**MitoTracker orange CMTMRos**	**MitoTracker red CMXRos**	**MitoTracker red FM**	**MitoTracker deep red FM**	
Readout	Active mitochondria exhibit brighter fluorescence compared to apoptotic mitochondria.				
Ex/Em (nm)	490/516	550/580	579/599	581/644	644/665
Suggested filter	FITC	TRITC	Texas red	Cy5	Cy5
Instrument platform	Fluorescence microscopy microplate reader				
Sample type	Live cells	Live cells	Live cells	Live cells	Live cells

	MitoTracker probes—flow cytometry			
MitoTracker green free mitochondria (FM)	**MitoTracker red CMXRos**	**MitoTracker red FM**	**MitoTracker deep red FM**	
Principle	Active mitochondria exhibit brighter fluorescence compared to apoptotic mitochondria.			
Ex/Em (nm)	490/516	579/599	581/644	644/665

Suggested filter	FITC	Texas red	Cy5	Cy5
Instrument platform	Flow cytometry			
Sample type	Live cells	Live cells	Live cells	Live cells

Detection of mitochondrial superoxide production.

	MitoSOX red (MSR) reagent	MitoSOX green reagent
Readout	Fluorescence increases with increased superoxide concentration. Highly selective detection of superoxide in the mitochondria of live cells. Note that two excitation peaks may be observed for MSR dye, as it is excited at both 510 nm and 396 nm. While 510 nm will excite the superoxide oxidation product, it can also excite nonspecific products, so we recommend using a 396 nm excitation for more selective detection of mitochondrial superoxide when using MSR dye.	
Ex/Em (nm)	396/610	488/510 nm
Suggested filter	Custom filter, UV/violet excitation filter and RFP emission filter	FITC
Instrument platform	Fluorescence microscopy, flow cytometry, microplate reader, high-content analysis	Fluorescence microscopy, high-content analysis
Sample type	Live cells	Live cells

Increases in cellular superoxide production have been implicated in multiple disease states. This increase, which is generated as a byproduct of oxidative phosphorylation, provides another method to assess the cells' apoptotic state.

Detection of mitochondrial calcium.

	Rhod-2, acetoxymethyl reagent
Readout	Calcium indicator targeted to mitochondria increases fluorescence with Ca^{2+} concentration
Ex/Em (nm)	552/577
Suggested filter	TRITC
Instrument platform	Fluorescence microscopy, flow cytometry
Sample type	Live cells

Detection of changes in mitochondrial permeability transition pore.

	Image-IT LIVE mitochondrial transition pore assay	MitoProbes transition pore assay kit
Application	Method of measuring mitochondrial permeability transition pore opening.	

Readout	In healthy cells, the mitochondria remain brightly fluorescent until mitochondrial pore activation permits quenching of the fluorescence.	
Ex/Em (nm)	494/517 (calcein) 579/599 (MitoTracker red stain) 361/497 (Hoechst 33342)	494/517 (calcein)
Suggested filter	TRITC	FITC
Instrument platform	Fluorescence microscopy	Flow cytometry
Sample type	Live cells	Live cells
Compatibility with fixation	Not compatible with fixation and detergent	Not compatible with fixation and detergent

Detailed information about mitochondria function assays

JC-1 dye and MitoProbe JC-1 assay kit: The often utilised dye for assessing mitochondrial membrane potential is the cell-permeant JC-1 dye, employed for identifying apoptotic cells by flow cytometry and imaging techniques (Fig. 1). JC-1 dye demonstrates accumulation in mitochondria that depends on potential. As the concentration of dye within the mitochondria reaches a sufficient level, J-aggregates are formed, resulting in a fluorescence emission shift from the monomer (green, ~529 nm), which predominates at lower dye concentrations, to J-aggregates (red, ~590 nm) that arise with increasing dye concentration. Thus mitochondrial depolarisation is signified by a reduction in the red/green fluorescence intensity ratio. JC-1 dye is offered as an independent reagent or within the MitoProbe JC-1 assay kit, which includes the JC-1 dye, carbonyl cyanide 3-chlorophenylhydrazone (CCCP) (a mitochondrial membrane potential disruptor in DMSO), 10x phosphate-buffered saline (PBS), and DMSO (see Fig. 1.8).

The MitoProbe JC-1 assay kit was utilised to label cells, which were subsequently analysed via flow cytometry employing 488 nm excitation and 530 nm and 585 nm bandpass emission filters. Green indicates apoptotic cells (decreased mitochondrial membrane potential), while red signifies normal cells. MitoProbe $DiOC_2(3)$ assay kit, in which $DiOC_2(3)$ is another dye that can serve as a ratiometric probe for mitochondrial membrane potential. The mechanism resembles that of JC-1 dye, wherein the monomer displays green fluorescence at lower concentrations, indicative of a diminished mitochondrial membrane potential, while the dye aggregates in more active mitochondria result in a shift of emission towards the red spectrum. In the instance of $DiOC_2(3)$, the shift occurs further into the red spectrum than JC-1, exceeding 650 nm. The MitoProbe $DiOC_2(3)$ assay kit is designed for flow cytometry applications and contains the mitochondrial membrane potential disruptor, CCCP. Tetramethylrhodamine, methyl ester (TMRM) and MitoProbe TMRM assay kit, in which TMRM and its analogue, tetramethylrhodamine, ethyl ester (TMRE), are diminutive, cell-permeable dyes that concentrate in metabolically active mitochondria. If the cells are healthy and possess functional mitochondria, the signal is luminous. When the mitochondrial membrane potential is lost, the accumulation of TMRM and TMRE halts, resulting in a diminished or absent signal from the mitochondria.

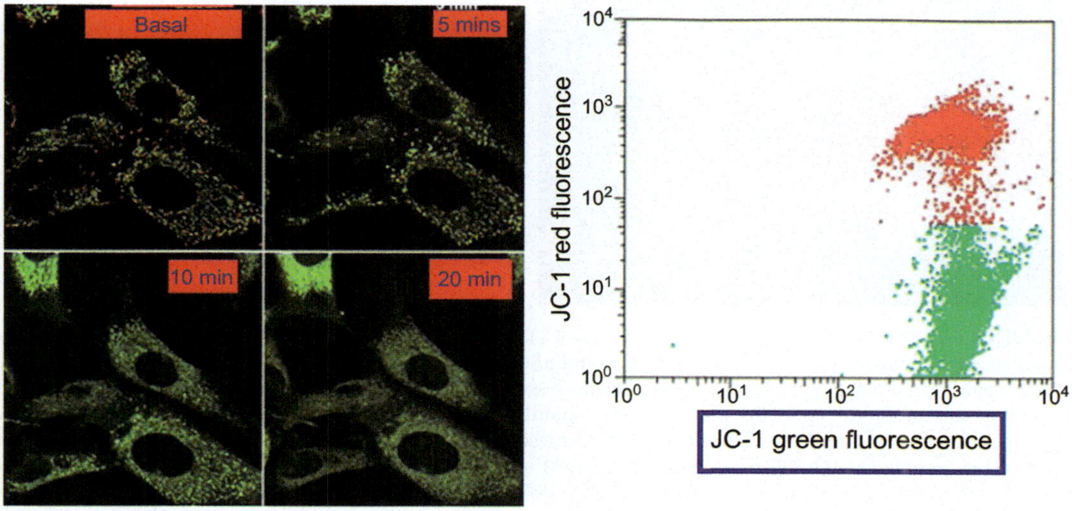

FIGURE 1.8 Assessment of mitochondrial membrane potential utilising JC-1 dye. NIH 3T3 fibroblasts labelled with JC-1 dye exhibit a gradual decline in red J-aggregate fluorescence and an increase in the cytoplasmic dispersion of green monomer fluorescence upon exposure to hydrogen peroxide. Images depict the identical field of cells observed prior to H_2O_2 treatment and at 5, 10, and 20 minutes posttreatment. *NIH*, National Institutes of Health.

The capability to actively monitor alterations in mitochondrial membrane potential is a notable advantage of this probe. TMRM and TMRE signals can be identified in live cells or isolated mitochondria using conventional and high-content fluorescence microscopy, microplates, or flow cytometry. TMRM is offered in many formats to accommodate your apoptotic assay requirements.

Single-emission reagents for detecting dynamic alterations in mitochondrial membrane potential via single-emission reagents. The MitoProbe TMRM assay kit for flow cytometry comprises TMRM and CCCP, a mitochondrial membrane potential disruptor, coupled with a comprehensive methodology for flow cytometry application (see Fig. 1.9).

MitoTracker probes: MitoTracker probes are diminutive (<1 kDa), cell-permeable, (see Fig. 1.10 mitochondrion-selective dyes featuring a thiol-reactive chloromethyl group that ensures the dye remains linked with the mitochondria postfixation. The probes establish covalent connections with mitochondrial thiols, hence they are suitable solely as endpoint assays for detecting mitochondrial membrane potential at the moment of loading in live cells and should not be utilised as sensors for dynamic alterations in mitochondrial membrane over time. **Influence of fixation on MitoTracker probes:** Although the MitoTracker signal can be stabilised postlabelling, it is highly recommended to utilise specific MitoProbe reagents for investigating mitochondrial membrane function in tests focused on mitochondrial structure. Only MitoTracker orange CMTMRos and MitoTracker red CMXRos have demonstrated a reliable disparity in signal attenuation when assessing mitochondrial membrane potential in control and treatment samples

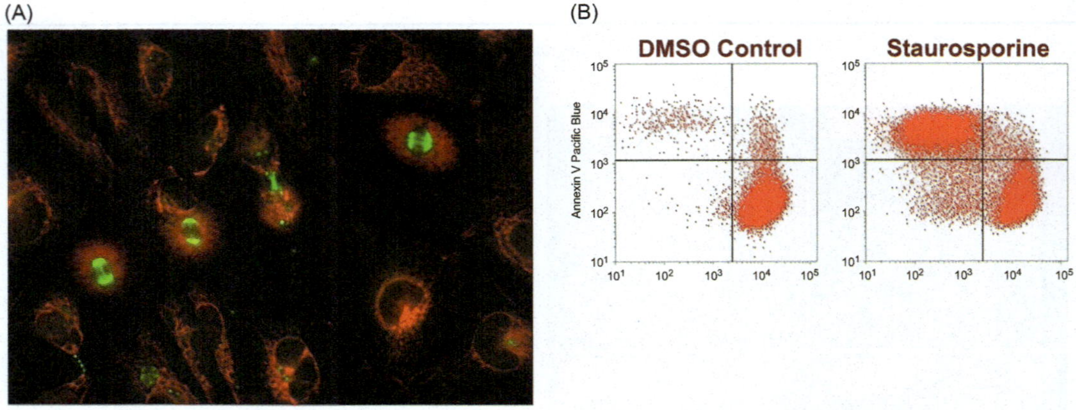

FIGURE 1.9 Tetramethylrhodamine, methyl ester (TMRM) utilised as a test for mitochondrial membrane potential in apoptosis. HeLa cells stained with tubulin tracker green and Image-iT TMRM reagent exhibit exceptional multiplexing performance and labelling specificity. The image depicts several live mitotic cells exhibiting microtubules organised into a mitotic spindle, with discernible microtubule filaments, alongside a cleavage furrow signifying the conclusion of cytokinesis. Jurkat cells, a human T-lymphocyte cell line, were subjected to treatment with DMSO (control) or 500 nM staurosporine for a duration of 2 hours. Cells were then stained with MitoProbe TMRM for 30 minutes at 37°C, followed by a wash and subsequent staining with Annexin V Pacific Blue conjugate. Staurosporine promoted apoptosis, leading to a heterogeneous cell population comprising healthy MitoProbe TMRM-positive cells and apoptotic, Annexin V Pacific Blue-positive, MitoProbe TMRM-low cells. **MitoProbe DiIC$_1$(5) fluorescent dye,** cationic carbocyanine dyes exhibit accumulation within cells in reaction to membrane potential. The MitoProbe DiIC$_1$(5) Kit is intended for flow cytometry applications and includes the far-red fluorescent DiIC$_1$(5) carbocyanine dye, as well as the mitochondrial membrane potential disruptor, carbonyl cyanide 3-chlorophenylhydrazone. At concentrations under 100 nm, the cationic DiIC$_1$(5) dye passively accumulates in the mitochondria of live cells exhibiting active membrane potentials. In contrast to DiOC$_2$(3), DiIC$_1$(5) dye displays a singular wavelength emission, with its fluorescence signal intensifying in the presence of more active mitochondria.

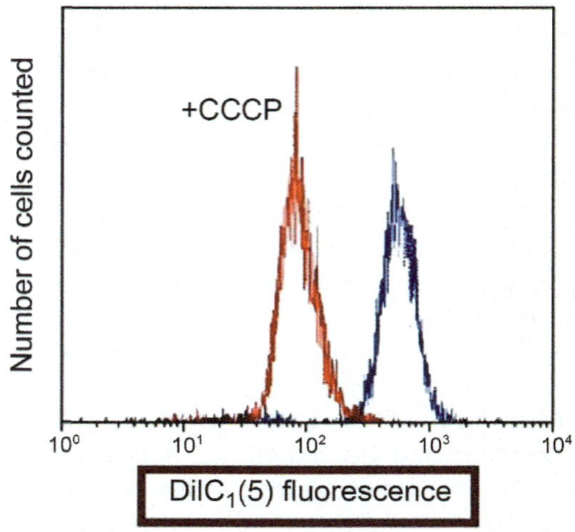

FIGURE 1.10 Assessment of alterations in mitochondrial membrane potential with the MitoProbe DiIC$_1$(5) assay kit for flow cytometry. Reduction in mitochondrial membrane potential seen by DiIC$_1$(5) fluorescence following the introduction of carbonyl cyanide 3-chlorophenylhydrazone (CCCP). Jurkat cells (human T-cell leukaemia) were stained with 50 nM DiIC$_1$(5) alone (*blue*) or in conjunction with 50 μM CCCP (*red*).

postfixation. These single-wavelength emission reagents are applicable to an experiment (see Figs. 1.11–1.18).

What methods can be employed to evaluate mitochondrial function? The respiratory control ratio is the most valuable overall indicator of function in isolated mitochondria. A high RCR signifies optimal function, whereas a low RCR typically denotes malfunction. The efficacy of RCR is contingent upon its intricacy: nearly any alteration in oxidative phosphorylation will affect RCR. **Origin of mitochondria:** Mitochondria are believed to have arisen from the engulfment of prokaryotes. The theory explaining the origin of mitochondria is known as the endosymbiotic theory. This idea posits that an aerobic, free-living prokaryotic cell was absorbed by a eukaryotic cell, leading to a symbiotic connection rather than its digestion. The prokaryotic cell supplied energy to the eukaryotic cell and ultimately evolved into an organelle. Conversely, the eukaryotic cell afforded space and shelter to the prokaryotic cell. This process spanned millions of years, with the emergence of prokaryotic cells or mitochondria serving as a defining characteristic of eukaryotic cells. This theory also elucidates the existence of DNA in mitochondria, distinct from nuclear DNA. The DNA found in mitochondria is believed to originate from the absorbed prokaryotic cell, residing within the cell, and is referred to as mtDNA. mtDNA is circular, akin to prokaryotic DNA. mtDNA is transmitted maternally to the subsequent generation, as sperm cells do not contain cytoplasm or mitochondria. New mitochondria are generated through binary fission of existing mitochondria. If all mitochondria are eliminated from a cell, that cell is incapable of regenerating mitochondria (see Figs. 1.19–1.22).

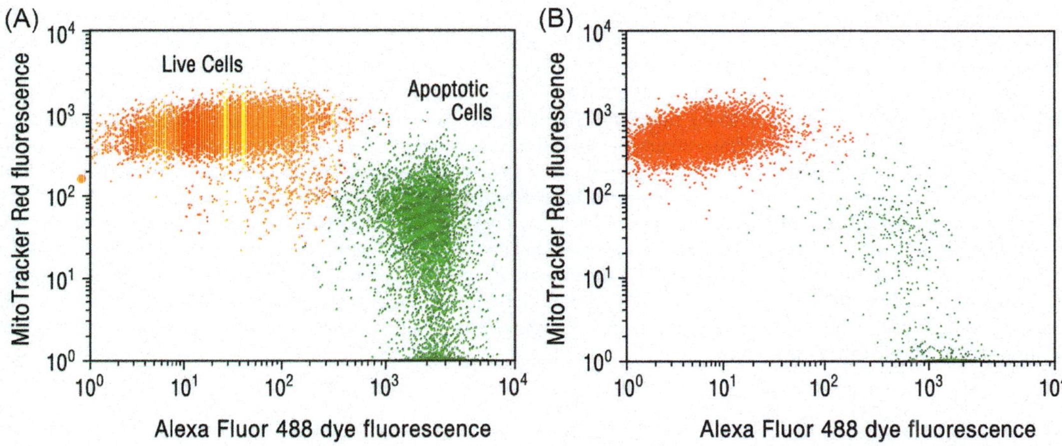

FIGURE 1.11 Detection of mitochondrial membrane potential using MitoTracker red CMXRos dye. Jurkat human T-cell leukaemia cells in complete media were (A) initially subjected to 10 µM camptothecin for 4 hours or (B) remained untreated. Subsequently, both cell populations were subjected to treatment with the reagents from the mitochondrial membrane potential/annexin V apoptosis kit and analysed using flow cytometry. Apoptotic cells exhibit increased reactivity for annexin V and diminished fluorescence of MitoTracker *red* dye compared to living cells.

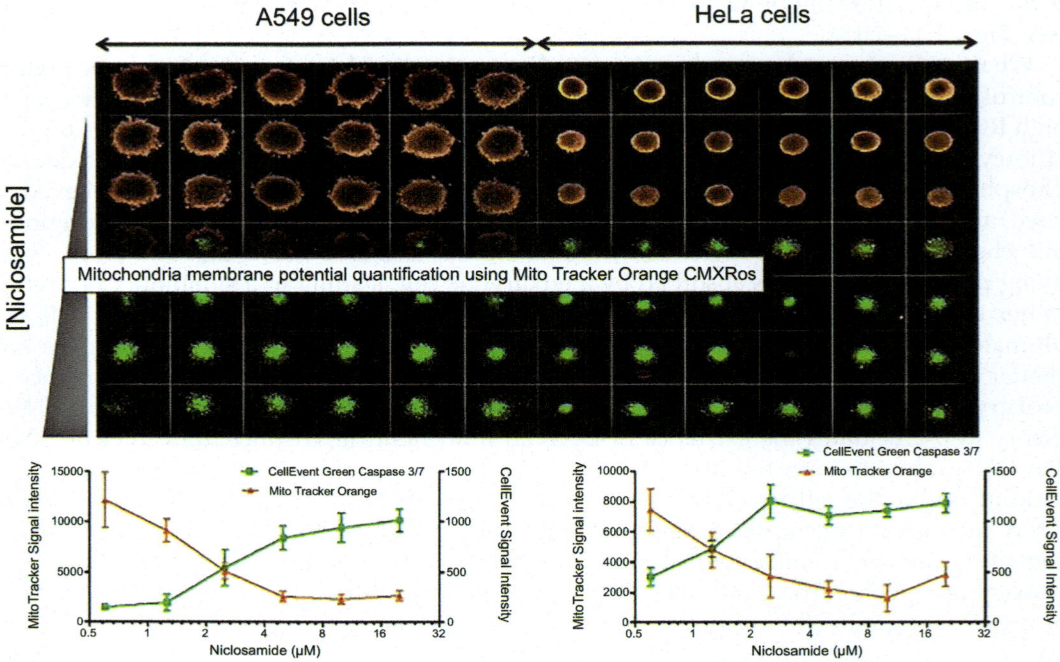

FIGURE 1.12 Quantification of mitochondrial membrane potential utilising MitoTracker orange CMTMRos. A549 or HeLa cells were cultivated in nunclon sphera plates in Grand Island Biological Company minimal essential medium (MEM) to generate spheroids over a period of two days. Samples were subjected to treatment with either DMSO vehicle or escalating concentrations of niclosamide for 24 hours, followed by labelling with 250 nM MitoTracker orange and 2.5 μM CellEvent caspase 3/7 green reagent for 30 minutes at 37°C under standard cell culture conditions. Confocal pictures were obtained with the Thermo Scientific CellInsight CX7 LZR high-content imaging system, subsequently analysed using HCS studio software V 2.0. Results indicate a niclosamide dose-dependent reduction in mitochondrial membrane potential (orange) and an increase in apoptotic cell death (green), both qualitatively (upper panels) and quantitatively as a dose-response relative to mean signal intensity in both channels (bottom graphs). **Assay for high-content mitochondrial health:** The HCS mitochondrial health kit was designed for concurrent quantitative assessment of two cellular health parameters—mitotoxicity and cytotoxicity—through high-content analysis inside a 96-well plate format. The MitoHealth stain accumulates in mitochondria of living cells in response to the mitochondrial membrane potential. Cytotoxicity is assessed using the Image-iT DEAD green viability stain. The Image-iT DEAD green viability stain exhibits a strong affinity for DNA, forming highly fluorescent and persistent dye-nucleic acid complexes; it remains nonfluorescent in the absence of DNA binding. *HCS*, high content screening.

Existence of circulating mitochondria in human and animal peripheral blood

It has been established that mitochondria reside in the cytoplasm of cells, where they produce ATP to facilitate cellular processes. We detected circulating mitochondria in the blood of humans and animals. Electron microscopy proved the presence of mitochondria in

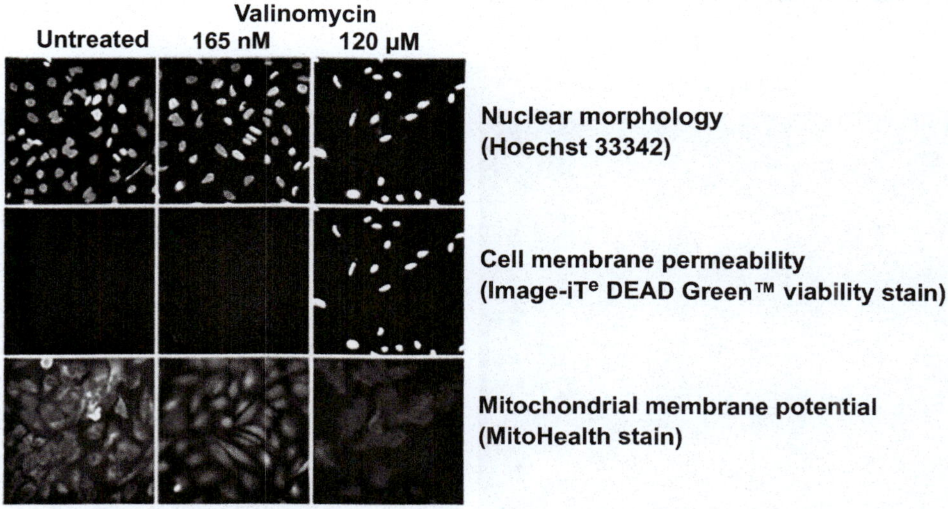

FIGURE 1.13 Assessment of mitotoxicity and cytotoxicity of valinomycin in HeLa cells utilising the HCS mitochondrial health kit. **MitoSOX red (MSR) and green indicators of mitochondrial superoxide:** MSR and green are fluorogenic reagents specifically engineered for the selective detection of superoxide within the mitochondria of live, healthy cells. The reagents are easily oxidised by superoxide, but not by other reactive oxygen species (ROS) or reactive nitrogen species producing systems, and superoxide dismutase inhibits the oxidation of the probe. The oxidation of MSR indicators by superoxide produces a fluorescence excitation peak at approximately 400 nm, which is not present in the excitation spectrum produced by other ROS. Consequently, fluorescence excitation at 400 nm with emission detection at approximately 590 nm ensures optimal differentiation of superoxide from other ROS. The MitoSOX green indicator is detectable using a conventional FITC/GFP filter combination. Selective detection of superoxide is achieved with 488 nm excitation and 510 nm emission.

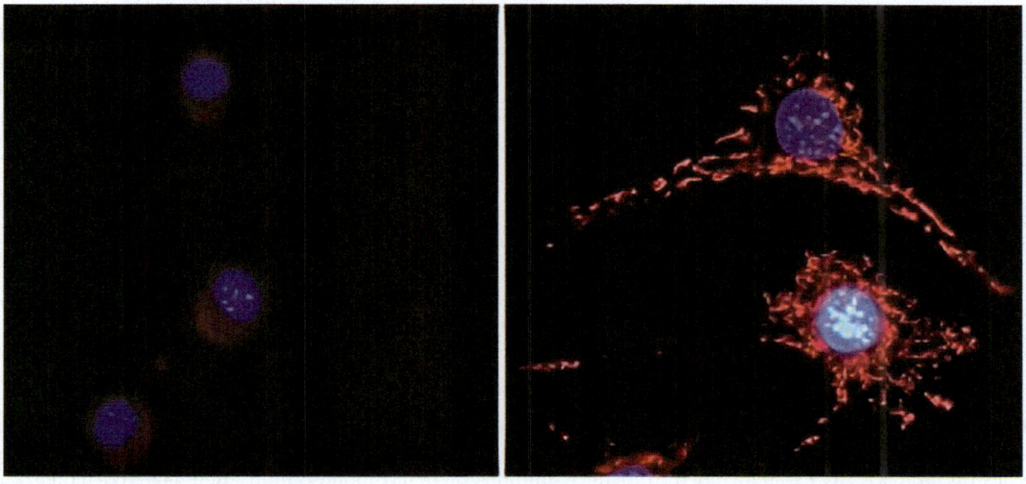

FIGURE 1.14 Detection of superoxide in live cells with MitoSOX red (MSR) superoxide indicator imaging. Live 3T3 cells were administered FeTCPP, a superoxide scavenger (left), or remained untreated (right). Cells were subsequently labelled with MSR reagent, which emits red fluorescence upon oxidation by superoxide, while nuclei were stained with the blue-fluorescent Hoechst 33342. The mitochondria of untreated cells displayed red fluorescence, signifying the existence of superoxide, while the mitochondria of treated cells exhibited negligible fluorescence.

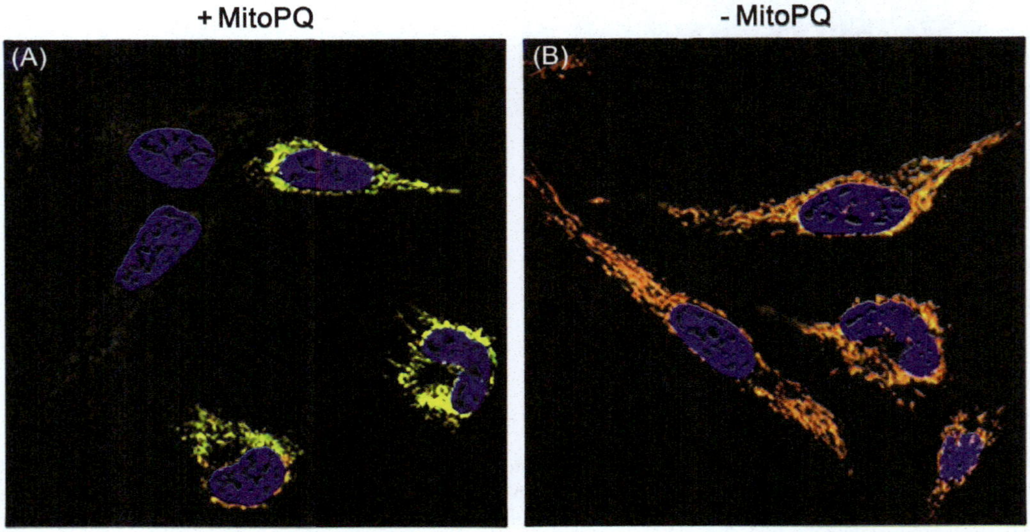

FIGURE 1.15 Detection of mitochondrial superoxide in live human osteosarcoma cells. Cells were subjected to overnight treatment with 30 μM MitoPQ to stimulate superoxide generation (A) or with ethanol as a vehicle control (B). Cells were subsequently stained with 1 μM MitoSOX green and 100 nM tetramethylrhodamine, methyl ester for 30 minutes to label the mitochondria. Nuclei were stained with Hoechst 33342 and subsequently photographed in HBSS using a Zeiss confocal microscope.

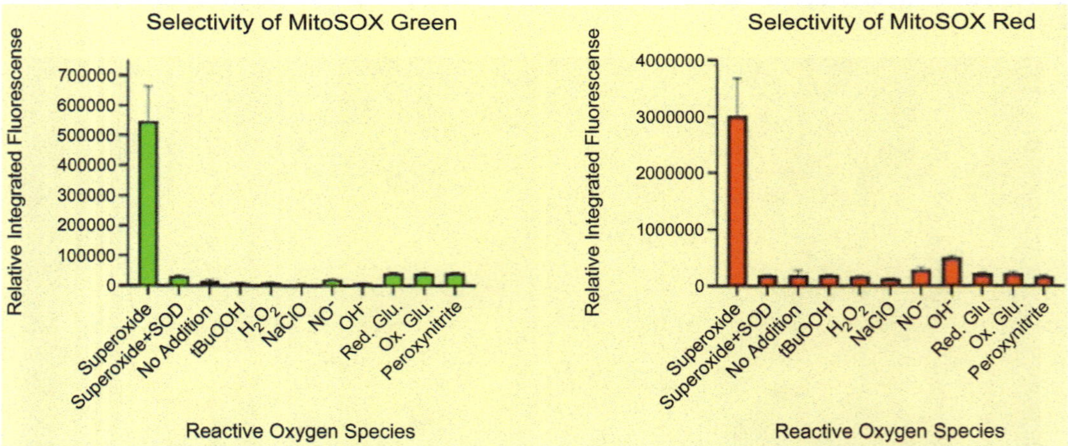

FIGURE 1.16 Assessing the selectivity of MitoSOX Green (MSG) and MitoSOX Red (MSR) as markers of mitochondrial superoxide in a cell-free environment. Cell-free systems were employed to produce reactive oxygen species (ROS). Each ROS was included in distinct solutions of 10 μM monosodium glutamate and monosodium aspartate and incubated for 30 minutes at room temperature. Excess DNA was incorporated into the solutions containing MSR. Superoxide dismutase served as a negative control for superoxide. Superoxide was produced utilising potassium superoxide for MSG, while the xanthine/xanthine oxidase system was employed to generate superoxide for MSR. The findings indicate that both MSG and MSR exhibit selectivity for superoxide, excluding other ROS.

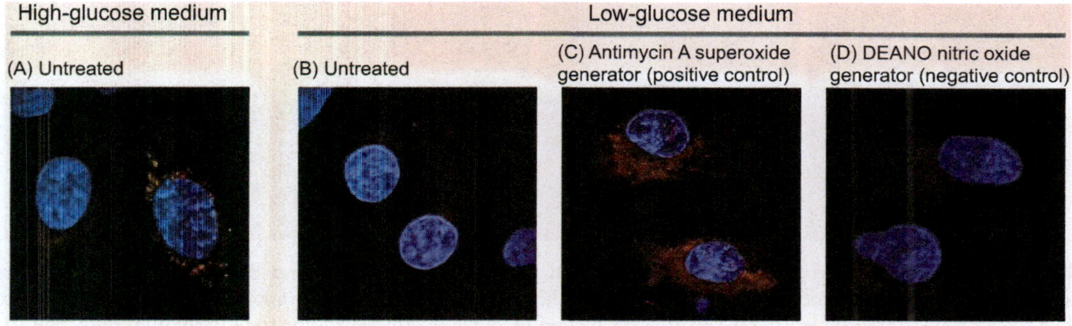

FIGURE 1.17 Visualisation of glucose-induced oxidative damage. Human osteosarcoma cells were cultured in minimum essential medium and incubated overnight at 37°C in a CO_2 atmosphere. To alleviate the impact of oxidative stress induced by elevated glucose levels, samples B–D were rinsed in PBS, submerged in low-glucose Dulbecco's modified eagle medium, and incubated overnight at 37°C in a CO_2 environment. Samples were subsequently rinsed in Hanks' balanced salt solution (HBSS) and subjected to the following treatments for 30 minutes: (A, B) cells remained in HBSS; (C) cells were incubated in HBSS supplemented with 100 μM antimycin A; (D) cells were incubated in HBSS supplemented with 100 μM DEANO. 5 μM MitoSOX red reagent was administered to each sample, followed by a 30-minute incubation, after which the cells were examined using confocal microscopy. **Rhod-2, acetoxymethyl (AM) reagent:** Increased mitochondrial Ca^{2+} is crucial for the onset of programmed cell death (apoptosis) and other cellular processes. Fluorescent probes exhibiting a spectrum response upon binding Ca^{2+} have facilitated researchers in examining alterations in intracellular free Ca^{2+} concentrations by fluorescence microscopy, flow cytometry, and fluorescence spectroscopy. Fig. 1.10 illustrates that Rhod-2 serves as a fluorescent calcium indicator capable of detecting mitochondrial Ca^{2+}. The AM ester variant of Rhod-2 AMfacilitates the efficient loading of the dye into living cells. Upon cell entry, intracellular esterases hydrolyse the AM group, liberating the Rhod-2 salt form to bind and emit fluorescence upon interaction with mitochondrial Ca^{2+}.

adult human blood plasma. Flow cytometry tests also verified the presence of circulating mitochondria in the plasma of human cord blood and adult peripheral blood. They similarly exhibited the immunological tolerance-related membrane markers, including CD270 and programmed cell death-ligand 1. Comparable data were acquired from foetal bovine serum and horse serum from various suppliers. Mitochondria were still identified following heat inactivation at 56°C. Expression profiles of many mitochondrial genes involved in human T- and B-cell activation were identified in animal serum using a real-time PCR array. Experiments demonstrated the migratory ability of mitochondria through their production of the chemokine receptor CXCR4 in response to its ligand, stromal-derived factor-1. Functional analysis demonstrated that human plasma mitochondria enhanced the proliferation of anti-CD3/CD28 bead-activated peripheral blood mononuclear cells (PBMC), increased the proportion of activated CD4+ T and CD8+ T cells, and reduced the production of inflammatory cytokines. These data indicate that circulating mitochondria in blood may serve as a novel mediator for intercellular communication and the maintenance of homoeostasis. Plasma-derived products must be utilised with caution in cell cultures due to mitochondrial contamination (see Fig. 1.23).

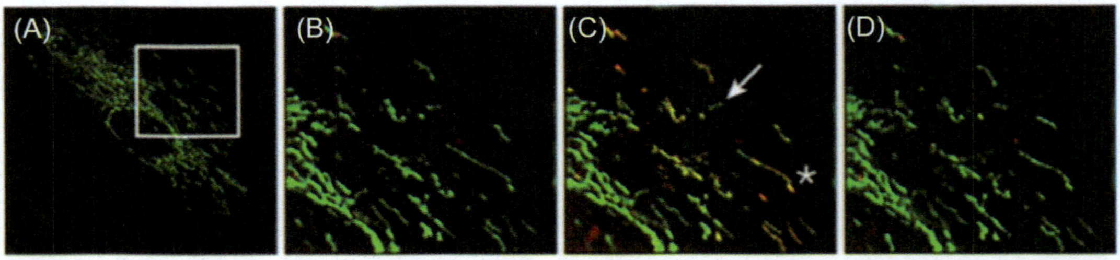

FIGURE 1.18 Multiplex imaging of mitochondrial calcium concentrations and dynamics. HeLa cells were stained with CellLight mitochondria-GFP and 5 μM Rhod-2, acetoxymethyl (AM) for 15 minutes at 37°C prior to live imaging for 100 seconds. (B–D) The area delineated in (A) is magnified to illustrate individual mitochondria within a single cell throughout time. Calcium is released from internal storage upon the administration of 10 μM histamine. The rise in orange-red fluorescence of Rhod-2 indicates the closeness of mitochondria to calcium release. The arrow in (C) indicates a mitochondrion that may exhibit compromised calcium uptake, a nuance that would have been overlooked with the only use of Rhod-2, AM. The asterisk indicates a mitochondrion exhibiting a temporary increase in calcium levels. ***Assays for mitochondrial permeability transition pore (MPTP)***: The MPTP is a generic channel composed of elements from both the inner and outer mitochondrial membranes, and is implicated in the release of mitochondrial constituents during cellular apoptosis. The opening of the hole significantly modifies the permeability of mitochondria and the mitochondrial membrane potential. The persistent activation of pores is due to mitochondrial Ca^{2+} excess, oxidation of mitochondrial glutathione, elevated reactive oxygen species in mitochondria, and several pro-apoptotic factors. We have created two kits for the detection of mitochondrial transition pore opening: one for image microscopy (Image-iT LIVE mitochondrial transition pore assay kit) and another for flow cytometry (MitoProbe transition pore assay kit). Both kits offer a more direct approach to assessing the opening of the MPTP compared to assays that depend solely on mitochondrial membrane potential. This assay utilises the AM ester of calcein, an uncoloured and nonfluorescent esterase substrate, along with $CoCl_2$ (cobalt), a quencher of calcein fluorescence, to specifically label mitochondria. Cells are infused with calcein AM, which passively diffuses into the cells and accumulates in cytosolic compartments, including the mitochondria. Upon entering cells, calcein AM is hydrolysed by intracellular esterases, releasing the highly polar fluorescent dye calcein, which minimally permeates the mitochondrial or plasma membranes during brief durations. The fluorescence of cytosolic calcein is diminished by the introduction of $CoCl_2$, whilst the fluorescence of mitochondrial calcein remains intact. As a control, cells loaded with calcein AM and $CoCl_2$ may also be subjected to a Ca^{2+} ionophore, such as ionomycin, facilitating the influx of excess Ca^{2+} into the cells, therefore inducing mitochondrial pore activation and the subsequent reduction of mitochondrial calcein fluorescence. The ionomycin response can be inhibited by cyclosporin A, a drug known to obstruct mitochondrial transition pore development by its binding to cyclophilin D.

Important or unique features for mitochondria— summarisation

Some interesting facts about mitochondria:

Mitochondria divide by simple fission, splitting in two just as bacterial cells do, and although the DNA **replication** strategies are a little different, forming displacement or D-loop structures, they partition their circular DNA in much the same way as bacteria do.

Mitochondria is considered as an **autonomous** cell organelle due to the following counts: **mitochondria** have their own DNA, which can **replicate** independently. However, most of the **mitochondrial** proteins are synthesised under instructions from cell nucleus (numtogenesis).

Mitochondria is a double-walled, DNA-containing, semi-autonomous, oxidative, energy-transducing organelle.

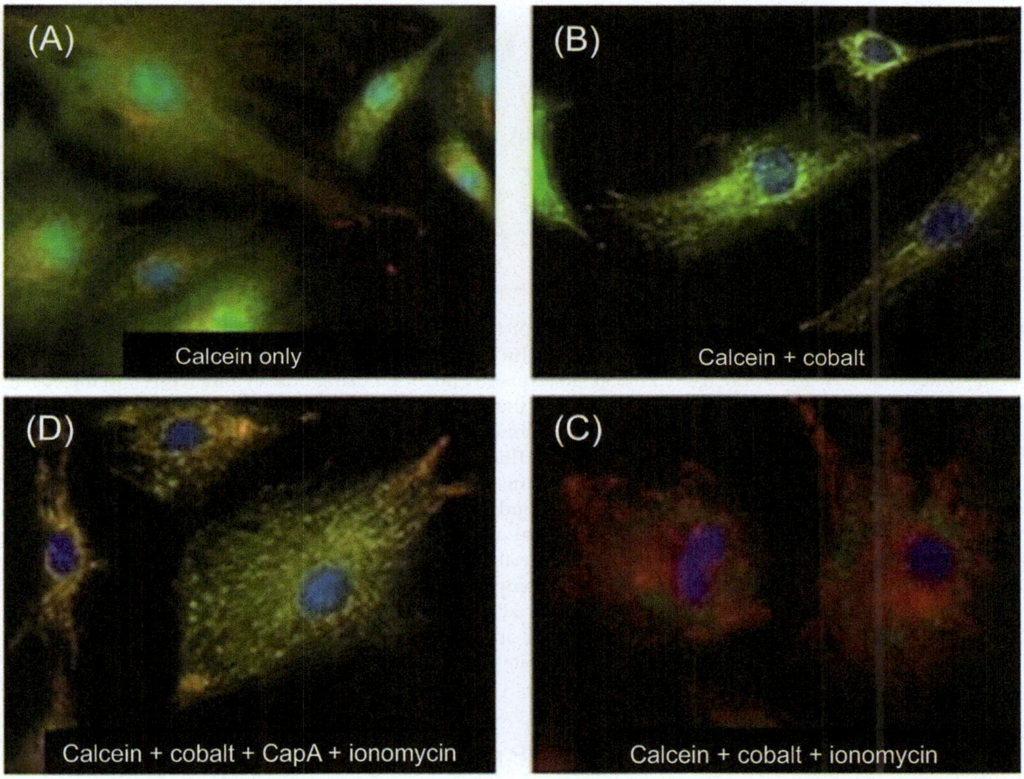

FIGURE 1.19 BPAE cells labelled utilising the Image-iT LIVE mitochondrial transition pore assay kit. BPAE cells are stained with components from the Image-iT LIVE mitochondrial transition pore assay kit. Cells are counterstained with MitoTracker red CMXRos to visualise mitochondria and with Hoechst 33342 to visualise nuclei. Panel A displays the consistent cellular fluorescence from unquenched calcein. Panel B illustrates the mitochondrial pattern following the addition of cobalt, which diminishes cytoplasmic calcein fluorescence while leaving mitochondrial calcein unaffected. Panel C illustrates the reduction of calcein fluorescence following the introduction of ionomycin, which facilitates the influx of cobalt and the efflux of calcein, but mitochondria remain discernible due to MitoTracker red. The activity of cyclophilin D is essential for the development of the mitochondrial permeability transition pore (MPTP) and is suppressed by cyclosporin A (CspA). Thus the inhibition of pore formation by CspA has been employed as evidence that a function is particular to the MPTP. Panel D demonstrates that the calcein pattern is preserved when CspA is administered prior to ionomycin, signifying that the ionomycin-induced alteration observed is an MPTP-mediated phenomenon. *BPAE*, bovine pulmonary artery endothelial cells.

It is found only in eukaryotic aerobic cells (*except mature RBCs totally absent in prokaryotes & anerobic eukaryotic cells).*

Mitochondria is ***semi autonomous*** *because*:

It contain their own DNA.

Having cytochromes electron transport system for ATP synthesis.

Mitochondria is arise from preexisting ones from division, that is *not de novo*.

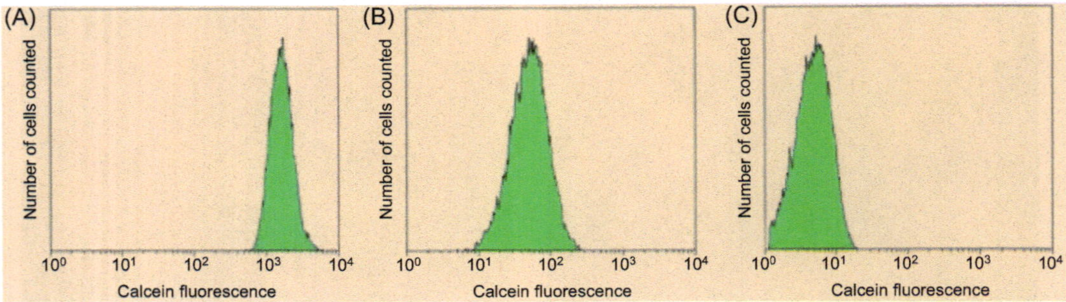

FIGURE 1.20 MitoProbe transition pore assay kit for flow cytometry analysis. The flow cytometry histograms illustrate the functions of the different kit components. Jurkat cells were subjected to incubation with the chemicals from the MitoProbe transition pore assay kit and subsequently analysed using flow cytometry. (A) In the absence of $CoCl_2$ and ionomycin, fluorescent calcein is localised in the cytosol and mitochondria, producing a luminous signal. (B) In the presence of $CoCl_2$, calcein within the mitochondria generates a signal, but the fluorescence of cytosolic calcein is muted, resulting in an overall reduction in fluorescence relative to calcein alone. When ionomycin, a calcium ionophore, and $CoCl_2$ are concurrently administered to the cells alongside calcein acetoxymethyl, the fluorescence signals from both the cytosol and mitochondria are significantly diminished. Tetramethylrhodamine, ethyl ester **(TMRE) mitochondrial membrane potential assay kit:** The TMRE-mitochondrial membrane potential assay kit ab113852 quantifies alterations in mitochondrial membrane potential in live cells using flow cytometry, microplate spectrophotometry, and fluorescence microscopy. Preliminary laboratory assessments, including lactate, pyruvate, urine organic acids, and plasma amino acids, can assist the clinician in identifying potential mitochondrial disease and shaping the genetic testing approach. Nonetheless, first biochemical assessments are not conclusive, necessitating genetic testing for a definite diagnosis.

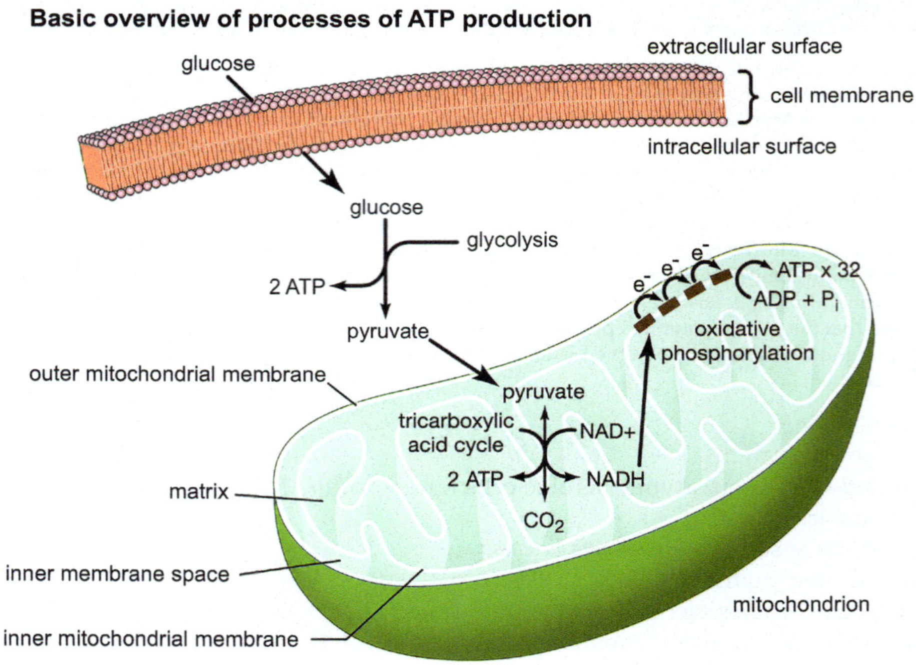

FIGURE 1.21 ATP production.

ETC

- ***In eukaryotic organism, the ETC, and site of oxidative phosphorylation, is found on the inner mitochondrial membrane.***
- ***ETC consists of 4 complexes and ATPase , in terms of molecular generator .***

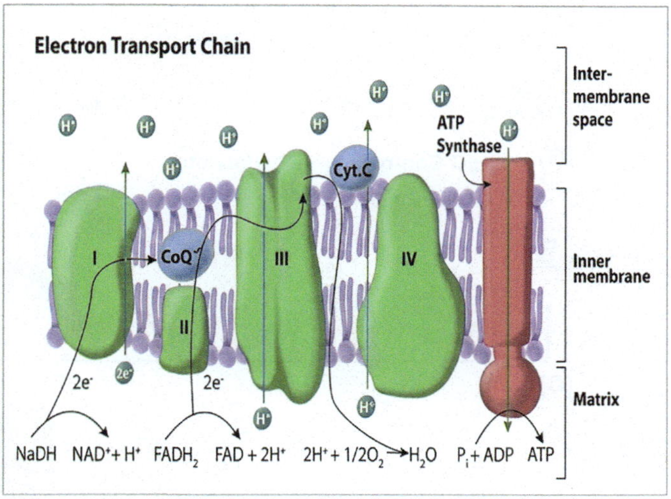

FIGURE 1.22 Electron transport chain.

Having DNA forming protien used in making their own membranes. But enzymes in mitochondria are **coded by nuclear DNA** (*therefore mitochondria is not fully autonomus it is semiautonomus*).

Mitochondria have their own **DNA,** which can replicate independently. The ***mtDNA*** produces its own ***mRNA, tRNA and rRNA***.

The organelles possess their own **ribosomes**, called ***mito ribosomes***.

Mitochondria synthesise some of their own structural proteins. However, most of the *mitochondrial proteins are synthesised under instructions from cell nucleus.*

The organelles synthesise some of the **enzymes** required for their functioning. *for example,* ***succinate dehydrogenase***.

They show **hypertrophy** *that is* internal growth.

However, both their structure and functioning are ***controlled by the nucleus of the cell and availability of materials from the cytoplasm***. Hence, they are termed as the"***semi-autonomous***" cell organelles.

Around 200 nuclear genes are responsible for maintaining mitochondrial function. However, some of the necessary proteins of mitochondria, especially proteins of outer membrane, transcribed by cell nucleus and prepared in cytosole as well as in rough endoplasmic reticulum and then transported to the mitochondria. Thus the reproduction is under the nucleus' control. This coordinates mitochondrial reproduction by responding to cell division and cellular physiological stresses, but not by the mitochondria's control.

Mitochondria have their own **DNA,** which can replicate independently. The ***mtDNA*** produces its own ***mRNA, tRNA and rRNA***.

The organelles possess their own **ribosomes**, called ***mito ribosomes***.

Mitochondria synthesise some of their own structural proteins. However, most of the *mitochondrial proteins are synthesised under instructions from cell nucleus.*

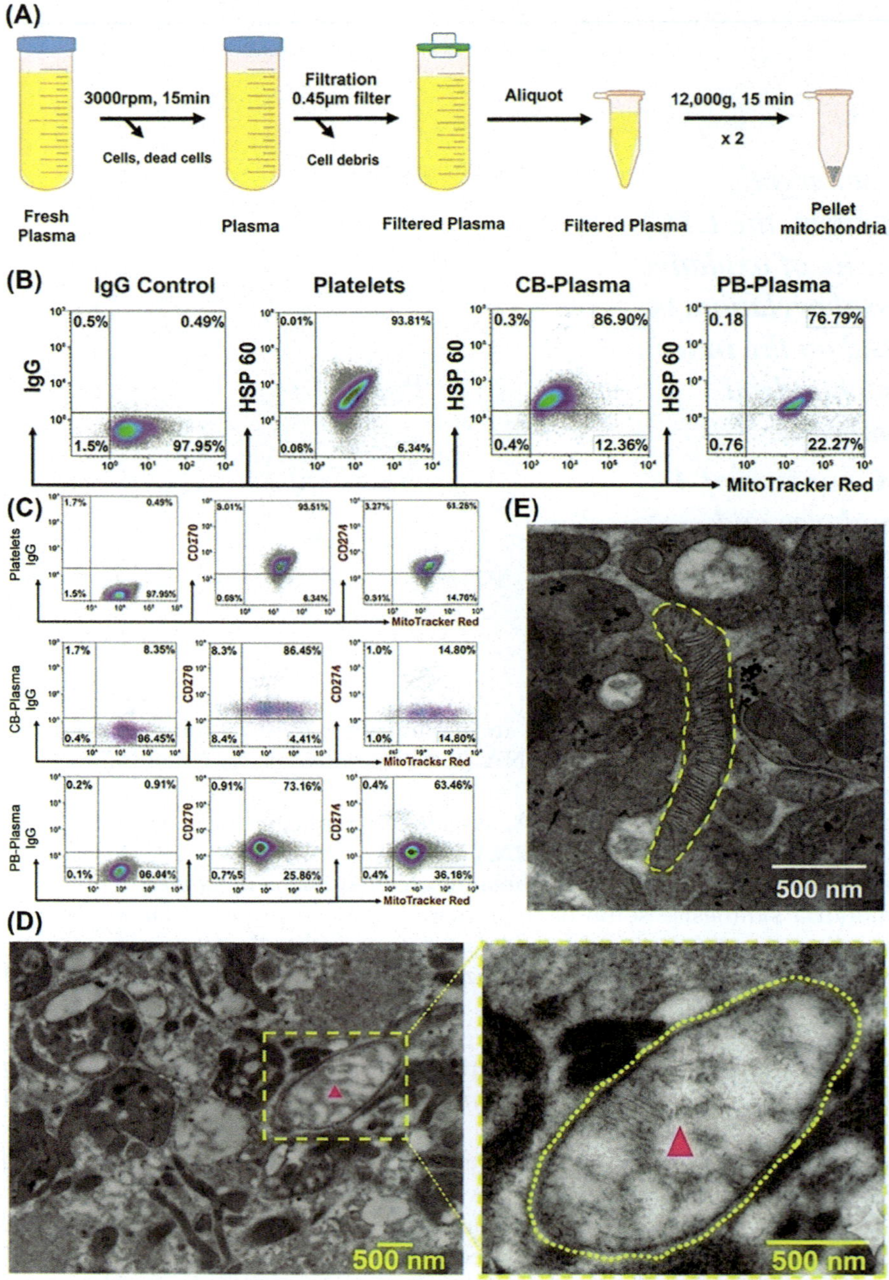

FIGURE 1.23 Mitochondrial characterisation in human blood. (A) Outline the process for isolating mitochondria from plasma. (B) Flow cytometry demonstrates that plasma-derived mitochondria are positive for particular mitochondrial markers (MitoTracker deep red) and heat shock protein 60 in the plasma of human cord blood (CB, N = 4) and adult peripheral blood (PB, N = 4). Mitochondria extracted from peripheral blood plasma (N = 4) served as the positive control. (C) Expression of immunological tolerance-associated markers CD270 and CD274 on mitochondria obtained from human cord blood and adult peripheral blood plasma, respectively. The positive control is established using mitochondria isolated from platelets obtained from peripheral blood. Isotype-matched immunoglobulins G served as the negative control. (D) The electron microscopic image illustrates free mitochondria (*red arrow, left*) within adult human blood plasma, captured at high magnification (*yellow dashed circle, right*). (E) Electron microscopy reveals a free elongated mitochondrion (*yellow dashed circle*) in adult human blood plasma.

The organelles synthesise some of the **enzymes** required for their functioning. *for example,* ***succinate dehydrogenase.***

They show **hypertrophy** *that is* internal growth.

It requires another (at least 200) nuclear genes for mitochondrial genome replication, transcription, maintenance and regulation. Even the mitochondrial translation machinery is incomplete; that is while rRNA and tRNA required for translation of mitochondrial protein-making genes are produced by mitochondria itself, the proteins required for mitochondrial translation complex are encoded by nuclear genes.

Mitochondria are called semi-autonomous cell organelle, due to the presence of mtDNA and s70S ribosomes in matrix. The **mtDNA**can replicate and code for mitochondrial proteins. The **mitochondria having 70S ribosomes and all the types of RNAs** synthesise some of the mitochondrial proteins, for example, the human mtDNA codes for 13 mitochondrial proteins.

Further Reading

Abraham, V. C., Towne, D. L., Waring, J. F., et al. (2008). Application of a High-Content Multiparameter Cytotoxicity Assay to Prioritize Compounds Based on Toxicity Potential in Humans. *Journal of Biomolecular Screening: the Official Journal of the Society for Biomolecular Screening, 13,* 527–537.

Dykens, J. A., Jamieson, J. D., Marroquin, L. D., et al. (2008). In Vitro Assessment of Mitochondrial Dysfunction and Cytotoxicity of Nefazodone, Trazodone, and Buspirone. *Toxicological Sciences: an Official Journal of the Society of Toxicology, 103,* 335–345.

Dykens, J. A., & Will, Y. (2007). The Significance of Mitochondrial Toxicity Testing in Drug Development. *Drug Discovery Today, 12,* 777–785.

Giorgi, C., Romagnoli, A., Pinton, P., et al. (2008). Ca2+ Signaling, Mitochondria and Cell Death. *Current Molecular Medicine, 8,* 119–130.

He, L., He, T., Farrar, S., et al. (2017). Antioxidants Maintain Cellular Redox Homeostasis by Elimination of Reactive Oxygen Species. *Cellular Physiology and Biochemistry: International Journal of Experimental Cellular Physiology, Biochemistry, and Pharmacology, 44,* 532–553.

O'Brien, P. J., Irwin, W., Diaz, D., et al. (2006). High Concordance of Drug-Induced Human Hepatotoxicity With in Vitro Cytotoxicity Measured in a Novel Cell-Based Model Using High Content Screening. *Archives of Toxicology, 80,* 580–604.

CHAPTER

2

Structural organisation of mitochondria and mitochondrial genetics

Introduction

In our earlier chapter, we discussed intact mitochondria in general. In this current chapter, we will be mostly concerned with the mitochondrial genome and genetics concerning to different species of animals, including humans.

Detailed structure of the mitochondrial genome

It is quite interesting to locate where mitochondrial DNA (mtDNA) is located in the mitochondria. It is located in the mitochondrial matrix. However, distinct information on how they are distributed, their pattern is not yet explored in detail. However, certain studies reveal stained mitochondria in the cell. mtDNA is circular in structure with light and heavy chains.

Can we stain the mtDNA within the mitochondria, an open question? This may explain how the mtDNA are organised within the mitochondrial matrix. We already have a clear idea how nuclear DNA is organised, with histone protein, through nucleosome formation, how they are winded.

Genomic organisation of mitochondria or mitochondrial genome mapping

How is mtDNA packed within mitochondria?

In contrast to nDNA, mtDNA is not associated with histones. However, mtDNA molecules are not 'naked', associated with areas called mt-nucleoids, along with different proteins responsible for mtDNA protection and stability. There exist multiple nucleoids per mitochondrion, while each nucleoid usually contains more than one mtDNA molecule.

A lot of studies are going on to explore the exact organisation and packaging of mtDNA. We attempt to discuss them.

The mitochondrial nucleoids in human cells were first time observed through epifluorescence by staining of DNA with the fluorochrome DAPI (19,090). Thereafter, further studies have

Mitochondrial Genomics
DOI: https://doi.org/10.1016/B978-0-443-26655-3.00015-9

documented that mtDNA is localised in discrete punctae—or mtDNA foci—within the mitochondrial network. They were localised with the inner mitochondrial membrane. These foci were named as mitochondrial nucleoids. The first nucleoid purification methodology identified two mtDNA binding proteins, which are the mitochondrial single-stranded DNA-binding protein (mtSSB) and transcription factor A (TFAM). Further studies were conducted to detect nucleoid-associated proteins through mass spectrometry, utilising either purified native or cross-linked nucleoids or immunoprecipitation of known nucleoid proteins.

Processes may be broadly classified as:

Mitochondrial replication machinery: acts through (DNA polymerase (POLG), Twinkle and mtSSB),

Mitochondrial transcription machinery: acts through (mitochondrial RNA polymerase (POLRMT), TFAM, TFB2M and mitochondrial transcription elongation factor (TEFM)).

Proteins associated with the nucleoid are listed as: ribosomal proteins, proteases, chaperones, RNA-binding proteins and RNA processing proteins.

The mitochondrial nucleoid has also been studied with the help of proximity-dependent labelling methods (biotin ligase-based (BioID) or peroxidase-based (APEX)) for tagging endogenous nucleoid-associated components, and simultaneous detection of their interacting partners in live cells.

A twinkle-APEX2 construct detected 37 nucleoid proteins through proteomic analysis. Among them, 30 were known nucleoid proteins and 12 had functions related to RNA. The technique could retrieve less cytosolic and nuclear contaminants in comparison to immunoprecipitation/fractionation techniques. More selective filters were used, as some proteins retrieved with immunoprecipitation/fractionation, such as LONP1 or MGME1, were not detected in the APEX study. As a result, this technique seems to be complementary to the immunoprecipitation/fractionation approaches and able to retrieve a specialised nucleoid proteins.

The detailed research of the nucleoid structure was quite challenging due to the small size of the nucleoid. Conventional fluorescence microscopy was not accurate enough to focus on structures smaller than ~200 nm. The recent development of superresolution techniques has enabled a better characterisation of nucleoid size, shape and organisation, and also a more precise counting of mtDNA molecules present within a nucleoid. Several high-resolution microscopy techniques, such as fluorescence photoactivation localisation microscopy, direct stochastic optical reconstruction microscopy, structured illumination microscopy and stimulated emission depletion microscopy, and also cryo-electron tomography and rotary-shadowing electron microscopy (EM), have been employed. These techniques have allowed the demonstration that nucleoids are rather small structures, with a mean size of ~110 nm. Moreover, a consensus has emerged regarding the shape of the mammalian mitochondrial nucleoid, which appears to be slightly ellipsoid, with an average shape varying between the studies from slightly elongated to truly ellipsoid.

Regarding the number of mtDNA molecules per nucleoid, nucleoids reconstituted in vitro and also nucleoids from mouse embryonic fibroblasts usually contain a single copy of mtDNA. However, reports supporting the idea of multiple mtDNA copies per nucleoid mtDNA are also available.

Mitochondrial nucleoids:

Mitochondrial nucleoids are large nucleoprotein complexes possessing mtDNA and regulatory factors, necessary for its packaging, replication, transcription, or repair. mtDNA encodes essential subunits of the electron transport chain complexes, which are crucial for the process of cellular energy production through oxidative phosphorylation. In most of the eukaryotic organisms, including animals, most of the complete copies of the mitochondrial genome are individually packed into the nucleoid complexes. They are distributed amongst the hundreds to thousands of individual mitochondria in each cell, through a dynamic and regulated process of mitochondrial fusion and fission. It is believed that mitochondria originate from an endosymbiont of alpha protobacterial origin and are being maintained through the cycles of genome duplication, membrane growth and division, similar to bacteria (see Fig. 2.1).

Compaction in mammals:

The molecular composition of the mammalian mitochondrial nucleoid is not completely known to date. Suggestions indicate that most of the proteins associated with different processes, like transcription and replication factors, are only temporarily linked to the nucleoid.

The important protein recognised as a core component of the mammalian nucleoid is the ***TFAM***. TFAM is a very commonly available protein, available in about 1000 molecules per mtDNA molecule or 1TFAM protein molecule per 16–17 bp of mtDNA in mammalian cells. TFAM is available to coat the entire mitochondrial genome and can introduce negative supercoils in DNA. TFAM is a 24 kDa basic protein with two domains, high-mobility group (HMG)-box (HMGA and HMGB), both of which intercalate into the groove of the DNA duplex. HMGA and HMGB are demarcated by a linker, and HMGB is linked to a C-terminal tail.

TFAM was earlier identified as a transcription factor that binds specifically to the promoter region of mtDNA. Along with the transcription factor B2, it increases transcription by the POLRMT. The structure of the initiation complex reveals that the HMGB domain of TFAM affects the interaction with POLRMT, which anchors POLRMT to the promoter.

On the contrary, Abf2p does not affect transcription, but is only responsible for the organisation and packaging of mtDNA. The C-terminal tail of TFAM is responsible for activation of transcription and is absent in Abf2p. Other than 'Initiation of transcription' TFAM also affects mtDNA metabolism, DNA repair and it can bind to RNA-containing

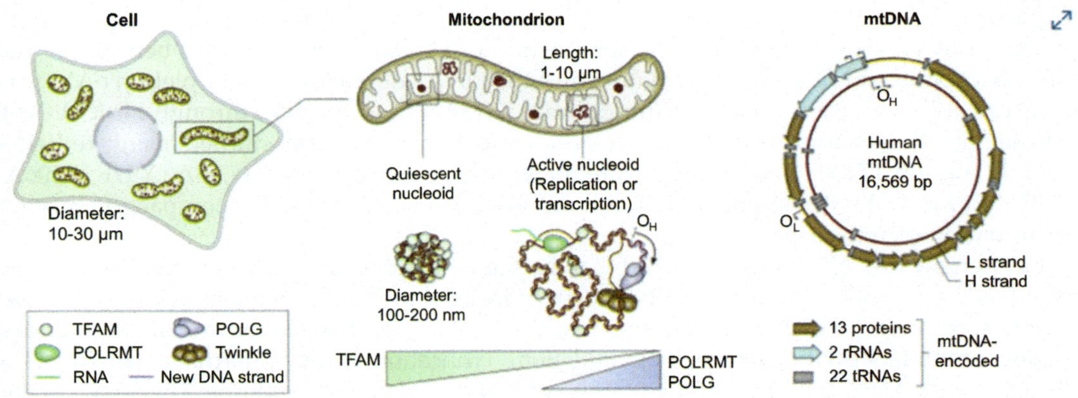

FIGURE 2.1 Organisation of mitochondrial DNA (mtDNA) within the mitochondrial matrix through mitochondrial nucleoids.

four-way junctions and to specific DNA structures—G quadruplexes. TFAM has an essential role in the organisation of the mitochondrial genome. As a result, it interacts specifically with the promoter regions; TFAM reveals strong nonsequence-specific DNA binding. This specific and unspecific binding to DNA is a unique feature of TFAM. This results in ordered binding of TFAM and DNA compaction throughout the entire mitochondrial genome. This aids in the control of mitochondrial gene expression and mtDNA replication.

To identify the compaction by TFAM, different techniques of visualisation and/or DNA manipulation have been employed. The mechanisms of DNA compaction by TFAM are yet to be explored. Abf2p, a mechanism proposed for DNA compaction by TFAM, is achieved by the bending of the DNA backbone. The structure of TFAM bound to the promoter region of mtDNA has been revealed that each of the two HMG-box domains of TFAM causes the DNA to bend by almost 90.

This leads to a complete 180 U-turn, which is a favourable arrangement for transcription initiation. This leads to the TFAM C-terminal tail, thereby recruiting the transcription machinery, to approach the initiation site. Accordingly, it has been reflected that TFAM also imposes a U-turn on DNA when bound unspecifically.

Single-molecule fluorescence resonance energy transfer (FRET) experiments on TFAM/light strand promoters (LSP) complexes have revealed that the DNA U-turn is induced by progressive and cooperative binding of the two TFAM HMG-box domains and the linker between them. FRET data also document a bending mechanism causing compaction of unspecific DNA by TFAM.

A combination of optical tweezers and fluorescence microscopy reveals that TFAM can bind cooperatively to DNA sequences nonspecifically and develops stable protein patches (filaments) in which each monomer covers about 30 bp of DNA. The binding of TFAM enhances the flexibility of the DNA through local base-pair melting of DNA (flexible-hinge mechanism), and leads to an effective means of compacting (CP) mtDNA.

EM and AFM have revealed that looping and cross-linking of DNA also affect mtDNA compaction. Low-angle rotary-shadowing EM depicted a progressive compaction of DNA at higher TFAM concentrations. Through this compaction process, they revealed cooperative binding of TFAM and isolated TFAM molecules, which were bridging neighbouring DNA duplexes.

The research also showed that the compaction of DNA by TFAM causes bending of the DNA backbone and DNA loop formation and the DNA compaction is complete. TFAM was observed to interact with itself, forming dimers in vitro and in vivo, through HMGA. The role of the dimerisation of TFAM in DNA compaction has been studied using tethered particle motion (TPM) and a dimer mutant with five substitutions (K95A, Y99F, E106A, E112A and R116A). It was concluded that dimerisation is necessary for compaction but not for initiation of transcription.

Electron microscopic images revealed that the same mutant could compact DNA to the same extent as the wild-type protein. The basic C-terminal tail of TFAM includes highly structured regions that interact with the transcription machinery. It is reported that the domain is necessary for the promoter-specific transcription initiation. Truncation of the C-terminal tail causes a decrease of the DNA-binding activity of TFAM by three orders of magnitude. This report states that the C-terminal tail of TFAM is important for the strong general binding to mtDNA and could influence DNA compaction.

TPM experiments also revealed that a truncated version of TFAM lacking the C-terminal tail possesses the capability to compact DNA, but to a lower degree than wild-type TFAM. Further studies need to be conducted.

Nucleoid organisation models

A 'multi-layer' model of mt-nucleoid organisation in human cells was suggested, with the central part—'core'—where DNA and proteins (DNA-packaging proteins and proteins involved in replication and transcription processes) are concentrated. The next layer possesses proteins temporarily appointed to carry out the special functions. These proteins are commonly retained in DNP by protein–protein interaction. On the circumference of a nucleoid, there exists a region where translation and other processes take place.

The nucleoid was proposed to be positioned on the internal mitochondrial membrane with the ATAD3 protein. This protein itself binds particularly to the D-loop of mtDNA. Other proteins of the inner mitochondrial membrane—prohibitin 1 and 2 (PHB1 and PHB2) were also predicted to be associated in keeping a nucleoid. Ribosomes (RS) were placed next to the second layer. This model was proposed for mammalian cells, but it can be extended to all metazoan cells.

Specific bodies called 'mitochondrial RNA granules' were observed in the mitochondrial matrix. Granules are complexes of RNase P in combination to the newly synthesised mtRNA. RNA-binding protein G-rich sequence factor 1 (GRSF1) was also detected in the complex. RNase P possesses three subunits (MRPP1, MRPP2 and MRPP3) and leads to the processing of the 5′-end of mitochondrial tRNAs in the in vitro system. The 3′-end of tRNAs is processed by ELAC2 enzyme (Z-type RNase). GRSF1 protein leads to the processing of mitochondrial RNA.

Subsequent to the identification of these granules, the model was slightly altered. A zone called 'RNA cloud' was placed at the periphery of DNP. Outside of this cloud of proteins, remodelling of a nucleoid was placed. Formation of macromolecular 'crowding' within the second layer was also included. A scheme of possible nucleoid organisation is shown in the figure (see Fig. 2.2).

Scheme of mitochondrial nucleoid organisation:

mtDNA with 'core' proteins is in the middle. The dashed line represents the boundaries of a nucleoid. The shadowed regions at the periphery of a nucleoid denote 'RNA granules'.

Legend: CP (packaging) proteins; RNA polymerase (RNApol); PolG; PHB1, PHB2; macromolecular crowding (region, where functional proteins of a nucleoid are localised); Rs; remodelling proteins.

A model of DNA organisation in an mt-nucleoid: (1) metazoan, fungi; (2) plants and prokaryotes. Legend: CP1, first-level packaging proteins; CP2, functional packaging proteins.

Is the nucleoid separated from the matrix?

Although there is an absence of membrane enveloping nucleoid (unlike nuclear membrane surrounding nucleus), recent research reveals structured 'crowding' by dispersed lipid droplets

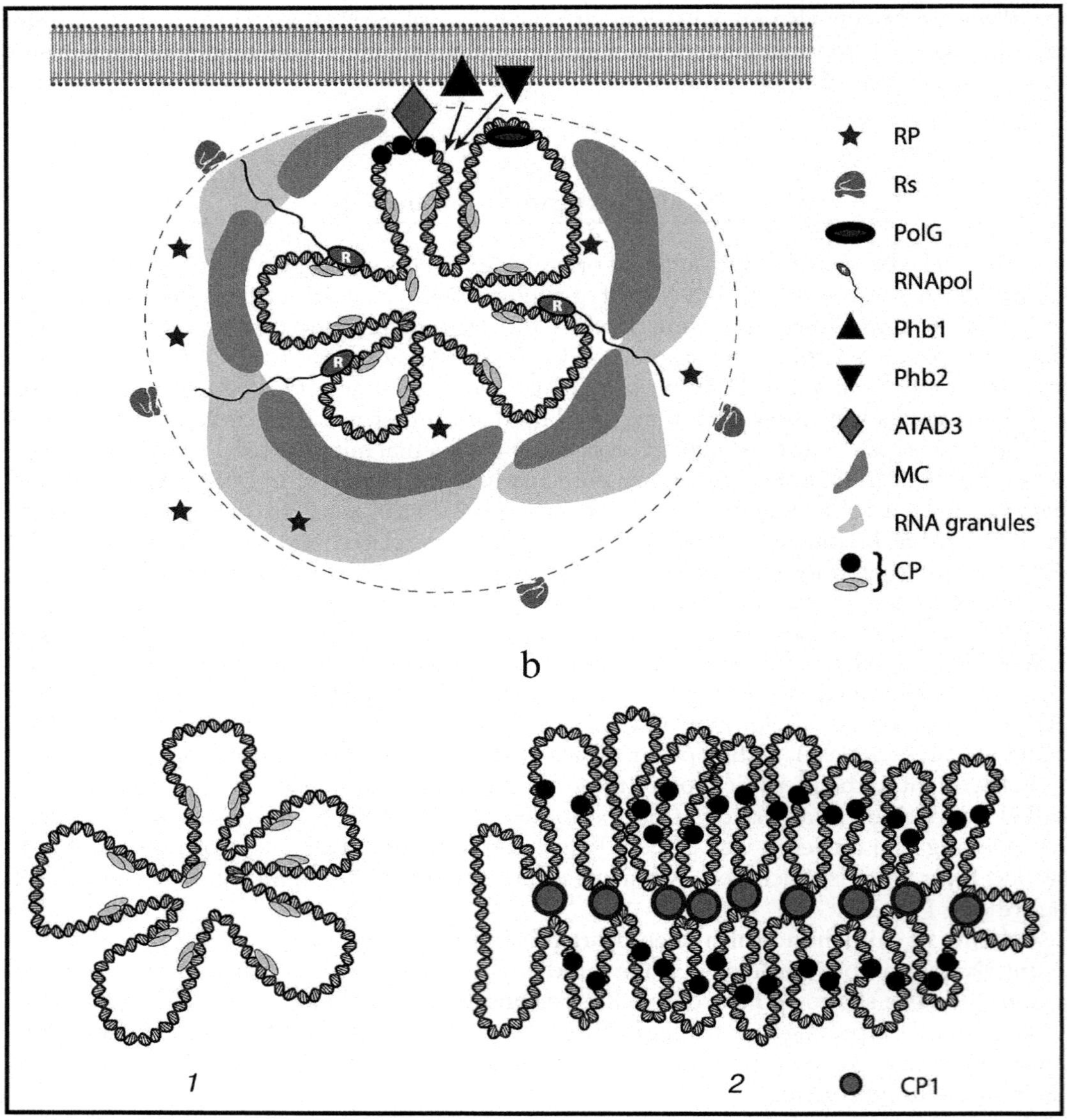

FIGURE 2.2 Nucleoid organisation of mitochondrial DNA.

and 'architectural RNA' (long noncoding RNA, longer than 150 nucleotides) existing as mitochondrial RNA granules, are involved in the formation of 'nucleoid backbone'.

Specific packaging of the loops would be conducted by another set of proteins. It is a two-level system of mtDNA packaging.

Since the mitochondrial genome size is less (for genomes smaller than 100 kb), the first level of DNA packaging is not a necessity, and HMG domains are present instead of histone-like proteins.

Replication, recombination and segregation of mtDNA are closely related to the structure of a mitochondrial nucleoid.

To date, there is a lack of complete information for a full comparative analysis of organisation of mitochondrial nucleoids in different groups of organisms. Throughout the evolution, a part of proteins was adjusted in mtDNP.

Mitochondrial DNA-37 genes

Function of mitochondrial genes:

The mitochondrial genome comprises 37 genes, which encode 13 proteins, 22 tRNAs and 2 rRNAs. The 13 proteins expressed by mitochondrial genes direct cells to synthesise protein components for the enzyme complexes of the oxidative phosphorylation system (OXPHOS), allowing mitochondria to function as the powerhouses of cells.

Mitochondrial codon: The initiation codon in mitochondria encodes N-formylmethionine. The start codon is the initial codon in a messenger RNA transcript that is translated by a mitoribosome. Mitochondrial genomes utilise alternative start codons more prominently (AUA and AUG in humans). The NCBI list of translation tables provides numerous instances, including codons, systematic range and citations. In human mitochondria, the conventional stop codon UGA is repurposed to encode tryptophan, while the codons AUA and AUU, which typically encode isoleucine, are redefined as start codons. Furthermore, the codons AGA and AGG, which specify arginine in the universal genetic code, lack corresponding tRNAs in mitochondria.

Mouse mtDNA is unique in that the translational start codon is AUN, with any of the four nucleotides in the third position, whereas the only translational stop codon is the orthodox UAA.

Mitochondrial genome in different species:

Whole mitochondrial genome sequencing is an important next-generation sequencing technique to explore the entire mitochondrial genome for different species:

Human: The human mitochondrial DNA has similar mitochondrial structure as other species.

Fig. 2.3

Mice: The complete mitochondrial genome (JN629047) of *A. agrarius* from Korea. *Apodemus agrarius* mitogenome is 16,263 bp in length and has a base composition of 33.9% A, 29.9% T, 23.6% C and 12.5% G. The total length of the 13 protein-coding genes is 11,398 bp long.

Sheep: We have studied whole mitochondrial genome sequences for different breeds of sheep, namely garole, bonpala, chotanagpuri and birbhum. We have also studied the whole mitochondrial genome for healthy and diseased Garole sheeps (see Fig. 2.4).

Pig: Whole mitochondrial genome sequencing conducted for Ghungroo pig. The mitochondrial map is being presented here.

Chicken: We have represented the whole mitochondrial genome sequence for the Haringhata Black chicken.

Duck: Duck mitochondrial DNA has similar structure to that of other species with more genetic similarity with chicken.

Fig. 2.5

Cattle: The complete mitochondrial genome sequence for *Bos taurus* and *Bos indicus* were observed.

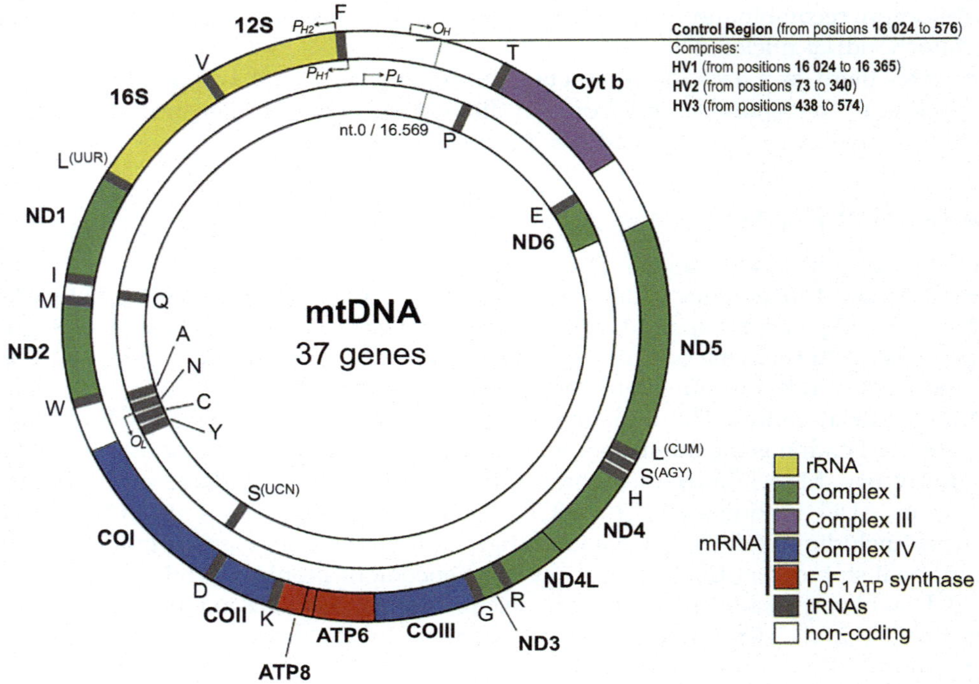

FIGURE 2.3 Human mitochondrial genome map.

Table 2.1 Goat: Complete mitochondrial genome sequence was observed to be 16637 bp. Remarks:

Overall base composition was estimated to be 33.68% for A, 27.36% for T, 25.86% for C and 13.10% for G, indicating an AT-rich (61.04%) feature in the *O. aries* mtgenome.

It is far interesting to note that even within a species, there exists mitochondrial genomic variation. The mitochondrial genome is very unique for a particular individual, which forms the basis for molecular phylogenetic analysis in evolutionary studies, with natural variations existing.

Any deviation from the normal mitogenome sequence, caused by mutations in terms of single nucleotide polymorphisms (SNPs) or deletions, may lead to diseases. Mutations causing defects in coded proteins, important domains may get changed; likewise, disease occurs.

Although the genome size varies from species to species, the genomic organisation is similar. In a later chapter, we have included whole mitochondrial genome sequencing in detail.

An important note:

There are reports that indicate that mtDNA can be synthesised artificially (ref). In that case, this may be treated as the easiest way to replace our healthy synthesised mitochondria. We have discussed mitochondria replacement therapy as well as gene therapy in later chapters.

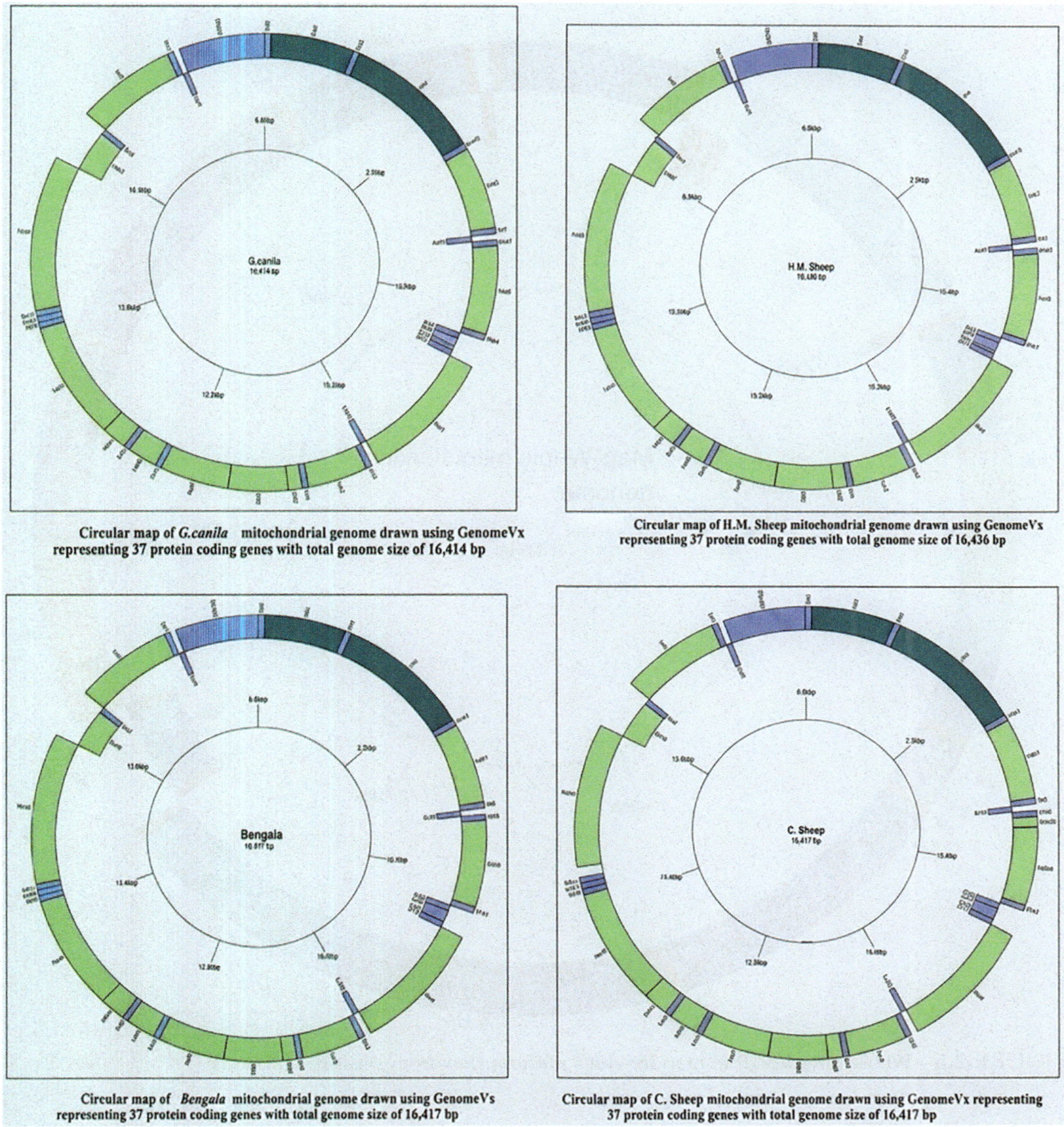

Circular map of *G.canila* mitochondrial genome drawn using GenomeVx representing 37 protein coding genes with total genome size of 16,414 bp

Circular map of H.M. Sheep mitochondrial genome drawn using GenomeVx representing 37 protein coding genes with total genome size of 16,436 bp

Circular map of *Bengala* mitochondrial genome drawn using GenomeVs representing 37 protein coding genes with total genome size of 16,417 bp

Circular map of C. Sheep mitochondrial genome drawn using GenomeVx representing 37 protein coding genes with total genome size of 16,417 bp

FIGURE 2.4 Complete mitochondrial genome map for sheep.

Mitochondrial genetics

The mitochondrial genome has been extensively used primarily for evolutionary studies for a long. We have also worked in our lab (ref sheep mitochondria, duck mitochondria). Mutations or copy number variations for the mitochondria have been studied for various diseases in human or mice. Very recently, our group has started working on association of

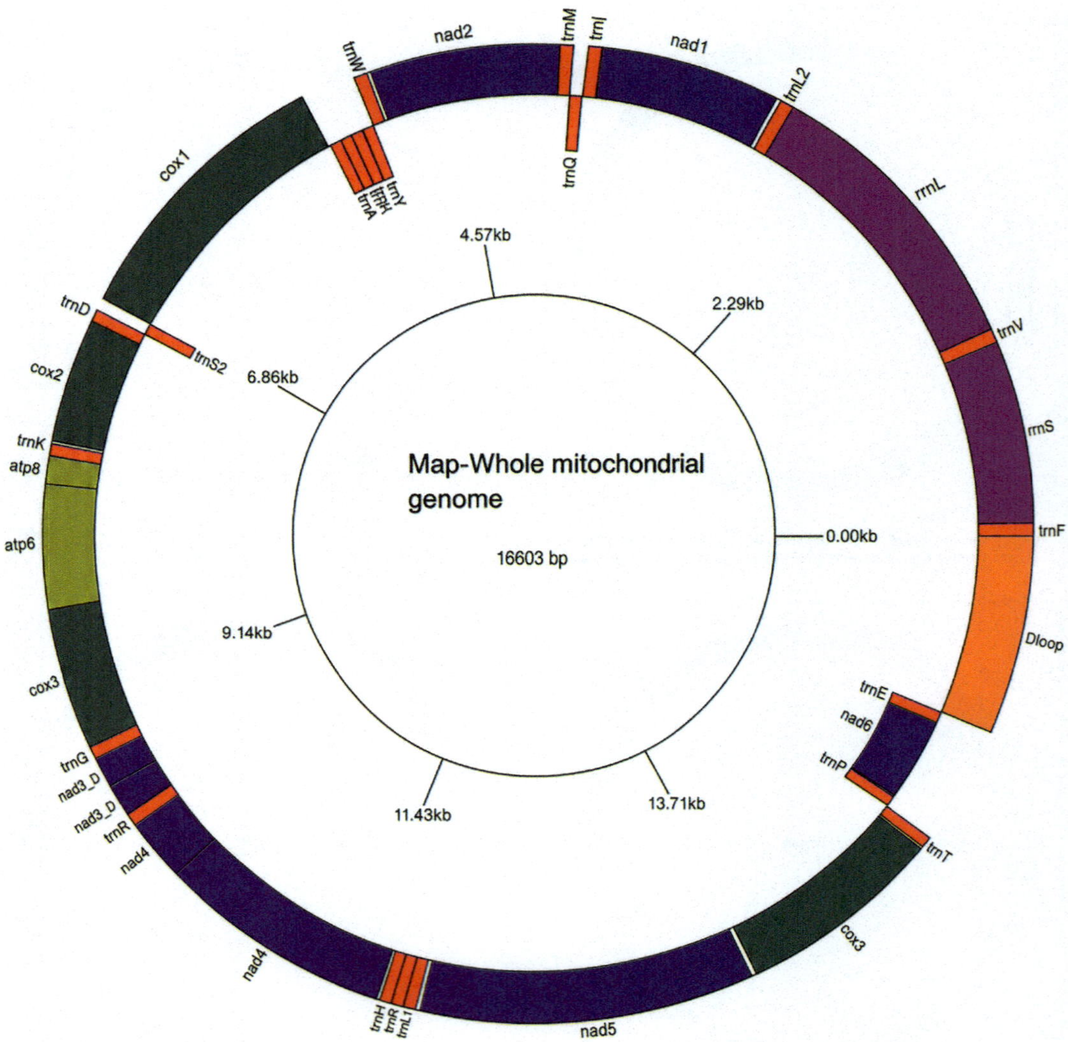

FIGURE 2.5 Whole mitochondrial map for ***Anas platyrynchos*** from West Bengal, India.

mitochondrial genome with respect to diseases (metabolic, viral or bacterial infections) in animal models, namely sheep, chicken and duck. We could also explore certain mutations in mitochondrial genes as well as copy number variations with respect to growth and reproduction in animal models (references).

Molecular interaction among the mitochondrial genes:

It is evident that for any biological function or pathway, there exists involvement of a lot of genes. Hence, it is quite important to study the molecular interaction among them. Very recent findings indicate molecular cross-talk between nuclear and mitochondrial genes (see Fig. 2.6).

TABLE 2.1 Complete mitochondrial genome size for different species of animals and humans (source: NCBI).

Species	Mitochondrial genome size
Human	16,569 bp
Mice	16,263 bp
Sheep	16,615 bp
Cattle	16,338 bp
Goat	16,637 bp
Pig	16 613 bp
Chicken	16,784 bp
Duck	16,603 bp
Dog	16,713 bp
Cat	17,009 bp

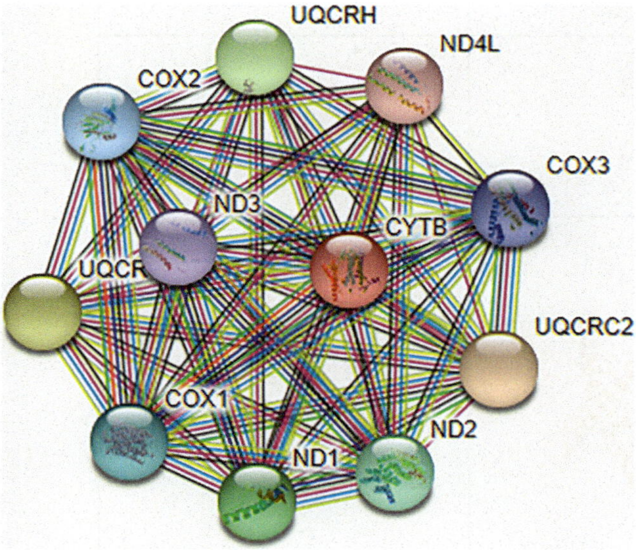

FIGURE 2.6 Molecular interaction studies among the ***Anas platyrynchos*** genes through STRING analysis.

3D structural organisation of the mitochondrial proteins:

For the appropriate biological function of a protein, the 3D structure of a protein arising out of polarity, hydrophilicity and other properties of respective amino acids, is essential. Here, we represent 13 polypeotide coding genes for mitochondria (see Fig.2.7).

Fig. 2.8

Genes	3 D Protein structure	Prosite
ND1 NADH dehydrogenase subunit 1 2807..3784	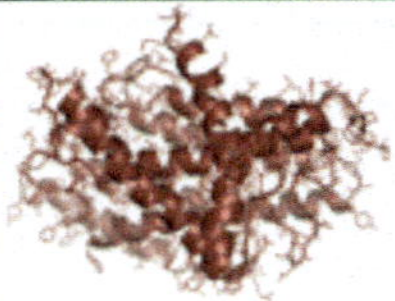	**51 - 66:** ***Respiratory-chain NADH dehydrogenase subunit 1 signatures*** **204 - 217:** **Respiratory-chain NADH dehydrogenase subunit 1 signature 2**
ND2 NADH dehydrogenase subunit 2 4001..5039	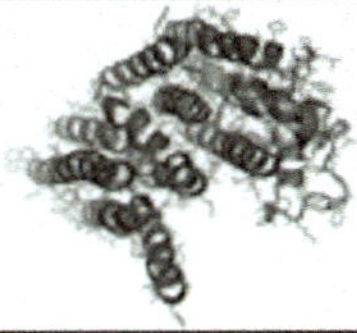	**7 – 512:** ***Cytochrome c oxidase subunit I signature and profile*** **237 - 292:**
COX1 cytochrome c oxidase subunit I	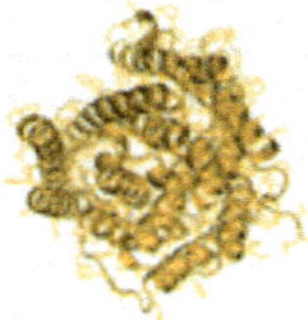	5401..6951
COX2 cytochrome c oxidase subunit II	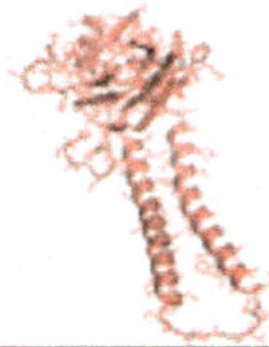	7088..7774
ATP8 ATP synthase F0 subunit 8		7846..8013
ATP6 ATP synthase F0 subunit 6	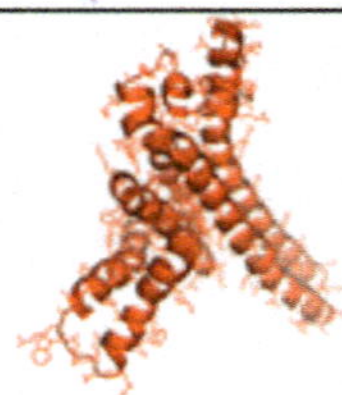	8004..8687
COX3 cytochrome c oxidase subunit III	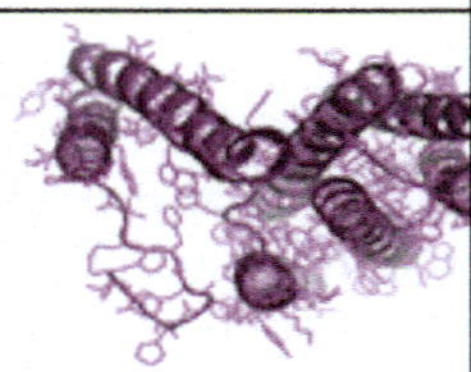	8687..9470

FIGURE 2.7 3D structural model for mitochondrial proteins.

ND3 NADH dehydrogenase subunit 3	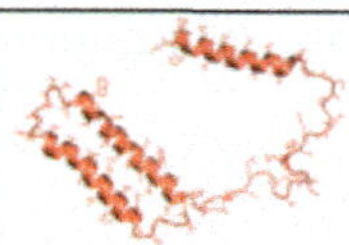	9540..9712, 9714..9891
ND4L NADH dehydrogenase subunit 4L	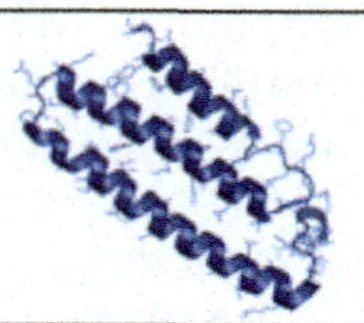	9963..10259
ND4 NADH dehydrogenase subunit 4	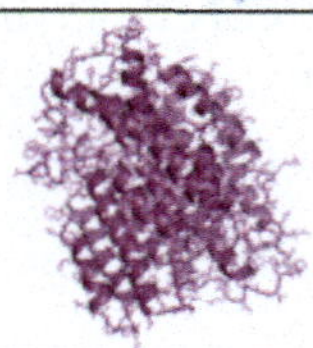	10253..11630
ND5 NADH dehydrogenase subunit 5	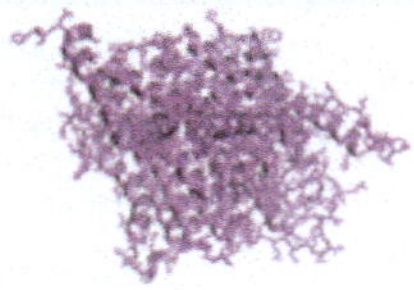	11836..13659
CYTB cytochrome b	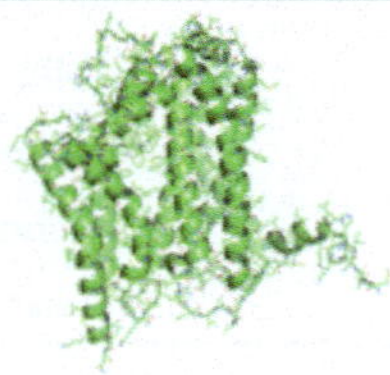	13659..14801
ND6 NADH dehydrogenase subunit 6	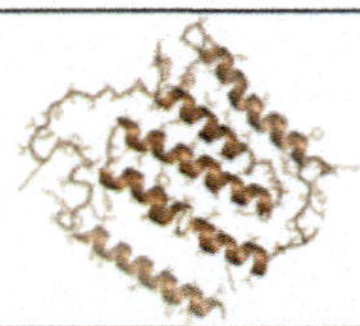	14963..15484

FIGURE 2.8 3D structural model for mitochondrial proteins.

Details about mitochondrial proteins and their predicted structures, diseases associated with mutations.

Mutation in mtDNA

A lot of studies have been conducted to detect mutations in the mtDNA. Several well-known diseases have already been documented as associated with mitochondrial mutations.

Association studies of disease traits with SNPs and other mutations of mtDNA.

Nad1: Nicotinamide adenine dinucleotide (NADH) dehydrogenase (ubiquinone) flavoprotein 1, mitochondrial

Functions of NADH dehydrogenase (ubiquinone) flavoprotein 1, mitochondrial
Tables 2.2, 2.3, 2.4.
Fig. 2.9
Table 2.5
NADH-ubiquinone oxidoreductase chain 2 MT-ND2
Involvement in disease due to mutations in mitochondrial ND2 gene.
Identified SNPs for ND2 (in red) are as follows. These lead to different disease conditions (see Fig. 2.10, Fig. 2.11, Fig. 2.12).
Leber hereditary optic neuropathy (LHON):
It is caused by detected mutations at N150D and G529S.
Alzheimer's disease mitochondrial:

TABLE 2.2 Functions of nicotinamide adenine dinucleotide (NADH) dehydrogenase (ubiquinone) flavoprotein 1, mitochondrial.

Molecular function	Four iron, four sulphur cluster binding
Molecular function	Flavin mononucleotide binding
Molecular function	Metal ion binding
Molecular function	NAD binding
Molecular function	NADH dehydrogenase (ubiquinone) activity
Molecular function	Aerobic respiration
Biological function	Mitochondrial ATP synthesis coupled to electron transport
Biological function	Mitochondrial electron transport, NADH to ubiquinone
Biological function	Proton motive force-driven ATP synthesis

TABLE 2.3 Binding site for nicotinamide adenine dinucleotide (NADH) dehydrogenase (ubiquinone) flavoprotein 1, mitochondrial.

Type	Position	Description
Binding site	87–96	NADH
Binding site	199–247	Flavin mononucleotide
Binding site	379	4Fe–4S
Binding site	382	4Fe–4S
Binding site	385	4Fe–4S
Binding site	425	4Fe–4S

TABLE 2.4 Variants arising due to nonsynonymous mutations leading to amino acid changes in NAD1 gene.

Positions	Change
214	E/K
341	A/V
423	T/M

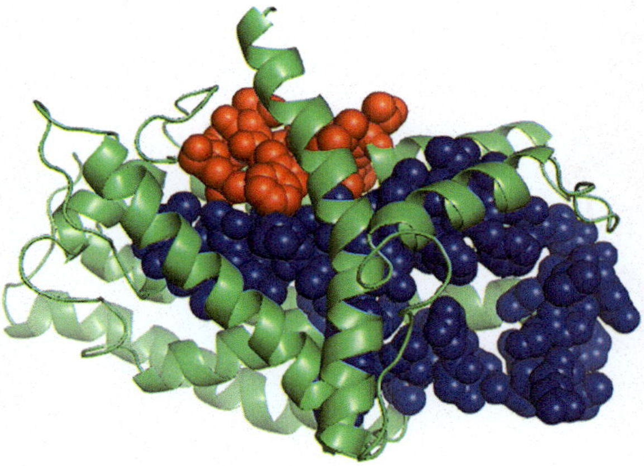

FIGURE 2.9 3D structure for NAD1, where red sphere (87–96 aa) represents nicotinamide adenine dinucleotide binding site, while blue sphere (199–247 aa) represents flavin mononucleotide binding site.

TABLE 2.5 Functions for NAD2.

Molecular function	Ionotropic glutamate receptor binding
Molecular function	Nicotinamide adenine dinucleotide dehydrogenase (ubiquinone) activity
Molecular function	Protein kinase binding
Biological function	Aerobic respiration
Biological function	Mitochondrial electron transport, NADH to ubiquinone
Biological function	Mitochondrial respiratory chain complex I assembly
Biological function	Proton motive force-driven mitochondrial ATP synthesis
Biological function	Reactive oxygen species metabolic process
Biological function	Response to hypoxia

It is caused by detected mutations at A331S.
Leigh syndrome (LS):
It is caused by detected mutations at L71P.

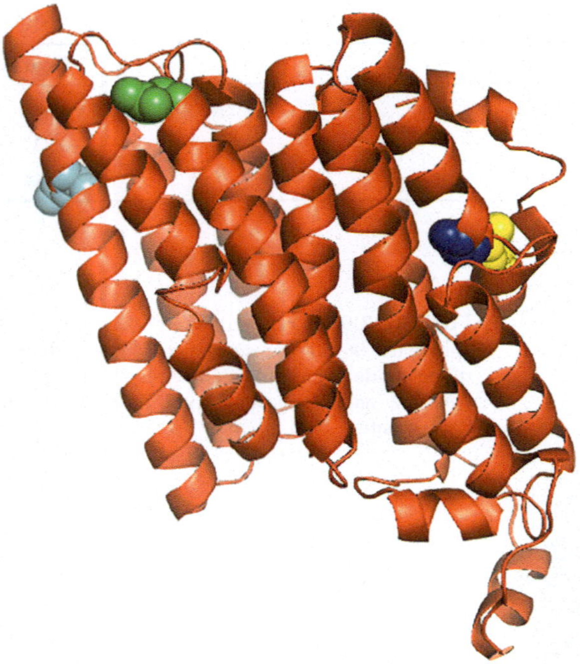

FIGURE 2.10 3D structure for ND2 with the sites for mutation at N150D: green sphere, G259S: blue sphere, A331S: yellow sphere and L71P: cyan sphere.

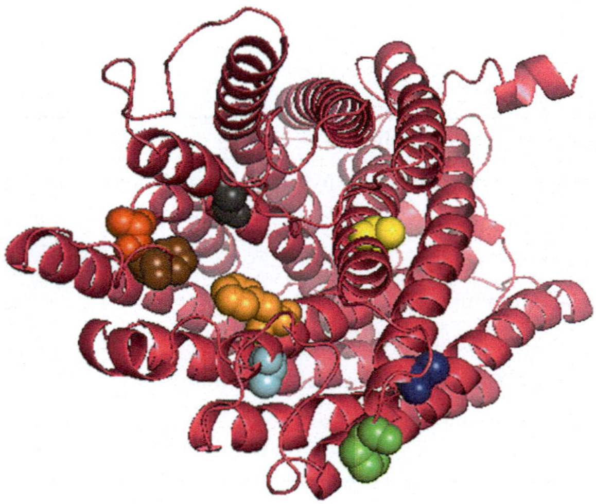

FIGURE 2.11 (A) 3D structure for cytochrome c oxidase subunit 1 (Cox1) with the binding sites predicted. Calcium binding sites: 40 (blue sphere), 45 (green sphere), iron binding site: 61 (yellow sphere), 376 (cyan sphere), 378 (orange sphere), copper cation binding site: 240 (deep grey sphere), 290 (red sphere), 291 (chocolate sphere). (B) 3D structure for Cox1 with the binding sites provided. Magnesium binding site: 368 (limon sphere), 369(deep blue sphere),oxygen binding site: 244 (bright orange sphere).

Cytochrome c oxidase subunit I (COX1)
Functions:
Table 2.6
Binding site

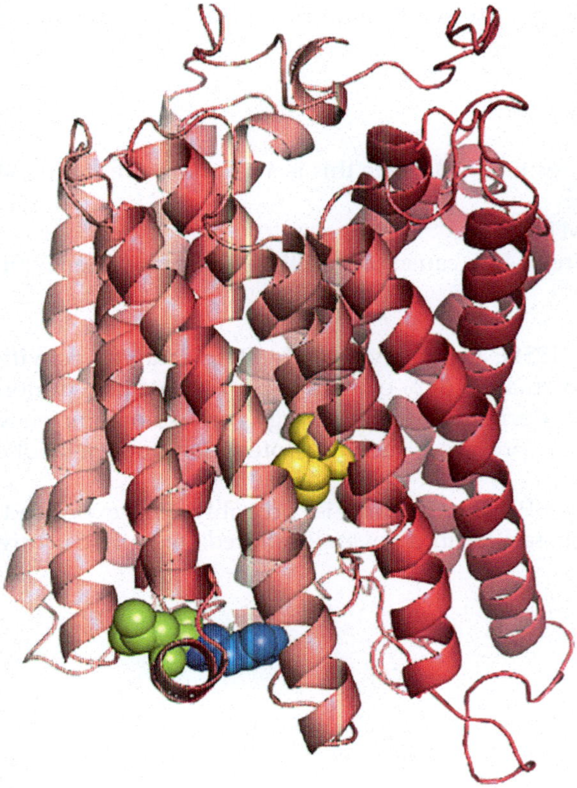

FIGURE 2.12 The 3D molecular structure of the Cox 1 protein is visualized using a ribbon representation. This protein consists of numerous alpha helices that form a dense transmembrane bundle. Within this helical framework, specific functional areas are marked by spherical clusters to denote binding sites. 3D structure for CoxI with the binding sites were predicted. Magnesium binding site: 368 (limon sphere), 369 (deep blue sphere). Oxygen binding site: 244 (bright orange sphere).

TABLE 2.6 Functions for cytochrome C oxidase I (COX1).

Molecular function	COX activity
Molecular function	Haem binding
Molecular function	Metal ion binding
Biological function	Aerobic respiration
Biological function	Cellular respiration
Biological function	Cerebellum development
Biological function	Mitochondrial electron transport, cytochrome C to oxygen
Biological function	Respiratory electron transport chain
Biological function	Response to copper ion
Biological function	Response to electric stimuli
Biological function	Response to hypoxia
Biological function	Response to oxidative stress

The binding site for mitochondrial Cox1 are very important for their functional significance. We are listing them as follows:

Table 2.7

Fig. 2.13

LHON:

It is caused by detected mutations at aa position 513. Since this is a secondary mutation, it does not seem to directly cause the disease.

Mitochondrial complex IV deficiency (MT-C4D):

The mutations identified in Cox1 gene are S142F (causes a significant decrease in enzyme activity) and L196I.

Colorectal cancer:

The identified mutation in Cox1 was G125D. It displays steady state catalytic activity linked to proton pumping, which is approximately 34% of wild type. An intrinsic proton leak is observed in the enzyme, that lead to decreased overall energy conversion efficiency of the respiratory chain, perturbing transport processes such as protein, ion and metabolic trafficking.

The other identified mutation was S458P. The mutant is probably not expressed, indicating that the amino acid substitution results in a severely altered overall structure of the enzyme.

COX2

Functions:

Table 2.8

Binding site:

Copper cation A1 binding site at amino acid positions 161, 196, 200 and 207.

TABLE 2.7 Binding site for cytochrome C oxidase I.

Type	Position	Description
Binding site	40	Ca^{2+}
Binding site	45	Ca^{2+}
Binding site	61	Fe of Fe(II)-haem a, axial binding residue
Binding site	240	Cu cation B
Binding site	244	O_2
Binding site	290	Cu cation B
Binding site	291	O_2
Binding site	368	Mg^{2+}
Binding site	369	Mg^{2+}
Binding site	376	Fe of haem a3, axial binding residue
Binding site	378	Fe of Fe(II)-haem a, axial binding residue

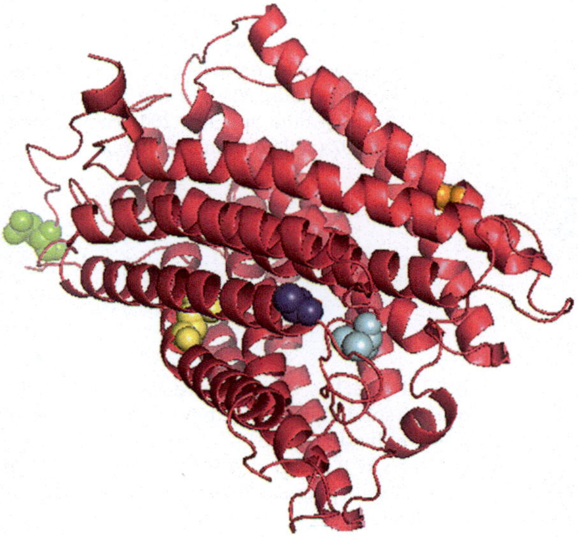

FIGURE 2.13 3D structure for cytochrome c oxidase subunit I (Cox1) with single nucleotide polymorphisms/mutations of Cox1 causing diseases. aa 513: split pea green sphere, leber hereditary optic neuropathy, aa 142: purple blue sphere mitochondrial complex IV deficiency (MT-C4D), aa 196: yellow orange sphere MT-C4D, aa 125: aquamarine (cyan) sphere colorectal cancer (CRC), aa 458: orange sphere CRC.

TABLE 2.8 Functions for cytochrome c oxidase subunit II (Cox2).

Molecular function	Copper ion binding
Molecular function	Cox activity
Biological process	ATP synthesis coupled to electron transport
Biological process	Cellular respiration
Biological process	Lactation
Biological process	Mitochondrial electron transport, cytochrome C to oxygen
Biological process	Response to hypoxia

Copper cation A2 binding site at amino acid positions 196, 198, 200 and 204.
Mg^{2+} binding site at amino acid position 198 (see Fig. 2.14).
Fig. 2.15
MT-C4D
Fig. 2.16

ATP synthase F(0) complex subunit 8

Function: Here, we describe the function of ATP8 at molecular level.
Table 2.9

Mitochondrial complex V deficiency, mitochondrial 2 (MC5DM2):
Identified mutation causing the disease was W55R.

Cardiomyopathy, infantile hypertrophic:
Identified mutation causing the disease was W55R.

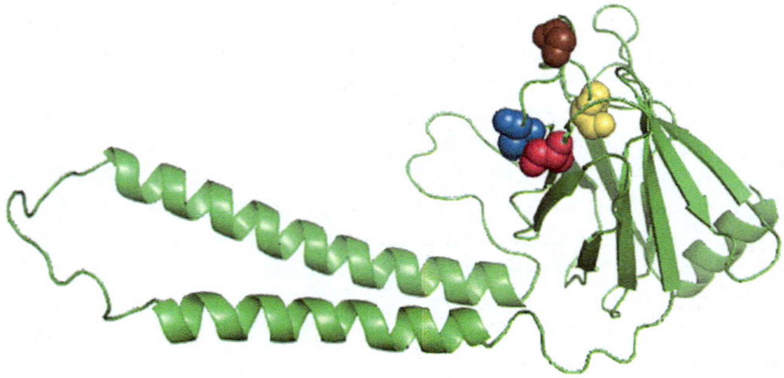

FIGURE 2.14 3D structure for cytochrome c oxidase subunit II with the binding sites predicted. Copper binding site (Cu cation A1): 161 (hot pink sphere), 196 (sky blue sphere), 200 (ruby sphere) and 207 (yellow orange sphere).

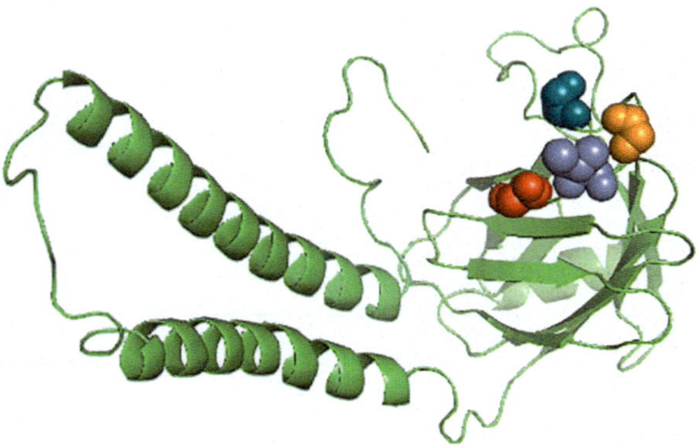

FIGURE 2.15 3D structure for cytochrome c oxidase subunit II with the binding sites predicted. Copper binding site (Cu cation A1), copper binding site (Cu cation A2): 196 (red sphere), 198 (slate bluish sphere), 200 (orange sphere) and 204 (deep teal sphere).

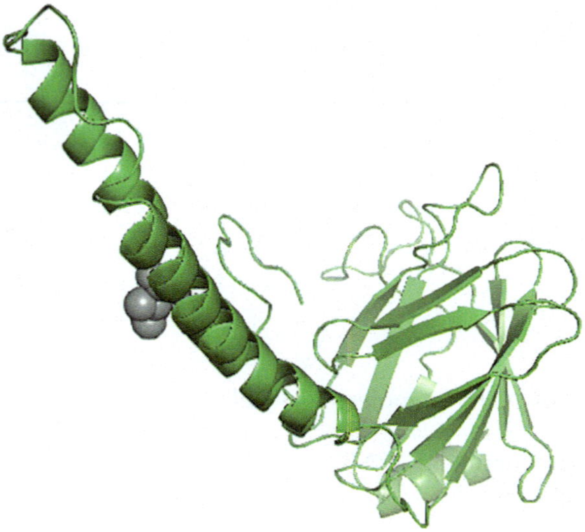

FIGURE 2.16 3D structure for cytochrome c oxidase subunit II mutations detected at M29K in grey sphere.

TABLE 2.9 Functions for ATP synthase F(0) complex subunit 8.

Molecular function	Proton transmembrane transporter activity
Biological process	Proton motive force-driven ATP synthesis
Biological process	Proton-motive force-driven mitochondrial ATP synthesis

TABLE 2.10 Functions for ATP synthase F0 subunit 6".

Molecular function	Proton channel activity
Biological process	Proton motive force driven ATP synthesis
Biological process	Proton motive force driven-mitochondrial ATP synthesis
Biological process	Proton transmembrane transport
Biological process	Response to hyperoxia

ATP synthase F0 subunit 6"

Function: We have tabulated the important functions of mitochondrial ATP6 at molecular level.

Table 2.10

Neuropathy, ataxia and retinitis pigmentosa:

Identified mutation causing the disease is L156R.

LHON:

Identified mutation causing the disease is I192T.

Mitochondrial infantile bilateral striatal necrosis:

Identified mutation causing the disease is L217P.

MC5DM1:

Identified mutation causing the disease is L156P.

Myopathy, lactic acidosis and sideroblastic anaemia 3:

Identified mutation causing the disease is S148N.

Ataxia and polyneuropathy, adult-onset:

Identified mutation causing the disease ia L156P (see Fig. 2.17).

COX3

Function: Following table details the function of Cox3.

Table 2.11

LHON:

Identified mutations causing the disease is G78S (secondary mutation, does not seem directly to cause the disease) and A200T (possible rare mutation).

MT-C4D:

Deletion mutation at 94–98.

SNPs/mutations detected:

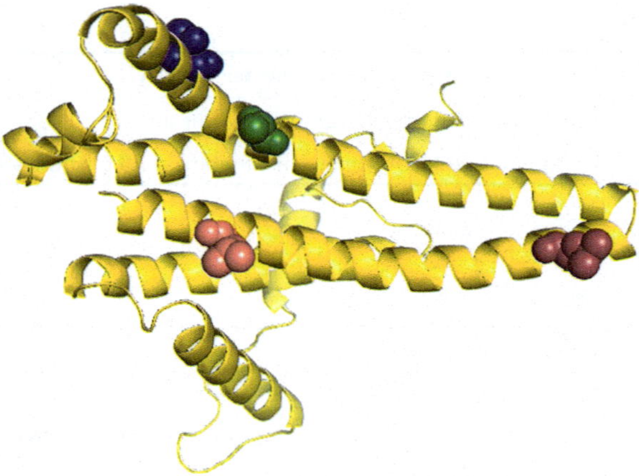

FIGURE 2.17 3D structure for ATP6 with the mutations or single nucleotide polymorphisms (SNPs) detected in ATP6 resulting into diseases. SNPs detected were L156R (forest green sphere), L156P forest green sphere, I192T (raspberry sphere), S148N (density blue) and L217P deep salmon sphere).

TABLE 2.11 Functions for cytochrome c oxidase subunit III (COX3).

Molecular function	COX activity
Biological process	Cellular respiration
Biological process	Mitochondrial electron transport, cytochrome C to oxygen
Biological process	Respiratory chain complex IV assembly

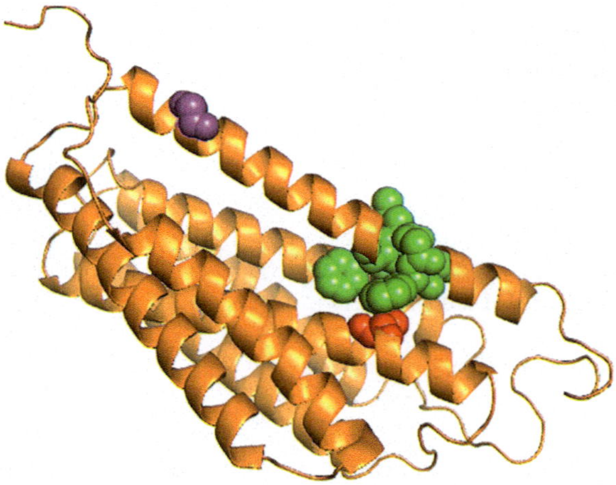

FIGURE 2.18 3D structure of cytochrome c oxidase subunit 3 with sites for mutation detected as G78S (magenta sphere), A200T (red sphere) and del 94–98 (green sphere).

Mitochondrial diiseases occur largely due to mutations in the mitochondrial genes that lead to the production of defective proteins.

Fig. 2.18

NADH dehydrogenase (ubiquinone) iron-sulphur protein 3, mitochondrial

Functions:

Table 2.12

PTM sites:

Post translation modification sites include sites for phosphothreonine at aa position 43, ubiquitinated lysine at aa 56 and acetyl lysine at aa 56 (see Fig. 2.19).

MC1DN8:

Identified mutations were R140W (causes a decrease in enzyme activity, impaired assembly of Complex I), T145I (decrease in enzyme activity, increased protein instability and aggregation), R199W (decrease in enzyme activity, impaired assembly of Complex I, increased protein instability and aggregation).

NADH dehydrogenase subunit 4 L

Function:

Table 2.13

TABLE 2.12 Functions for nicotinamide adenine dinucleotide (NADH) dehydrogenase (ubiquinone) iron-sulphur protein 3, mitochondrial.

Molecular function	Electron transfer activity
Molecular function	NADH dehydrogenase (ubiquinone activity)
Molecular function	NADH dehydrogenase activity
Biological process	Aerobic respiration
Biological process	Mitochondrial electron transport, NADH to ubiquinone
Biological process	Mitochondrial respiratory chain complex I assembly
Biological process	Proton motive force-driven mitochondrial ATP synthesis
Biological process	Reactive oxygen species metabolic process
Biological process	Substantia nigra development

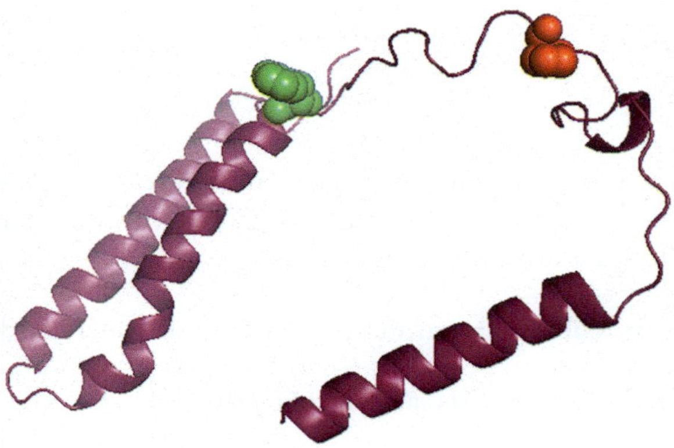

FIGURE 2.19 3D structure for ND3 with sites for mutation detected as aa 43 (red sphere) and aa 56 (green sphere).

TABLE 2.13 Functions for nicotinamide adenine dinucleotide (NADH) dehydrogenase subunit 4 L.

Molecular function	NADH dehydrogenase (ubiquinone) activity
Biological process	Aerobic respiration
Biological process	Mitochondrial electron transport, NADH to ubiquinone
Biological process	Proton motive force driven mitochondrial ATP synthesis

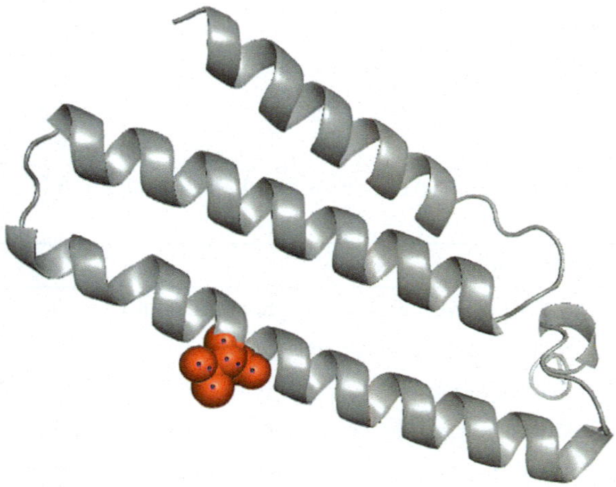

FIGURE 2.20 3D structure for ND4L with mutational sites detected as V65A (red sphere).

Transmembrane sites were predicted for 1–21 (helical), 29–49 (helical) and 58–78 (helical).

Involvement in disease

LHON

Possibly rare primary mutations identified as V65A (see Fig. 2.20).

NADH dehydrogenase subunit 4

Functions:

Table 2.14

Involvement in disease

SNPs/mutation in ND4: Fig. 2.21

LHON:

Identified mutation was R340H. It is the primary and most frequent mutation identified with almost no vision recovery. The mutation seems to have no effect on electron transfer activity of the complex in inner mitochondrial membrane preparations.

LHON with dystonia (LDYT):

Identified mutation was V313I. This is possibly rare mutation, leading to decreased enzyme activity.

Mitochondrial encephalomyopathy with lactic acidosis and stroke-like episodes syndrome (MELAS):

TABLE 2.14 Functions for nicotinamide adenine dinucleotide (NADH) dehydrogenase subunit 4.

Molecular function	NADH dehydrogenase (ubiquinone) activity
Molecular function	Ubiquinone binding
Biological process	Aerobic respiration
Biological process	Cerebellum development
Biological process	Electron transport coupled proton transport
Biological process	In utero embryonic development
Biological process	Mitochondrial electron transport, NADH to ubiquinone
Biological process	Mitochondrial respiratory chain complex I assembly
Biological process	Proton motive force driven mitochondrial ATP synthesis
Biological process	Response to ethanol
Biological process	Response to hypoxia

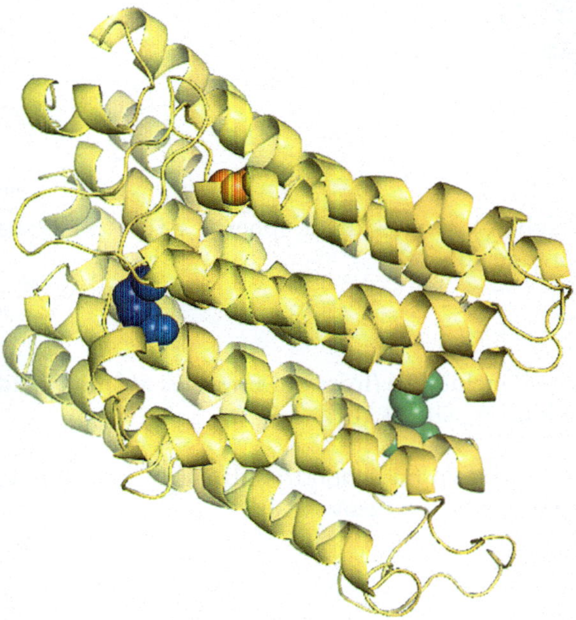

FIGURE 2.21 3D structure for ND4 with sites for mutation detected at R340H (green sphere), V313I (blue sphere) and T109A (red sphere).

The identified mutation was T109A.

NADH dehydrogenase subunit 5

NADH dehydrogenase (ubiquinone) 1 beta subcomplex subunit 5, mitochondrial

Function: Important roles are being listed below.

Table 2.15

TABLE 2.15 Functions for nicotinamide adenine dinucleotide (NADH) dehydrogenase subunit 5.

Molecular function	NADH dehydrogenase (ubiquinone) activity
Biological process	Aerobic respiration
Biological process	Mitochondrial electron transport, NADH to ubiquinone
Biological process	Proton motive force driven mitochondrial ATP synthesis

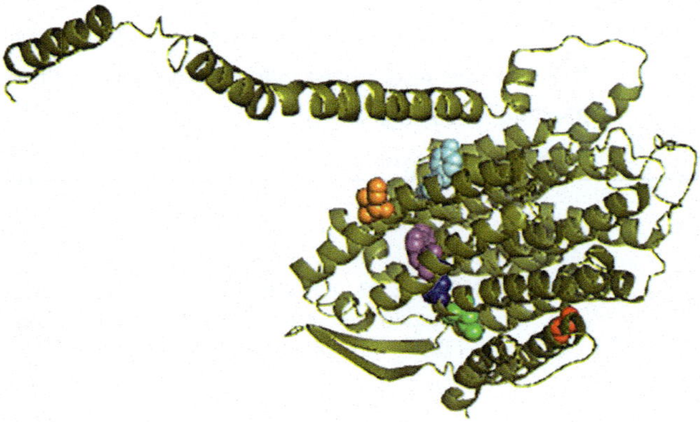

FIGURE 2.22 3D structure for ND5 with sites for post translation modification.

Fig. 2.22

ND5: aa 39 (red sphere), aa131 (green sphere), aa 133 (blue sphere), aa146 (magenta sphere), aa 176 (cyan sphere) and aa 182 (orange sphere).

Cytochrome b

Functions:

Table 2.16

The important binding sites for cytochrome B include Fe (iron) binding sites at amino acid positions 83, 97, 182, 196 and a ubiquinone binding site at 201.

Fig. 2.23

Disease occurrence due to SNPs/mutations: Certain mitochondrial diseases occur due to mutations in mitochondrial gene Cytochrome B as listed below:

CMIH:

The identified mutation was G521D.

LHON:

The identified mutations were D171N (uncertain significance, secondary mutation), V356M (secondary mutation, does not seem to directly cause the disease).

SNPs/mutations for cytochrome B causing diseases (see Fig. 2.24).

NADH-ubiquinone oxidoreductase chain 6

Function: We have listed the important functions for the gene tabulated below.

Table 2.17

Fig. 2.25

TABLE 2.16 Function for cytochrome B.

Molecular function	**Metal ion binding**
Molecular function	Protein-containing complex binding
Molecular function	Ubiquinol-cytochrome C reductase activity
Biological function	Animal organ regeneration
Biological function	Cellular respiration
Biological function	Electron transport coupled proton transport
Biological function	Mitochondrial electron transport, ubiquinol to cytochrome C
Biological function	Response to cadmium ion
Biological function	Response to calcium ion
Biological function	Response to cobalamin
Biological function	Response to copper ion

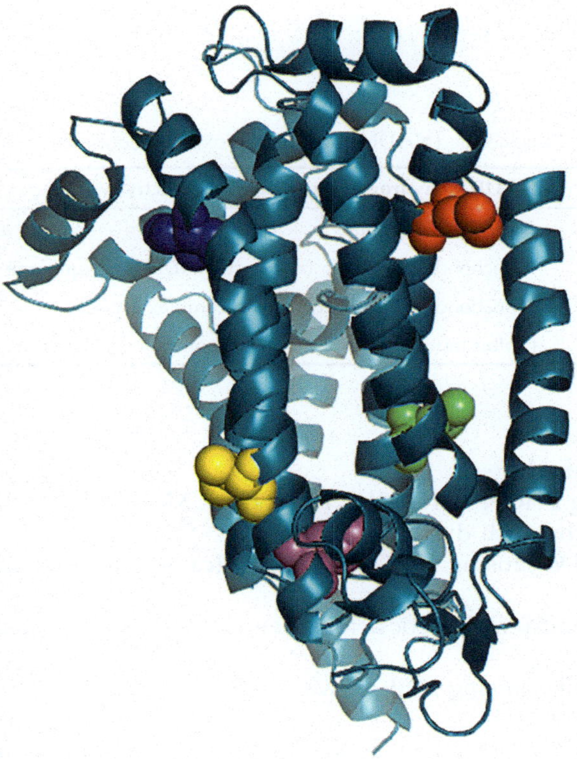

FIGURE 2.23 3D structure for cytochrome B with detected binding site for cytochrome B: aa83 (red sphere), aa 97 (green sphere), aa 182 (blue sphere), aa 196 (yellow sphere) and aa 201 (magenta sphere).

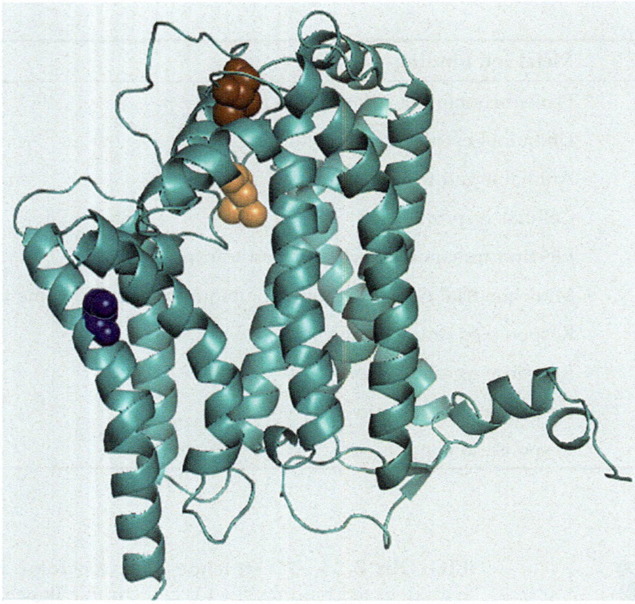

FIGURE 2.24 3D structure for cytochrome B with sites for mutation predicted as G251D (orange sphere), D171N chocolate sphere) and V356M (density blue sphere).

TABLE 2.17 Function for nicotinamide adenine dinucleotide (NADH)-ubiquinone oxidoreductase chain 6.

Molecular function	NADH dehydrogenase (ubiquinone) activity
Biological process	Aerobic respiration
Biological process	Mitochondrial electron transport, NADH to ubiquinone
Biological process	Mitochondrial respiratory chain complex I assembly
Biological process	Proton motive force-driven mitochondrial ATP synthesis

Involvement in disease

LHON (Leber hereditary optic neuropathy**):**

The identified mutations in ND6 causing LHON were G36S, Y59C, L60S, M64I and M64V.

LDYT (Leber hereditary optic neuropathy with dystonia**):**

The identified mutations detected in ND6 causing LDYT were I26M (causes decrease in enzyme activity), A72V.

MELAS (Mitochondrial encephalomyopathy with lactic acidosis and stroke-like episodes syndrome**):**

The identified mutations detected in ND6 causing MELAS was A74V.

LS (Leigh syndrome**):**

The identified mutations detected in ND6 causing LS was M63V. The detected mutation causes decrease in enzyme activity and impaired assemby of complex I.

Fig. 2.26

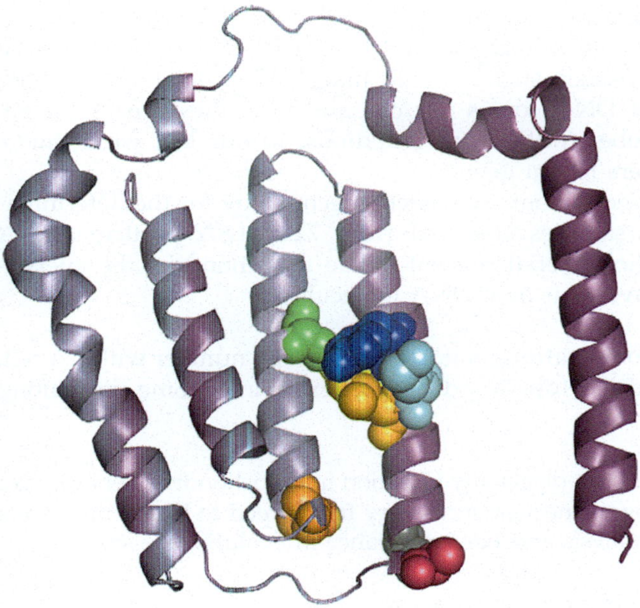

FIGURE 2.25 3D structure for ND6 with sites for mutation predicted as I26M (red sphere), G36S (green sphere), Y59C (blue sphere), L60S (yellow sphere), M63V (cyan sphere), M64I (orange sphere), M64V (orange sphere), A72V (grey sphere) and A74V (warm pink sphere).

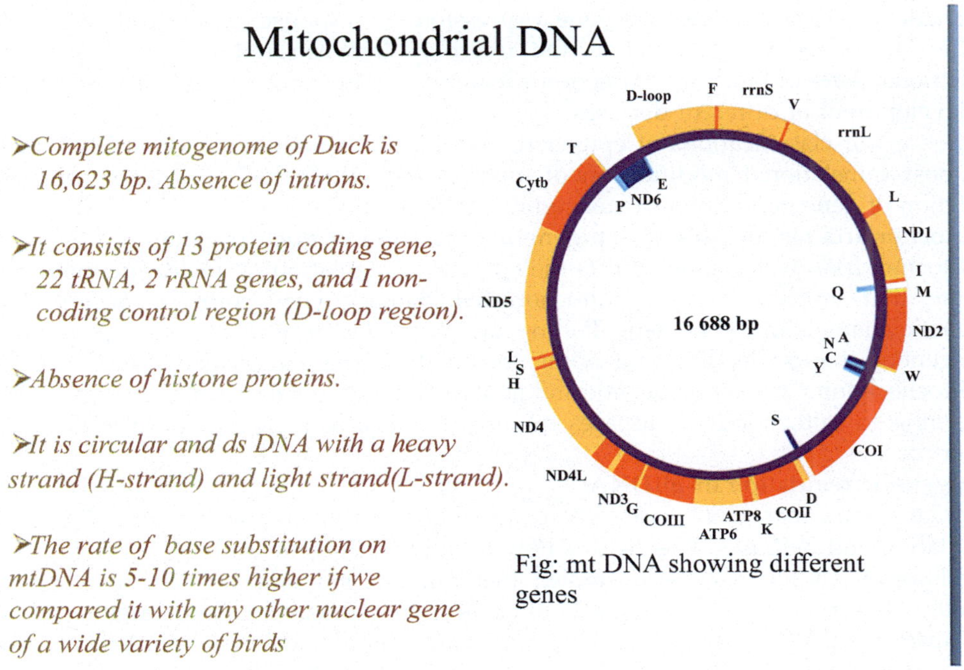

FIGURE 2.26 Schematic mitochondrial DNA (mtDNA) map.

There are two major considerations:

Firstly, the copy number variations in mitochondria. There exists a larger number of mitochondria in metabolically active tissues, such as heart, lungs and skeletal muscle. Still, we lack evidence for the number of DNA copies per cell and how they are regulated. Determination of mtDNA copy number (mtDNAcn) is tedious. In our lab, we estimate through quantitative PCR with primers for mt genes.

The other researchers have employed genome sequencing technology for the determination of the mtDNAcn from blood samples collected from 722 BLSA participants. A mtDNAcn is an estimation of genetic material present inside mitochondria, the energy-producing structures of a cell that have their own DNA molecules.

Between species:

Although there exists no information regarding mitochondrial copy number within a cell, not even mtDNAcn within mitochondria. However, difference exists in genome size among the species.

Between breeds:

So far, reports are available, we could not detect any variation in mitochondrial genome size among the breeds within species. Mitogenome sequences vary from breed to breed, this forms the basis for molecular phylogenetic analysis and has importance in evolutionary studies.

Among different body organs:

Mitochondrial copy number varies from organ to organ.

Mitochondrial epigenetics:

mtDNA is controlled by epigenetic mechanisms, a process called mitoepigenetics. Epigenetic mechanisms regulate gene expression in response to environmental stimuli and are key regulatory events at the interplay between mitochondrial and nuclear interactions. Alterations in the mt genome leading to mitochondrial dysfunction may affect the development of different diseases.

Three major classifications of epigenetic modifications include genomic DNA methylation, post translational modification of histone tails within higher-order chromatin and regulation of gene expression by noncoding RNAs (ncRNAs).

Mitochondria responsible for epigenetic regulation of nuclear gene:

'Mitochondrial dysfunction' affects the epigenetic regulation of nuclear genome. Studies indicate that 'mitocheckpoint' (mitochondrial damage checkpoint) is responsible for epigenetic changes in the nucleus. Pathogenic mutations in mtDNA may also have a role in epigenetic changes leading to genomic instability in the nuclear genome.

'Mitocheckpoint' causes epigenetic and genetic changes, resulting in phenotypic variation concerning to mitochondrial diseases or diseases in which mitochondrial defect plays a primary role.

Epigenetic regulation of mtDNA:

Similar to nuclear DNA, mtDNA is regulated by epigenetic factors, due to DNA methylation and ncRNAs. This field is now called mitoepigenetics.

Additionally, it has also been studied that nucleus and mitochondria are continuously communicating to each other to regulate different cellular pathways. Knowledge regarding the mechanisms underlying mitoepigenetics and nuclei–mitochondria communication, and also about the involvement of the ncRNAs in mitochondrial functions and related diseases, is not completely explored yet. ncRNAs also play a significant role in mitoepigenetic regulation.

The term 'epigenetics' was first termed by Conrad H. Waddington in 1942 to describe a whole complex of developmental processes connecting genotype to phenotype. Currently, epigenetics is referred to as heritable changes in the genome resulting in altered gene expression patterns without affecting the DNA sequence. Gene regulation conducted by epigenetic mechanisms is critical for cellular identity and homoeostasis, and is associated with vital cellular processes, including transcription, replication and DNA repair.

These epigenetic mechanisms occur through three deeply associated layers: DNA methylation, histone modifications and ncRNAs. DNA methylation and histone modifications are chemical events arranged by enzymes responsible for adding chemical residues. Both mechanisms are chromatin-related processes that act synergistically to ensure or block the DNA accessibility to the transcription machinery.

DNA methylation is carried out by the DNA-methyl transferases family of enzymes (DNMT1, DNMT2, DNMT3A, DNMT3B and DNMT3L). They catalyse the addition of a methyl group to position 5 of cytosines using S-adenosyl-l-methionine as a methyl donor. Nuclear DNA cytosine methylation is related not only to transcriptional repression but also to chromosomal stability. In mammals, 5-methylcytosine (5mC) is mainly found in the context of nuclear cytosine-guanine dinucleotides (CpG sites) in promoter regions, leading to gene silencing.

ncRNAs are RNAs that serve as several regulatory and structural functions as a protein template. In humans, they represent 70% of the genome, evidencing their abundance.

There exist two main classes of ncRNAs:

Structural ncRNAs: It includes transfer and ribosomal RNAs, which are not usually considered in the epigenetics context.

Regulatory RNAs: They are mostly epigenetic-related. As per their length, the epigenetic-related regulatory ncRNAs can be grouped into:

1. Long ncRNAs (>200 nt)
2. Circular ones
3. Small ncRNAs
4. miRNAs
5. Piwi-interacting RNAs

Fig. 2.27

Table 2.18

It was previously mentioned that though mitochondria lack histone packing, mtDNA is not 'naked', but forms a nucleoid (mitochromosome), where DNA is packed in TFAM-associated nucleoids. TFAM functions in mtDNA packing, replication, transcription and general maintenance of mtDNA. Also includes posttranslational modification as acetylation, phosphorylation and ubiquitination. TFAM in mitochondria is analogous to histone in nucleus. They alter mitochondrial gene expression and ubiquitination, though details yet to be explored. It is presumed that modifications in TFAM modify mitochondrial gene expression. It is interesting to note that several histone family members, particularly H2A and H2B, were found within mitochondria by a shotgun proteomics approach. These proteins were identified mainly in the mitochondrial membrane. Other than being directly bound to DNA and their function in the organelle is unclear (see Fig. 2.28).

Mitochondrial genetics—heretability, repeatability and genetic correlation for mitochondrial genes.

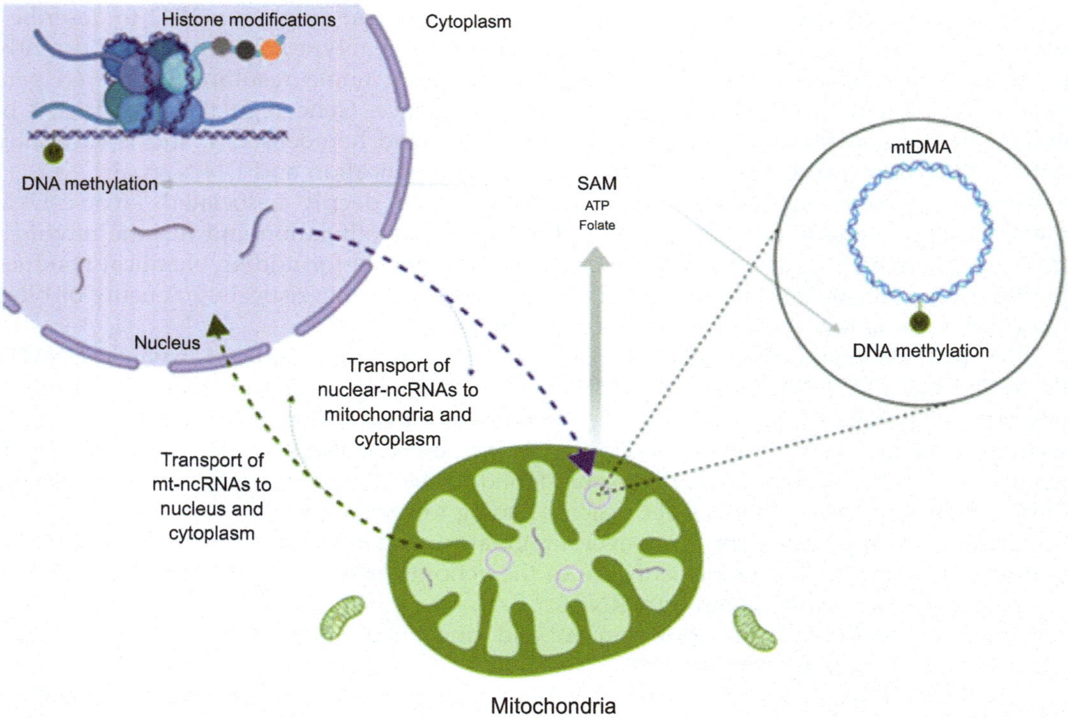

FIGURE 2.27 Mitochondrial epigenetics. Scheme showing the occurrence of DNA methylation in mtDNA and of noncoding RNAs (ncRNAs) inside mitochondria. These ncRNAs are encoded. Mitochondrial epigenetics. Scheme showing the occurrence of DNA methylation in mtDNA and of ncRNAs inside mitochondria. These ncRNAs are encoded by both nucleus and mitochondrial genomes and participate in the nuclei–mitochondria communication. Mitochondrial metabolism is responsible for the production of the universal methyl donor (S-adenosyl-l-methionine (SAM)), which is used in all DNA methylation events in a cell.

Genes are inherited from generation to generation, offsprings obtain their genes from their parents. The term 'gene' is used for both nuclear and mitochondrial genes. Although sperms possess mitochondria in their mid piece that allow them to travel a long distance to the ovum prior to fertilisation, the mid piece is sheded out prior to fertilisation.

Mitochondrial traits are inherited exclusively from the mother through the female germline, or oocyte. This is unlike nuclear DNA, which is inherited equally from both parents. Accordingly, mtDNA mutations and diseases are also always inherited through maternal inheritance. For instance, if a mother has a mutation in a mitochondrial gene, she inherits it onto all of her children, but her sons will not be able to pass it on to their children. However, men with mtDNA mutations will be affected by the conditions.

Such a kind of recombination is absent in the mitochondrial genome.

Fig. 2.29

Fig. 2.30

Mitochondrial inheritance empty pedigree disorders resulting from mutations in mtDNA have atypical patterns.

TABLE 2.18 Characterisation of mtncRNA and their related diseases and functions.

ncRNA in mitochondria	Type	Genomic origin	Probable signalling	Function	Related diseases	Organism
1 lncRNA	Chimeric	Mitochondrial	Retrograde	NS	NA	Mouse
3 lncRNAs	Simple	Mitochondrial	Retrograde	NS	Cervical cancer	Human
1 lncRNA	Chimeric	Mitochondrial	Retrograde	NS	NA	Mouse
1 lncRNA	Chimeric	Mitochondrial	Retrograde	NS	NA	Mouse, human
1 lncRNA	Chimeric	Mitochondrial	Retrograde	Cell proliferation	NA	Human
2 lncRNAs	Chimeric	Mitochondrial	Retrograde	Cell cycle regulation	Multiple cancers	Human
2 lncRNAs	Chimeric	Mitochondrial	Retrograde	Cell survival	Renal cell carcinoma	Human
2 lncRNAs	Chimeric	Mitochondrial	Retrograde	Apoptosis	Multiple cancers	Human
2 lncRNAs	Chimeric	Mitochondrial	Retrograde	NS	Melanoma	Mouse, human
1 lncRNA	Chimeric	Mitochondrial	Retrograde	Cell cycle regulation	NA	Mouse, human
2 lncRNAs	Chimeric	Mitochondrial	Retrograde	Cell proliferation	HPV infection and cervical cancer	Human
9 lncRNAs	NS	Mitochondrial	Retrograde	NS	Heart failure	Human

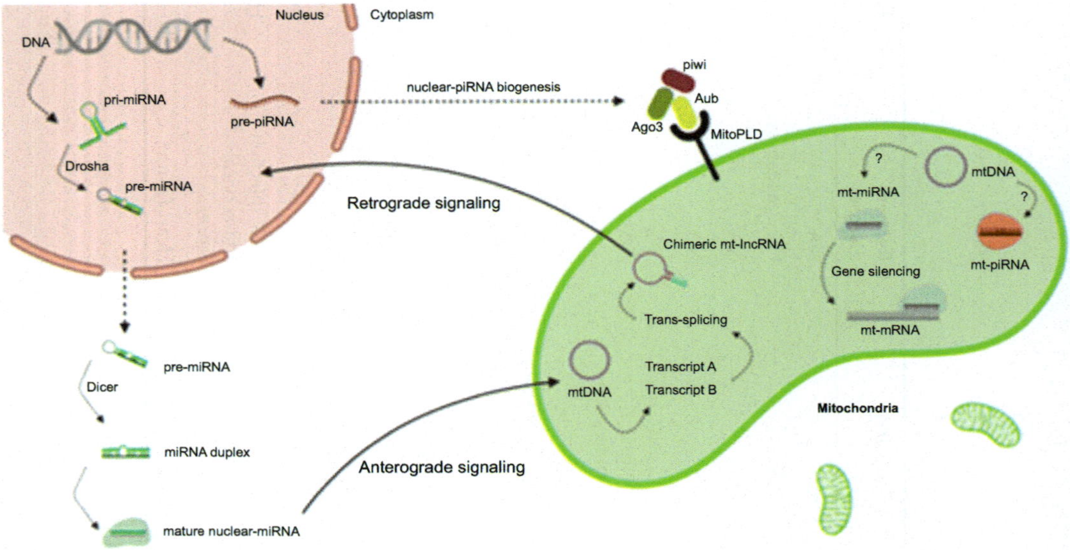

FIGURE 2.28 Mitochondria and noncoding RNAs (ncRNAs). Pathway showing the complex communication between nucleus and mitochondria through ncRNAs. Nuclear-ncRNAs (e.g. miRNAs) control mitochondrial gene expression through anterograde signalling. On the other hand, mitochondrial-ncRNAs (such as linear ncRNAs (lncRNAs)) regulate nuclear functions through retrograde signalling. Moreover, mitochondria secrete its own piwi-interacting RNAs (piRNAs) and participate in nuclear-piRNAs' biogenesis. nuclear functions with retrograde signalling.

- Both males and females are impacted.
- The trait is inherited from the female to her progeny.
- If a man possesses the trait while his partner does not, the trait will be absent in their children. In the initial pedigree, it is observed that when a female exhibits the characteristic, all of her progeny inherit the trait; however, in the subsequent pedigree, this is not invariably true. The manifestation of mitochondrial disorders is inconsistent. Each cell contains multiple mitochondria. Copy number variation exists in the quantity of mitochondria that harbour the mutation. A phenomenon referred to as heteroplasmy. A specific fraction of defective mitochondria in a cell can be accommodated without manifesting the disease in the organism. A greater proportion of mutant mitochondria, however, may result in the manifestation of the disease in the organism. The initial pedigree illustrates the transmission of inheritance via the female's cytoplasm within her ovum. The second pedigree illustrates the difference in expression contingent upon the proportion of mitochondria harbouring the mutation. Illustrative instance: LS.

Instances of mtDNAcn variation affecting diseases. A case study: mtDNA content—its genetic heritability and correlation with renal cell carcinoma.

Context: Research is underway to determine how genetic factors influence mtDNA content, also known as mtDNAcn, in normal human cells. Furthermore, it is uncertain whether inherited variations in mtDNA content in normal cells are linked to cancer susceptibility. Renal cell carcinoma constitutes 85% of all renal cancers. No research has

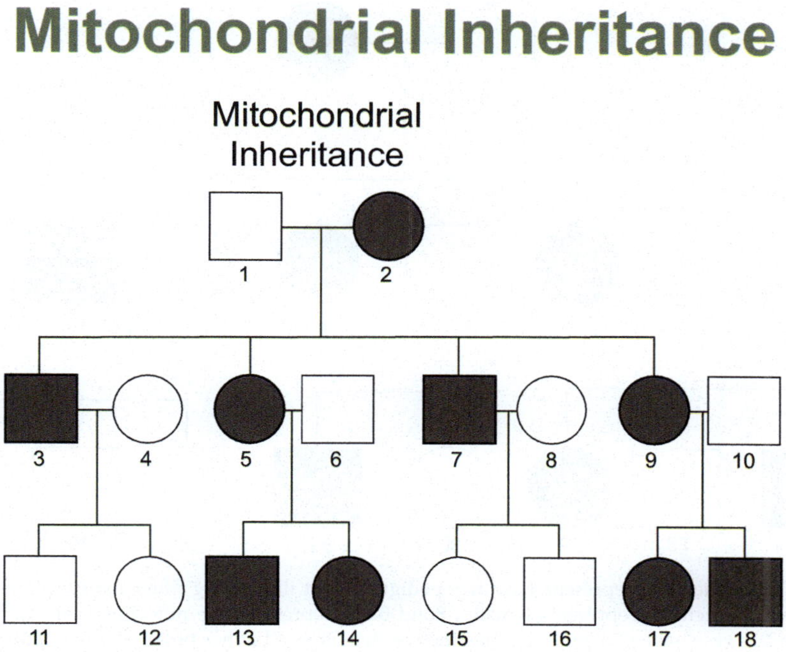

FIGURE 2.29 A schematic representation of a pedigree chart illustrating mitochondrial inheritance. Squares represent males, while circles represents females. Solid black represents the mitochondrial traits to be inherited from mother. It is observed that when a female exhibits the characteristic, all of her progeny inherit the trait, not the male.

demonstrated the association between mtDNA content and the risk of renal cell cancer. Approaches: A conventional twin research design was utilised to assess the genetic contribution to mtDNA content determination. The mtDNA content was quantified using real-time polymerase chain reaction in peripheral blood lymphocytes from 250 monozygotic twins, 92 dizygotic twins and 33 siblings (i.e. individual siblings of a twin pair). Biometric genetic modelling was utilised to assess the heredity of mtDNA content. A case-control study was conducted involving 260 patients with renal cell carcinoma and 281 matched control subjects, utilising multivariable logistic regression analysis to investigate the relationship between mtDNA content in peripheral blood lymphocytes and the risk of renal cell carcinoma. All statistical tests employed were bilateral. Results: The heritability of mtDNA content was found to be 65% (95% confidence interval (CI) =50% to 72%; $P<.001$). Heritability is defined as the proportion of phenotypic variation in a population that can be attributed to genetic diversity among individuals. Case patients possessing renal cell carcinoma were discovered to have a statistically substantially lower mtDNA content (1.18 copies) than control subjects (1.29 copies) (difference =0.11, 95% CI =0.03 to 0.17; P=.006). Individuals with low mtDNA content (i.e. below the median of control participants) exhibited a statistically significant elevation in the risk of renal cell carcinoma compared to those with high mtDNA content (odds ratio =1.53, 95% CI =1.07 to 2.19). While trend

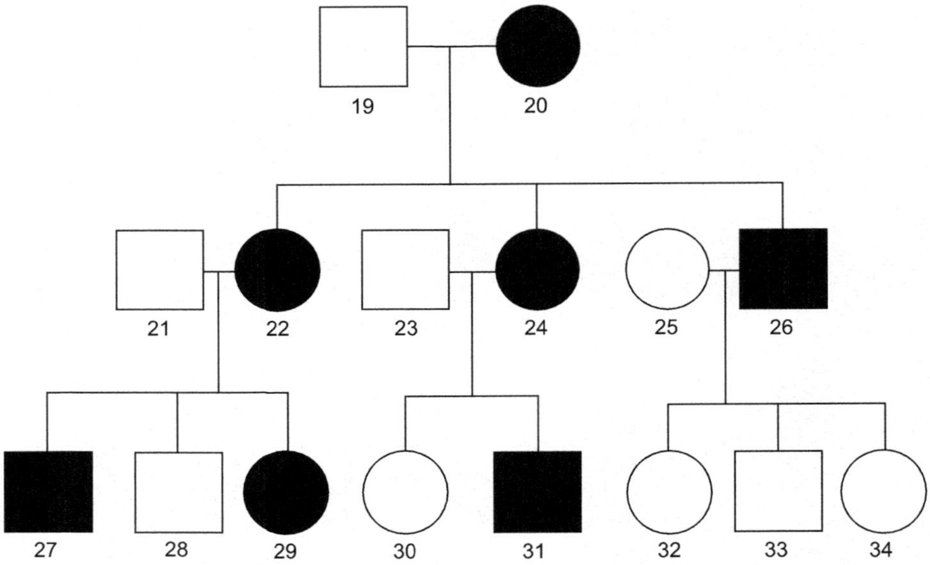

FIGURE 2.30 A schematic representation of a pedigree chart illustrating mitochondrial inheritance. Squares represent males, while circles represents females. Solid black represents the mitochondrial traits to be inherited from mother. In this case, when a female exhibits the characteristic, all of her progeny do not inherit the trait in F1 generation since the male number 26 is not able to transmit the trait to next F2 generation.

analysis was done, a statistically significant dose–response connection was identified between decreasing mtDNA content and rising risk of renal cell cancer (*P* for trend <.001). Thus mtDNA content was measured to exhibit strong heredity. Another noteworthy result is that diminished mtDNA content seems to correlate with an elevated risk of renal cell cancer.

Biparental inheritance of mtDNA in humans: a case study: Controversies exist about the potential coexistence of paternal mtDNA transfer alongside maternal mtDNA transmission. Nonetheless, it is widely accepted that mitochondria and mtDNA are inherited solely from the maternal lineage in humans. In this instance, three distinct multigenerational families were identified exhibiting a significant degree of mtDNA heteroplasmy (varying from 24% to 76%) among a total of 17 individuals. Heteroplasmy of mtDNA was independently assessed using high-depth full mtDNA sequencing analysis conducted simultaneously in different facilities. These were identified through laboratories approved by the Laboratory Improvement Amendments and the College of American Pathologists employing various methodologies. An extensive analysis of mtDNA segregation in these families indicates biparental mtDNA transmission with a pattern of inheritance resembling autosomal dominance. The results indicate that, while the core dogma of maternal inheritance of mtDNA is valid, there are unusual instances of paternal mtDNA transmission to kids. The investigation of the molecular mechanism behind this atypical manner of inheritance will yield fresh insights into the transmission of mtDNA from parent to progeny. This could potentially result in the development of novel approaches aimed at therapeutic interventions for pathogenic mtDNA transmission.

Hence, the penetrance and expressivity for the traits governed by genes vary. The factors complicating the process include:

1 Copy number variation among the mitochondria within the cell
2 Heteroplasmy
3 mtDNAcn variation within the cell
4 Size and shape of mitochondria—also determine mtDNA content
5 Mutations for mtDNA

The mitochondrial mutation is one of the most important factors. To date, a lot of mitochondrial mutations have been detected mostly in the coding region and also in D-loop region, leading to a lot of diseases. But the condition of heteroplasmy further creates complications.

It is to be noted that heteroplasmy is a condition where some of the mitochondria are mutant while others are wild type. Whenever the proportion of mutant mitochondria that is containing DNA mutations, increases compared to the wild type, disease occurs.

Does any mechanism exist in the body to correct the condition of heteroplasmy? There exists a natural mechanism for neutralisation of mutated cells through apoptosis and mitophagy. There exist certain open questions: when and how these are regulated?

In the next chapter, we have explained in detail about mitophagy. Certain regulating factors have been explained. Still, it is not completely clear if the entire heteroplasmy is being corrected, or a part of it.

We can summarise certain interesting facts about mitochondria:

Generally, we learn that mitochondria is present in cells in multiple copies.

Copy number variation exists depending upon the energy requirement of the tissue, for example, muscle, heart, brain and kidney require more energy. How copy numbers for mitochondria are regulated is not known.

There exists variation in the shape and size of mitochondria. Elongated shape of mitochondria may be formed by mitochondrial fusion.

mtDNA exists in the matrix of mitochondria in multiples, within the inner mitochondrial membrane. The exact organisation of mtDNA, the mechanism of packaging is not properly known. mtDNA is double-stranded, circular in shape.

There exists heteroplasmy, that is there is the presence of both mutated and wild type mtDNA within the cell. Whenever the mutated mtDNA exceeds that of the wild type, an abnormal phenotype is exhibited.

There exists a process of mitophagy, in which the abnormal/mutated mitochondria are destroyed.

Recent reports indicate the transfer of mitochondria from one cell to another.

There is a complex interaction between the nucleus and mitochondria. Around 200 nuclear genes code for mitochondrial proteins, apart from 37 own genes of mitochondria.

Nuclear mitochondrial DNA exists, mtDNA often integrates with the nucleus.

In the most recent report, there is the existence of circulating mtDNA in blood, which may integrate into any organ.

A case analysis: Correlation of personality traits with blood mtDNAcn—analysed from genomic sequences in two genetic cohort studies. Context: mtDNAcn in tissues and blood is known to be altered in illnesses such as diabetes, serious depression, ageing and lifespan. To date, there is scant research regarding the correlation between mtDNAcn and personality traits, as well as their connections to emotional states, metabolic health and lifespan. The study aims to investigate blood mtDNAcn and its link with personality traits, as well as to demonstrate the correlation between personality and mortality. Methods: The five major personality domains and facets were assessed using the revised NEO personality inventory, depressive symptoms were evaluated with the centre for epidemiologic studies depression scale, mtDNAcn levels were estimated through whole-genome sequencing, and mortality was monitored in participants from the Baltimore longitudinal study of ageing. Results: mtDNAcn exhibited a negative correlation with the neuroticism domain and its facets, while demonstrating a positive correlation with features from the other four domains. Moreover, elevated depressed symptoms correlated with reduced mtDNAcn levels. mtDNAcn was found to be associated with personality traits and mortality risk. mtDNAcn may serve as a biomarker elucidating a portion of the relationship between personality and mortality.

Presence of mtDNA in blood cells: a guess on how plasma acquires mtDNA. ***The fate of mtDNA upon cellular demise.*** During apoptosis, mtDNA is fragmented into minute pieces that subsequently circulate in the bloodstream. Cell loss resulting from injury, sickness, or disruptions in blood flow can lead to the observation of bigger fragments, which may incite persistent inflammation.

- **Erythrocytes:** mtDNA is located within human red blood cells and associates with them rather than with plasma.
- **Thrombocytes:** Platelets possess the capability to release undamaged mitochondria into the bloodstream under specific conditions. The quantity of mtDNA in blood cells may indicate mitochondrial health and an individual's capacity to metabolise glucose into energy. Alterations in mtDNA levels were shown to be correlated with illnesses such as diabetes and cancer, in addition to ageing and death. Individuals exhibiting elevated mtDNA levels in their blood correlated with extended lifespans, diminished incidences of depression and neuroticism, and enhanced manifestations of specific personality traits. Consequently, mtDNAcn may function as a molecular indicator of mitochondrial activity. Epidemiologic research has demonstrated that reduced mtDNAcn in blood correlates with diminished health, increased all-cause mortality and several cardiometabolic features, including elevated LDL cholesterol levels.

A case study: Correlation between whole blood-derived mtDNAcn, low-density lipoprotein, cholesterol and cardiovascular disease risk.

The association between mtDNAcn and cardiovascular disease represents a recent discovery. Cross-sectional and prospective association analyses of blood-derived mtDNAcn and cardiovascular disease outcomes were conducted in 27,316 adults across eight cohorts representing various racial and ethnic groupings, utilising whole-genome sequencing. Mendelian randomisation was conducted to investigate the causal associations between mtDNAcn and coronary heart disease, as well as cardiometabolic risk variables including obesity, diabetes, hypertension and hyperlipidaemia. We confirmed the majority of the previously documented correlations between mtDNAcn and cardiovascular disease outcomes. Mendelian randomisation analysis indicated no causal relationship between decreased mtDNAcn and increased coronary heart disease risk. Bidirectional Mendelian randomisation analyses indicate that low-density lipoprotein cholesterol exerted a causal influence on mtDNAcn, however, the opposite relationship was not statistically significant. No causal relationships were detected between mtDNAcn and obesity, diabetes, or hypertension, in either direction. Multivariable Mendelian randomisation analysis indicated no causal influence of coronary heart disease on mtDNAcn, but a significant direct causal effect of elevated low-density lipoprotein cholesterol on reduced mtDNAcn was seen, after correcting for coronary heart disease status.

Replication of mtDNA:

The maintenance of mitochondrial genome is independent of the nuclear genome through the proteins which are transported to the organelle in case of DNA replication, transcription, translation and repair.

Apart from the 13 polypeptides encoded by mtDNA, the mitochondrial proteome consists of an estimated 1158 proteins which are encoded by nuclear genes and transported to mitochondria through a variety of import pathways. The expression of these proteins was controlled through epigenetic, posttranscriptional, or posttranslational modifications either within the nucleus or the mitochondrion.

The replication of mtDNA is facilitated by the DNA polymerase gamma (Polγ) complex, consisting of a 140 kDa catalytic PolG encoded by the **POLG gene** and two 55 kDa auxiliary subunits expressed by the **POLG2 gene**.

The replisome complex comprises,

1 PolG,
2 TWINKLE and
3 mitochondrial SSB proteins.

The three major proteins involved with the process of mtDNA replication are Polγ, the mitochondrial helicase, TWINKLE and the mitochondrial single-stranded DNA binding protein, mtSSB. The Polγ holoenzyme is a heterotrimer comprising of two identical accessory subunits (Polγ-B) and one catalytic subunit (Polγ-A), when the accessory subunits enhance the processivity and catalytic function of Polγ-A.

The DNA helicase, TWINKLE, possesses a hexameric donut-shaped structure and is needed at the replication fork when it functions ahead of Polγ and unwinds the DNA template for replication.

Polγ was believed to be the sole polymerase within the mitochondrion, responsible for both mtDNA replication and repair; however, four other PolGs affect mtDNA maintenance, namely,

1. PolG beta,
2. PrimPol,
3. polymerase theta and
4. polymerase zeta.

The replication of mtDNA is a nonrestrictive process that transpires continuously during the cell cycle and throughout the lifespan of an organism.

It entails the replication of mtDNA molecules during mitosis, in addition to autonomous replication outside the cell cycle.

The process is facilitated by Polγ, which possesses 5′ → 3′ PolG activity and 3′ → 5′ exonuclease activity.

mtDNA replication relies on numerous nuclear-encoded proteins that are transported into the mitochondria to preserve and safeguard the mitochondrial genome. These proteins encompass **TFAM** that compacts DNA into dense nucleoid complexes. Deficiencies in these pathways might result in mtDNA maintenance problems, manifested as mtDNA depletion or deletions. The existing therapeutic alternatives for these problems are constrained. mtDNA is a circular, double-stranded, multicopy DNA genome that encodes numerous important protein components of the OXPHOS. It also encodes the RNA molecules necessary for the creation of these proteins within mitochondria. ***Upon release into the cytoplasm, extracellular space, or circulation, mtDNA can function as a damage-associated molecular pattern that activates innate immune sensors***. This may lead to antiviral resistance and can induce harmful inflammation in many illness situations.

Two-way process for mtDNA replication:

Strict replication: mtDNA molecules *replicate during mitosis*, analogous to the replication of nuclear DNA.

Relaxed replication: Continuous replication is going on in mtDNA molecules, independent of the cell cycle, *analogous to the replication process of bacterial genomes*.

The dominant perspective is that mtDNA molecules undergo unidirectional replication from two distinct, spatially and temporally separate, strand-specific origins. Replication initiates in the ***D-loop*** at the origin of H-strand replication (OH) with the synthesis of a daughter H-strand, resulting in the displacement of the parental H-strand. The primary reason for the rejection of the replication-rate hypothesis was that the duration needed for mtDNA replication is merely 1–2 hours, while mitochondrial division occurs over approximately 1–3 weeks.

The noncoding region (NCR) within mtDNA contains an origin of replication on the heavy strand (OH) and a second origin on the light strand (OL), which is located approximately 11 kb away from OH. The exact mechanism of mitochondrial replication is still not completely known and the theories conflict among them.

mtDNA transcription related to initiation, elongation and termination needs to be explored further. Since, the genetic information is present within both strands of mtDNA, the bidirectionality of mtDNA transcription was reported. In vitro reconstitution of mtDNA transcription requires TFAM, mitochondrial transcription factor B2 (TFB2M), and POLRMT, however, the exact role of TFAM has been debated.

Transcription and translation for mtDNA:

Out of 37 mitochondrial genes, there exist 13 polypeptide coding genes, one D-loop coding gene and the rest are tRNA coding genes that are responsible for oxidative phosphorylation processes. We have already discussed the mitochondrial codon and how it differs from nuclear codon. Although transcription is present for all 37 genes, translation is processed for only 13 genes.

LSP and heavy strand promoters (HSP)-promoter for mitochondrial gene expression:

The human mitochondrial genome contains two main LSP and HSP that are crucial for gene expression. These promoters are located within a NCR and are responsible for the initiation of transcription from the light and heavy strands of mtDNA. The LSP is situated upstream of the origin of H-strand mtDNA replication (OriH), while the HSP is positioned upstream of the origin of light-strand mtDNA replication (OriL). The transcription from these promoters is polycistronic, meaning that multiple genes can be transcribed from the same promoter. The transcripts synthesised from these promoters are rapidly processed by endoribonucleases to release 5′-monophosphorylated mRNAs, tRNAs and rRNAs for mitochondrial translation.

Transcription of mtDNA is started at one of two promoter regions, the heavy strand promoters (HSP1 and HSP2) and the light strand promoter (LSP) (see Fig. 2.31),

1. located on the heavy strand (HSP1 and HSP2) and
2. located on the LSP, where an overwhelming majority of transcripts are produced by transcription from HSP1.

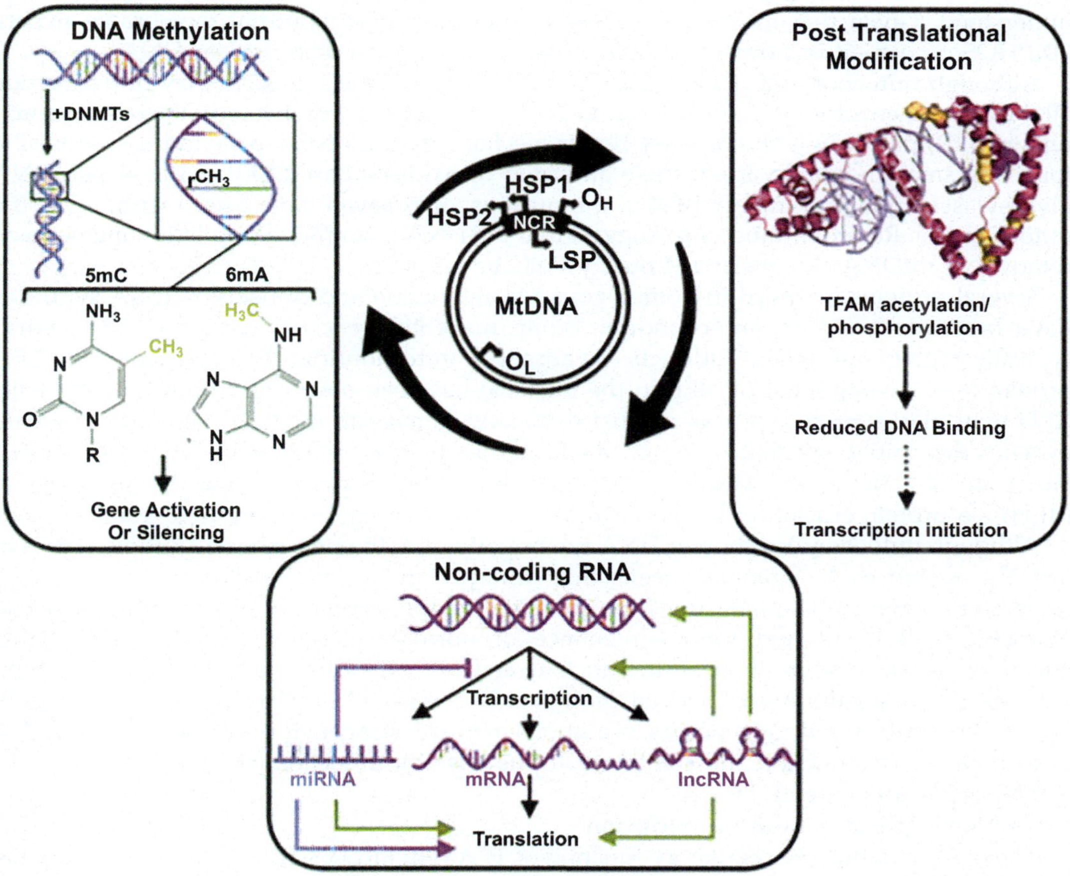

FIGURE 2.31 The 16.569 kbp mitochondrial genome is replicated from two origins OH and OL located 11 kb apart on the heavy and light strands, respectively. Mitochondrial DNA (mtDNA) codes for 13 polypeptide sequences, 22 tRNAs and 2 rRNAs. Transcription process is regulated by the heavy strand promoters (HSP1 and HSP2) and the light strand promoter (LSP).

Most of these control elements are present within the NCR. The main alterations that govern gene expression (gene activation or silencing) in the mitochondria are DNA methylation (cytosine or adenine where 5mC and 6 mA is 6-methyl adenine), nrRNAs and post translational modifications of nucleoid proteins.

In ***the process of posttranslational modification,*** the crystal structure of TFAM bound to DNA is denoted by the acetylation and phosphorylation sites within the HMG 1 domain, denoted by yellow and red spheres, respectively. The dotted arrow represents an unknown outcome. In the noncoding RNA panel, black arrows represent cellular processes, green arrows represent increased regulation, red arrow indicates a decrease in regulation, and inhibition is denoted by a red line.

Among the components involved in mitochondrial transcription, TFAM is known to be regulated through posttranslational modifications. However, the exact effects of these modifications on the epigenetic control of mitochondrial transcription are still not well understood. Given that mtDNA lies close to the inner mitochondrial membrane—where OXPHOS occurs—it is more vulnerable to oxidative damage than nuclear DNA.

Although mitochondria possess their own DNA repair mechanisms, they don't include all the components found in the nuclear DNA repair systems. For instance, the base excision repair (BER) pathway is the primary repair mechanism in mitochondria and is responsible for fixing small, nonbulky lesions like alkylated or oxidised bases. Out of the eleven DNA glycosylases known to initiate BER in the nucleus, only seven have been identified in the mitochondria. Recent findings also suggest that the DNA resynthesis step following damage removal in mtDNA can be carried out by PolG beta.

Several proteins involved in homologous recombination and translesion DNA synthesis have been identified in mitochondria. While nucleotide excision repair (NER)—which typically repairs bulky DNA adducts—is absent in mitochondria, the presence of the XPD protein (a key component of NER in the nucleus) has been observed within the organelle. XPD may play a protective role against oxidative damage to mtDNA. Similarly, the core proteins that initiate mismatch repair (MMR) in the nucleus are missing from mitochondria. However, low levels of MMR-like activity have been reported, and the presence of MLH1—a protein crucial for nuclear MMR—has also been detected in the organelle.

Although mitochondria contain DNA repair pathways, the presence of multiple copies of mtDNA within each organelle means that mutated and normal mtDNA can coexist—a condition known as heteroplasmy. This creates a complex scenario for both mitochondrial disease development and the maintenance of normal cellular function. Additionally, effective communication between the nucleus and mitochondria is essential for mitochondrial health. Since the mitochondrial proteome is composed of proteins encoded by both nuclear and mitochondrial genomes, mutations in nuclear-encoded proteins can also lead to mitochondrial dysfunction and contribute to the onset of mitochondrial diseases.

The regulatory role of TFAM

mtDNA replication and transcription

Prior to evaluating the regulatory function of TFAM in mtDNA transactions, it is essential to examine the processes of mtDNA replication and transcription. In mammalian cells, mtDNA replication is mediated by a distinct set of proteins that are functionally and evolutionarily separate from those involved in nuclear DNA replication, with several components sharing homology with bacteriophage replication factors. In human mitochondria, the primary replicative enzyme is POLγ, a heterotrimer composed of a single catalytic subunit (POLγA) and a dimer of accessory subunits (POLγB). Although additional PolGss—including PrimPol, PolG β, PolG θ and PolG ζ—have been proposed to function within mitochondria, none have been shown to functionally replace POLγ in mtDNA replication.

At the mitochondrial replication fork, POLγ operates in coordination with the mtDNA helicase Twinkle, which translocates ahead of POLγ to unwind the double-stranded DNA template. Twinkle functions as a hexamer or heptamer and requires a replication fork structure to be properly loaded and to initiate DNA unwinding. A critical accessory factor at the replication fork is the mtSSB, which binds to the nascent single-stranded DNA to protect it from nuclease degradation and to inhibit secondary structure formation. Additionally, mtSSB enhances replication efficiency by stimulating the helicase activity of Twinkle and by increasing the processivity of POLγ.

Mammalian mitochondria possess a dedicated transcriptional apparatus distinct from that of the nucleus. Transcription is initiated by the POLRMT in coordination with TFAM and TFB2M. Together, these components initiate transcription from the HSP and LSP in vitro. Transcription elongation by POLRMT is facilitated by the TEFM, which enhances the stability of the elongation complex by reinforcing the interaction between POLRMT and the DNA template.

Multiple models have been proposed to explain the mechanism of mtDNA replication. The most widely accepted is the strand displacement model. According to this model, replication initiates independently at two distinct origins: the origin of the heavy strand (OriH) and the origin of the light strand (OriL). DNA synthesis begins at OriH, resulting in the unidirectional synthesis of the nascent heavy (H) strand. As replication progresses and approximately two-thirds of the H-strand is synthesised, the OriL region becomes single-stranded and adopts a stem-loop secondary structure. This conformational change permits POLRMT to synthesise an RNA primer from a poly-dT tract within the loop region. These primers serve as initiation sites for POLγ to begin replication of the light (L) strand, using the parental H-strand as a template. Replication of both strands continues until two complete daughter mtDNA molecules are generated. The resolution of the replicated molecules is mediated by a topoisomerase 3-dependent decatenation process.

The precise molecular mechanisms governing the initiation of mtDNA replication at the origin of the heavy strand (OriH) remain incompletely understood. It is established that the RNA primers required for initiation at OriH are synthesised by the POLRMT. Transcription initiated from the LSP can yield full-length transcripts extending to the TERF1 binding site; however, the majority of transcription events initiated at LSP terminate prematurely after approximately 120 nucleotides, resulting in the formation of stable RNA–DNA hybrid structures known as R-loops. These nascent R-loops are subsequently processed by RNase H1, which generates 3' hydroxyl termini that serve as primers for POLγ to initiate DNA synthesis.

The balance between premature transcription termination leading to R-loop formation and the generation of full-length transcripts is modulated by the transcription elongation factor TEFM. Despite these advances, several critical aspects of OriH-dependent replication initiation remain unresolved, including the precise mechanisms of primer removal and replisome assembly. Furthermore, the role of TFAM in these processes is not yet fully defined. It has been hypothesised that TFAM-mediated compaction of mtDNA may influence primer processing and replication initiation, suggesting a potential epigenetic regulatory role for TFAM in mtDNA maintenance.

The role of TFAM in the regulation of mtDNA replication and transcription:

Across all domains of life, DNA compaction serves as a fundamental mechanism for regulating DNA transactions such as transcription and replication. As previously noted, the histone-like nucleoid structuring protein functions as a global transcriptional regulator in prokaryotes, modulating the expression of hundreds of genes. It also acts as an environmental sensor, facilitating physiological adaptations to changing environmental conditions. In eukaryotic nuclei, the organisation of DNA into chromatin through its association with histone proteins significantly impacts transcriptional regulation and DNA replication. Nucleosomes—structural units of chromatin composed of DNA wrapped around histone octamers—are often precisely positioned at gene promoters and origins of replication. Chromatin remodelling is mediated by specialised proteins that reposition nucleosomes or alter histone posttranslational modifications (e.g. acetylation or methylation), thereby enabling access to DNA by transcriptional and replicative machinery.

Emerging evidence indicates that mitochondrial nucleoids are highly organised structures, potentially regulated by mechanisms analogous to nuclear chromatin. In yeast, transcriptionally active nucleoids exhibit a lower abundance of the Abf2 protein per mtDNA molecule. Conversely, overexpression of Abf2p leads to rapid mtDNA loss, a phenomenon attributed to excessive DNA compaction that hinders access of the replication machinery to the mitochondrial genome.

The influence of TFAM concentration on mtDNA replication and transcription has been investigated using in vitro reconstituted nucleoid-like particles. These studies demonstrate that even modest increases in TFAM levels substantially reduce the proportion of mtDNA molecules accessible for transcription and replication, suggesting that TFAM-mediated compaction plays a critical regulatory role in mitochondrial gene expression and genome maintenance (see Fig. 2.32).

A proposed model illustrates how TFAM concentration and mtDNA compaction coordinate the regulation of mtDNA replication and transcription. In this model, increased levels of TFAM (depicted in green) lead to enhanced compaction of the mtDNA (represented as a grey duplex ring), ultimately resulting in the formation of a highly compacted nucleoid structure. Under conditions of low TFAM abundance, mtDNA remains less compacted, thereby providing a more permissive environment for the initiation and progression of both replication (indicated by black arrows) and transcription (represented by orange lines).

Conversely, elevated TFAM concentrations promote the formation of densely compacted nucleoids that are less accessible to the transcriptional and replicative machinery. As a result, mitochondria with high TFAM levels exhibit a reduced number of actively replicating and transcribing mtDNA molecules. This suggests that TFAM not only functions as a structural component of the nucleoid but also serves as a dynamic regulator of mitochondrial genome accessibility and gene expression.

Within compacted mitochondrial nucleoids, TFAM has been shown to form stable protein–DNA filaments that inhibit local DNA strand separation, thereby obstructing the progression of both the replication and transcription machinery. These findings support the hypothesis that relatively small changes in the TFAM-to-mtDNA ratio may serve as a regulatory mechanism controlling mitochondrial gene expression and genome replication.

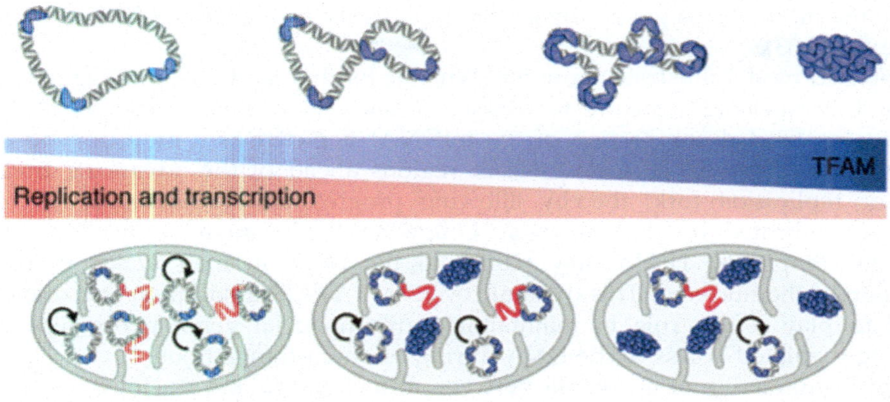

FIGURE 2.32 Regulatory role of mitochondrial transcription factor in mitochondrial DNA transactions.

In vivo studies have revealed heterogeneous patterns of colocalization between TFAM and bromodeoxyuridine (BrdU)-positive foci, a marker of active DNA synthesis. Notably, a subset of BrdU-labelled foci lacks detectable TFAM signal, suggesting that these may represent actively replicating mtDNA templates. It has been proposed that such regions, initially depleted of TFAM to permit replication, may later acquire TFAM and undergo compaction following the completion of DNA synthesis. This dynamic redistribution of TFAM may thus play a critical role in coordinating the temporal regulation of mtDNA accessibility during the replication cycle.

Super-resolution microscopy studies conducted in vivo have revealed the existence of distinct structural forms of mammalian mitochondrial nucleoids. These observations support the hypothesis that highly compacted nucleoids may serve as storage forms of mtDNA, whereas larger, more relaxed nucleoids are associated with active transcription and/or replication. This spatial and structural heterogeneity suggests that TFAM may act as an epigenetic regulator, modulating the availability of mtDNA molecules for transcription and replication through differential compaction.

In addition, posttranslational modifications of TFAM have been proposed as a mechanism for regulating its DNA-binding activity. Specifically, acetylation and phosphorylation events within the HMGA domain of TFAM may influence its interaction with mtDNA. Functional studies using TFAM mutants that mimic acetylated or phosphorylated states have demonstrated alterations in DNA binding affinity, compaction efficiency and sliding dynamics along DNA when compared to wild-type TFAM. These findings suggest that such modifications could play a regulatory role in controlling mtDNA transcription in vivo.

Analogous to the nucleus, where chromatin remodelling complexes modulate nucleosome structure to facilitate transcription, it is conceivable that mitochondria also possess specific remodelling factors that regulate TFAM-mediated packaging of mtDNA during replication and transcription. These may include mitochondrial-specific chaperones or enzymes that modulate TFAM-DNA interactions. Alternatively, the inherently high on/off dynamics of TFAM binding to DNA might obviate the need for such remodelling factors. These possibilities remain speculative and warrant further experimental investigation to elucidate the mechanisms governing nucleoid plasticity and mitochondrial genome regulation (see Fig. 2.33).

A proposed model illustrates how the dynamic binding and release of TFAM modulate mitochondrial nucleoid structure to facilitate transcription and replication. During these processes, localised decompaction of the mtDNA (depicted as a grey duplex) occurs as TFAM (shown in green) dissociates from regions ahead of the advancing transcription bubble or replication fork, thereby allowing progression of the respective molecular machinery (indicated by dark arrows). Following the passage of the transcription or replication complexes, TFAM may rebind to the DNA in their wake, promoting recompaction of the nucleoid. This dynamic cycling of TFAM association and dissociation is thought to play a key role in maintaining mitochondrial genome organisation while permitting regulated access to the DNA during essential transactions.

Complex interaction or cross-talk between mitochondria and nucleus:

In any physiological process, mitochondria can not work alone. It acts through a complex interaction between both nuclear and mitochondrial proteins through a series of biochemical pathways.

A case study:

We had conducted a study involving innate immunity against bacterial pathogen (*Pasteurella multocida*) in *Anas platyrynchos* (duck) model. We could explore the complex role of certain nuclear gene, such as NRLP3, IL18 and stimulator of interferon genes, with mitochondrial genes, such as cytochrome B and cytochrome C. This has been explained with biochemical KEGG pathways (see Fig. 2.34).

Fig. 2.35

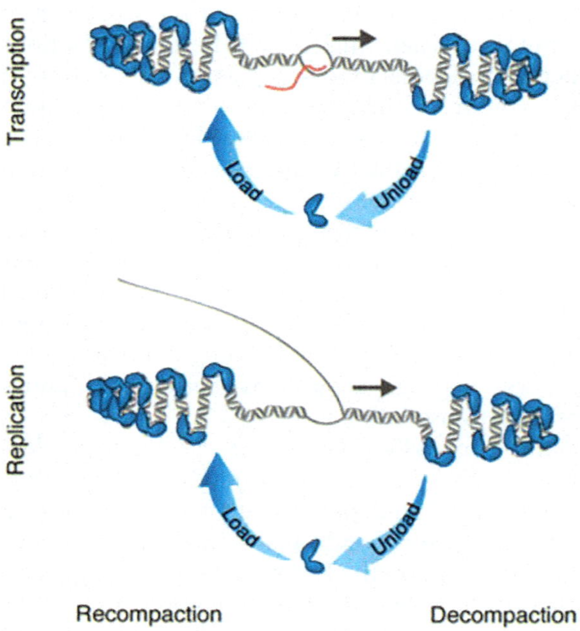

FIGURE 2.33 Dynamics of mitochondrial transcription factor binding during transcription and replication.

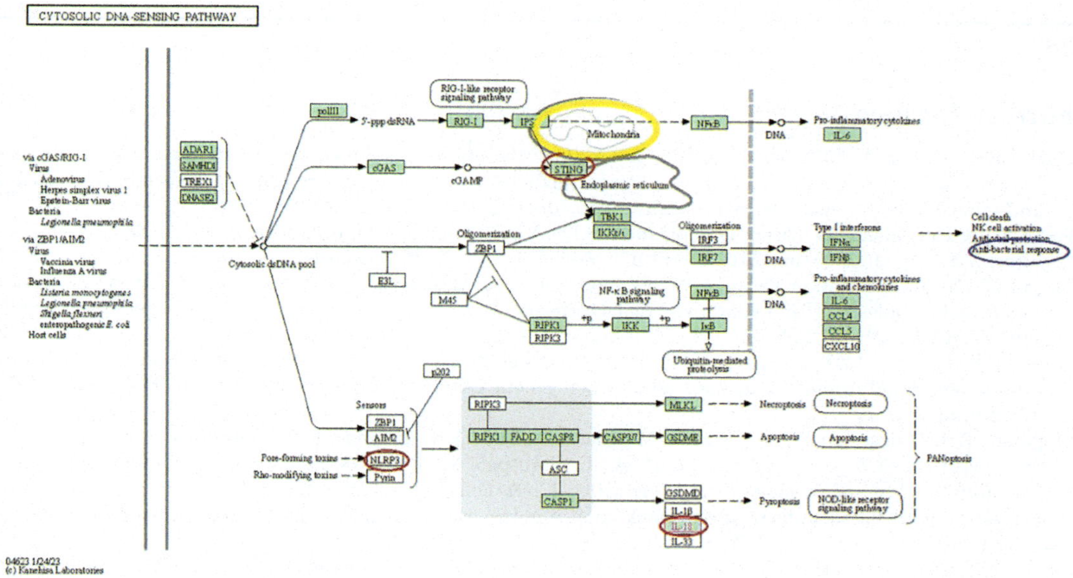

FIGURE 2.34 Role of mitochondria and other nuclear genes/proteins in cytosolic DNA sensing pathway.

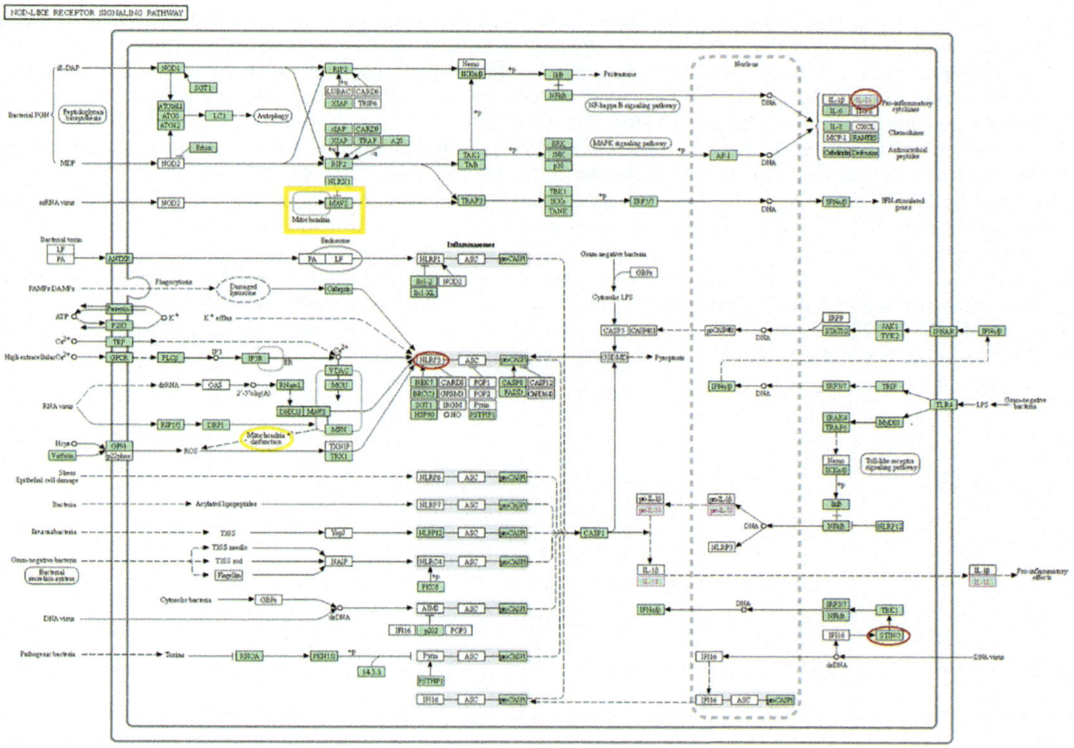

FIGURE 2.35 Role of mitochondria along with other nuclear protein/gene in NOD-like receptor signalling pathway.

Further Reading

Chakraborty,A. Pal, A. 2023. Mitochondrial genes affect the immune response against bacterial infection (Duck Pasteurellosis) through a cascade of mechanisms mediated by nuclear immune genes- a cross talk with nuclear and mitochondrial gene. bioRxiv preprint https://doi.org/10.1101/2023.08.30.555483

Farge, G., & Falkenberg, M. (2019). Organization of DNA in Mammalian Mitochondria. *Int J Mol Sci, 20*(11), 2770. https://doi.org/10.3390/ijms20112770 PMID: 31195723; PMCID: PMC6600607.

Liu, et al. (2023). Association Between Whole Blood–Derived Mitochondrial DNA Copy Number, Low-Density Lipoprotein Cholesterol, and Cardiovascular Disease Risk. *J Am Heart Assoc, 12*, e029090. https://doi.org/10.1161/JAHA.122.029090 2023.

Oppong, R. F., et al. (2022). Personality traits are consistently associated with blood mitochondrial DNA copy number estimated from genome sequences in two genetic cohort studies. *eLife*. https://doi.org/10.7554/eLife.77806 Epub Dec. 20.

Oppong1, R. F., Terracciano2,3, A., Picard4, M., Qian1, Y., Butler1, T. J., Tanaka1, T., Moore1, A. Z., Simonsick1, E. M., Opsahl-Ong1, K., Coletta5, C., Sutin6, A. R., Gorospe5, M., Resnick3, S. M., Cucca7, F., Scholz8,9, S. W., Traynor9,10, B. J., Schlessinger5, D., Ferrucci1*, L., & Ding1, J. (2022). Personality traits are consistently associated with blood mitochondrial DNA copy number estimated from genome sequences in two genetic cohort studies. * *eLife, 11*, e77806. https://doi.org/10.7554/eLife.77806.

Pal, A., Banerjee, S., Pathak, K., Das, M.K., Chatterjee, P.N. 2022. Molecular phylogenetic analysis for *Ovis aries* with whole mitochondrial genome sequencing. https://doi.org/10.1101/2022.07.29.502105.

Pal, A., Banerjee,S., Rawat, K., Deb, D. and Chatterjee, P.N. 2025. Copy number variations in mitochondrial DNA has a potent role in parasitic immunity against Haemonchus contortus-A novel report. https://doi.org/10.1101/2025.09.10.675310.

Pal,A., Chakraborty, A., Debnath,M. 2022. Whole mitochondrial genome sequencing- a novel approach for studying Phylogenomics for *Anas platyrynchos*. https://doi.org/10.1101/2022.07.11.499548.

Pal,A., Chakraborty, A., Debnath,M. 2022. Molecular evolution and characterization of domestic duck (Anas platyrynchos) and Goose (Anser indicus) with reference to its wild relatives through whole mitochondrial genome sequencing. https://doi.org/10.1101/2022.07.19.500621.

Pal, A., Pal, A. B., Banerjee, S., Batobyal, S., & Chatterjee, P. N. (2019). Mutation in ***Cytochrome B gene*** causes debility and adverse effects on ***health***of sheep. *Mitochondrion, 46*, 393–404 https://doi.org/10.1016/j.mito.2018.10.003.

Pradhan, M., Pal, A., Samanta, A. K., Banerjee, S., & Samanta, R. (2018). Mutations in ***cytochrome B gene*** effects female reproduction of Ghungroo pig. *Theriogenology, 119*, 121–130. http://doi.org/10.1016/j.theriogenology.2018.05.015.

Sahu, J., Pal, A. et al., 2023. Role of NUMTs (Nuclear mitochondrial DNA) genes in affecting disease resistance in duck against Pasteurellosis. bioRxiv preprint https://doi.org/10.1101/2023.08.30.555528.

Xing, et al. (2008). Mitochondrial DNA Content: Its Genetic Heritability and Association With Renal Cell Carcinoma. *Journal of the National Cancer Institute, 100*, 1104–1112 2008.

CHAPTER

3

Mitophagy

Introduction

It is always necessary to have a good excretory system for our body to excrete the undigested material as faeces, metabolic waste as nitrogenous waste for example, urea through urine. Cellular apoptosis is a similar phenomenon to destruction of damaged cells. In case of disruption, cells may be tumorogenic. Similarly, damaged mitochondria need to be removed through a well-regulated process.

What is mitophagy?

Mitophagy is a form of macroautophagy that specifically targets and degrades impaired mitochondria. Autophagosomes are double-membraned structures that may encapsulate entire mitochondria or selectively target damaged regions. Subsequently, autophagosomes merge with lysosomes for destruction.

Significance of mitophagy

- Maintains cellular health by preventing the accumulation of dysfunctional mitochondria that may result in cellular degeneration.
- Regulates mitochondrial quantity in response to fluctuating cellular metabolic requirements.
- Facilitates steady-state mitochondrial turnover, and
- Occurs during specific cellular developmental phases, such as the differentiation of red blood cells.

Mitochondrial Genomics
DOI: https://doi.org/10.1016/B978-0-443-26655-3.00007-X

Can mitophagy destroy mitochondria with a higher proportion of mutant mtDNA relative to wild-type mtDNA, in the context of heteroplasmy?

'Mitophagy of mitochondria' is equivalent to 'apoptosis of cells' or cell death. Oxidative phosphorylation results in the production of several reactive oxygen species (ROS) within the mitochondria and submitochondrial particles. ROS generated as mitochondrial byproducts would induce cytotoxicity and cellular apoptosis via mitochondrial impairment. Mitochondrial damage results in diminished ATP synthesis and the release of cytochrome C, subsequently activating caspases and initiating apoptosis. Mitochondrial damage is not solely attributable to oxidative stress or pathological processes. Normal mitochondria progressively acquire oxidative damage, rendering them detrimental to both the mitochondria and the cell. These faulty mitochondria will generate less ATP, elevate the formation of ROS, and consequently release pro-apoptotic proteins such as caspases. Therefore it is essential to eliminate damaged and old mitochondria to preserve cellular integrity. Consequently, there is sequestration and hydrolytic breakdown by the lysosome, a process referred to as mitophagy. A reduced quantity of mitochondria leads to a decrease in senescence effectors and phenotypes, while maintaining ATP synthesis via augmenting glycolysis.

Mitophagy is directly associated with specific disorders such as cancer, Parkinson's disease (PD), and tuberculosis. The pathways for mitophagy differ between mammals and yeast. PTEN-induced putative kinase 1 (PINK1) and PARKIN pathway in mammals, ATG32 for *Saccharomyces cerevisiae*.

Genetic regulation of mitophagy

According to the requirements of the book, we must investigate the genetic regulation of mitophagy. Mitophagy is initiated by stressors such as hypoxia, food deprivation, mtDNA damage, loss of mitochondrial membrane potential (MMP), and accumulation of ROS. Six mitophagy-related genes, proto-oncogene tyrosine-protein kinase (SRC), ubiquitin B (UBB), PINK1, FUN14 domain-containing protein (FUNDC1), microtubule associated protein 1 light chain 3 beta (MAP1LC3B), and CSNK2A1, were identified as closely connected with the onset and progression of non-small cell lung cancer. The principal routes found are the PINK1–Parkin-mediated ubiquitin pathway and the FUNDC1 receptor-mediated system.

Mitophagy is a key regulatory mechanism for mitochondrial quality control. It is regulated by PINK1/Parkin and hypoxia-inducible factor 1-alpha (HIF1α)- mediated adaptors.

BNIP3, NIX and FUNDC1: In the process of tumorigenesis, mitophagy is arrested via mutations in parkin. Mutations in BCL2/adenovirus E1B 19 kDa protein-interacting protein 3 (BNIP3) also lead to mitophagy dysregulation.

The salient features associated with mitochondrial dynamics and mitophagy affecting the mitochondrial quality control are shown in the figure. Normal physiological conditions have tubular mitochondria.

1. Mitochondrial injury caused by physiological stress leading to mitochondrial impairment.

2. Segregation of the damaged part of the mitochondria through mitochondrial fission process.
3. The impaired or damaged mitochondria are selectively flagged by the PINK1.
4. Facilitation of Parkin recruitment and mitophagy.
5. All repaired mitochondria are integrated into the functional mitochondrial network through fusion with other mitochondria.
6. The factors that affect fission, fusion and mitophagy are depicted.

The mitochondrial dynamics and mitophagy play vital roles in metabolism, cellular differentiation, cancer, neurodegeneration and several other pathologies. Earlier, the aspects of mitochondrial dynamics were studied in detail in the context of neurodegenerative disorders such as Parkinson's, Alzheimer's and Huntington's diseases. The function of mitochondrial dynamics in viral infections is rare and explained only for a very small number of viruses. Our comprehension of mitochondrial dynamics and mitophagy during viral infections is limited, although it undoubtedly signifies a burgeoning area of molecular research in virology. The roles of the mitochondria in innate immune signalling and metabolism depict its role in viral pathogenesis.

The mitochondria are affected in two ways:

1. Either directly targeted by viral proteins or
2. Influenced by the physiological alterations to cellular environment during viral pathogenesis, such as deregulated calcium homoeostasis, endoplasmic reticulum stress, oxidative stress and hypoxia.

Recent work has revealed the mechanism of hepatitis B virus and hepatitis C virus (HCV) and their role in the utilisation of alterations in mitochondrial dynamics for the maintenance of persistent infection. Studies on the effect of mitochondrial dynamics in viral infections will lead to a better understanding of virus–host interactions and their role in pathogenesis. This will lead to the therapeutic design of new antiviral strategies. The process of virus-regulated mitochondrial dynamics and their function in viral pathogenesis are important.

Mitochondrial dynamics and mitophagy

Mitochondrial homoeostasis is broadly maintained by the two closely interlinked and crucial processes, that is mitochondrial dynamics and mitophagy. Molecular and morphological examination of the mitochondrial network demonstrates that a single mitochondrion experiences several instances of fission and fusion throughout its life cycle, and when the cellular machinery detects damaged mitochondria, they are removed through a process called mitophagy. Fig. 3.1 delineates a sequence of pivotal events related to this process. While mitophagy is widely regarded as a primary mechanism for the removal of malfunctioning mitochondria, it also actively contributes to the elimination of fully functional mitochondria in several biological processes, including starvation, fertilisation, and erythroid differentiation. Fission, fusion, and mitophagy are interconnected yet different processes that collectively enable the segregation of dysfunctional mitochondria

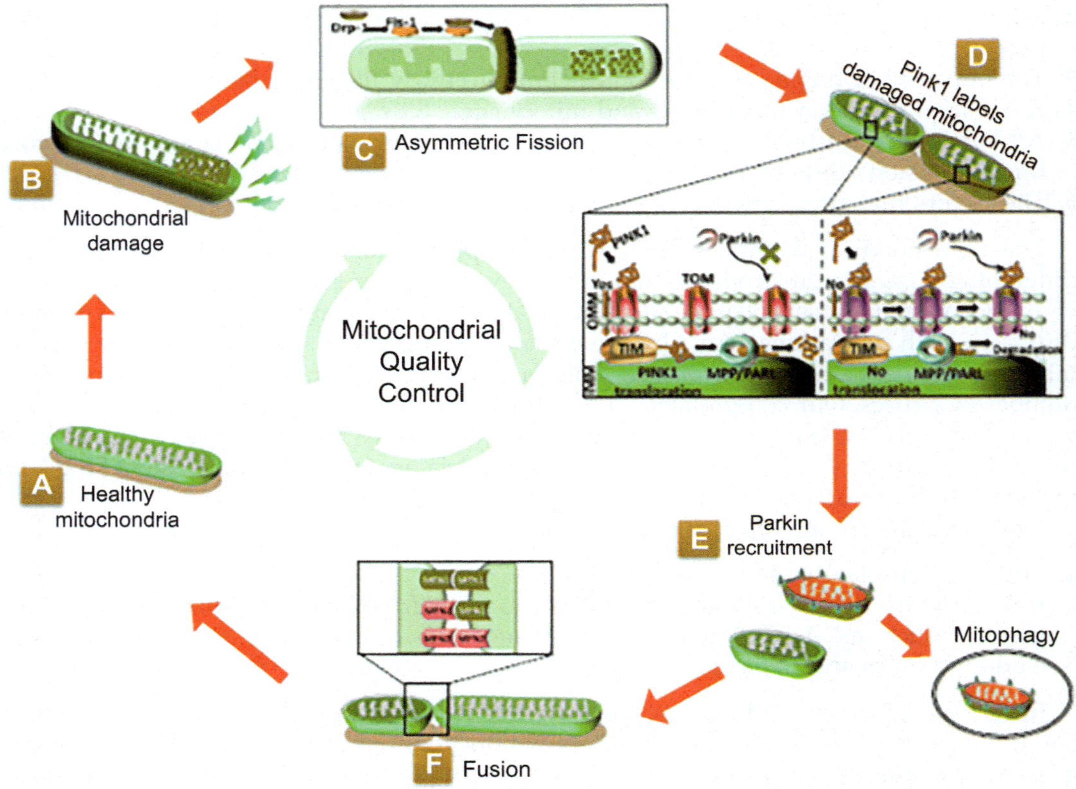

FIGURE 3.1 Mitochondrial quality control. The key events associated.

and their subsequent transport to lysosomes for full degradation. Fission has been associated with the segregation of mitochondria containing mutant mtDNA copies. In contrast to fission, fusion is a mechanism via which mitochondria rapidly exchange and equilibrate matrix metabolites, intact mtDNA copies, and mitochondrial membrane constituents. This procedure mitigates damage by merging a partially damaged mitochondrion with a fully functional, tubular mitochondrial network through the straightforward interchange of matrix proteins (see Fig. 3.1).

Initially, the aforementioned explanation elucidates the inverse link between mitochondrial fusion and mitophagy; yet, within a biological context, both activities are complementary and together contribute to the preservation of homoeostasis. Highly depolarised and permanently damaged mitochondria are designated to constitute a premitophagic pool and are preferentially removed through mitophagy. The mitochondria that participate in fusion are selectively salvaged from the premitophagic pool and reintegrated into the functional mitochondrial population. The MMP ($\Delta\Psi m$) is crucial for the sorting and segregation of faulty mitochondria. Based on the degree of damage, the dysfunctional and isolated daughter mitochondria are either repaired and reintegrated or entirely removed through mitophagy, should the $\Delta\Psi m$ be irreparable. The processes of fusion and fission are exceptionally dynamic. During the biological cycle, mitochondria exhibit a kiss-and-run mechanism, characterised by a transient fusion event lasting seconds, succeeded by fission. In eukaryotic cells, the mitochondrial life cycle is characterised by two distinct phases: the prefusion phase and the postfusion phase. In the prefusion stage, mitochondria exist as isolated entities, whereas in the postfusion stage, they are organised into a complex network. Occasionally, during the solitary phase of mitochondria, asymmetric fission can significantly affect the $\Delta\Psi m$ because of the uneven distribution of membrane surface and nucleoid structure between the two daughter mitochondria. Consequently, it is inferred that the depolarised mitochondria may possibly arise via spontaneous depolarisation during the solitary phase. Regardless of whether it is a spontaneous or stress-induced occurrence, such depolarised mitochondria are fated for either repair or elimination. Research correlating mitochondrial size with mitophagy rates indicates that elongated mitochondria demonstrate a decreased rate of mitophagy, while mitochondria within autophagosomes performing mitophagy are smaller in size (b1 μM). The mitophagy of larger, elongated mitochondria seems improbable. Fission enables the segregation of dysfunctional mitochondria from the healthy mitochondrial reservoir. If the damage in a sorted mitochondrion is reparable, it may proceed to the fusing step; nevertheless, irreversibly damaged mitochondria are removed via mitophagy. The mitochondrial dynamics and mitophagic machinery collaborate to inhibit the influx of highly damaged mitochondria into the active and healthy pool (see Fig. 3.1C–E).

Regulation of mitochondrial fusion and fission mechanisms in relation to mitophagy

Mitochondrial dynamics, specifically fission and fusion, are regulated by *dynamin-related proteins*. Dynamin can oligomerize into spirals and mechanically alter membranes by harnessing energy from GTP hydrolysis. They typically facilitate membrane scission; nevertheless, their function in mitochondrial fusion is distinctly unique. The functionality and quality of mitochondria are governed by the delicate equilibrium between ongoing fission and fusion processes. We will examine the key participants in these processes.

Chimeric proteins

The fuzzy onion (Fzo) was the inaugural fusion-regulating protein found in *Drosophila*. Fzo governs mitochondrial fusion, and its absence leads to fragmented mitochondria. In Fzo mutant flies, mitochondria congregate and do not undergo fusion, resulting in the aggregation of mitochondria in spermatids. The molecules regulating the fusion process in mammals include Mitofusins (Mfn1 and Mfn2) and optic atrophy 1 protein (OPA1). Silencing any of these proteins results in embryonic mortality and mitochondrial malfunction. Mfn1 and Mfn2 are homologues of Fzo, possessing analogous yet distinct functions in regulating mitochondrial morphology. Mfn1 and Mfn2 are members of the membrane-anchored dynamin family, expressed in diverse mammalian cells, and consistently localised to the outer mitochondrial membrane (OMM). Mitochondrial fusion entails the amalgamation of the outer and inner mitochondrial membranes (IMMs), leading to the convergence of two distinct mitochondria and the simultaneous blending of their contents. The OMM fusion is orchestrated by Mfn1 and Mfn2. Mutations in any of the Mfns result in mitochondrial morphological defects linked to a significant decrease in mitochondrial fusion. The depletion of both Mfn1 and Mfn2 in cells leads to impaired cell growth and reduced cellular respiration, correlated with the variability of MMP. The fusion of the IMM is regulated by OPA1. OPA1 is a dynamin-related GTPase found in various mammalian cell types and is localised on the IMM. OPA1 is regarded as a multifunctional protein essential for mitochondrial cristae structure, mitochondrial bioenergetics, and apoptosis.

Mitochondrial fusion is a multistep process, wherein Mfns facilitate outer membrane fusions, followed by inner membrane fusions mediated by OPA1. The presence of Mfns on both fusing mitochondria is essential for outer membrane fusion. The Mfns on neighbouring mitochondria oligomerize, drawing the opposing membranes into close proximity and ultimately facilitating fusion through membrane merger. Mfn1 and Mfn2 engage in both homotypic and heterotypic interactions, leading to the formation of Mfn1 homotypic oligomers, Mfn2 homotypic oligomers, and Mfn1–Mfn2 heterotypic oligomers (Fig. 3.1F). Notably, OPA1-mediated inner membrane fusion does not necessitate its localisation on the membranes that are fusing.

Both Mfns and OPA1 experience posttranslational changes and proteolytic cleavage. OPA1 manifests as diverse isoforms owing to variable RNA splicing and precursor protein processing. The long isoform possesses a membrane anchor, whereas the short isoform lacks one, yet does interact with membranes. Typically, both long and short isoforms facilitate membrane fusion; however, under stress conditions, only the long isoform exclusively stimulates mitochondrial fusion. OPA1 activity is modulated by sirtuin 3 (SIRT3)-mediated deacetylation. Furthermore, recent findings indicate that a reduction in MMP initiates the proteolysis of OPA1 by the isoenzymes of the ATP-dependent matrix AAA protease and overlapping with m-AAA protease 1 (OMA1). OMA1 exhibits constitutive activity under standard cellular circumstances. Nonetheless, several stressors can significantly augment its catalytic activity, resulting in increased turnover. Baker et al. recently identified the N-terminal region of OMA1 as a stress sensor responsible for the stress-induced activation of OMA1. This method likely facilitates the selective fusing of normal mitochondria, leading to the segregation of faulty or dysfunctional mitochondria. The phosphorylation of Mfn2 by c-Jun N-terminal kinases (JNK) during genotoxic stress facilitates its association with Huwe1, an E3 ubiquitin ligase, culminating in the proteasomal destruction of Mfn2, mitochondrial fragmentation, and subsequent apoptotic cell death. The recruitment of parkin to injured mitochondria leads to the ubiquitination and subsequent destruction of Mfns, hence inhibiting the fusion of impaired mitochondria with the healthy mitochondrial reservoir.

Fission proteins

The primary component in mitochondrial fission is the cytosolic dynamin-related GTPase, dynamin-related protein 1 (Drp1). Drp1 is predominantly situated in the cytosol, and in mammalian cells, the accessory proteins Mid49, Mid51, and ***mitochondrial fission factor (Mff)*** facilitate its recruitment to the mitochondria by serving as ligands for Drp1 on the mitochondrial surface. Upon recruitment to the mitochondria, Drp1 oligomerizes into rings on mitochondrial tubules, resulting in the scission of both the outer and IMMs. Our current comprehension of the mitochondrial fission mechanism is inadequate; nevertheless, recent research indicates that ER tubules encircle and restrict the mitochondria, designating them for further division by Drp1. The ER-mitochondrial interface sites are abundant in Drp1, indicating that the interaction between the ER and mitochondria regulates mitochondrial division. Mitochondrial constriction at ER contact sites was observed in cells lacking Drp1 and Mff, indicating that the ER-mediated signalling for mitochondrial division occurs before the recruitment of Drp1. The mechanical force necessary for the constriction of mitochondrial tubules by the endoplasmic reticulum is believed to be produced by actin assembly at the ER–mitochondria contact sites through interaction with the ER-associated formin inverted formin-2 (INF2). Genetic or Chemical suppression of Drp1 results in enlarged mitochondria. The initial molecule recognised as the Drp1 receptor was fission1 (Fis1) in yeast. Fis1 is a diminutive C-tail protein affixed to the OMM and exhibits a homogeneous distribution. The function of mammalian Fis1 in mitochondrial fission remains unclear, as mitochondrial morphology and the recruitment of Drp1 were unchanged in Fis1 mutant cells, casting doubt on the significance of mammalian Fis1 in this process. The Mff is additionally a C-tail attached protein on the OMM and a prominent receptor of Drp1. It temporarily engages with Drp1 via its N-terminal cytoplasmic domain. Unlike the homogeneous localisation of Fis1 on the OMM, Mff predominantly colocalises with Drp1 foci on the OMM. The overexpression of Mff enhances the recruitment of Drp1 to mitochondria, leading to mitochondrial fragmentation, whereas the silencing of Mff induces mitochondrial elongation. The MiDs can facilitate mitochondrial fission without the presence of Fis1 and Mff. Knockdown of either gene leads to an increase in mitochondrial length and connectedness, indicating that these proteins positively control mitochondrial fission.

Notably, the overexpression of MiDs leads to the elongation of mitochondria. The mitochondrial elongation resulting from MiD overexpression correlates with increased phosphorylation of Drp1 at the S637 location, which inhibits Drp1 function. Despite the mutant Drp1 cells resisting mitochondrial fragmentation caused by depolarisation, they display significant mitochondrial fragmentation upon treatment.

Apoptotic stimuli, including actinomycin D and etoposide, indicate the existence of Drp1-independent pathways of mitochondrial fission. The mitochondrial fragmentation caused by the pore-forming toxin listeriolysin O, released by the bacteria *Listeria monocytogenes*, is not reliant on the conventional fission protein Drp1 but is contingent upon the actin cytoskeleton. Multiple posttranslational changes of Drp1 govern mitochondrial fission in response to diverse physiological signals. In typical cells, Drp1 is predominantly cytoplasmic, with approximately 3% of total Drp1 linked with the mitochondria. The overexpression of Drp1 often does not result in increased fission; nevertheless, the stimulation of Drp1's mitochondrial recruitment, GTPase activity, and spiral assembly directly affects mitochondrial fission (see Fig. 3.1A). During mitosis, Cdk1/cyclin B phosphorylates Drp1 at the S616 residue (amino acid numbering pertains to human Drp1), leading to enhanced mitochondrial recruitment of Drp1 and increased fission activity. This control is crucial in cells undergoing mitosis to guarantee the equitable distribution of mitochondria to daughter cells. During mitosis, Aurora A kinase facilitates mitochondrial fission by phosphorylating the small Ras-like GTPase RalA, resulting in its translocation to the mitochondria, subsequently recruiting its effector RalBP1 to the mitochondria. RalBP1 subsequently promotes CDK1-mediated phosphorylation of Drp1 at S616 and its translocation to the mitochondria.

A study reveals that high-glucose stimulation of liver-derived cells activates Ca2+-mediated microtubule-associated proteins (MAP) kinase signalling, resulting in extracellular signal-regulated kinase (ERK)1/2-dependent phosphorylation of Drp1 at S616 and increased mitochondrial fission. Phosphorylation of Drp1 at the S637 residue inhibits fission activity. Protein kinase A (PKA), also known as cAMP-dependent protein kinase, phosphorylates Drp1 at S637 within the GTPase effector domain (GED), thereby disrupting the intramolecular connection between the GED and GTPase domains. In hypoxia, the phosphorylation of Drp1 mediated by PKA is influenced by the proteasomal degradation of mitochondrial A kinase anchoring protein 121, which anchors PKA to the mitochondria. Calcineurin, a calcium-dependent phosphatase, dephosphorylates Drp1 at S637, promoting the mitochondrial recruitment of Drp1 and subsequent fission. The same position is likewise dephosphorylated by the phosphatase protein phosphatase 2A/Bcl-2-associated X protein (PP2A/Bβ 2). Conversely, phosphorylation of the same residue in response to elevated intracellular calcium by Ca2+/calmodulin-dependent protein kinase α (CaMKIα) in neural cells or by rho-associated, coiled-coil-containing protein kinase 1 (ROCK1) leads to increased fission.

Recent results unexpectedly highlight the significance of upstream kinases, rather than the phosphorylation location, as a crucial determinant of Drp1 activity. The phosphorylation of S616-Drp1 by CDK1 or protein kinase C Delta (PKCΔ) is documented to augment fission through the activation of Drp1; conversely, phosphorylation at the same site by CDK5 is described to promote the dissociation of Drp1 from its oligomeric to monomeric state, significantly diminishing its fission-inducing activity. The dephosphorylation of S637 mediated by calcineurin, together with the phosphorylation of S637 by CaMKIα, paradoxically promotes increased fission, indicating that varying physiological settings may provide distinct effects of Drp1 phosphorylation on the same serine residue. Further research is necessary to elucidate how phosphorylation at the identical location by various kinases might influence Drp1 function with divergent outcomes. Furthermore, ubiquitination, sumoylation, and s-nitrosylation have been demonstrated to regulate Drp1 function. Parkin and mitochondria ubiquitin ligase promote the degradation of Drp1 through proteasomal mechanisms.

Regulation of mitophagy machinery

Autophagy is a catabolic process wherein a designated cargo is actively engulfed by a phagophore, followed by the fusion of the autophagosome with lysosomes to facilitate the destruction of the cargo within the lysosome. Autophagy is a nonselective mechanism; nevertheless, new developments have confirmed the existence of selective autophagy, wherein specific cargo or damaged organelles are targeted for elimination. This text addresses the selective autophagy of mitochondria, known as mitophagy. Mammalian mitophagy can be categorised into two separate classes. **(1) Mitophagy reliant on Parkin; and (2) Mitophagy independent of Parkin**. Parkin-dependent mitophagy entails the selective identification of damaged mitochondria by PINK1, which is succeeded by the recruitment of parkin to the compromised mitochondria and the ensuing mitophagy. The serine/threonine kinase PINK1 possesses a mitochondrial targeting sequence (MTS) and is imported into healthy mitochondria in a sequential manner through the translocase of the outer mitochondrial membrane (TOM) and the translocase of the inner mitochondrial membrane. At the IMM, PINK1 undergoes processing by the mitochondrial processing protease (MPP) to remove the MTS, followed by cleavage by the presenilin-associated rhomboid-like protease (PARL) to produce a 52 kDa protein, which is thereafter destroyed swiftly. In damaged mitochondria, the loss of membrane potential ($\Delta\Psi m$) swiftly disrupts the import of PINK1 to the IMM and inhibits its destruction by MPP, PARL, and other proteases (see Fig. 3.1D). This results in the presentation of PINK1 on the mitochondrial surface, thereby designating the impaired mitochondria for parkin recruitment to commence the mitophagy process. Numerous investigations have elucidated the mechanism by which PINK1 attracts parkin to the mitochondria. PINK1 auto-phosphorylation on depolarised mitochondria is essential for parkin recruitment, and direct phosphorylation of parkin by PINK1 at the S65 residue has been demonstrated to enhance parkin's E3 ligase function. PINK1 phosphorylates the S65 residue of ubiquitin, which subsequently alleviates parkin's autoinhibition at the active cysteine site. The ubiquitin mutant S65A obstructed parkin's transport to dysfunctional mitochondria. Small compounds that emulate S65 phospho-ubiquitin may serve as possible treatment agents for PD. Collectively, these data suggest a feed-forward mechanism for the recruitment of parkin to the mitochondria, which involves PINK1-mediated phosphorylation of both parkin and polyubiquitin chains produced by Parkin at the IMM. Certain investigations indicate a direct physical interaction between Parkin and PINK1. Other substrates of PINK1 kinase activity are likely implicated in the recruitment of Parkin to the mitochondria. While PINK1 aids in the recruitment of Parkin to the mitochondria, it is not the sole mechanism for Parkin recruitment in flies. Mitochondrial morphology in PINK-1-deficient *Drosophila* can be restored with Parkin overexpression, but Parkin-deficient *Drosophila* exhibit a more severe phenotype. than the latter, indicating PINK1-independent activities of Parkin in *Drosophila*. It remains to be confirmed if Parkin can be recruited to the mitochondria independently of PINK1 in mammalian cells. In a particular investigation, PINK1-mediated phosphorylation of Mfn2 was demonstrated to facilitate Parkin recruitment to depolarised mitochondria. Parkin ubiquitinates and promotes the degradation of OMM proteins, including Mfn1, Mfn2, and the TOM complex, which appears to be crucial for the subsequent autophagic clearance of damaged mitochondria. The suppression of proteasomal activity disrupts Parkin-dependent mitophagy. The aggregation of proteasomal subunits on the OMM after Parkin's recruitment indicates that the proteasomal machinery is directly summoned to the mitochondria to enable destruction. The AAA + ATPase p97, which structurally remodels or unfolds ubiquitinated client proteins designated for degradation, is demonstrated to be recruited to the mitochondria alongside Parkin, hence regulating protein degradation at the OMM.

The mechanisms by which Parkin targets proteins and the specific targets essential for the onset of mitophagy remain ambiguous. Parkin not only facilitates K-48-linked polyubiquitination, resulting in proteasomal degradation, but also induces the synthesis of diverse lysine-linked polyubiquitin chains attached to its target proteins. Given Parkin-mediated K-63-linked ubiquitination, it is probable that the p62 adaptor protein, which associates with K-63-linked polyubiquitin-conjugated proteins, contributes to the recruitment of mitochondria to autophagosomes. Nonetheless, mitochondrial elimination has been documented even without p62 aggregation at the mitochondria. Histone deacetylase 6 (HDAC6), which also binds ubiquitinated proteins, is recruited to the mitochondria in a Parkin-dependent way. HDAC6 has been demonstrated to contribute to autophagosome maturation. PINK1 and Parkin have been demonstrated to directly engage with beclin-1 and the phosphatidylinositol-4,5-bisphosphate 3-kinase complex to facilitate autophagy. Parkin also aids in the recruitment of the activation molecule in beclin-1-regulated autophagy (Ambra-1) to the mitochondria, potentially activating beclin-1 and enhancing the nucleation of isolation membranes (phagophores) around the mitochondria. Mutations in both Parkin and PINK1 are the predominant cause of autosomal recessive genetic illness Parkinsonism, typically linked to the accumulation of dysfunctional mitochondria resulting in neurodegeneration in PD. This supports the notion that PINK1/Parkin-mediated mitophagy is crucial for mitochondrial quality regulation. The PINK1/Parkin pathway not only triggers the mitophagic elimination of impaired mitochondria but also promotes the segregation of damaged mitochondria by inhibiting their fusion with healthy, functional mitochondria. This segregation is accomplished through two mechanisms: inducing fusion defects in impaired mitochondria via Parkin-mediated degradation of Mfns, and limiting the cytosolic mobility of damaged mitochondria by the targeting of Miro (Rho-GTPase), a mitochondrial adaptor that anchors the kinesin motor complex to the outer mitochondrial surface.

PINK1 phosphorylates Miro, facilitating its recognition and degradation by Parkin. The breakdown of Miro liberates the kinesin motor from the mitochondria, hence impeding their mobility. These events clearly demonstrate that mitochondrial dynamics and mitophagy are interconnected and regulated in a synchronised manner. Numerous aberrations in the PINK1–Parkin route indicate the presence of PINK1–Parkin-independent mitophagy pathways. The PINK1–Parkin route may not be the sole mechanism for the steady-state turnover of mitochondria, as Parkin absence does not result in a significant reduction of mitochondrial mass, and Parkin-deficient mice exhibit normalcy unless subjected to stress.

Allen et al. noted that iron chelation prompts mitophagy independently of PINK1 stabilisation and Parkin activation in primary human fibroblasts and those derived from PD patients with Parkin mutations. AMBRA-1 is said to promote Parkin-independent mitophagy through direct interaction with LC3 via its LC3-interacting region (LIR). AMBRA-1 can stimulate mitophagy in Parkin-deficient cells and enhance Parkin-mediated mitophagy. In reaction to energy shortage, AMP-activated protein kinase (AMPK) activates autophagy by phosphorylating unc-51 like autophagy activating kinase 1 (ULK1). The absence of AMPK or ULK1 in mammalian cells results in the buildup of the autophagy adaptor p62 and impaired mitophagy, indicating that ULK1 is essential for mitochondrial and cellular homoeostasis during starvation. A recent study indicates that ULK1 translocates to the mitochondria and phosphorylates the FUNDC1, which directly interacts with LC3 through its LIR motif to regulate mitophagy. BNIP3L, also known as Nix, promotes the removal of mitochondria from reticulocytes during erythrocyte maturation by directly binding to LC3 through its LIR motif. A separate study identifies FUNDC1, an OMM protein, as a modulator of hypoxia-induced mitophagy. In neurones, the externalisation of cardiolipin to the OMM acts as a signal for the removal of damaged mitochondria. Cardiolipin, a phospholipid found in the IMM, when externalised to the OMM, interacts with LC3 to promote mitochondrial breakdown. The functionality of the parkin-dependent and -independent pathways in parallel during depolarisation-induced mitophagy remains uncertain.

Role of mitochondrial dynamics and mitophagy in cell physiology

In the last 10 years, it has become clear that mitochondrial dynamics and mitophagy are the two primary mechanisms necessary for sustaining mitochondrial quality control and cellular homoeostasis. Disruptions in these pathways have physiological implications and are a prevalent cause behind the pathogenesis of numerous neurological illnesses, cancer, inflammation, metabolic syndrome, and cardiac dysfunctions. We examine the functions of mitochondrial dynamics and mitophagy in cellular physiology below.

Mitochondrial quality control, cellular viability and apoptosis

Disruption in mitochondrial dynamics and mitophagy typically results in a compromised mitochondrial population, creating an energy-deficient state in the cytosol that may ultimately induce death. Mitochondrial fragmentation has been associated with apoptosis. Augmented fragmentation heightens cellular susceptibility to apoptotic stimuli, whereas increased fusion correlates with a delay in Bax activation, cytochrome C release, and subsequent cell death. Drp1-mediated fission has been demonstrated to promote the release of cytochrome C from the mitochondria. The Drp1-mediated restructuring of the OMM promotes the oligomerisation and incorporation of Bax into the mitochondria. In contrast, both Bax and Bak enhance Drp1 sumoylation and stabilise Drp1 on the mitochondria, hence facilitating fission. In the neuronal cell model of nitric oxide-induced oxidative stress, mitochondrial fission occurred prior to death, and the suppression of fission provided protection. Augmented mitochondrial fusion has demonstrated the capability to safeguard cells from mitochondria-induced apoptosis, although it does not confer protection against mitochondria-independent apoptotic triggers. Challenging the prevailing belief, cytochrome C release was noted in Drp1-deficient cells during apoptosis induction, indicating that mitochondrial fission and mitochondrial outer membrane permeabilisation are distinct processes. The exact function of mitochondrial fission in apoptosis remains ambiguous; nevertheless, the antiapoptotic B-cell lymphoma 2 (Bcl2) family members may influence mitochondrial dynamics to enhance cellular resilience to diverse stressors. Impaired mitochondria may generate ROS, potentially causing harm to healthy mitochondria. Consequently, disruption in the swift elimination of impaired mitochondria would trigger a detrimental cycle of ROS production and mitochondrial injury, ultimately compromising other cellular constituents and facilitating cell death. Impaired mitochondrial quality has been linked to the initiation of inflammatory responses through inflammasome activation. Preconditioning cells to upregulate mitochondrial quality control pathways has demonstrated an increase in resistance to stressors. Interorganelle interactions are crucial for mitochondrial functions and are involved in mitochondrial dynamics.

An example is the role of the ER–mitochondria encounter structure in preserving mitochondrial shape in budding yeast. Mitochondrial-associated membranes (MAMs) in mammals denote the contact locations between the endoplasmic reticulum and mitochondria. Mfn2 is abundant in MAMs and is posited to be essential for stabilising the connection between the ER and mitochondria by forming homotypic and heterotypic complexes with

Mfn1/2 on the mitochondrial surface. Changes in Mfn2 expression influence ER-mitochondria interaction, therefore impeding lipid exchange and calcium signalling between the ER and mitochondria. Numerous connections between endoplasmic reticulum and mitochondrial proteins have been documented in mammals; however, it is yet to be established whether these interactions constitute an essential component of a physical complex.

Role of mitochondrial dynamics in innate immunity

The significance of mitochondrial dynamics in innate immune signalling is an emerging area of research. Pattern recognition receptors (PRRs), integral to the innate immune system, identify pathogen-associated molecular patterns (PAMPs) and initiate a series of intricate pathways to eradicate infection, occasionally resulting in the apoptosis of infected cells. Numerous PRRs have been identified, each associated with a distinct ligand. Numerous viral PAMPs are detected by the retinoic acid-inducible gene I (RIG-I) and RIG-I-like receptors (RLRs), subsequently activating the mitochondria-associated antiviral signalling protein (MAVS) through binding mediated by the caspase activation recruitment domain adaptor domains of each protein. Recent investigations indicate that mitochondrial dynamics and mitophagy influence this inherent defence mechanism.

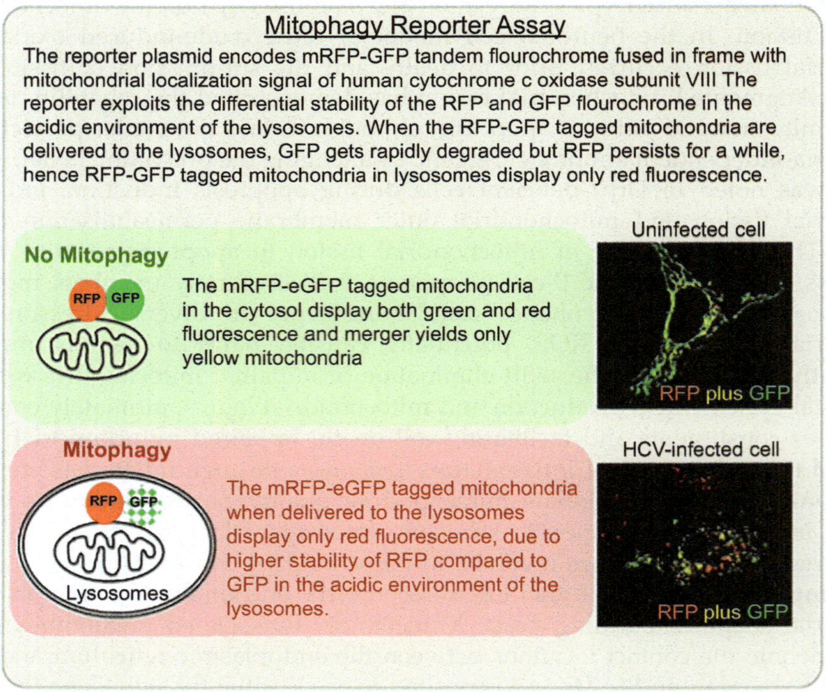

FIGURE 3.2 Mitophagy reporter assay.

Mitochondrial fusion and fission are pivotal in modulating RIG-I-like receptor signalling intensity

Mitochondrial fusion enhances MAVS interactions with downstream signalling molecules, leading to increased interferon (IFN) generation in response to polyI: C stimulation or Sendai virus infection. Mitochondrial fission inhibits MAVS downstream signalling, leading to decreased IFN production. Consistent with this concept, fibroblasts deficient in both Mfn1 and Mfn2 exhibited diminished synthesis of IFN and pro-inflammatory cytokines in reaction to viral infection. The lack of fusion results in the depletion of MMP in these double-knockout cells, which is associated with diminished MAVS signalling, indicating that MMP

TABLE 3.1 Posttranslational modifications of the protein involved in mitochondrial dynamics.

Protein	Modification	Induced by	Effect
Drp1	Phosphorylation	CDK1	Enhanced mitochondrial fission
		Erk1/2	Enhanced mitochondrial fission
		PKC6	Enhanced mitochondrial fission
		CDK5	Reduced mitochondrial fission
	Phosphorylation	PKA	Reduced mitochondrial fission
		CaMK1α	Enhanced mitochondrial fission
		ROCK1	Enhanced mitochondrial fission
	Dephosphorylation	Calcineurin	Enhanced mitochondrial fission
	S-Nitrosylation	Nitric oxide	Enhanced mitochondrial fission
	Sumoylation	MAPL	Enhanced mitochondrial fission
	Desumoylation	SENP5	Reduced mitochondrial fission
Fis1	Ubiquitination	MITOL	Degradation
Mfn1	Ubiquitination	Parkin	Degradation
Mfn1	Ubiquitination	MITOL	Degradation
	Ubiquitination	Parkin	Degradation
	Ubiquitination	MARCH5	Degradation
Mfn2	Ubiquitination	PARKIN	Degradation
	Phosphorylation	PINK1	Parkin recruitment
OPA1	Acetylation	Stress	Reduced GTPase activity
	Deacetylation	SIRT3	High GTPase activity

List of important posttranslational modifications of the key molecules of mitochondrial dynamics and their upstream modulators, with its final consequences on cell physiology.
Throughout the chapter, we have discussed about varieties of proteins directly or indirectly related to one of the most important phenomenon as mitophagy. These proteins were mostly regulated by nuclear originated genes, but functioning for mitochondrial structure and function. We focus on the genomics for such proteins in subsequent chapters.

may be essential for MAVS signalling. Mitochondrial fusion and the subsequent creation of a tubular mitochondrial network around viral replication sites are anticipated to facilitate the concentration of the MAVS signalosome at these sites of viral replication. A recent publication asserts that Mfn1 facilitates significant redistribution of MAVS on the mitochondria upon RLR activation and that Mfn1 is essential for the recruitment of MAVS-enriched mitochondria to the RIG-I-enriched regions of viral replication. Mitochondrial shape also influences the signalling of the ER-associated innate immunity protein stimulator of interferon genes (STING). Elongated tubular mitochondria are more likely to engage with the endoplasmic reticulum, hence facilitating STING's subsequent interaction with MAVS. We noted that suppressing Drp1 to decrease mitochondrial fission during HCV infection augmented IFN production, suggesting that HCV-induced alterations in mitochondrial dynamics disrupt antiviral defence mechanisms. This may function alongside the HCV NS3/4a-mediated proteolytic cleavage of MAVS to undermine host innate immunity. Fig. 3.2; Table 3.1.

Further reading

Chen, G., Kroemer, G., & Kepp, O. (2020). Mitophagy: An Emerging Role in Aging and Age-Associated Diseases. *Front Cell Developmental Biol, 8*. https://doi.org/10.3389/fcell.2020.00200.

Ding, W. X., & Yin, X. M. (2012). Mitophagy: Mechanisms, pathophysiological roles, and analysis. *Biol Chem, 393*(7), 547–564. https://doi.org/10.1515/hsz-2012-0119.

Killackey, S. A., Philpott, D. J., & Girardin, S. E. (2020). Mitophagy pathways in health and disease. *J Cell Biol, 219*(11). https://doi.org/10.1083/jcb.202004029.

Onishi, M., Yamano, K., Sato, M., Matsuda, N., & Okamoto, K. (2021). Molecular mechanisms and physiological functions of mitophagy. *EMBO J, 40*(3). https://doi.org/10.15252/embj.2020104705.

Zhang, J. (2013). Autophagy and mitophagy in cellular damage control. *Redox Biol, 1*(1), 19–23. https://doi.org/10.1016/j.redox.2012.11.008.

CHAPTER

4

Introduction to mitochondrial antiviral signalling protein in different livestock species

Introduction

Generally, we come across the immune response molecules mostly synthesised from the lymphoid organs, coded by nuclear genes. Later research reports indicate the role of mitochondria in innate immunity. There have been many pathways showing how mitochondrial proteins are helpful for immune function. Mitochondrial antiviral-signalling protein (MAVS) is one of such mitochondrial proteins, although it is encoded by a nuclear gene. MAVS serves as a central hub for signal transduction initiated by retinoic acid-inducible gene I (RIG-I)-like receptors, which predominantly recognise viral RNA.

What is mitochondrial antiviral-signalling protein?

RIG-I like receptors (RLRs) have been thoroughly examined as detectors of cytoplasmic viral dsRNA, and their significance in regulating RNA virus infections is now evident. The RLR family has three members: RIG-I, melanoma differentiation-associated gene 5 (MDA5), and laboratory of genetics and physiology protein 2 (LGP2). RIG-I and MDA5 are quintessential pattern recognition receptors, while LGP2 is regarded as a modulator of RIG-I- and MDA5-mediated signalling pathways. Following viral detection, RIG-I and MDA5 engage with the mitochondrial antiviral-signalling protein (MAVS, also referred to as IFN-β promoter stimulator 1 (IPS-1), virus-induced signalling adaptor (VISA), and caspase activation recruitment domain adaptor inducing interferon-β (CARDIF)), leading to the activation of the transcription factors interferon regulatory factors 3 and 7 (IRF3/7) and nuclear transcription factor-kB (NF-κB). This process culminates in the production of several proinflammatory factors and antiviral genes, including interferon (IFN) and IFN-stimulated genes (ISGs), which impede viral replication and transmission. MAVS is the essential adaptor protein necessary for defence against RNA viruses recognised by RLR. The regulation of MAVS expression and signalling functions, together with its manipulation

Mitochondrial Genomics
DOI: https://doi.org/10.1016/B978-0-443-26655-3.00009-3

by viruses, is a significant domain of study. IPS-1 and MAVS refer to the same protein, MAVS.

- IPS-1: Commonly referred to as IFN-beta promoter stimulator I MAVS, sometimes referred to as CARDIF or VISA. MAVS is a protein composed of 540 amino acids that is integral to the body's innate immune response to viral infections.

Mitochondrial antiviral-signalling protein comprises three components

1. A caspase activation recruitment domain (CARD) at the N-terminus,
2. A proline-rich domain, and
3. A transmembrane (TM) C-terminal domain. MAVS stimulates the NF-κB and IRF3 signalling pathways in response to viral stimuli.

Genomic organisation of mitochondrial antiviral-signalling protein

MAVS cDNA is comprised of 1625bp nucleotide. MAVS protein comprises of three domains as CARD (amino acid position 10-77), PRR (amino acid position 103 to 173) and TM domains (amino acid position 514 to 535) from N to C terminals (Fig. 4.1, Fig. 4.2).

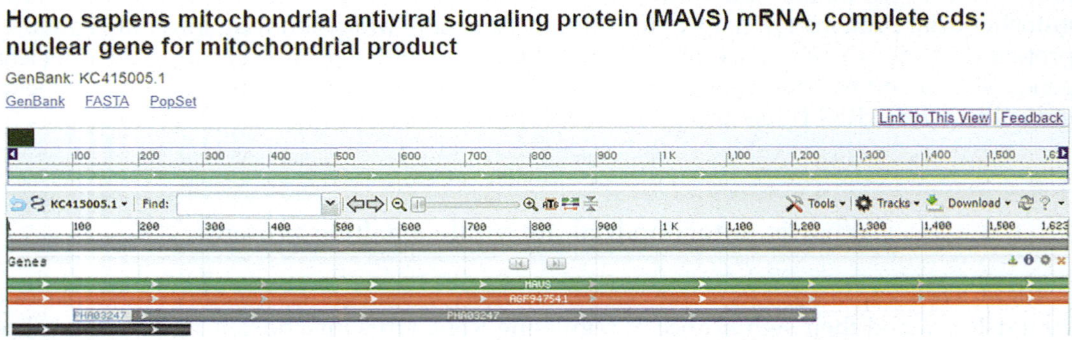

FIGURE 4.1 Genomic organisation of mitochondrial antiviral-signalling protein (MAVS).

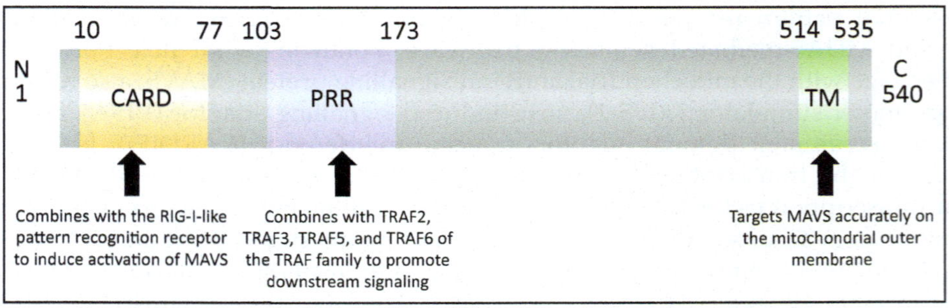

FIGURE 4.2 Important structural domains for mitochondrial antiviral-signalling protein (MAVS).

Secondary protein structure for MAVS (see Fig. 4.3).
3D protein structural organisation of MAVS (see Fig. 4.4
Fig. 4.5).
Molecular interaction of MAVS with other proteins (see Fig. 4.6).

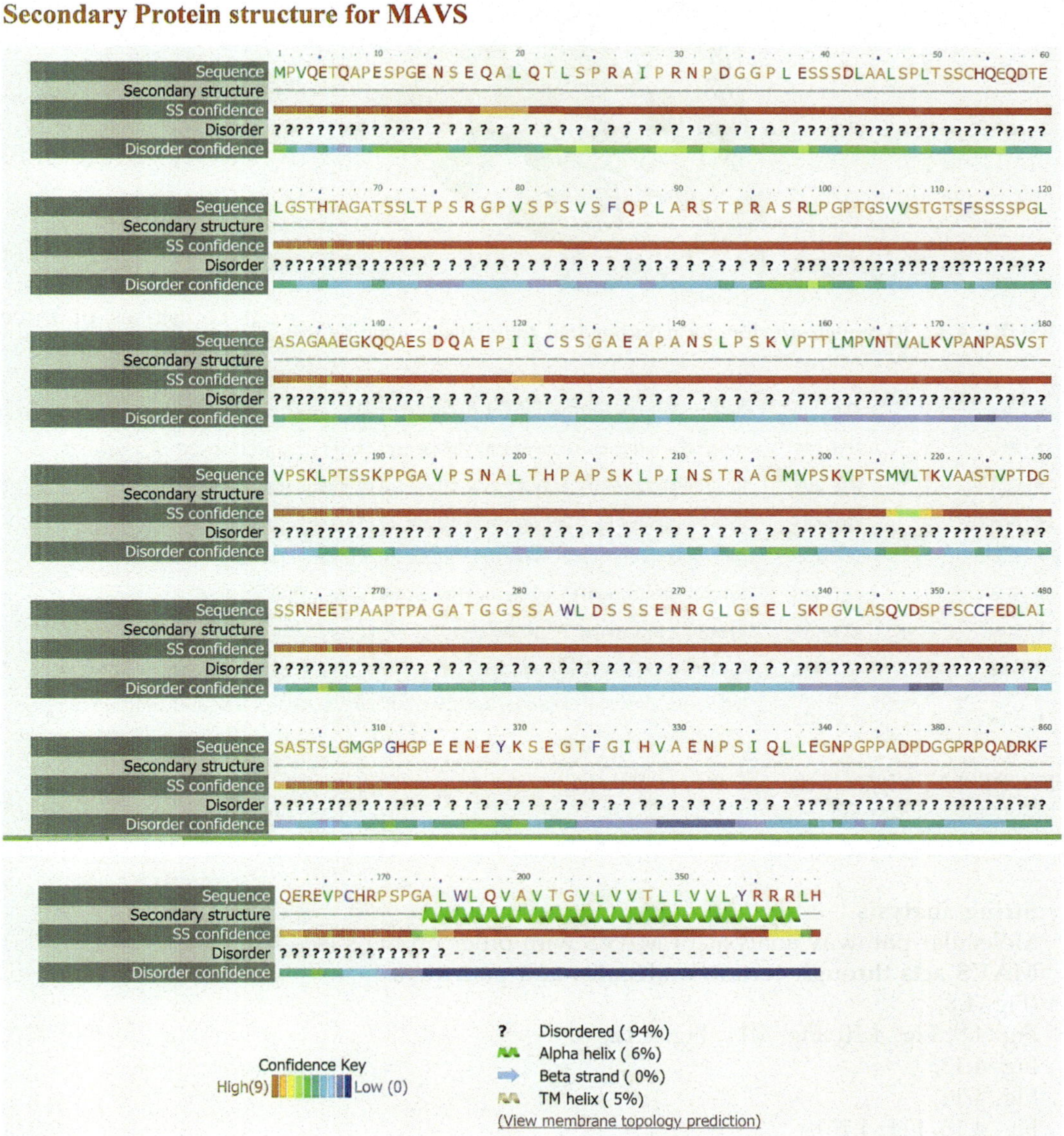

FIGURE 4.3 Secondary structure and disorder prediction for human mitochondrial antiviral-signalling protein (MAVS) derived protein.

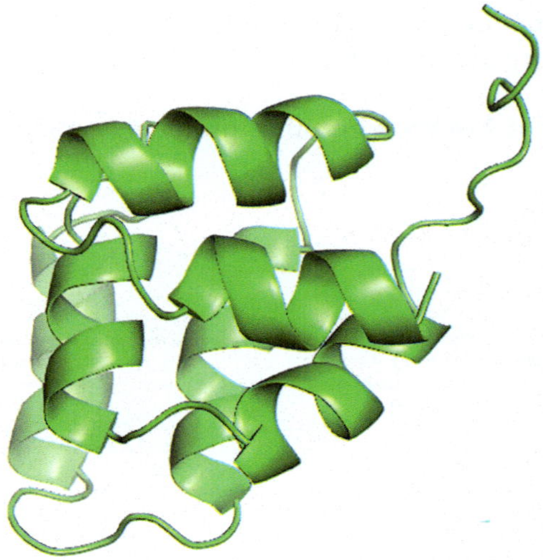

FIGURE 4.4 3D protein structure for mitochondrial antiviral-signalling protein.

▶ Site	148	(Microbial infection) Cleavage by CV3B 1 Publication
▶ Site	148-149	(Microbial infection) Cleavage; by viral Seneca Valley virus protease 3C 1 Publication
▶ Site	208-209	(Microbial infection) Cleavage by protease 2A of enterovirus 71 1 Publication
▶ Site	250-251	(Microbial infection) Cleavage by protease 2A of enterovirus 71 1 Publication
▶ Site	264-265	(Microbial infection) Cleavage by protease 2A of enterovirus 71 1 Publication
▶ Site	427-428	(Microbial infection) Cleavage; by HAV protein 3ABC 1 Publication
▶ Site	429-430	Cleavage; by CASP3 1 Publication
▶ Site	490-491	Cleavage; by CASP3 1 Publication
▶ Site	508-509	(Microbial infection) Cleavage; by HCV and hepatitis GB virus B NS3/4A protease complex 2 Publications

FIGURE 4.5 Important domains for mitochondrial antiviral-signalling protein.

String analysis

Molecular pathway analysis of MAVS with other proteins (see Fig. 4.7).

MAVS acts through certain molecules and pathways.

(Fig. 4.8,
Fig. 4.9, Fig. 4.10, Fig. 4.11, Fig. 4.12,
Fig. 4.13,
Fig. 4.14,
Fig. 4.15, Fig. 4.16).

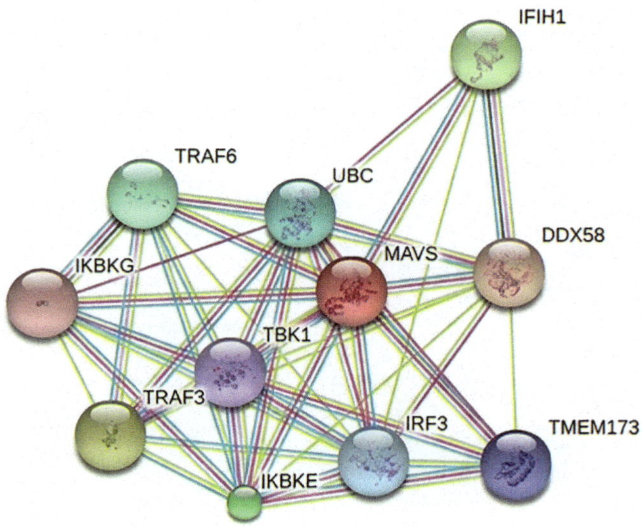

Your Input:

● MAVS mitochondrial antiviral signaling protein: Required for innate immune defense aglenst viruses. Acts downstream of DDX58 and IFIH1/MDAS, which detect intracellular dsRNA produced during viral regilcation, to coordinate pathways leading in the activation of NF-kappa-B, IRF3 and IRF7, and to the subsequent induction of antiviral cytokines such as IFN-beta and RANTES (CCL5). May activate the same pathways following detection of extracellular dsRNA by TLR3. May protect cells from synptoms (540 aa) (*Homo sapiens*)

Predicted Functional Partners:

		Neighborhood	Gene Fusion	Cooccurrence	Cooccurrence	Experiments	Databases	Databasing	[Homology]	Score
DDXS8	DEAD (Asp-Glu-Ala-Asp) box polypeptide 58 Sequired for innate immune defense agense viruses. U [...] (925 aa)					•	•	•		0.999
TRAF3	TNF receptor-associated factor 3: Adapter protein and signal transducer that haks members of t [...] (685 aa)					•	•	•		0.999
IFIH1	interferon induced with helicase C domain 1: RNA belicase that, through its ATP-aspponent unwin [...] (1025 aa)					•	•	•		0.998
IKBKE	inhibitor of kappa light polypeptide gene enhancer in B-cells, kinase epsilon: Phosphorylates i [...] (716 aa)					•	•	•		0.997
TRAF6	TNF receptor-associated factor 6: Adapter protein and signal transducer that haks members of t [...] (522 aa)					•	•	•		0.997
UBC	ubiquitin C (685 aa)					•	•			0.997
IRF3	interferon regulatory factor 3: Mediates interferon-stimulated response element (ISRE) prometer [...] (427 aa)					•	•	•		0.997
TMEM173	transmembrane protein 173: Acts as a facilitator of innate immune signaling. Able to activate b [...] (379 aa)					•	•	•		0.997
TBK1	TANK-binding kinase 1: Serine/feneenine protein involved in the signaling ceacalic converning to [...] (729 aa)					•	•	•		0.997
IKBKG	inhibitor of kappa light polypeptide gene enhancer in B-cells, kinase gamaty: Regulatory subunit [...] (487 aa)					•	•	•		0.991

FIGURE 4.6 Molecular interaction of mitochondrial antiviral-signalling protein (MAVS) with other proteins.

Mechanism of action of mitochondrial antiviral-signalling protein—a link for antiviral activity for mitochondria

MAVS is a protein comprising 540 amino acids, encoded by the nuclear genome. MAVS is mostly situated on the mitochondrial outer membrane, while it has also been identified on peroxisomal and mitochondrial-associated endoplasmic reticulum membranes.

MAVS comprises three domains: (1) N-terminal CARD, (2) a central proline-rich region (PRR) and (3) C-terminal TM domain. The MAVS CARD interacts with analogous CARDs found in RIG-I and MDA5, resulting in the activation of MAVS. The MAVS PRR domain interacts with tumour necrosis factor receptor-associated factor (TRAF) family members

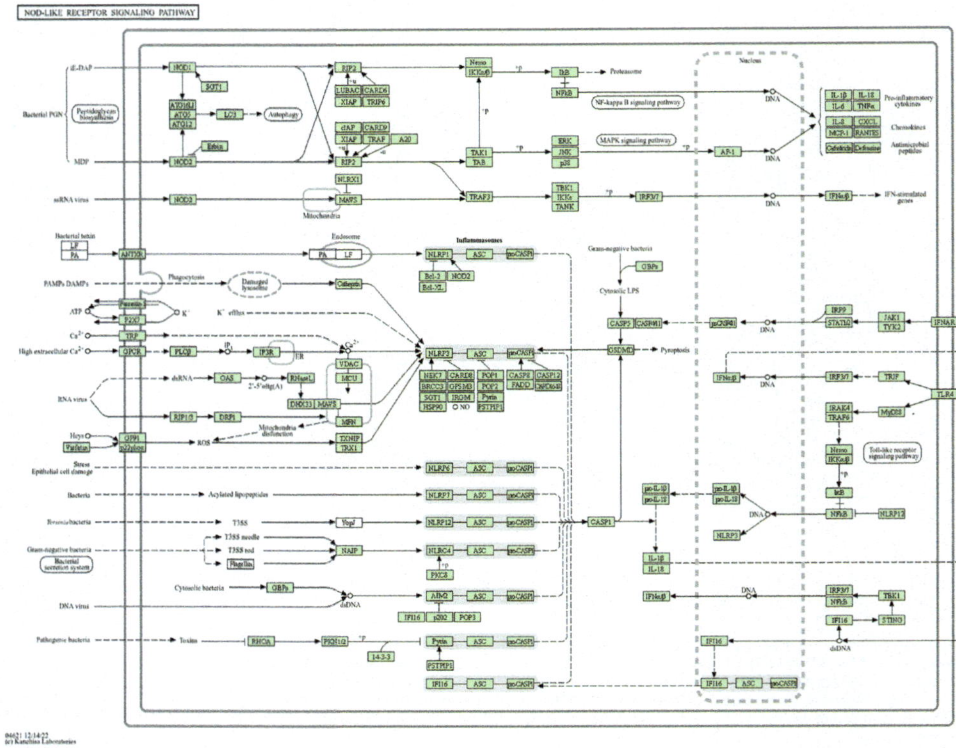

FIGURE 4.7 Molecular pathway analysis involving mitochondrial antiviral-signalling protein in nucleotide-binding oligomerization domain (NOD)-like receptor signalling pathway.

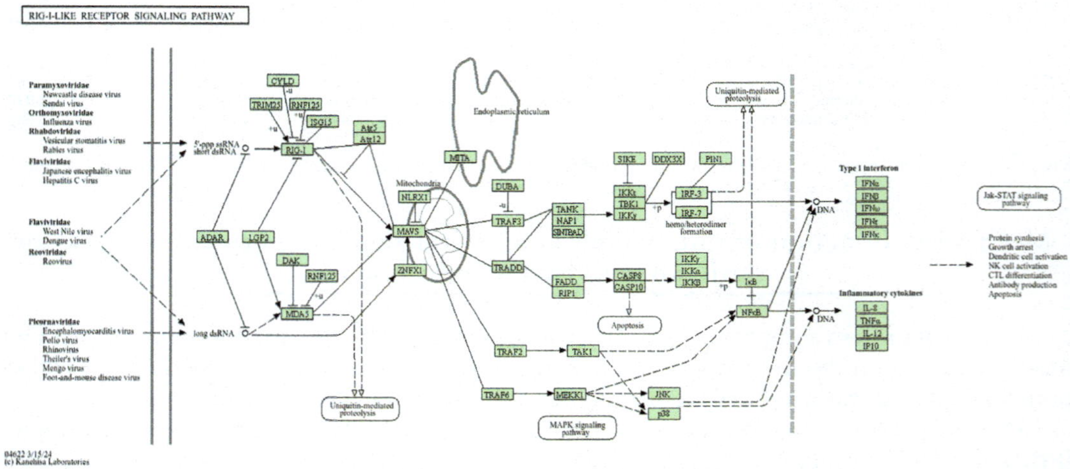

FIGURE 4.8 Molecular pathway analysis involving mitochondrial antiviral-signalling protein in retinoic acid-inducible gene I-like receptor signalling pathway.

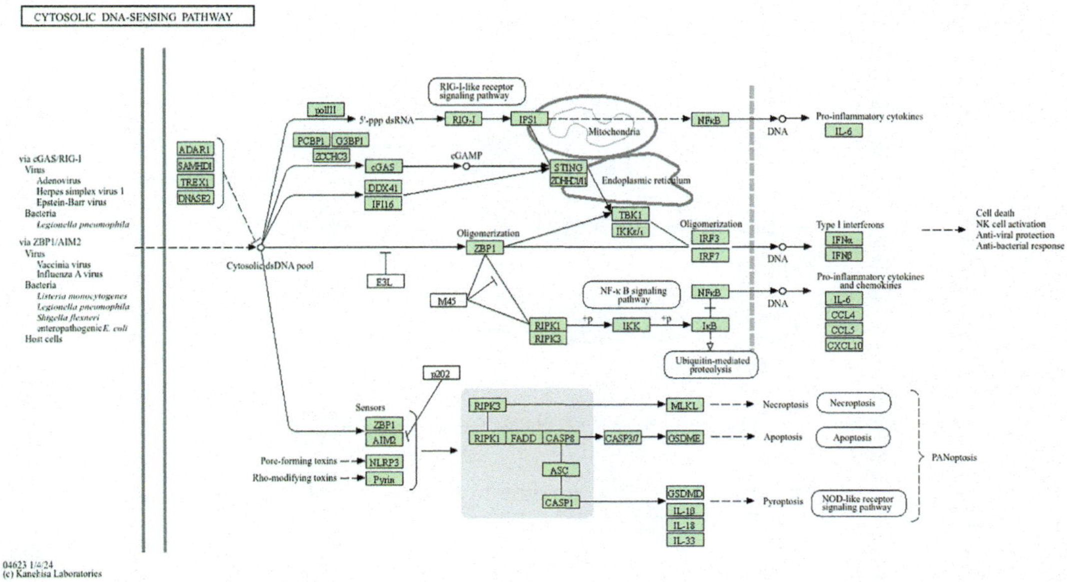

FIGURE 4.9 Molecular pathway analysis involving mitochondrial antiviral-signalling protein in cytosolic DNA sensing pathway.

TRAF2, TRAF3, TRAF5, and TRAF6 to facilitate downstream signal transduction, whereas the TM domain guarantees the localisation of MAVS to the mitochondrial outer membrane. Upon the interaction of viral components, RIG-I and MDA5 experience a conformational alteration that reveals their N-terminal tandem CARDs and facilitates tetramer formation through these CARDs, which are then modified with K63-polyubiquitin chains by the E3 ubiquitin ligases Tripartite Motif-containing (TRIM)25, RNF125, and RIG-I–like receptor E3 ubiquitin ligase (RIPLET). K63-linked ubiquitination facilitates the interaction of RIG-I and MDA5 with MAVS through their CARDs. The MAVS CARD swiftly generates prion-like aggregates that transform other MAVS on the mitochondrial outer membrane into prion-like aggregates. The aggregation process is crucial for the biological functions of MAVS.

MAVS subsequently associates via its PRR domain with TRAF2, TRAF3, TRAF5, or TRAF6. It thereby facilitates the activation of the TRAF family member-associated NF-κβ activator binding kinase I (TBK1) complex (comprising TBK1, IkB kinase [IKK]i/+, and NEMO) when interacting with TRAF2/3/5/6, and the IKK complex (consisting of IκB kinase alpha/beta (IKKa/b) and NEMO) when engaging with TRAF2/5/6. The TBK1 complex facilitates the phosphorylation and homodimerization of IRF3 and/or IRF7, which then translocate to the nucleus to bind to IFN-stimulated response elements, therefore inducing the transcription of target genes, including type I IFNs. Likewise, the TRAF2/5/6-activated IKK complex stimulates NF-κB to enhance the production of proinflammatory cytokines. Consequently, the two MAVS-mediated signalling pathways serve unique yet essential functions in antiviral innate immunity.

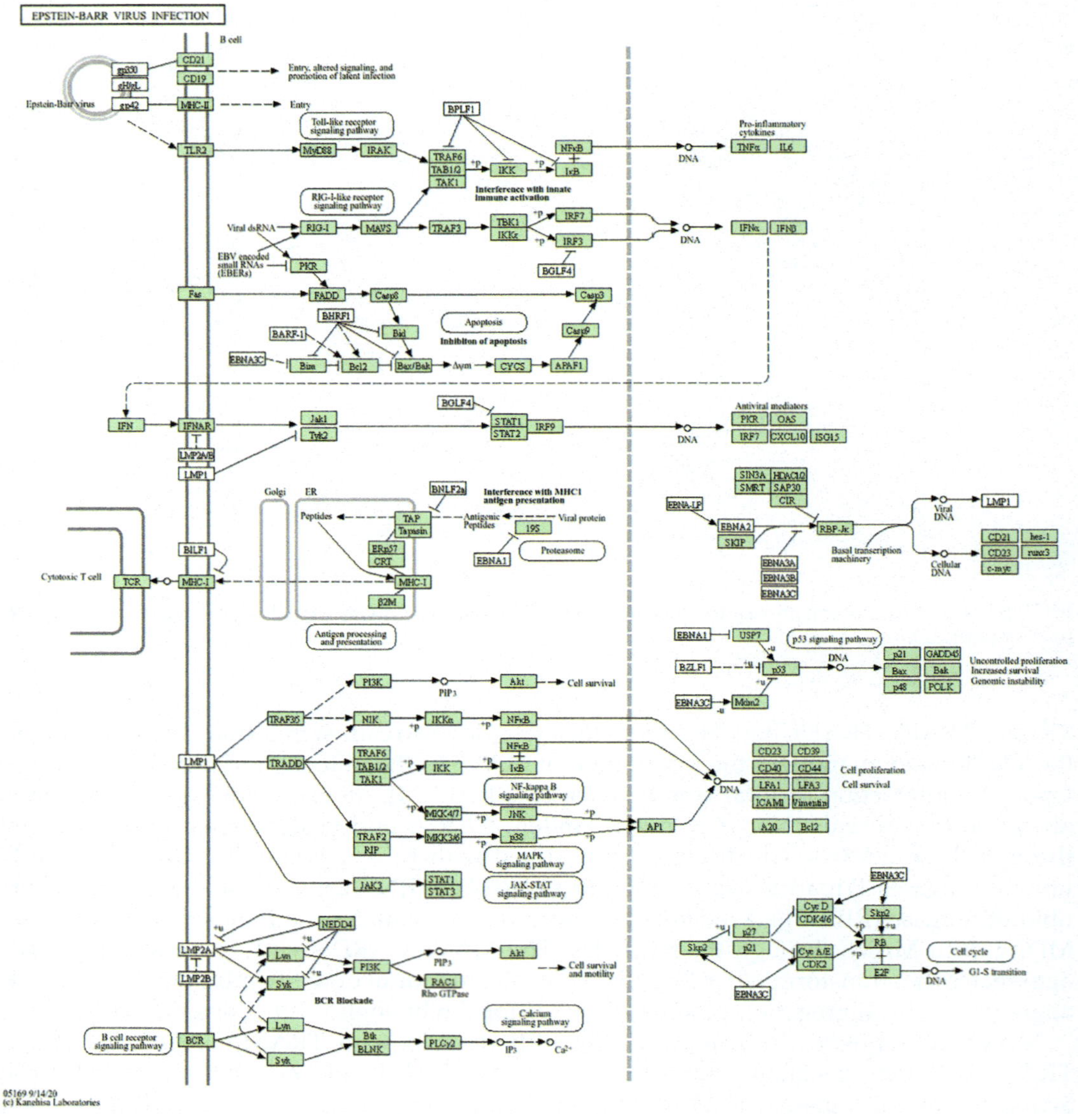

FIGURE 4.10 Mitochondrial antiviral-signalling protein acts on certain antiviral immune pathways against Epstein–Barr virus infection.

Regulation of mitochondrial antiviral-signalling protein function under physiological conditions and during viral infection

The expression and function of MAVS are meticulously regulated at both posttranscriptional and posttranslational levels, ensuring that RLR signalling pathways are swiftly

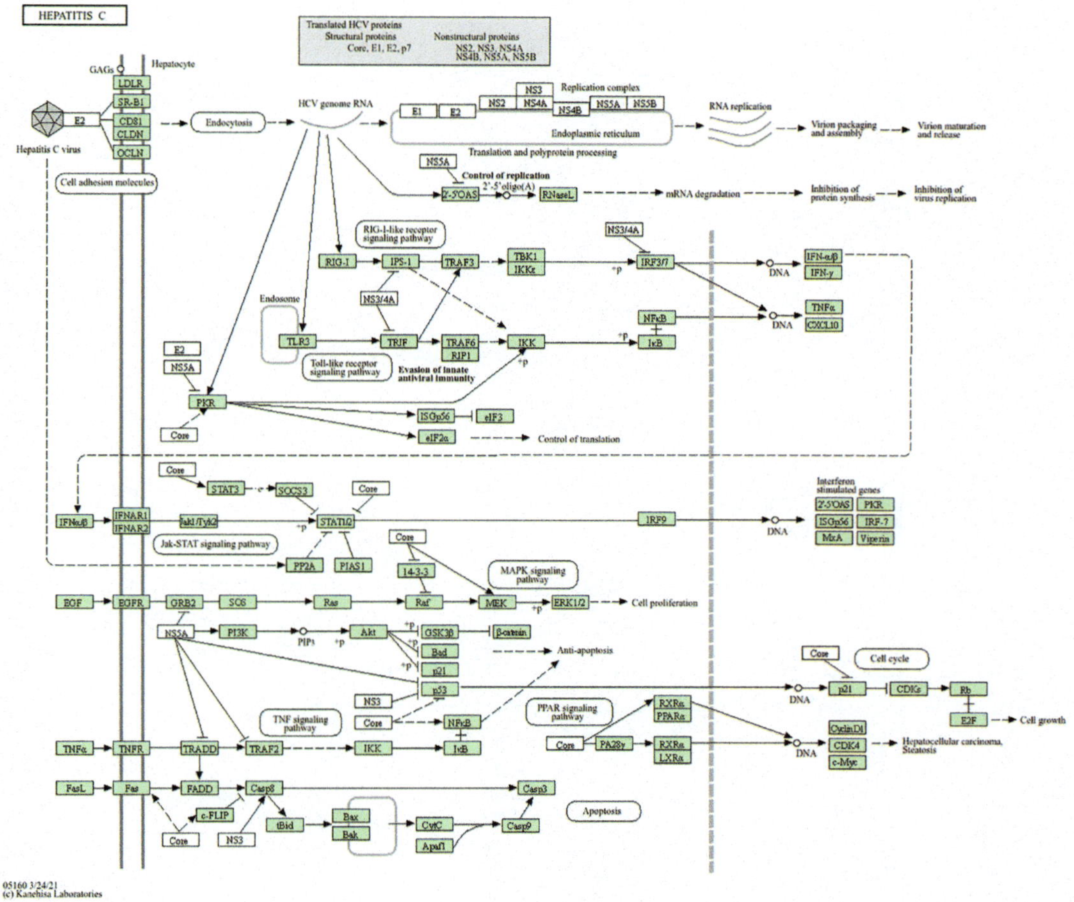

FIGURE 4.11 Mitochondrial antiviral-signalling protein acts on certain antiviral immune pathways against hepatitis C virus infection.

activated upon viral detection and promptly attenuated following viral eradication to prevent potential tissue damage.

Posttranscriptional regulation of mitochondrial antiviral-signalling protein

MAVS mRNA is polycistronic, characterised by the ability for protein translation to initiate not only from the N-terminal methionine but also from interior methionine residues, hence producing isoforms by alternate translation at several start locations. MAVS mRNA was shown to generate a full-length protein and five CARD-deleted variants (FL MAVS and MAVS-M142/303/358/367/449) from six separate translational start sites determined by methionine distribution. Under physiological settings, the binding of shortened MAVS occurs. Isoforms of FL MAVS limit its spontaneous aggregation, whereas the resultant

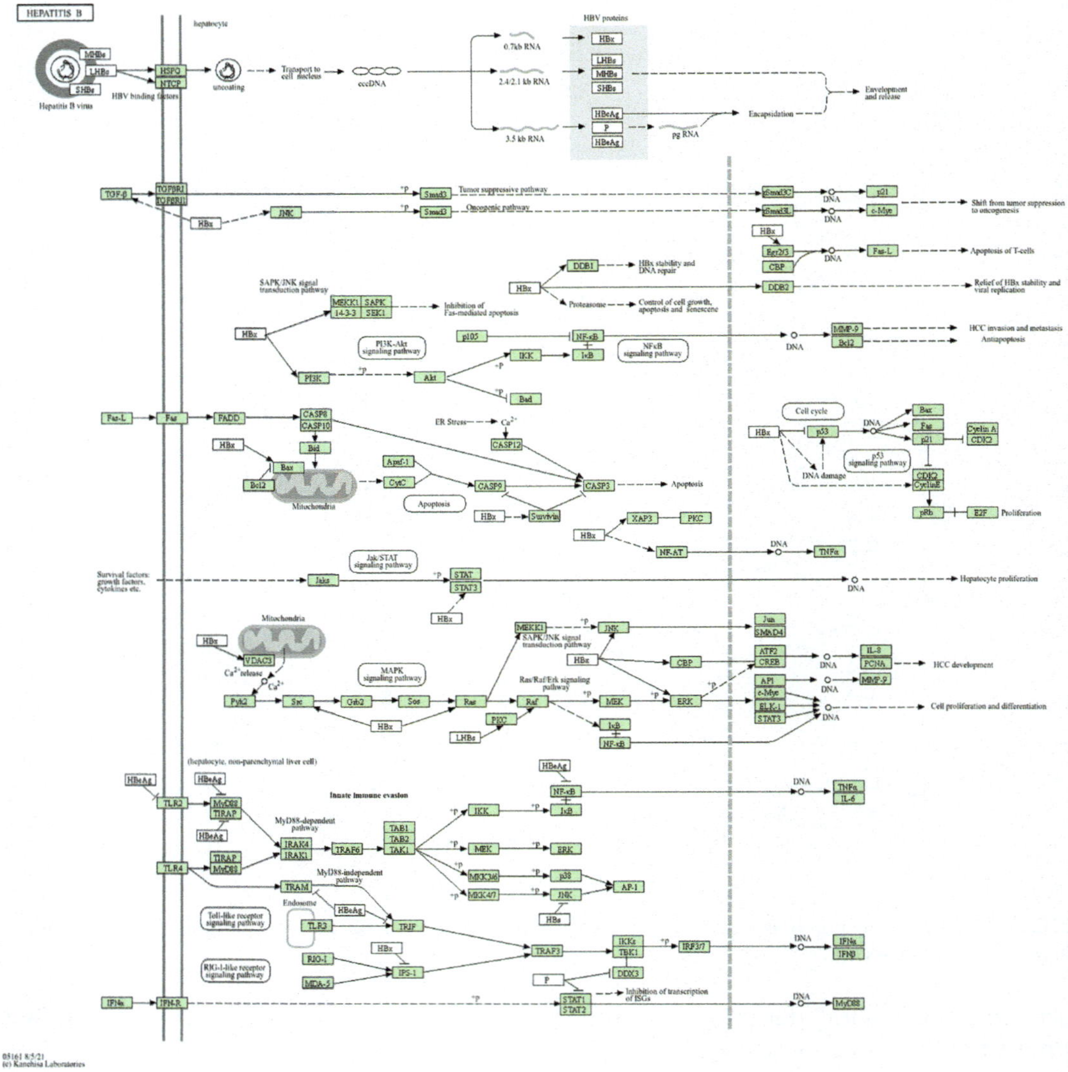

FIGURE 4.12 Mitochondrial antiviral-signalling protein acts on certain antiviral immune pathways against hepatitis **B** virus infection.

aggregates are destroyed by mitochondrial autophagy, thus preserving autoimmune homoeostasis. MAVS-M142 does not inhibit the aggregation of FL MAVS during viral infection, but it can obstruct the generation of IFN following the binding of MAVS-M142 to TRAF. The control of MAVS and MAVS-M142 can be accomplished through an upstream open reading frame (uORF) located in the 5′ untranslated region (UTR). The translation of uORF may bypass the initiation site of full-length MAVS and commence the translation of

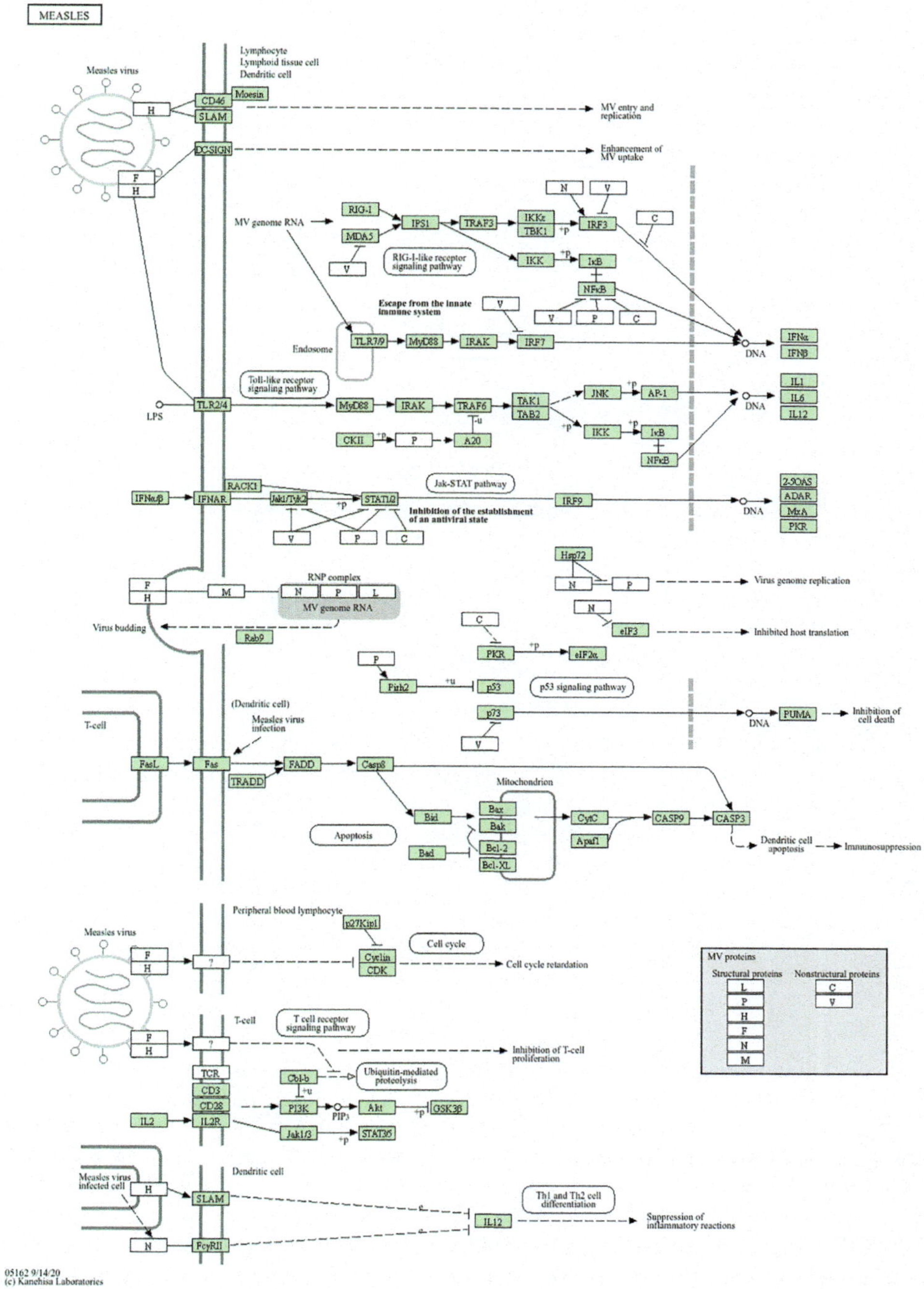

FIGURE 4.13 Mitochondrial antiviral-signalling protein acts on certain antiviral immune pathways against measles virus infection.

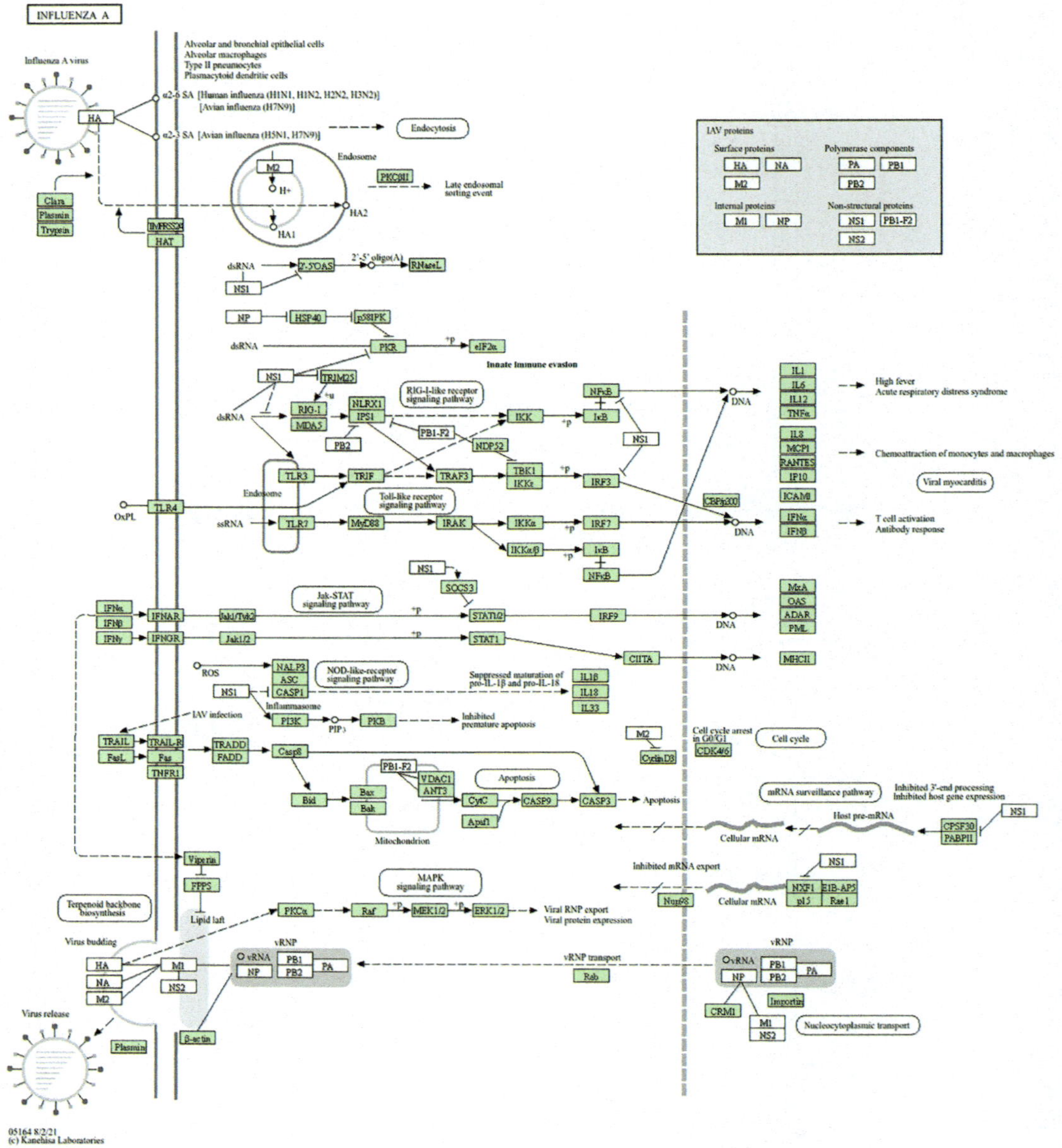

FIGURE 4.14 Mitochondrial antiviral-signalling protein acts on certain antiviral immune pathways against influenza A virus infection.

MAVS-M142. Consequently, the result of MAVS signalling is influenced to some degree by the cumulative effects of full-length MAVS and shortened MAVS isoforms. Currently, only a limited number of genetic studies have examined the control of FL MAVS. The functional

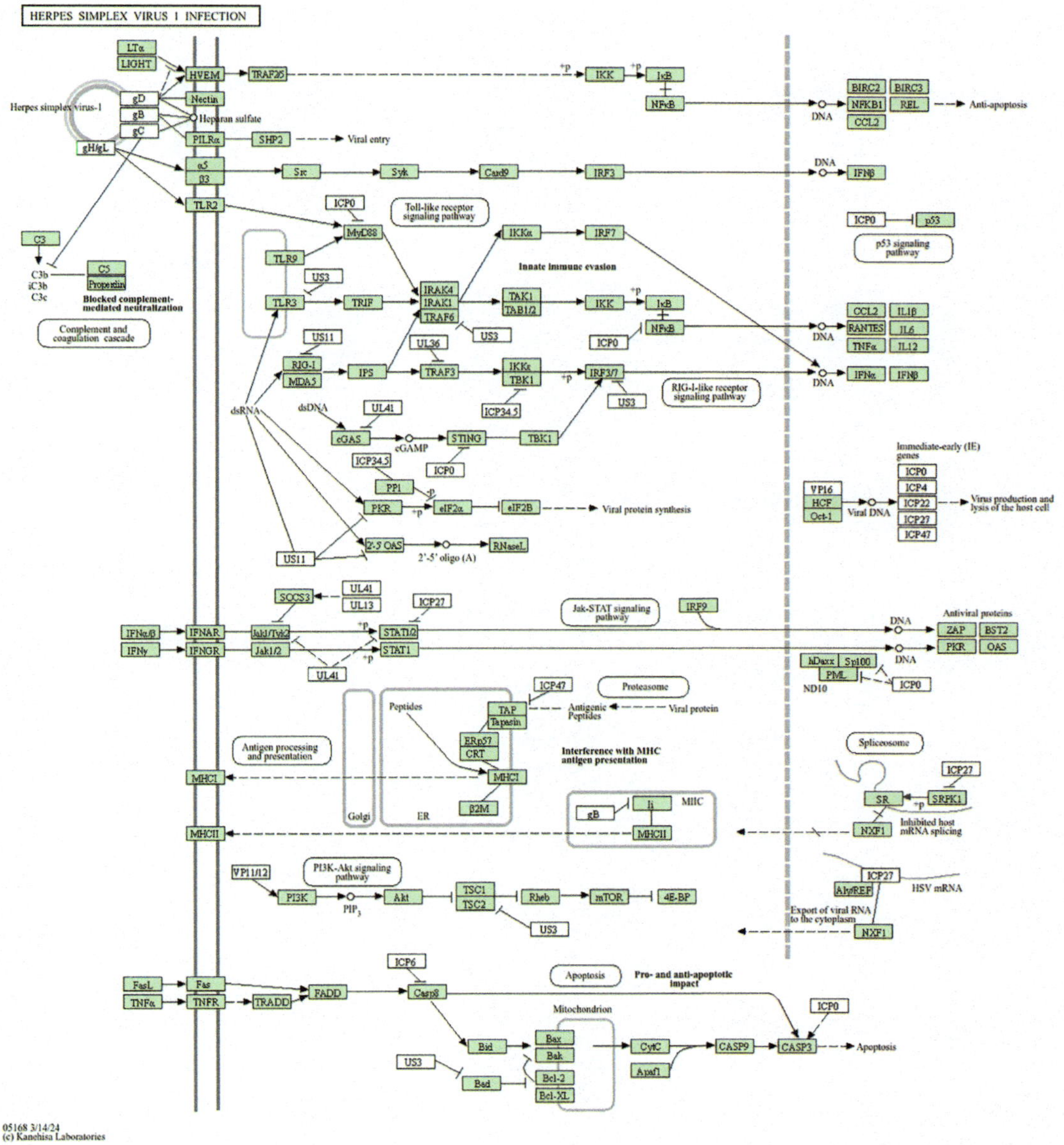

FIGURE 4.15 Mitochondrial antiviral-signalling protein acts on certain antiviral immune pathways against herpes simplex virus infection.

rivalry among MAVS shortened isoforms and the mechanism driving it remains ambiguous. mRNA is recognised for including many regulatory elements in the 3′ and 5′ UTRs. The attachment of regulatory components, including proteins and complementary short RNA molecules, to these sites is crucial for regulating protein translation. Three AU-rich regions

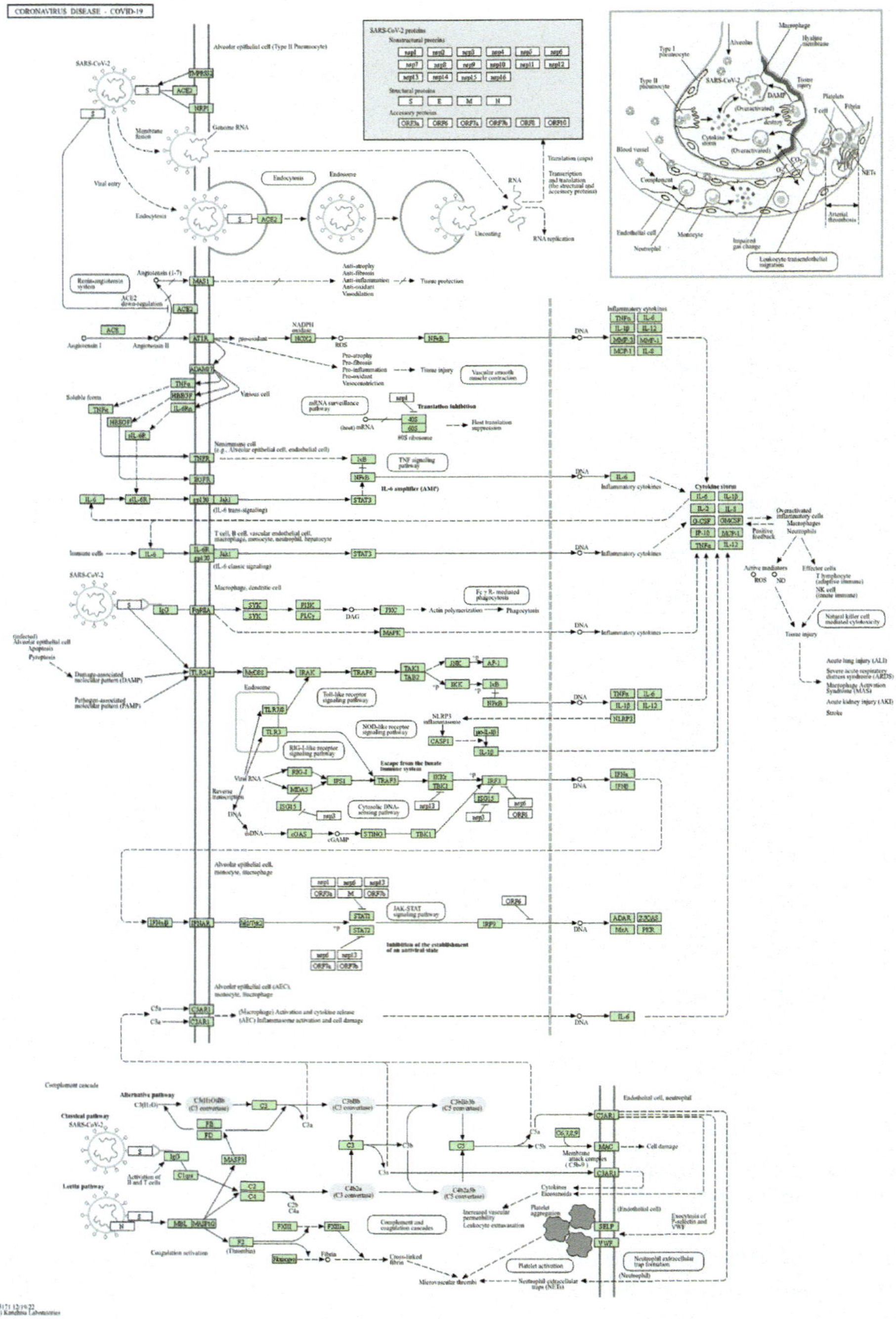

FIGURE 4.16 Mitochondrial antiviral-signalling protein acts on certain antiviral immune pathways against COVID-19 virus infection.

in the 3' UTR of MAVS mRNA facilitate the binding of human antigen R, which destabilises the mRNA and sustains low levels of MAVS protein under normal physiological conditions. MicroRNAs (miRNAs) are diminutive noncoding RNAs that modulate gene expression by facilitating the degradation and/or obstructing the translation of target mRNAs. Four miR-27a-binding sites have been discovered in the 3' UTR of MAVS mRNA, and the upregulation of miR-27a can suppress MAVS expression.

The Aspartate–Glutamate–Alanine–Aspartate (DEAD)-box (DDX) family of highly conserved RNA helicases is crucial to RNA synthesis, processing, transport, and destruction. A recent study demonstrated that, in virus-infected cells, DDX46, situated in the nucleus, binds to many conserved CCGGUU regions in MAVS, TRAF3, and TRAF6 mRNAs, thereby inhibiting their translocation. The cytoplasm facilitates translation, therefore diminishing the protein levels of MAVS, TRAF3, and TRAF6.

Posttranslational modulation of mitochondrial antiviral-signalling protein

MAVS can experience various posttranslational changes (Table 4.1) that affect its role in enhancing innate immune responses. Viral infection results in K63-linked ubiquitination of MAVS in the mitochondrial outer membrane, subsequently inducing MAVS aggregation, an indicator of its activation. Currently, the sole identified initiator of MAVS aggregation and activation is ubiquitination by the E3 ubiquitin ligase TRIM31 (Table 4.1). TRIM31-mediated MAVS aggregation is absent in RIG-I-defective or nonviral infection scenarios, indicating that RIG-I activation is essential for TRIM31-mediated MAVS aggregation following viral infection. The glycosylation of MAVS by O-linked N-acetylglucosamine transferase (OGT) augments TRIM31-mediated K63-linked ubiquitination of MAVS (Table 4.1). A viral infection enhances glucose metabolism in the host cell, involving the activation of the hexosamine production pathway and an increase in OGT levels. Despite just 2%–5% of intracellular glucose entering the hexosamine production route under physiological settings, OGT's capacity to enhance RLR signalling underscores the significance of glucose metabolism in antiviral innate immunity. Research indicates that sustained MAVS aggregation may correlate with the development of autoimmune disorders, including systemic lupus erythematosus. The prolonged production of IFN and its regulatory genes, along with unsuppressed signals, will ultimately lead to tissue damage. The K27-linked ubiquitination mediated by the E3 ubiquitin ligase TRIM21 positively regulates MAVS. A viral infection increases the expression of TRIM21, facilitating the recruitment of TBK1 to MAVS and amplifying downstream signalling. Phosphorylation is a significant posttranslational change that regulates MAVS signalling. Upon activation of TBK1 complexes by MAVS, the components TBK1 and IKK kinases phosphorylate MAVS, enabling it to bind IRF3. Recruited IRF3 is subsequently phosphorylated by TBK1, leading to IRF3 homodimerization, nuclear translocation, and enhancement of IFN production. K48-linked ubiquitination of MAVS is recognised to facilitate its proteasomal breakdown, and proteins that impede MAVS K48-linked ubiquitination can therefore enhance RLR-mediated signalling. The endogenous protein cyclophilin A, which typically competes with TRIM25 for MAVS binding, is increased during viral infection, thereby effectively blocking TRIM25-mediated MAVS ubiquitination and destruction. Ubiquitination is a reversible mechanism, and viral infection enhances the expression of ovarian tumour family deubiquitinase 4 (OTUD4),

TABLE 4.1 The posttranslational modifications (PTMs) of mitochondrial antiviral-signalling protein (MAVS).

Regulator of PTMs	PTMs	Site	Regulatory effect
-	K63-linked ubiquitination	K500	Promotes the recruitment of IKKε to MAVS, then IKKε detaches TRAF3 from mitochondria and promotes K63-linked deubiquitination of TRAF3 and its subsequent K48-linked ubiquitination degradation
TRIM31	K63-linked ubiquitination	K10/K311/K461	Promotes the aggregation and activation of MAVS during viral infection
OGT	Glycosylation	S366	Promotes the TRIM31-mediated K63-linked ubiquitination of MAVS and RLRs signalling
TRIM21	K27-linked ubiquitination	K325	Promotes TBK1 recruitment and thereby MAVS-mediated IFN production
MARCH8	K27-linked ubiquitination	K7	Promotes NDP52 to recognise K27-linked ubiquitination signal on MAVS and induces lysosomal autophagy of MAVS
TRIM25	K48-linked ubiquitination	K7/K10	Induces proteasomal degradation of MAVS
Atrophin-1 Interacting Protein 4 (AIP4)/Itchy E3 ubiquitin protein ligase (ITCH)	K48-linked ubiquitination	K371/K420	Induces proteasomal degradation of MAVS
PCBP1	K48-linked ubiquitination	NA	Promotes the K48-linked ubiquitination degradation of MAVS by recruiting AIP4 under normal conditions, thus preventing spontaneous activation of immune response
PCBP2	K48-linked ubiquitination	NA	Promotes the K48-linked ubiquitination degradation of MAVS by recruiting AIP4 at the late stage of viral infection, thus avoiding excessive immune response
TAX1BP1	K48-linked ubiquitination	NA	Promotes the K48-linked ubiquitination degradation of MAVS by recruiting AIP4 under normal and viral infection conditions, thus playing a role similar to PCBP1/2
Smurf1 and Smurf2	K48-linked ubiquitination	NA	Induces proteasomal degradation of MAVS
Ndfip1	K48-linked ubiquitination	NA	Promotes the K48-linked ubiquitination degradation of MAVS by recruiting Smurf1

(Continued)

TABLE 4.1 (Continued)

Regulator of PTMs	PTMs	Site	Regulatory effect
OTUD1	K48-linked ubiquitination	NA	Promotes the K48-linked ubiquitination degradation of MAVS by recruiting Smurf1 at the early stage of viral infection, inhibiting the RLRs pathway
MARCH5	K48-linked ubiquitination	K7/K500	Induces proteasomal degradation of MAVS
RNF5	K48-linked ubiquitination	K362/K461	Induces proteasomal degradation of MAVS
OTUD4	Removal of K48-linked ubiquitination	NA	Stabilises MAVS and thereby induces IFN production during viral infection
TBK1 and IKKβ	Phosphorylation	S442	Gives phosphorylated MAVS the ability to recruit IRF3 for its phosphorylation by TBK1
NLK	Phosphorylation	S121/S212/ S258/S329	Inducing the degradation of MAVS in the later phase of infection, thereby inhibiting further signalling transduction
PPM1A/PP2Cα	Dephosphorylation	NA	Dephosphorylates MAVS, TBK1, and IKKε, thus blocking signalling conduction

which eliminates K48-linked ubiquitin from MAVS, hence diminishing its breakdown. The findings indicate that the host cell has developed mechanisms to suppress K48-linked ubiquitination and degradation of MAVS during viral infection, hence preserving a minimum level of signalling for the antiviral response. These studies collectively demonstrate that MAVS signalling is modulated through three primary posttranslational changes that function synergistically.

1. TRIM31-induced K63-linked ubiquitination activates MAVS. (2) TRIM21-mediated K27-linked ubiquitination facilitates the recruitment of TBK1 to MAVS. (3) Phosphorylation of MAVS by TBK1 and IKK facilitates the recruitment of IRF3 to MAVS. Moreover, the suppression of K48-linked ubiquitination and degradation is essential for preserving sufficient cellular levels of MAVS to guarantee antiviral protection.

Posttranslational inhibition of mitochondrial antiviral-signalling protein

Currently, the adverse posttranslational regulation of MAVS primarily occurs through K48-linked ubiquitination modification, signal inhibition, autophagy, and death.

Regulation of mitochondrial antiviral-signalling protein by ubiquitination

K48-linked ubiquitination of MAVS induces its proteasomal degradation, hence constraining RLR signalling. Several E3 ubiquitin ligases, including TRIM25, AIP4, Smurf1, Smurf2, MARCH5, and RNF5, have been demonstrated to facilitate MAVS K48-linked ubiquitination (see Table 4.1 for further details). Moreover, accumulating evidence indicates that the interaction of several adaptor proteins with MAVS influences the ubiquitination-mediated destruction of MAVS. K48-linked ubiquitination of MAVS, facilitated by the E3 ubiquitin ligase AIP4 (sometimes referred to as Itch), is enhanced by the poly (C)-binding protein 2 (PCBP2), which is increased following viral infection. PCBP2 associates with the TM domain of MAVS and recruits AIP4, facilitating the degradation of MAVS (see Table 4.1).

Recent research has shown a comparable function for the highly related PCBP1; however, the expression level of PCBP1 does not seem to change during viral infection. MAVS expression may be sustained at a low level under physiological conditions partly due to PCBP1-mediated degradation promotion, while the upregulation of PCBP2 during viral infection facilitates MAVS degradation, serving as a negative feedback mechanism to restore homoeostasis and prevent excessive immune signalling. AIP4 is controlled by the adaptor protein Tax1-binding protein 1 (TAX1BP1), which interacts with the MAVS CARD and recruits AIP4. TAX1BP1 modulates MAVS degradation both under physiological settings and during viral infection, analogous to PCBP1/2. The E3 ubiquitin ligase Smurf1, which facilitates K48-linked degradation of MAVS, is favourably regulated by two adaptor proteins: OTUD1 and Nedd4 family-interacting protein 1 (Ndfip1). OTUD1 associates with Smurf1 and acts as a deubiquitinase, eliminating K48-linked ubiquitination of Smurf1, thereby inhibiting its self-ubiquitination degradation and elevating its protein levels. Conversely, Ndfip1's binding to MAVS recruits Smurf1 to MAVS, and the interaction between Ndfip1 and Smurf1 amplifies Smurf1's self-ubiquitination (potentially K63-linked) and enzymatic activity. The variety of E3 ubiquitin ligases implicated in MAVS degradation indicates that they may offer additional targets for the treatment of viral infections. Numerous proteins decrease antiviral innate immunity by obstructing TRIM31-mediated K63-linked ubiquitination and activation of MAVS. Fas-associated factor 1 (FAF1) is a protein that facilitates Fas-induced apoptosis and has recently been demonstrated to play a role in the MAVS response. Under physiological settings, FAF1 aggregates generated by its ubiquitin-like proteins (UBL) domain bind to and antagonise MAVS by competing with TRIM31 for MAVS interaction, thereby limiting MAVS aggregation. Subsequently, upon viral infection, FAF1 undergoes phosphorylation by active IKK+, facilitating its depolymerisation and lysosomal breakdown, hence alleviating MAVS inhibition. Rac1 is recognised for its role in facilitating tumour cell proliferation. Viral infection induces geranylgeranylation and palmitoylation of Rac1, hence inhibiting the TRIM31–MAVS interaction and suppressing innate defence; notably, under physiological settings, Rac1 can suppress MAVS aggregation as well. TRAF3, an E3 ligase situated downstream of MAVS signalling, undergoes self-ubiquitination via K63 linkages, which enhances the interaction between TRAF3 and TBK1. K63-linked ubiquitination at Lys500 of MAVS, facilitated by an unidentified enzyme, can induce the dissociation of TRAF3 from MAVS, leading to K63-linked deubiquitination of TRAF3 and its subsequent K48-linked ubiquitin-mediated destruction, hence impeding TRAF3-mediated signalling. These studies collectively indicate that MAVS K48-linked ubiquitination under physiological settings facilitates MAVS breakdown and inhibits unprovoked activation of the immune response. The same procedure transpiring throughout the advanced phases of viral infection functions to re-establish cellular baseline activity and avert the overproduction of possibly detrimental immune mediators. Likewise, suppression of TRIM31-mediated K63-linked ubiquitination of MAVS inhibits its aggregation and activation, while still preserving MAVS homoeostasis.

Inhibition of mitochondrial antiviral-signalling protein through direct protein interaction

Another mechanism regulating MAVS signalling under healthy settings involves the direct binding of tubulin-tyrosine ligase-like protein 12 (TTLL12) to MAVS, TBK1, and IKK +. In the course of a viral infection, the expression of TTLL12 diminishes, thereby removing the inhibition on MAVS-mediated immune response activation. Protein phosphatase magnesium-dependent 1A (PPM1A, or PP2Ca) is a constituent of the TBK1/IKK+ complex and typically preserves MAVS, TBK1, and IKK+ in a dephosphorylated condition, hence inhibiting MAVS downstream signalling. Nonetheless, the elevation of MAVS expression during viral infection circumvents the PPM1A-mediated obstruction in signalling and establishes a threshold for the activation of antiviral signalling. Lactate, the final result of anaerobic glycolysis, has traditionally been seen as a metabolic waste, although current research indicates its significant involvement in immunomodulation. Lactate serves as a principal negative regulator of RLR signalling by directly binding to the MAVS TM domain, hence inhibiting MAVS mitochondrial localisation, aggregate formation, and signalling activity. Lactate levels are intimately correlated with glycolysis. Under physiological conditions, the interaction between hexokinase II (HK2), a crucial rate-limiting enzyme for glycolysis situated on the mitochondrial outer membrane, and MAVS sustains HK2 activity at its basal level, thereby facilitating glycolytic metabolism and inhibiting MAVS mediated by lactate. During viral infection, the binding of active RIG-I to MAVS results in the dissociation of HK2, which diminishes glycolysis and lactate synthesis, mitigates lactate-mediated MAVS inhibition, and enhances downstream signalling for IFN production. The relationship between MAVS and HK2 is linked to its activation and functionality. Notably, the downregulation of glucose metabolism and lactate-mediated inhibition of MAVS transpires solely during the initial phases of viral infection. This aligns with the discovery that viral infection leads to a general increase in glucose metabolism to guarantee sufficient energy resources for replication and proliferation. In the advanced phases of viral infection, MAVS activity is inhibited by UBX-domain-containing protein 1 (UBXN1), which belongs to the ubiquitin-binding protein family. UBXN1 expression is significantly elevated late in viral infection and competes with TRAF3/TRAF6 for binding to MAVS (amino acids 455–460), hence inhibiting MAVS signalling. Gpatch domain-containing protein 3 (GPATCH3), a ubiquitously expressed protein, functions analogously to UBXN1.

GPATCH3 interaction with MAVS inhibits the formation of the MAVS/TRAF6/TBK1 complex during viral infection.

Mitochondrial antiviral-signalling protein regulation concerning apoptosis and autophagy

When alternative mechanisms are ineffective, apoptosis may serve as a technique to impede viral replication. Nonetheless, the cell has developed multiple feedback systems to guarantee prompt regulation of immunological signalling, so preventing excessive tissue damage. NLRP11, a type I IFN-induced nucleotide-binding oligomerization domain (Nod)-like receptor, is increased upon viral infection and can suppress type I IFN production. Recent findings indicate that a portion of NLRP11 translocates from the cytoplasm to the mitochondria and interacts with MAVS; however, NLRP11 does not interfere with the interaction between MAVS and RIG-I, nor does it affect the ubiquitination of MAVS. NLRP11 significantly interacts with TRAF6, facilitating K48-linked ubiquitination and subsequent degradation of TRAF6, which inhibits RLR signalling. Notably, the over-expression of NLRP11 suppressed the cleavage of poly-ADP-ribose polymerase, a protein implicated in apoptosis associated with DNA damage. The deletion of TRAF6 negated the NLRP11-mediated inhibition in cells. Given that TRAF6 is implicated in apoptosis following viral infection, NLRP11 may have the capacity to prevent apoptosis. It also guarantees the survival of the host cells. The deletion of MAVS abolished the interaction between NLRP11 and TRAF6. The findings indicate that MAVS functions as a platform for NLRP11 to degrade TRAF6, hence inhibiting TRAF6-dependent apoptosis. NLRP11's dual blocking function in the antiviral immune response acts as a negative feedback loop that inhibits RLR signalling, IFN production, and apoptosis. Cysteinyl aspartate specific proteinase-3 (caspase-3), primarily recognised as a pivotal executor of apoptosis, is garnering heightened interest for its nonapoptotic roles.

In the course of viral infection, activated caspase-3 cleaves MAVS at Asp429 and Asp490, thereby inhibiting excessive signalling. This result enhances our comprehension of caspase-3 actions and offers novel insights for medication development aimed at preventing abnormal immune responses. Furthermore, the hemagglutinating virus of Japan envelope (HVJ-E) is produced from inactivated replication-defective Sendai virus exhibited antitumor action reliant on RIG-I/MAVS signalling. HVJ-E promotes the RIG-I/MAVS signalling pathway to enhance the production of its downstream pro-apoptotic genes TNF-related apoptosis-inducing ligand (TRAIL) and Noxa, thereby triggering selective apoptosis in tumour cells, including prostate, lung, and breast cancer cells. This procedure notably does not take place in typical immune cells. Autophagy, a cellular breakdown process, functions as an antiviral defence mechanism by eliminating intracellular pathogens and engaging with the innate immune response. Tetherin, often referred to as bone marrow stromal cell antigen 2 (BST2) or CD317, is an IFN-induced membrane protein that is increased in response to viral infection. Tetherin facilitates K27-linked ubiquitination of MAVS through the recruitment of the E3 ubiquitin ligase. The autophagy protein NDP52 subsequently associates with K27-linked ubiquitinated MAVS, resulting in lysosomal destruction. The autophagic degradation of MAVS mediated by tetherin might be regarded as an additional negative feedback mechanism that mitigates excessive signalling. Further negative regulatory mechanisms in MAVS-mediated antiviral immunity involve the phosphorylation of MAVS by Nemo-like kinase (NLK) which transpires in the later stages of viral infection and prompts MAVS degradation. The mitochondrial fusion proteins mitofusin 1 and 2 (Mfn1 and Mfn2), situated on the outer mitochondrial membrane, exert antagonistic functions in the localisation of MAVS during RLR-mediated signalling. Mfn2 engages with the C-terminal, specifically the TM domain, of MAVS interacts with the HR1 region, thereby inhibiting the activation of IRF3 and NF-κB, a mechanism reliant on an Mfn2-dependent complex (~600) kDa. The depletion of endogenous Mnf2 diminished the relative molecular mass of the MAVS complex from a high to a lower molecular mass, hence augmenting the antiviral efficacy of MAVS. Nonetheless, data suggest that the MAVS aggregate exceeds 700 kDa, indicating that the MAVS complex may consist of MAVS and its downstream signalling components. The Mfn2-dependent form of molecules is not conducive to signalling transmission. Unlike Mfn2, Mfn1 exerts a beneficial influence via modulating the mitochondrial redistribution of MAVS establishing its speckled staining pattern within cells, which may be associated with the development of MAVS prion-like structures. Homologous proteins Mfn1 and Mfn2 collectively modulate MAVS-mediated signalling while regulating mitochondrial fusion. Moreover, Mfn1/Mfn2-mediated mitochondrial fusion can augment RLR signalling, the precise mechanism is ambiguous (see Fig. 4.17, Fig. 4.18).

1. Under physiological conditions, MAVS activity is inhibited to prevent aberrant activation of the immune response.
2. Early after viral infection, MAVS is activated by TRIM31-mediated k63-linked ubiquitination, and the inhibition of MAVS is relieved, thus initiating the RLR signalling pathway.
3. At later stages of viral infection, MAVS is cleaved and degraded, which effectively arrests RLR-stimulated signalling.

How viruses overcome the host innate immunity—role of MAVS

Host possess an innate immune mechanism, whenever it weakens or damaged, the pathogens like virus find a way out how to enter.

Regulation of mitochondrial antiviral-signalling protein by viral proteins

Viruses can evade the host's antiviral immune response by facilitating the cleavage or degradation of MAVS and by directly disrupting components of the RLR-activated signalling pathway. Significantly, there is less evidence that viral proteins contribute to the augmentation of the host's innate immune response.

Cleavage of MAVS protein: A multitude of virus-encoded proteins function as proteases, capable of cleaving MAVS without necessitating proteasomal destruction or apoptosis, hence obstructing RLR signalling (Fig. 4.4). The initial viral protein identified to colocalise with and cleave MAVS in the mitochondria was the hepatitis C virus (HCV)

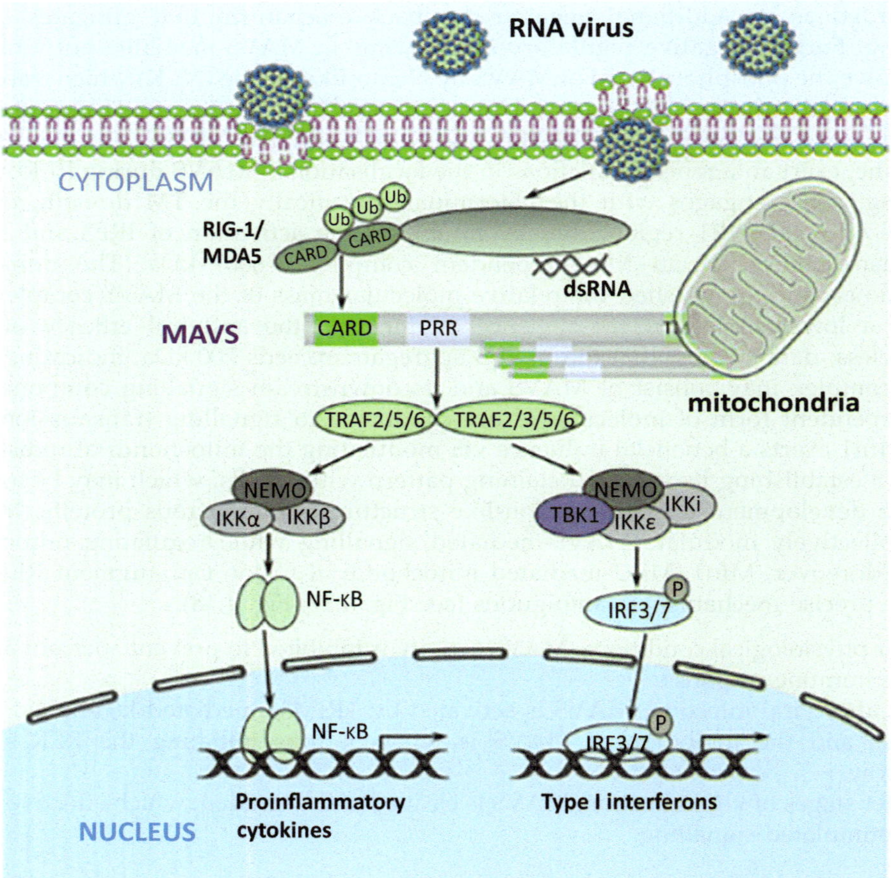

FIGURE 4.17 RLR signalling pathways. Binding of viral components to retinoic acid-inducible gene I (RIG-I) and melanoma differentiation-associated gene 5 (MDA5) induces their interaction with mitochondrial antiviral-signalling protein (MAVS) via their common caspase activation recruitment domains (CARDS). MAVS activates two signalling cascades leading to the production of immune factors. The tumour necrosis factor receptor-associated factor (TRAF) 2/5/6-IkB kinase (IKK) complex-nuclear transcription factor-κB (NF-κB) pathway induces transcription of proinflammatory cytokines, whereas the TRAF2/3/5/6-TRAF family member-associated NF-κβ activator binding kinase I (TBK1) complex-interferon regulatory factors (IRF3/7) pathway induces the expression of type I interferon (IFN) genes.

serine protease NS3/4A. NS3/4A cleaves MAVS at Cys508, displacing the N-terminal portion of MAVS from the mitochondria, diminishing downstream signalling, and facilitating sustained viral infection. Seneca valley virus, human rhinovirus C, and coxsackievirus B3 (CVB3) are small RNA viruses that each encode a cysteine protease, 3Cpro, which cleaves MAVS at Gln148, hence inhibiting its function. CVB3 additionally generates a second MAVS-cleaving protease, 2Apro, although the precise cleavage location remains ambiguous. Furthermore, a 3C-like serine protease, analogous to 3Cpro, is synthesised by the porcine reproductive and respiratory syndrome virus and

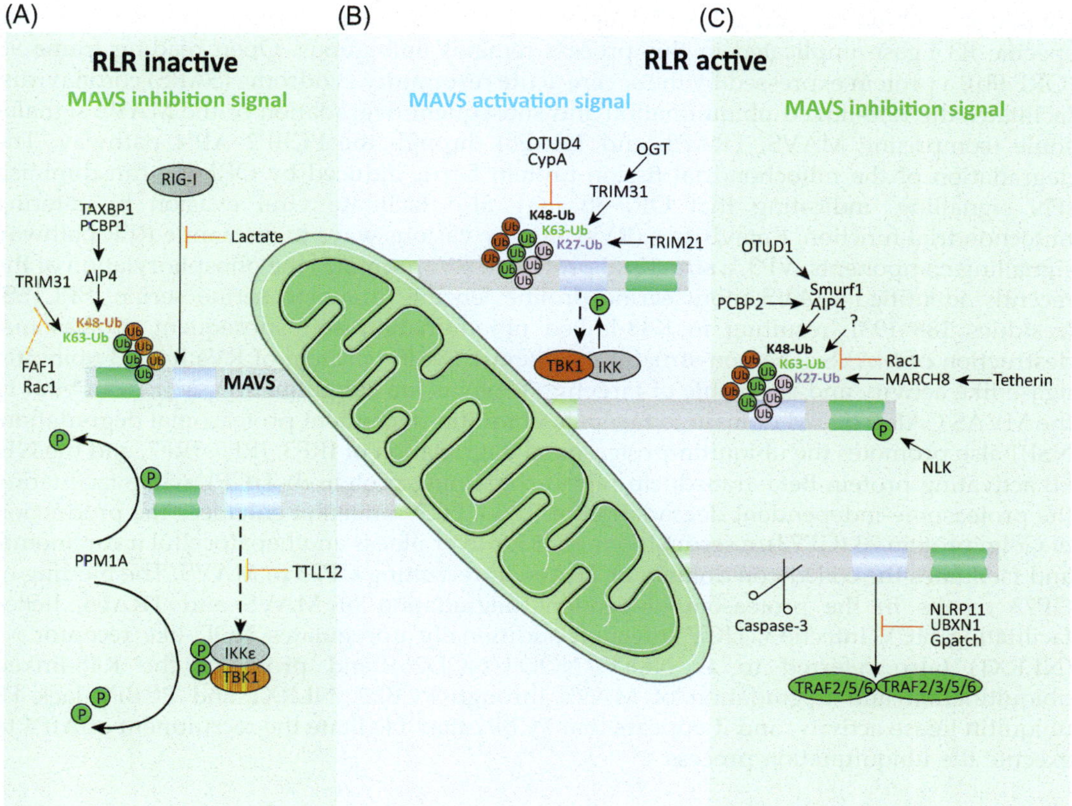

FIGURE 4.18 Positive and negative regulation of mitochondrial antiviral-signalling protein (MAVS) activity in antiviral immunity.

cleaves MAVS at Glu268. The proteases 3ABC, encoded by the hepatitis A virus (HAV), and 2Apro, from enterovirus 71 (EV71), are especially significant. 3ABC, generated by bat and human HAV, cleaves human MAVS at Glu463 and Gln427, respectively, indicating that cross-species interference with MAVS signalling may facilitate the transfer of HAV between species. The initial viral protein identified to cleave MAVS at various amino acid residues was the cysteine protease 2Apro, generated by EV71, the primary aetiological agent of hand, foot, and mouth disease. 2Apro cleaves MAVS at Gly209, Gly251, and Gly265.

Degradation of mitochondrial antiviral-signalling protein

Besides direct cleavage, several viral proteins promote the proteasomal degradation of MAVS (Fig. 4.4). The hepatitis B virus (HBV) protein X (HBX) interacts with MAVS and facilitates its ubiquitination at Lys136, resulting in proteasomal destruction. Nonetheless, the specific E3 ligase implicated in this process remains ambiguous. Open reading frame 9b (ORF-9b), a protein expressed by the severe acute respiratory syndrome (SARS) coronavirus, facilitates the K48-linked ubiquitination and subsequent degradation of the MAVS signalosome (comprising MAVS, TRAF3, and TRAF6) through the PCBP2–AIP4 pathway. The degradation of the mitochondrial fission protein Drp1, induced by ORF-9b, can diminish IFN signalling, indicating that ORF-9b may also facilitate viral evasion by altering mitochondrial function. Rotaviruses (RVs) employ various ways to dismantle RLR pathway signalling components. VP3, a structural protein of RV, promotes the phosphorylation of the recently identified MAVS motif serine proline leucine threonine serine serine (SPLTSS) (residues 188–193), resulting in K48-linked ubiquitination and subsequent proteasomal destruction of MAVS. The non-structural protein 1 (NSP1) protein of RVs has E3 ubiquitin ligase-like activity and is capable of targeting proteins for ubiquitination. NSP1 can bind to the MVAS CARD or TM domain to facilitate ubiquitin-dependent proteasomal degradation. NSP1 also promotes the ubiquitin-proteasomal degradation of IRF3, IRF5, IRF7, and the NF-κB activating protein Beta–transducin repeat-containing protein (b-TrCP), while facilitating the proteasome-independent degradation of RIG-I. HCV infection enhances the production of Golgi protein 73 (GP73), a serum biomarker for liver illness and hepatocellular carcinoma, and facilitates the coiled-coil domain of TRAF6 in recruiting GP73 to MAVS. The binding of GP73 results in the proteasome-dependent degradation of MAVS and TRAF6, hence facilitating HCV infection. HCV infection additionally upregulates NOD-like receptor X1 (NLRX1) (also referred to as NOD5/NOD9/NOD26) and promotes the K48-linked ubiquitination and degradation of MAVS through PCBP2. NLRX1 and PCBP2 lack E3 ubiquitin ligase activity, and it appears that PCBP2 may facilitate the recruitment of AIP4 to execute the ubiquitination process.

Regulation of mitochondrial antiviral-signalling protein signalling

Viral proteins can obstruct MAVS-mediated signalling by direct or indirect suppression of specific pathway components (Fig. 4.4). Numerous viral proteins attach to and inhibit the interaction among MDA5, RIG-I, and MAVS. This encompasses NS1 and NS2 of respiratory syncytial virus, NS4A of Zika virus, and 3Cpro proteins of coxsackievirus (CV)-A16, CV-A6, and EV-D68. The glycoprotein US9, encoded by human cytomegalovirus, compromises mitochondrial integrity and causes a reduction in membrane potential, resulting in the detachment of MAVS. The virulence factors M2-2 and glycoprotein G of human metapneumovirus (hMPV) inhibit RLR signalling. M2-2 interacts with MAVS, inhibiting the assembly of MAVS/TRAF signalosomes, whereas glycoprotein G attaches to the CARD of RIG-I, hence obstructing MAVS binding. The capsid protein VP16 of herpes simplex virus 1 obstructs the synthesis of early ISGs facilitated by peroxisome MAVS, hence impeding the initial peroxisome-mediated response to viral infection. The influenza A virus (IAV), the predominant strain of influenza virus, has instigated numerous global pandemics due to its diversity and elevated pathogenicity. The H5N1 strain of IAV encodes the PB1-F2 protein and PB21, a faulty yet functional polypeptide, which have contrasting effects on MAVS-mediated antiviral innate immunity. PB1-F2's interaction with the MAVS TM domain diminishes mitochondrial membrane potential and impairs

MAVS signalling, while PB21 directly engages with MAVS to enhance signalling for type I IFN production, thus reducing viral multiplication. IAV NS1 impedes TRAF3–MAVS interactions by eliminating K63-linked ubiquitination on TRAF3, hence negatively modulating the antiviral response. IAV also suppresses MAVS expression at the posttranscriptional stage. IAV enhances the expression of miRNA-125a and miRNA-125b, which interact with the MAVS mRNA 3' UTR and suppress its translation. Furthermore, the M2 protein of IAV stimulates calcium-dependent reactive oxygen species generation, hence augmenting MAVS aggregation and activity. Viral proteins may potentially induce intracellular proteins to suppress MAVS activity (Fig. 4.4). For instance, HCV non-structural protein 5A (NS5A) facilitates the binding of the mitochondrial protein leucine-rich PRR motif containing protein to MAVS, hence inhibiting the interaction between MAVS and TRAF3/6. HBX can facilitate the recruitment of the E3 ubiquitin ligase linear ubiquitin chain assembly complex to the mitochondria, and the ensuing linear ubiquitination of MAVS inhibits MAVS–TRAF3 interactions. The V protein of the Nipah virus obstructs the degradation of the negative regulatory factor UBXN1, thereby augmenting the inhibition of MAVS–TRAF interactions by UBXN1. Numerous viruses utilise various techniques to undermine antiviral immunity. For instance, HCV not only cleaves MAVS through N3/4A but also promotes the proteasomal degradation of MAVS via GP73 or NLRX1. While the majority of contemporary research concentrates on hepatitis viruses (HAV, HBV, and HCV) and respiratory viruses (IAV, human rhinovirus C, SARS coronavirus, respiratory syncytial virus, EV-D68, and hMPV), there is a growing focus on viruses such as EV71, CV-A16, CV-A6, Nipah virus, Seneca valley virus, and Zika virus. This research will certainly enhance our comprehension of the methods by which viruses circumvent the human antiviral response. The acknowledgement of MAVS's critical function in antiviral immunity has resulted in an increase in research focused on this molecule. Cells and viruses now evidently regulate MAVS at both posttranscriptional and posttranslational stages through several methods. Ubiquitination is essential for the initiation, maintenance, and termination of RLR-mediated signalling pathways, with K27-, K48-, and K63-linked ubiquitination collaboratively regulating MAVS signalling. The impact of different forms of ubiquitination (e.g. K6-, K11-, K29-, K33-, and M1-linked ubiquitin chains) on the biological functions of MAVS remains uncertain. The influence of changes such as phosphorylation and glycosylation, together with byproducts of anaerobic glycolysis (lactate) and elements of the apoptosis and autophagy pathways, on MAVS is garnering heightened attention. Viral multiplication necessitates significant food absorption from the host organism, and the potential regulatory role of other elements of carbohydrate and protein metabolism on the antiviral immune response represents an emerging area of research. MAVS, as the principal adaptor protein in RLR-mediated signalling, is essential to the innate immune response. Notwithstanding considerable progress in recent years, as demonstrated in this review, the mechanisms governing MAVS-mediated signalling in other species remain ambiguous. Subsequent research in this domain is anticipated to elucidate MAVS function in antiviral immunity and to catalyse the development of novel therapeutics for viral infections (see Fig. 4.19).

We will discuss it further in the section for Chapter 6, mitogenomics and viral disease.

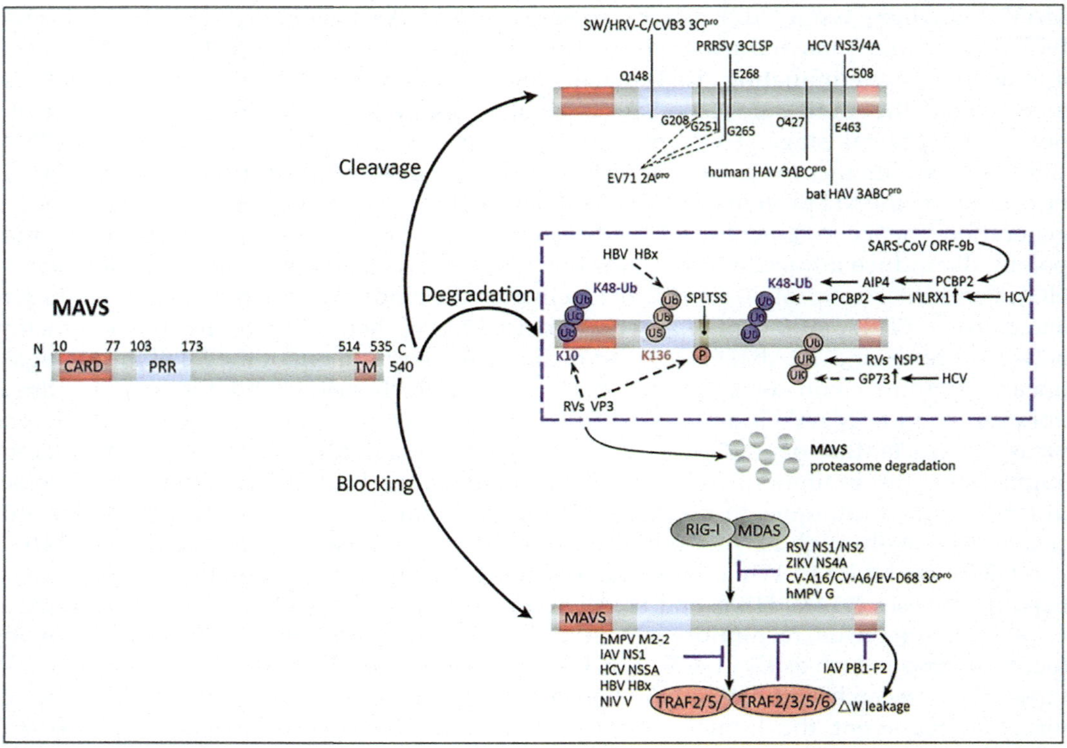

FIGURE 4.19 Negative regulation of mitochondrial antiviral-signalling protein (MAVS) signalling by viral proteins. Viruses employ various strategies to escape the host antiviral immune response.

Further Reading

Brun, A. M., Leser, G. P., Lamb, R. A., & Horvath, C. M. (2014). The innate immune sensor LGP2 activates antiviral signaling by regulating MDA5- RNA interaction and filament assembly. *Molecular Cell, 55*, 771–781. http://doi.org/10.1016/j.molcel.2014.07.003.

Ren, Z., Ding, T., Zuo, Z., Xu, Z., Deng, J., & Wei, Z. (2020). Regulation of MAVS Expression and Signaling Function in the Antiviral Innate Immune Response. *Frontiers in Immunology*. http://doi.org/10.3389/fimmu.2020.01030.

Yoneyama, M., Kikuchi, M., Matsumoto, K., Imaizumi, T., Miyagishi, M., Taira, K., et al. (2005). Shared and unique functions of the DExD/H-Box helicases RIG-I, MDA5, and LGP2 in antiviral innate immunity. *Journal of Immunology, 175*, 2851–2858. http://doi.org/10.4049/jimmunol.175.5.2851.

CHAPTER

5

Mitogenomics and metabolic disease

Introduction

Normal physiology of the body relies on a series of metabolic processes. Energy utilisation and disposal of metabolic waste form the basis of life. Our body is associated with a series of metabolic pathways, namely glycolysis, tricarboxylic acid (TCA) cycle, oxidative phosphorylation (OXPHOS), etc. Metabolic pathways involving carbohydrate metabolism, lipid metabolism, protein metabolism and others like gluconeogenesis, urea cycle (waste disposal) are the most important metabolic pathways. These pathways involve numerous enzymes, coenzymes and cofactors.

Any defect in any one of these components leads to metabolic disease. Some of the most common metabolic diseases in humans include diabetes, while in animals, they include ketosis, grass tetany and hypocalcemia.

Mitochondria play a vital role in such metabolic processes

Fundamental functions of mitochondria in metabolic processes

The mitochondrion, in its classical definition, is the cellular organelle that generates energy, facilitating the final stages of organic substance metabolism to produce adenosine triphosphate (ATP) for the cell. In mammalian cells, the majority of the redox potential utilised for ATP synthesis is delivered to the mitochondrion as nicotinamide adenine dinucleotide (NADH) and flavin adenine dinucleotide (FADH2), which are reduced coenzymes produced through the acceptance of electrons from the degradation of organic compounds in the TCA cycle. The electron transport chain (ETC), comprising four protein complexes in the inner membrane, transforms the redox energy held in NADH and FADH2 into chemical energy as ATP. Complexes I and II accept electrons from NADH and FADH2, respectively, then transfer them to complexes III and IV, culminating in their donation to an oxygen molecule, resulting in the formation of H_2O. Supplementary proteins can facilitate electron transfer from donors to the ETC, including electron-transferring flavoprotein dehydrogenase, which is part of a pathway that acquires electrons from β-oxidation and amino acid catabolism and CHCHD4 (Coiled-Coil-Helix-Coiled-Coil-Helix Domain Containing 4), which accepts electrons from disulphide bridges. During the transfer of electrons through complexes I, III and IV via electron carriers,

Mitochondrial Genomics
DOI: https://doi.org/10.1016/B978-0-443-26655-3.00013-5

protons (H^+) are transported from the mitochondrial matrix to the intermembrane space. This results in a net positive charge in the intermembrane space and a net negative charge in the matrix. The electrochemical gradient leads to the necessary driving power for ATP synthesis.

F_1F_O-ATP synthase is a substantial enzymatic complex composed of more than 22 subunits. The predominant route for protons accumulated in the intermembrane space to re-enter the mitochondrial matrix, thereby dissipating the electrochemical gradient and re-establishing equilibrium, is through the ATP synthase enzyme complex. Protons traverse the channel domain of ATP synthase, generating a motive force that rotates a substantial catalytic domain oriented towards the matrix, facilitating the coupling of adenosine diphosphate (ADP) with inorganic phosphate (Pi) to produce ATP. Research utilising isolated enzymes from bovines and yeast demonstrated that the proton requirement for ATP synthase to produce a single ATP molecule is nearly double in lower eukaryotes compared to mammals. Consequently, the efficiency of ATP generation by the ATP synthase enzyme has markedly improved throughout the evolution of multicellular creatures from their early eukaryotic predecessors. Mitochondria primarily generate ATP through the oxidation of NADH and FADH2 derived from the catabolism of carbohydrates like glucose. In mammals, a secondary energy source is derived from the catabolism of fatty acids, which are released from lipid reserves, such as those found in adipocytes. Fatty acids, commonly present as CoA-esters in chains, are initially transported across the outer mitochondrial membrane by palmitoyl transferase I, leading to their translocation to the intermembrane gap and conversion into an acyl-carnitine form. Subsequently, acyl-carnitines are transported across the inner mitochondrial membrane via carnitine acyl translocase. Within the matrix, acyl-carnitines are reverted to their original acyl-CoA form by palmitoyl transferase II, which can subsequently undergo oxidation to provide energy. β-Oxidation is facilitated by four enzymes located in the matrix, leading to the entry of acyl-CoA into the citric acid cycle, which produces NADH and FADH2, subsequently employed by the ETC for ATP synthesis. In humans, β-oxidation serves as a primary energy source and impairments in this process, or in the ETC overall, may result in ATP deficit, as shown further below.

Common identified mutations in genes or gene groups resulting in mitochondrial-associated metabolic diseases.

Mutations affecting oxidative phosphorylation are the foundation of miochondrial-associated metabolic disorders (see Fig. 5.1).

Mutations (purple) in genes critical for mitochondrial processes lead to mitochondrial-associated diseases (green). Human mtDNA encodes 13 proteins (along with subunits of ETC complexes) and 22 tRNAs on a circular genome. A D-loop region is the site of mtDNA polymerase gamma binding. Mutations within the D loop region, or in the POLG gene, lead to defects in mtDNA replication and eventually mtDNA depletion. In the case of mitochondrial neurogastrointestinal encephalomyopathy (MNGIE), mutations in the nuclear TYMP gene, which encodes a protein involved in nucleotide synthesis, also lead to defects in mtDNA replication due to an unbalanced nucleotide pool. Mitochondria that have no mtDNA copies can no longer synthesise ETC subunits encoded by the mtDNA, leading to defects in OXPHOS. In another example of a mutation in a nuclear gene affecting mitochondrial function, mutations in genes encoding proteins required for coenzyme Q10 (CoQ) synthesis lead to CoQ deficiency and, therefore defects in electron transfer in the ETC, resulting in defects in OXPHOS. Finally, mutations in mtDNA genes themselves, including

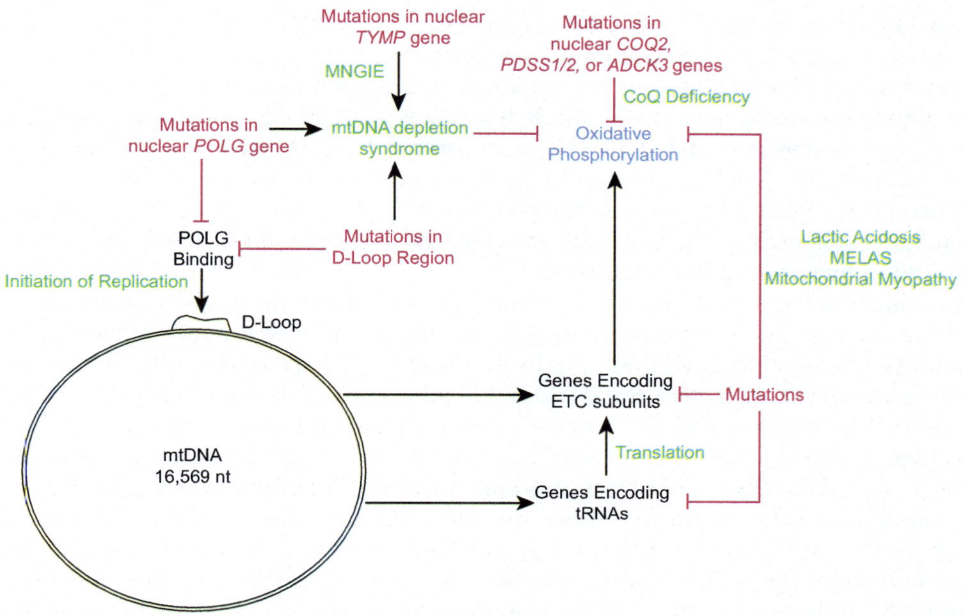

FIGURE 5.1 Mutations affecting oxidative phosphorylation cause mitochondrial-associated metabolic disorder.

genes encoding ETC subunits or tRNAs essential for mitochondrial gene expression, lead to defects in the synthesis of ETC protein subunits or nonfunctional complexes. Such mutations comprise the majority of mutations associated with mitochondrial-associated diseases (excluding POLG mutations) and have been associated with lactic acidosis (LA), MELAS and mitochondrial myopathies.

The preservation of mtDNA is essential for metabolic activity

Although mitochondria are believed to have assimilated into host cells approximately 2 billion years ago, the incorporation of mtDNA components into the nuclear genome seems to be a continuous process. Mitochondria typically possess 2–10 mtDNA molecules, contingent upon their phase in the fusion-fission cycle. Due to the absence of mutation-suppressing components like intron sequences in human mtDNA, it is estimated that mtDNA accumulates mutations at a rate 10 times greater than that of nuclear DNA. Furthermore, the diminutive size of mtDNA molecules (about 16.5kb in humans) elevates the incidence of initial mutations in the mtDNA sequence, potentially resulting in defects in the ETC and thus impairing mitochondrial function. The mechanisms governing mtDNA maintenance are crucial for sustaining mitochondrial integrity and avoiding illness. Damage to mtDNA is primarily induced by reactive oxygen species (ROS) produced within the mitochondrion. The elevated local concentration of ROS, along with the substantial effects of mtDNA mutations, constitutes a perilous combination. Although the mitochondrion lacks its own DNA maintenance machinery, it has been found that numerous nuclear-encoded proteins associated with DNA maintenance and repair are also directed to mitochondria. The most well-defined DNA repair mechanism in mitochondria is base excision repair, in which a mispaired or damaged nucleotide base is excised, the resulting gap is filled through DNA repair synthesis and the newly incorporated base is subsequently ligated to the adjacent nucleotide. Nuclear proteins that are essential for nucleotide excision repair (NER) have also been observed to localise to mitochondria under conditions of cellular stress or dysfunction, suggesting the potential presence of a similar NER-like repair mechanism within the mitochondrial compartment.

The mitochondrion must also address single- or double-strand DNA breaks, which can result from ROS-induced sugar damage or from the inability of topoisomerases to alleviate torsional stress after transcription. Single-strand DNA breaks attract ubiquitous enzymes. Base excision repair addresses the damaged 5′ and 3′ nucleotides, in conjunction with the tyrosyl-DNA phosphodiesterase 1 protein, which rectifies breaks that arise during DNA replication, a period when the majority of single-strand breaks transpire. Double-stranded fractures in nuclear DNA are highly detrimental to the cell; thus the nucleus has evolved three mechanisms for repairing these breaks, including end joining and homologous recombination. Initial evidence indicates that these repair mechanisms may also operate within mitochondria. Mitochondria has the capability to rectify insertion-deletion loops resulting from base-base mismatches through a mismatch repair pathway. When the mitochondrion fails to repair mtDNA damage, it may sequester the impaired mtDNA within a specific region, which, upon fission, will promptly exhibit diminished respiratory activity and will be earmarked for mitophagy, leading to its degradation the compromised mtDNA molecules inside. In extreme and uncommon circumstances, high quantities of mutant and/or unrepaired mtDNA may induce apoptosis of the entire cell.

Mechanisms for mtDNA maintenance and repair are essential for preserving mitochondrial integrity. Mitochondrial metabolism differs between tissues. Mitochondria, similar to differentiated cells, are specialised to execute distinct functions inside tissues. Mitochondria in the liver are predominantly biosynthetic, whereas those in the heart are chiefly ATP-generating. The coevolution of mitochondria inside tissues has resulted in methods whereby nuclear-encoded proteins regulate numerous mitochondrial processes, significantly aiding metabolic balance in humans. Although evidence robustly indicates that the nucleus regulates numerous functions of the mitochondrion, the mechanisms underlying many proposed pathways remain unclear. Certain nuclear-encoded proteins, such as mitochondrial transmembrane carrier proteins for glutamate and ATP-Mg/Pi exchange, exhibit variability among species and tissues, frequently aligning with the specialised functions of specific cell types. Another discovered alteration is the hyperphosphorylation of proteins within mitochondrial respiratory complexes in cardiac muscle cells, which enhances ETC activity. Consequently, ATP synthesis. Moreover, the differential or tissue-specific expression of proteins, such as transporters or gene regulatory proteins, is a mechanism via which mitochondria in various cell types can exhibit specialisation. Mitochondria also engage in signalling through calcium buffering inside tissues. In addition to ATP synthesis, a crucial physiological function of mitochondria is the storage of calcium (Ca^{2+}) ions. Calcium ions ingress into mitochondria via a calcium ion uniporter, where they are isolated. The rate and process of calcium ion release from mitochondria into the cytosol exhibit significant variability across different tissue types.

Minor changes in cytosolic calcium ion concentration, even at nanomolar levels, can significantly impact signalling pathways and modify cellular physiology; thus the sequestration of calcium ions within mitochondria plays a crucial role in regulating intracellular signalling. As previously mentioned, elevated mitochondrial calcium levels also promote ATP synthesis, establishing a clear connection between calcium sequestration in mitochondria and mitochondrial respiratory function at the level of the ETC.

Interactions between the endoplasmic reticulum and mitochondria facilitate biological functions

Mitochondria and the endoplasmic reticulum (ER) do not operate as autonomous, independently functioning organelles. Their tight association is essential for various activities, including mitochondrial fission, calcium buffering, lipid flow and organelle inheritance. Mitochondria and the ER interchange various lipids, some of which facilitate mitochondrial membrane elongation, while others undergo chemical modification within the mitochondrion before being translocated back to the ER. Dysfunctions in lipid flow are characteristic of various cardiovascular illnesses, which are strongly associated with deficiencies in ETC activity. Consequently, discovering additional components involved in ER mitochondrial lipid exchange is of significant interest. One of the crucial roles of ER-mitochondria contact sites is the modulation of mitochondrial fission. The ER is believed to envelop mitochondria, creating an initial constriction prior to the Drp1 protein exerting the mechanical force necessary for fission. The ER may also govern the recruitment of fission-required proteins, such as Drp1, from the cytosol to the mitochondria. Recent research has significantly enhanced our comprehension of the physical and functional relationship between the ER and mitochondria. The subsequent focus will be to further explore the molecular mechanisms that facilitate these interactions and how they may be modified in mitochondrial-associated diseases. Besides lipid flow, ER-mitochondria contact sites are essential for calcium buffering between the two organelles. Initial research indicated that augmented calcium outflow from the ER directly led to elevated mitochondrial calcium concentrations, thus confirming that the ER and mitochondria can exchange calcium ions in a controlled manner. Mitochondria and the ER both sequester calcium ions, which can subsequently be released into the cytosol, triggering calcium-dependent signalling pathways involved in numerous processes. Mitochondria serve as both storage organelles for calcium ions and targets for these ions, functioning as signals to modulate mitochondrial metabolic activity and other functions. A primary role of calcium exchange is believed to be the activation of calcium-dependent mitochondrial activities, including the TCA cycle, facilitated by the elevated concentration of calcium ions in the ER compared to the cytosol. Mitochondria and the ER may coregulate calcium ion release for cellular signalling, particularly during instances requiring elevated calcium concentrations, such as the initiation of the apoptotic cascade. Variations in calcium concentration serve as an activator of mitochondrial fission at ER-mitochondria contact sites, hence requiring a proximate calcium supply. In conclusion, the interactions between mitochondria and the ER have been associated with nearly all mitochondrial functions, hence increasing the intricacy of mitochondrial activity in both health and disease. The key processes outlined below are likely collaborating to synergistically regulate mitochondrial and ER function. Calcium buffering at ER-mitochondria contact sites significantly affects ETC function, thereby impacting mitochondrial respiration.

Diet and nutrition affect mitochondrial functions

Diet and nutrition can impact mitochondrial activity, hence affecting mitochondrial-associated disorders. As previously mentioned, dietary energy sources comprise carbs and lipids. Carbohydrates are metabolised via the glycolysis route and TCA cycle to produce reduced NADH and FADH2, which are transported to the ETC for ATP synthesis. Fats are metabolised through β-oxidation, producing reduced NADH and FADH2, along with acetyl-CoA, which subsequently enters the TCA cycle. Furthermore, ketone bodies are synthesised in the liver and kidneys as a consequence of fatty acid β-oxidation. Ketone bodies are high-energy molecules that can be carried through the bloodstream to various tissues, where they are metabolised and join the TCA cycle. Ketone bodies can function as an alternate energy source for tissues when glucose oxidation is compromised. Although clinical data examining the impact of diet on mitochondrial function are currently scarce, the ketogenic diet is a well-researched dietary regimen linked to alterations in mitochondrial function. The ketogenic diet is composed of high fat and low carbohydrate intake, which promotes the production of ketone bodies through the substitution of lipids for carbs as the main energy source. As elaborated below, numerous mitochondrial diseases stem from deficiencies in mitochondrial metabolism. In certain illness instances, mutations impact proteins essential for glucose oxidation, resulting in the blockage of the ETC and the consequent inability of glucose to function as a major carbon source. The ketogenic diet has been reported as an effective treatment for impaired glucose oxidation in patients, as it stimulates fatty acid oxidation in mitochondria, circumventing the glucose oxidation pathway and supplying a glucose-independent energy source. In this therapeutic approach, ketone bodies can function as an energy source when glucose metabolism is impaired. A well-documented instance is the efficacy of a ketogenic diet in managing epilepsy, a manifestation of mitochondrial ETC abnormalities in hippocampus neurones.

Most research examining the influence of nutrition on mitochondrial function has been conducted using rat, fly and cell line models possessing wild-type mtDNA or mtDNA with mutations linked to recognised metabolic disorders. In laboratory models, ketone bodies have demonstrated the ability to enhance the replication and translation of wild-type mtDNA under heteroplasmic circumstances, where a cell or individual has both wildtype and mutant mtDNA copies. A mouse model of mitochondrial myopathy has shown that a ketogenic diet enhances mitochondrial function and partially alleviates certain symptoms of mitochondrial myopathies. Recent laboratory research has examined the impact of dietary composition, specifically the quantities of protein, fat and carbohydrates, on mitochondrial function. In a Drosophila model exhibiting compromised ETC activity, a high glucose-to-protein ratio induced mitochondrial abnormalities. Consequently, there is data suggesting that dietary modifications may mitigate specific mitochondrial conditions.

Phenotypes in individuals with sickness

Although there is an increasing array of clinical and experimental evidence regarding the impact of diet on mitochondrial function, the molecular mechanisms behind this influence remain largely unexplored. Further investigation into the impact of nutrition on mitochondrial function is expected to yield significant advancements in the development of innovative therapeutics. As outlined below, almost all mutations and abnormalities linked to mitochondrial disorders lead to diminished activity of ETC complexes. At now, to address these disorders, researchers and physicians must initially identify the precise mutations and proteins implicated in the pathology before endeavouring to develop a medication, which will be effective solely for that particular variant of the condition. Conversely, nutrition is capable of directly affecting mitochondrial metabolism, potentially alleviating or mitigating metabolic abnormalities linked to mitochondrial disorders. Consequently, by formulating a dietary regimen that enhances metabolic performance in instances of mitochondrial dysfunction, a singular diet plan might potentially address a broad spectrum of mitochondrial disorders.

Pathologies linked to functional or hereditary mitochondrial abnormalities mitochondrial myopathy

Mitochondrial myopathy, also known as mitochondrial encephalomyopathy, is a condition characterised by dysfunctional mitochondria in muscle fibres, resulting in muscle weakness and the accumulation of mitochondria within the fibres, which produces the distinctive 'ragged red' fibres observed in *Gömöri trichrome* staining. Certain patients exhibit mitochondrial myopathy as persistent muscle weakness, whilst others only manifest weakness during vigorous activity; in other variants, this condition is associated with additional severe symptoms, including seizures. Mitochondrial myopathies mostly result from mtDNA abnormalities that cause impairments in OXPHOS and ATP synthesis in muscle fibres. Myopathy-related mtDNA mutations predominantly consist of large-scale deletions or point mutations in genes that encode tRNAs, rRNAs, or proteins. Typically, only a subset of mtDNA molecules within mitochondria harbours the relevant mutations, whereas other copies remain nonmutated. Moreover, mutations in gene products responsible for regulating mitochondrial dynamics cause disruptions in mitochondrial synthesis and fusion, leading to the accumulation of mitochondria inside fibres and the manifestation of the 'ragged red' phenotype. Mitochondrial myopathies are challenging to diagnose, as the clinical manifestations of mitochondrial illnesses frequently overlap with various other conditions, including hepatic failure, stroke-like symptoms, diabetes, and more symptoms. Consequently, accurately determining the precise number of patients with mitochondrial myopathy and the proportion of inherited vs spontaneous cases is challenging. It is predicted that around 12.48 per 100,000 individuals are affected by mitochondrial disorders or are at risk of developing one, with a considerable proportion anticipated to exhibit symptoms of mitochondrial myopathies. Certain therapies for mitochondrial myopathies have undergone clinical investigation, including the integration of satellite cells with wild-type mtDNA into myocytes during wound healing and the transplantation of embryonic myocytes containing wild-type mtDNA to reduce mutant copy number. Alternative therapy options, including transitioning to a ketogenic diet to mitigate the consequences of myopathies, are presently being investigation. Notwithstanding these efforts, a clinical therapy for mitochondrial myopathies remains unavailable today. Mitochondrial myopathies impact practically all organ systems, making the characterisation and therapy of these myopathies a priority in managing mitochondrial-associated illnesses. Mutations in numerous genes that directly or indirectly influence mitochondrial respiratory activity frequently result in myopathy-like symptoms, marked by muscular weakening and malfunction throughout various organs, including the heart, kidneys and gut.

Persistent lactic acidosis

Mutations in genes that encode mitochondrial proteins essential for the assembly and operation of ETC complexes, together with the administration of specific prescription medications and various physiological stressors, have all been linked to the onset of LA. LA is defined by the accumulation of lactate resulting from diminished mitochondrial respiration, potentially causing tissue acidification. In healthy human cells, ATP is mostly produced through aerobic mitochondrial respiration. In response to genetic or functional impairments in the mitochondrial ETC, cells undergo a metabolic shift from OXPHOS to glycolysis in order to compensate for the resulting deficit in ATP production associated with mitochondrial dysfunction. Consequently, patients have elevated lactate levels, potentially resulting in tissue acidification. The precise aetiology of mitochondrial dysfunction remains unidentified; however, the problem seems to be inherited maternally, as numerous instances are congenital. However, certain instances are either sporadic or contrived, as in patients administered metformin, a medication indicated for diabetes treatment. Untreated patients with elevated lactate levels may experience tissue acidification. Tissue acidification can induce systemic disruptions in physiological function, making the detection and treatment of persistent LA imperative. Available treatments presently involve the administration of ion-containing fluids to elevate systemic pH to near-normal values. The majority of people with chronic LA are effectively managed and no acidification is observable, despite systemic lactate levels being up to five times more than those in healthy individuals. Nevertheless, with acute illnesses such as infections or metformin ingestion, blood pH may decrease, resulting in symptoms such as vomiting, tachypnea, and abdominal pain (attributable to compromised intestinal function).

Melas

Mitochondrial myopathy, encephalopathy, LA and stroke-like episodes (MELAS) are distinguished by stroke-like episodes and seizures, accompanied by additional myopathy-related symptoms, including fatigue and muscle weakness. The onset of MELAS occurs prior to the age of 40 and it is both progressive and lethal. In 80% of instances, individuals possessed a singular mtDNA point mutation impacting complex I and/or tRNA functionality, which is therefore maternally inherited. Patients with MELAS also shows the symptoms of LA. Similar to numerous mitochondrial-related disorders, clinical identification is challenging and no therapy is presently available.

Coenzyme Q deficiency

The mitochondrial CoQ is a crucial element of the ETC, facilitating electron transfer between mitochondrial complexes I and II, from electron-transferring-flavoprotein dehydrogenase (ETFDH) to complex III and acts as an antioxidant and necessary cofactor for pyrimidine biosynthesis. Mutations in the nuclear COQ2, decaprenyl diphosphate synthase subunit 1 (PDSS1/2) and aarF domain containing kinase 3 (ADCK3)/coenzyme Q10 carrier protein 1 (CABC1) genes, which encode enzymes essential for CoQ production, may result in CoQ deficits. CoQ insufficiency can be identified using high-performance liquid chromatography of skeletal muscle tissue and manifests symptoms including skeletal muscle degradation, muscular weakness, seizures, migraines, ptosis and LA. In contrast to most other mitochondrial-associated illnesses, CoQ insufficiency can be treated with oral supplementation of CoQ for COQ2 mutations, or riboflavin supplements for ETFDH mutations. Patients with CoQ deficits attributed to PDSS2, COQ9 and several other related mutations have not exhibited favourable responses to CoQ supplementation. Consequently, although the management of CoQ insufficiency has proven effective in several instances, there remain cases where other treatments must be discovered.

Mitochondrial DNA depletion syndrome

mtDNA depletion syndrome (MDS) is defined by a significant reduction in mitochondrial DNA levels within the mitochondria of diverse organs. The primary gene products of mtDNA are essential for the correct assembly and operation of the respiratory chain complexes; hence, their loss leads to significant ETC malfunction across nearly all tissues, resulting in a profound systemic reduction of ATP. Congenital mutations become evident at birth, coinciding with the newborn's metabolic transition from anaerobic to aerobic processes. Mutations in any of the 10 known genes can lead to MDS and further genes are anticipated to be discovered. The majority of altered genes responsible with MDS are implicated in nucleotide synthesis and can be detected using PCR or Southern blotting. MDS is a prevalent mitochondrial-associated metabolic illness, identified as the underlying cause of mitochondrial malfunction in 11% of symptomatic children under 2 years of age. The majority of cases are lethal and presently no treatment is available; however, certain symptoms may be alleviated by CoQ therapy. The examination of altered genes associated with MDS, along with potential therapies, is currently a prominent area of research due to its incidence and symptom severity.

Mitochondrial neurogastrointestinal encephalomyopathy

MNGIE is a well-defined mitochondrial disorder, alongside CoQ deficiency, for which a clinical treatment has been established. MNGIE is an uncommon mitochondrial disorder, documented in approximately 100 instances. This occurs in patients with a recessive mutation in the TYMP gene, resulting in a deficit of thymidine phosphorylase. MNGIE, akin to other mitochondrial disorders, manifests a diverse array of symptoms in patients, such as myopathy, ptosis, peripheral neuropathy, gastrointestinal dysfunction, leukoencephalopathy and cachexia and frequently culminates in mitochondrial DNA depletion. The disease advances swiftly, typically manifesting about age 19 and frequently results in mortality, with an average lifespan of 37 years if untreated. Mutations in TYMP leads to a shortage of thymidine phosphorylase. Thymidine phosphorylase is necessary for nucleotide assembly, that is, converting thymidine and deoxyuridine into thymine and uracil, respectively. MNGIE-induced thymidine phosphorylase deficit causes the buildup of thymidine and deoxyuridine, which can be detected via HPLC. This accumulation is proposed to disrupt the balance of the mitochondrial nucleotide pool required for mtDNA replication and transcription, thereby contributing to mitochondrial DNA instability and degradation. Investigations into the management of this condition have concentrated on eliminating the surplus concentrations of thymidine and deoxyuridine from the bloodstream of MNGIE patients. Therapies like haemodialyses and platelet injection have shown helpful in decreasing blood plasma concentrations of these two nucleosides, resulting in symptom relief. It has recently been suggested that allogenic hematopoietic stem-cell transplantation (AHSCT) is the most successful treatment for MNGIE, as the transplanted cells with a healthy genetic background restore the appropriate dinucleotide triphosphate (dNTP) pool balances. Clinical trials for the treatment of MNGIE utilising AHSCT are currently in the planning stage.

Mutations in the POLG gene

As previously stated, due to the vital metabolic roles of its gene products, the maintenance and replication of mtDNA are crucial for optimal mitochondrial function. Initially, it was believed that due to the molecule's diminutive size and pronounced vulnerability to loss-of-function mutations, changes in mtDNA would be the primary cause of most mitochondrial disorders. Pathology resulting from POLG/POLG2 mutations exemplifies a mitochondrial-associated disease caused by the mutation of a nuclear gene. The mitochondrial replicative polymerase is mtDNA polymerase γ, composed of two subunits encoded by the POLG and POLG2 genes. In humans, POLG and POLG2 are situated on nuclear chromosomes 15 and 17, respectively. Mutations in POLG/POLG2 may be inherited or arise spontaneously, while further chemical modifications or inhibition of the polymerase, such as those generated by anti-HIV medications, can also result in analogous symptoms. As of yet, over 200 disease-related POLG/POLG2 mutations resulting in ETC malfunction have been documented in a POLG mutation database. Mutations in POLG/POLG2 frequently result in mtDNA depletion, presenting symptoms akin to MDS or MNGIE. Mutations impacting genes that encode other proteins essential for replication, such as components near the replication fork, have also been demonstrated to result in mtDNA depletion. An estimated 12.48 per 100,000 adults and children have or are at risk of developing a mitochondrial disease, with 25% of those individuals identified as having POLG mutations. Considering that approximately 2% of the population harbours deleterious POLG mutations, along with the disease's considerable prevalence and severity, forthcoming investigations into potential therapeutics are garnering significant attention.

Mitochondrial impairment in type 2 diabetes

As of 2011, it was stated that 8.3% of the U.S. The American population is afflicted with diabetes, with around 90% diagnosed with adult-onset type 2 diabetes. The defining feature of type 2 diabetes is the increased blood glucose level, resulting from insulin insufficiency and insulin resistance. Insulin, produced by pancreatic β-cells, functions to promote glucose absorption from the bloodstream into target cells that necessitate carbs for energy, such as those in the liver and skeletal muscles. This condition results from a complex interplay of factors, including genetic predisposition and lifestyle, with mitochondrial dysfunction recognised as a significant element in its pathogenesis. Numerous chemicals have been linked to the development of insulin resistance, wherein target cells fail to respond to insulin signals for glucose uptake from the bloodstream, leading to elevated blood glucose levels and energy-deficient tissues. Fatty acids serve as significant inhibitors of glucose absorption. In skeletal muscle, elevated fatty acid levels in blood plasma lead to heightened intracellular acyl-CoA concentrations, triggering a signal cascade that culminates in the inhibition of insulin-dependent glucose transport. Mitochondrial functional abnormalities are thought to significantly contribute to the onset of insulin resistance. Deficiencies in mitochondrial β-oxidation likely lead to elevated fatty acid levels, which contribute to insulin resistance produced by fatty acids. Defects in β-oxidation are proposed to result from spontaneous mtDNA mutations that accumulate with ageing. Moreover, these mutations may hinder the ATP-producing capacity of mitochondria, resulting in an ATP deficiency essential for glucose and insulin transport. A study revealed that mitochondria in skeletal muscle cells are diminished in both quantity and size and demonstrate decreased ETC activity in individuals with type 2 diabetes, corroborating the notion of hereditary mitochondrial abnormalities in this condition. Impaired mitochondrial β-oxidation resulting in insulin resistance is arguably the most substantiated. In addition to mitochondrial involvement in type 2 diabetes, numerous other ideas are proposed. Some investigations indicate that pancreatic β-cell malfunction, namely reduced insulin generation, plays a more significant role in pathogenesis than insulin resistance. This is endorsed by a diabetes-related mtDNA mutation in a gene that encodes a tRNA, resulting in diminished insulin production. Mitochondria in type 2 diabetes are believed to have a role in various facets of disease aetiology and contribute to certain downstream effects following onset. Central to mitochondrial involvement in type 2 diabetes, as well as most mitochondrial-related metabolic disorders, are impairments in ETC function or fatty acid oxidation. Consequently, continuous investigation into treatments for other mitochondrial-associated disorders are likely to influence the treatment of specific facets of type 2 diabetes and conversely.

Modified mitochondrial function in neoplasia

Besides strictly metabolic diseases, mitochondrial function is impaired in cancer patients. Tumour progression is frequently linked to enhanced mitochondrial respiration, linked to the accelerated proliferation of cells, leading to more ROS generation. Elevated ROS production correlates with a heightened risk of mtDNA mutation. MtDNA mutations are hence a characteristic feature of numerous cancer patients. A significant number of mutations occur in the noncoding D-loop region of mtDNA, which is important for the initiation of replication. mtDNA mutations may not only result from cancer but may also facilitate it, as indicated by a strong link between advancement stages and specific mtDNA mutations. Given that the D-loop area is essential for polymerase binding and the regulation of replication start, it is believed that these alterations often impede mtDNA replication. It has been recorded that a low mtDNA copy number is a defining trait of several malignancies, including renal cell carcinoma, lung cancer and breast cancer. The relationship between reduced mtDNA copy number and its potential advantages for cancer cell survival has yet to be clarified. Recent investigations into mitochondria within cancer cells have revealed other alterations in mitochondrial function, including the suppression of mitochondrial calcium uptake to enhance cancer cell viability. Recent reports indicate that mitochondrial miRNAs may influence mitochondrial function and energy metabolism in cancer contexts. miRNA regulation is modified. Additional investigation into mitochondrial diversity in tumour progression is expected to yield new biomarkers for monitoring carcinogenesis and innovative therapeutic targets for the clinical community.

We have discussed the role of mitochondria and cancer in a separate chapter too, later in detail.

The importance of mitochondrial malfunction in heart failure

Mitochondrial dysfunction has been associated with heart failure, similar to its role in cancer. The significance of mitochondria in cardiac muscle is critical due to their function in high-energy demand. Prior research on individuals with chronic cardiac disease demonstrated diminished ETC complex activity and discovered a nonpathological point mutation in the mtDNA-encoded cytochrome b gene. A critical factor in the initiation of heart failure is the compromised blood flow to and within the heart. Impaired cardiac blood flow results in diminished oxygen availability at the ETC of mitochondria; as oxygen serves as the ultimate electron acceptor in the ETC, a lower level of oxygen is responsible for the buildup of electrons at ETC complexes. The accumulation of electrons in ETC complexes triggers the creation of ROS, which functions as a stress response indicating limited oxygen supply. Nevertheless, ROS generation results in mtDNA damage and mutations. Consequently, mtDNA mutations are commonly detected in heart failure, resulting in diminished ETC activity and in certain instances, mtDNA depletion and reduction of mitochondrial mass. Mitochondrial malfunction resulting from the buildup of ROS is indicative of heart failure and other mitochondrial disorders. Mitochondrial dysfunction may be mitigated through the administration of antioxidant chemicals to patients, which work as ROS scavengers, hence diminishing their detrimental effects and decreasing the incidence of mtDNA mutations. Only a single clinical experiment with antioxidants has been conducted, aimed at mitigating mitochondrial abnormalities in Parkinson's disease. The experiment proved ineffective, mostly because of the challenges in targeting the chemical to mitochondria. Recent research in animal models and cell cultures has effectively administered antioxidant chemicals or enzymes to mitochondria, resulting in a reduction of ROS-related abnormalities. Consequently, the use of antioxidants appears to be a potential strategy for addressing ROS-induced mitochondrial dysfunction; nevertheless, more research remains necessary before human therapeutics can be developed. Although the significance of mitochondria in heart failure is more peripheral compared to other metabolic disorders addressed here, mitochondrial malfunction is a critical factor in the terminal phases of heart failure. Therapies aimed at mitochondrial biogenesis and oxidative stress for the treatment of heart disease have progressed markedly relative to other mitochondrial-related disorders, with mitochondria emerging as viable therapeutic targets. According to the classical definition, the primary role of the mitochondrion is to serve as the metabolic centre of the cell. Indeed, as previously noted, when mitochondrial respiratory function is compromised, systemic symptoms manifest. Recent research has identified that mitochondria participate in numerous more activities, which can also affect mitochondrial function and dynamics, potentially leading to disease. Novel treatments and therapies for mitochondrial-associated diseases are currently in development; although significant effort is still required, the discovery of new mediators of metabolic disease has revitalised interest in creating treatments for numerous ailments. Future research into mitochondrial dynamics and noncanonical functions is expected to establish new correlations between mitochondrial deficiencies and abnormalities in other cellular processes, hence offering novel avenues for therapeutic development Fig. 5.2

Exercise and calorie restriction lead to an increase in AMP:ATP and NAD+:NADH ratios, leading to activation of AMP-activated protein kinase (AMPK) and Sirtuin-1 (SIRT-1) pathways, respectively. These activate Peroxisome proliferator-activated receptor (PPAR) gamma coactivator-1α (PGC-1α), the master regulator of mitochondrial biogenesis. PGC-1α activation leads to activation of several transcription factors, including nuclear respiratory factors (NRF-1 and 2) and oestrogen related receptors (ERRs). NRFs and ERRs cause transcription of TCA cycle enzymes and OXPHOS subunit genes among several others. Transcription of genes for mitochondrial import proteins is also upregulated. These also promote the transcription of mitochondrial transcription factor A (TFAM). TFAM leads to mitochondrial DNA replication and transcription. Thus enhanced synthesis of nuclear DNA- derived mitochondrial proteins, their import into the mitochondria and mitochondrial DNA replication and transcription are mediated by PGC-1α. Moreover, PGC-1α also activates, (PPARs) which cause enhanced fatty acid β-oxidation and increased antioxidant defence.

Therapeutic implications

Strategy 1: augmenting mitochondrial biogenesis

Augmenting mitochondrial biogenesis facilitates effective nutrition management and enhances antioxidative defence mechanisms. Mitochondrial biogenesis can be enhanced using pharmacological and nonpharmacological methods by activating Sirtuin or AMPK pathways.

1. **Sirtuin pathway**: Sirtuins are deacetylases regulated by cellular NAD+ levels. Caloric restriction and heightened energy demand result in a reduction of NADH and a subsequent elevation of NAD+ levels. An elevated NAD+/NADH ratio results in heightened levels of sirtuins (Sirtuin 1 and 3). Sirtuin-1 is a crucial regulator of numerous intracellular activities. An elevation in sirtuin expression results in a rise in PGC-1α, hence promoting mitochondrial biogenesis, substrate oxidation and a decrease in oxidative stress.
2. **AMPK**: AMPK acts as an energy sensor and is activated by an increase in the AMP/ADP or AMP/ATP ratio. AMPK seeks to re-establish energy balance by suppressing energy-consuming anabolic processes, including lipid synthesis. Furthermore, it enhances mitochondrial function through the stimulation of PGC-1α, leading to improved mitochondrial biogenesis and antioxidant defence. Caloric restriction and physical activity result in elevated AMP/ATP and NAD+/NADH ratios, hence activating both AMPK and Sirtuin pathways.

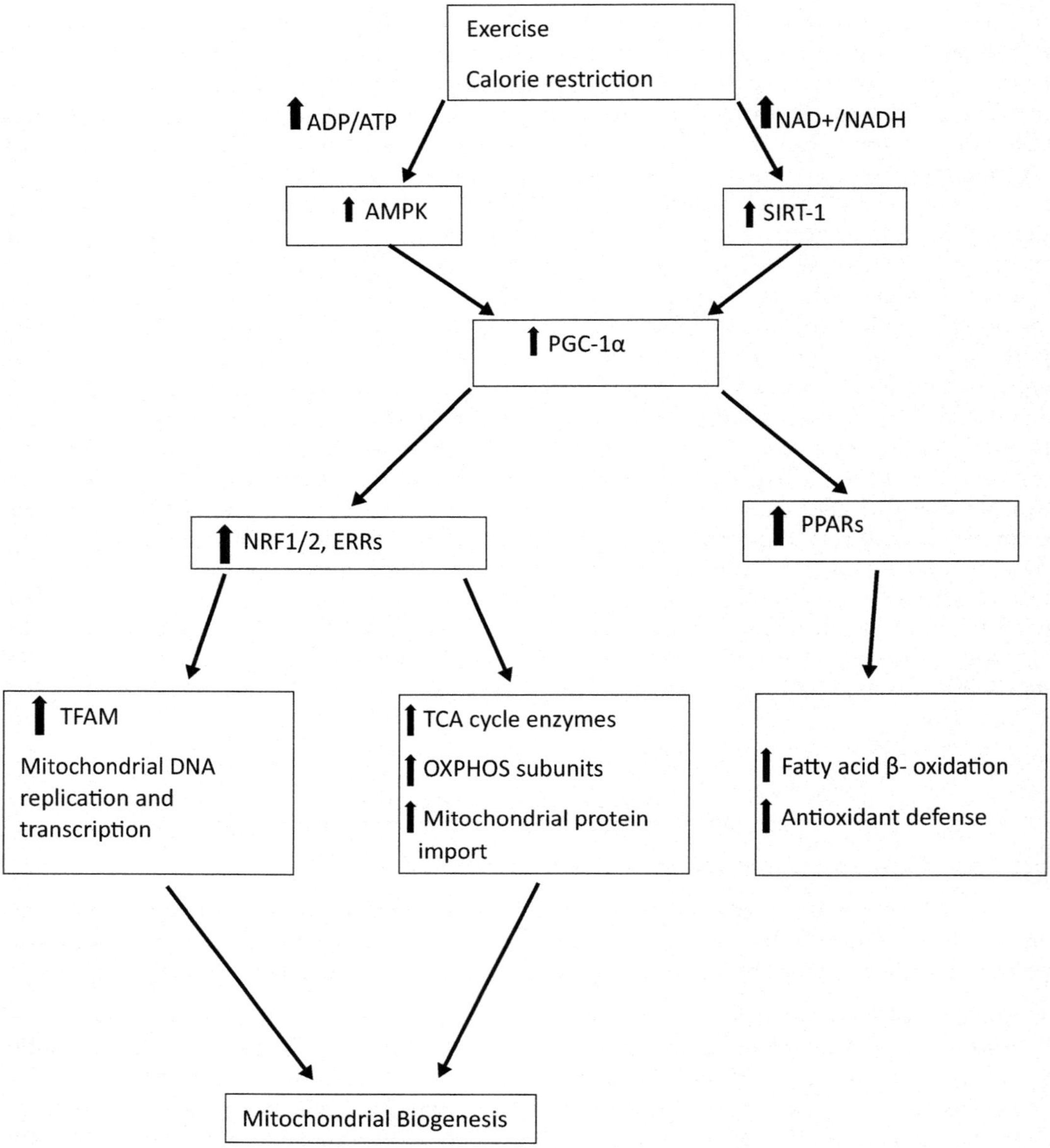

FIGURE 5.2 Exercise and caloric restriction lead to Mitochondrial biogenesis.

Mitochondrial biogenesis induced by energy deficiency from exercise and caloric restriction is documented. Resveratrol is an SIRT-1 activator present in grape skins. It has been beneficial in the control of insulin resistance in metabolic syndrome (MetS). Metformin is an antihyperglycemic agent. It is an activator of AMPK and many of its advantageous effects on insulin resistance are deemed secondary to AMPK activation. Besides promoting mitochondrial biogenesis, PGC-1α augments mitochondrial fatty acid oxidation through the activation of PPAR agonists. PPAR agonists may enhance insulin sensitivity by effectively managing fatty acids in adipose tissue, skeletal muscle and the liver. A number of PPAR agonists are presently undergoing clinical trials (Table 5.1).

Strategy 2: augmenting antioxidative capacity

Oxidative stress is a significant contributor to the pathophysiology of MetS. The use of both natural and synthetic antioxidant therapy is being investigated in the treatment of MetS. Commonly utilised antioxidants include vitamin C, lipoic acid, ubiquinone, vitamin E and N-acetyl cysteine. Nonetheless, extensive metaanalyses have not demonstrated a beneficial effect from these supplements. Mitochondrial-targeted antioxidants, including MitoQ and Szeto-Schiller peptides, are presently under investigation as prospective therapeutic treatments (Table 5.2).

Prospective considerations MetS, while primarily regarded as a lifestyle disease associated with sedentary behaviour and excessive diet, possesses a notable hereditary component. Family studies demonstrated a heritability of roughly 25%–30%. Furthermore, there exists a racial disparity in the prevalence of MetS. The analysis of data from 8814 men and women aged 20 years or older, derived from the Third National Health and Nutrition Examination Survey (1988–94) in the United States, revealed an overall age-adjusted prevalence of 23.7%. Within Mexican Americans, the age-adjusted frequency was 31.9%. Consequently, a hereditary predisposition to MetS may exist. Genome-wide association studies have primarily demonstrated associations with genes implicated in lipid metabolism or plasma lipid levels. This may indicate a correlation with hyperlipidaemia and may not accurately represent the fundamental pathophysiology of MetS (Table 5.3).

There is a scarcity of research about the association between MetS and polymorphisms in the mitochondrial DNA. A study has shown that pathogenic mutations in mitochondrial tRNA genes are linked to insulin resistance in women with polycystic ovarian syndrome. The association of the A3243G variant in the mitochondrial genome with diabetes mellitus is well established.

An alternative view is that the cytoplasm of the tumour cell could reprogramme the nucleus to become tumorigenic. Some researchers state that cancer is a mitochondrial metabolic disease.

Metabolic defect as manifested in animal model sheep:

In our lab, we have earlier worked on certain metabolic disorders manifested through debility, cardiac impairment, disorders in kidney function test for sheep, arising due to certain detected mutations in the cytochrome B gene.

Cytochrome B is a mitochondrial protein that functions as an integral component of the ETC and constitutes the principal subunit of the transmembrane cytochrome bc_1 and b_6f complexes, thereby playing a critical role in energy metabolism through OXPHOS. The present study was undertaken to investigate the impact of mutations in the *cytochrome B* gene on the health status of sheep. This work represents the first report establishing an association between a mitochondrial gene mutation and disease-related traits in a livestock

TABLE 5.1 Common mutations in genes or gene groups result in mitochondrial-associated metabolic diseases.

Gene(s)	Location	Product	Function	Phenotype	Associated diseases
COQ2, COQ9, PDSS1/2, ADCK3	Nuclear	Coenzyme Q10 (CoQ)	Required for synthesis of CoQ; it is an essential electron shuttle, acts as an antioxidant, cofactor of pyrimidine synthesis	Decreased ETC activity; skeletal muscle breakdown	CoQ deficiency
ETC subunits	Mitochondrial	Mitochondrial respiratory complex subunits	Oxidative phosphorylation	Respiratory complex deficiency	Mitochondrial myopathy; lactic acidosis; MELAS: (complex I)
ETFDH	Nuclear	Electron-transferring-flavoprotein dehydrogenase	Required for delivering electrons to complex III; derived from fatty acid and amino acid oxidation	Accumulation of toxic intermediates from fatty acid and amino acid oxidation; decreased ETC efficiency	Symptoms of glutaric acidemia type 2
POLG, POLG2	Nuclear	Mitochondrial DNA polymerase gamma	mtDNA replication	mtDNA, depletion	mtDNA depletion syndrome; symptoms of MNGIE
12S rRNA, 16S rRNA	Mitochondrial	Mitochondrial ribosomal RNAs	Required for synthesis of mtDNA-encoded proteins	Respiratory complex deficiency	Mitochondrial myopathy; lactic acidosis
tRNAs	Mitochondrial	Mitochondrial transfer RNAs	Required for synthesis of mtDNA-encoded proteins	Respiratory complex deficiency	Mitochondrial myopathy; lactic acidosis; MELAS
TYMP	Nuclear	Thymidine phosphorylase	Required for processing thymidine and deoxyuridine into thymine and uracil	Accumulation of thymidine and deoxyuridine; imbalance in mitochondrial (d)NTP pool; impaired mtDNA replication and transcription	MNGIE

TABLE 5.2 Identified mutations with sheep cytochrome B gene.

Identified mutation	Amino acid change	Healthy sheep	Diseased/ debilitated	Variants of identified cytochrome B
Wild type				Variant A KU 246230 and KU 246229.
T99G SNP	F33L	T	G	Variant B KY 110724
C189T SNP		C	T	
T309C SNP		T	C	
G511A SNP	D171N	G	A	Variant C KU 246231
Deletion at a position 290				Variant D KU 246228

species. Nonsynonymous substitutions (F33 L and D171N) and Indel mutations were observed for cytochrome B gene, leading to a truncated protein, where anaemia, malfunctioning of most of the vital organs such as the liver, kidney and mineral status associated with clinical manifestation of debility with exercise intolerance and cardiomyopathy in extreme cases were detected. These findings were confirmed by bioinformatics analysis, haematological and biochemical data analysis and other phenotypical physiological data pertaining to different vital organs. The molecular mechanism of cytochrome B mutation was that the mutant variant interferes with the site of haem binding (iron-containing) domain and calcium binding essential for ETC. Mutation at amino acid site 33 is located within transmembrane helix A, a hydrophobic environment at the Qi site and close to haem binding domain and mutation affects these domains and diseases occur. Thermodynamic stability was also observed to decrease in mutant variant. Sheep cytochrome B, being genetically more similar to humans, may be used as a model for studying human diseases related to cytochrome B defects. Future prospects involve the use of recombinant proteins as therapeutics, gene therapy associated with gene editing and marker-assisted/genomic selection of disease-resistant livestock.

Characterisation of cytochrome B gene in Garole sheep

Garole sheep cytochrome B gene is characterised by 1140bp nucleotide and 379 amino acids. The sequences submitted for diseased sheep with mutated cytochrome B were KY 110724, KU 246231, KU 246228, while the wild types were KU 246230 and KU 246229.

Mutation detection in Cyt B gene

Four single nucleotide polymorphisms (SNPs) (T99G SNP, C189T SNP, T309C SNP, G511A SNP) and one site for deletion had been detected in Garole sheep. Identified nonsynonymous SNPs between healthy and debilitated sheep were F33L and D171N.

The sites for posttranslational modification vary for different mutant forms from that of the wild type and have been tabulated in a table. It has been observed that the variant D of sheep cytochrome B had most of the depressed posttranslational modification processes, in comparison to the wild type variant A.

TABLE 5.3 Association of SNPs of cytochrome B gene with respect to different functional parameters in sheep.

Physiological parameters					
	Wild type/ variant A	Variant B	Variant C	Variant D	Remarks
Body temperature (°F)	102.5 ± 0.54 [a]	98.0 ± 0.57 [b]	100.00 ± 0.62 [a]	97.0 ± 0.65 [b]	Subnormal temp. for debilitated animals.
Heart rate (per minute)	75.5 ± 0.43 [a]	120 ± 0.56 [b]	110 ± 0.58 [b]	123 ± 0.62 [b]	Tachycardia was observed for diseased sheep
Pulse rate (per minute)	70 ± 0.45	65 ± 0.47	70 ± 0.56	64 ± 0.59	Weak and feeble pulse rate was observed
Respiration (per minute)	16.5 ± 0.12 [a]	40 ± 0.23 [b]	35 ± 0.28 [b]	45 ± 0.32 [b]	Tachypnoea was observed for debilitated sheep
Muscle weakness	Muscle weakness not detected	Increased muscle weakness	Moderate muscle weakness	Extreme muscle weakness-sheep is unable to rise	The debilitated sheep was not able to stand of its own, general weakness of muscle
Exercise intolerance	Not detected	Detected	Detected	Detected	
Haematological properties					
Complete hemogram test					
Haemoglobin	11 ± 0.32[a]	06 ± 0.28 [b]	10 ± 0.34 [a]	05 ± 0.31 [b]	gm/dl
PCV (haematocrit)	40.34 ± 0.35 [a]	25 ± 0.48 [b]	35 ± 0.44 [b]	24 ± 0.56 [b]	%
Total erythrocyte count	5.4×10^{6} [a]	4.92×10^{6} [a]	4.67×10^{6} [a]	3.28×10^{6} [b]	Cu mm
Total leucocyte count	6300 ± 23.34 [a]	4700 ± 34.56 [b]	6230 ± 23.34 [a]	4300 ± 28.56 [b]	Cu mm
Differential leucocyte count					
Neutrophil	29.5 ± 0.034	29.5 ± 0.235	29.3 ± 0.034	28.5 ± 0.352	%
Eosinophil	9.4 ± 0.383	11.0 ± 0.435	10.5 ± 0.383	11.0 ± 0.523	%
Basophil	-	-	-	-	%
Lymphocyte	51.5 ± 0.234	50.0 ± 0.324	51.4 ± 0.234	51.0 ± 0.367	%

Monocyte	9.6 ± 0.23	9.5 ± 0.435	8.8 ± 0.23	9.5 ± 0.435	%
Platelet count	450,000 ± 34.67 [a]	135,000 ± 56.54 [b]	200,000 ± 40.34 [a]	97,000 ± 45.34 [b]	%
MCV	54 ± 0.45 [a]	40.34 ± 0.34 [b]	38. 34 ± 0.47 [b]	35.56 ± 0.54 [b]	Fl
MCH	10 ± 0.034 [a]	5.5 ± 0.034 [b]	9.5 ± 0.023 [a]	4.67 ± 0.078 [b]	Pg
MCHC	36 ± 0.53	35 ± 0.64	37 ± 0.34	38 ± 0.45	%/dl
ESR (1st hour)	02 ± 0.04	2.35 ± 0.05	2.3 ± 0.03	2.38 ± 0.06	Mm
Kidney function test					
Glucose (fasting) mg/dL	52 ± 0.24 [a]	120 ± 0.45 [b]	100 ± 0.56 [b]	245 ± 0.68 [b]	mg/dL
Urea	39 ± 0.47 [a]	72 ± 0.59 [b]	44 ± 0.58 [a]	74 ± 0.59 [b]	mg /dL
BUN	18.5 ± 0.13 [a]	25.3 ± 0.21 [b]	22.89 ± 0.23 [b]	27.0 ± 0.32 [b]	mg/dL
Creatinine	1.5 ± 0.034 [a]	1.9 ± 0.045 [a]	1.9 ± 0.063 [a]	2.5 ± 0.072 [b]	mg/dL
IV. Liver function test					
Total protein	7.2 ± 0.12 [a]	8.5 ± 0.14 [a]	8.3 ± 0.23 [a]	10.7 ± 0.34[b]	gm/dL
Albumin	3.0 ± 0.03 [a]	1.8 ± 0.05 [b]	2.2 ± 0.07 [c]	1.3 ± 0.09[b]	gm/dL
Globulin	4.2 ± 0.16 [a]	5.1 ± 0.08 [b]	4.5 ± 0.13 [a]	5.5 ± 0.17 [b]	gm/dL
Total bilirubin	0.4 ± 0.02 [a]	7.0 ± 0.08 [b]	5.0 ± 0.12 [b]	7.8 ± 0.15 [b]	mg/dL
Direct bilirubin (conjugated)	0.1 ± 0.01 [a]	0.5 ± 0.04 [a]	0.4 ± 0.06 [a]	1.0 ± 0.09 [b]	mg/dL
SGOT (ALT)	100 ± 2.45 [a]	352 ± 3.64 [b]	300 ± 3.56 [b]	382 ± 4.5 [b]	IU/L
SGPT (AST)	24 ± 0.16 [a]	35 ± 0.19 [b]	30 ± 0. 18 [b]	45 ± 0.20 [b]	IU/L
V. Mineral status					
Serum calcium (Ca)	11.75 ± 0.13 [a]	8.23 ± 0.23 [b]	7.86 ± 0.32 [b]	6.78 ± 0.34 [b]	mg/dL
Phosphorus(P)	7.1 ± 0.34	5.2 ± 0.45	6.3 ± 0.37	4.1 ± 0.47	mg/dL
Magnesium	2.6 ± 0.23	1.8 ± 0.32	2.0 ± 0.34	1.6 ± 0.37	Mmol/L

The wild type variant A, possessing nine transmembrane helices, results in increased biological potentiality compared to that of variant D with a deleted mutation. The entropy of variants B, C and D was increased to that of variant A, which reflects better stability of variant A. The sites for posttranslational modification sites have been studied through various bioinformatics software. The secondary structure of cytochrome B had been depicted for different variants. Gene 3D result analysis reveals stereochemical quality of the model and Procheck reveals the index of thermodynamic stability as rG being -14,747 kCal/mol. Ramachandran plot was depicted for both the wild type and mutant variants of cytochrome B.

Effect of mutation of cytochrome B gene on health parameters (physiological, haematological and biochemical)

A deletion mutation was observed in Garole sheep number, which leads to a truncated cytochrome B protein, with reduced biological efficiency. The sheep were observed to be extremely debilitated, unable to stand, with extreme muscle weakness. The case of cardiomyopathy was confirmed by electrocardiogram (ECG). Respiratory disorders were detected with examination by a stethoscope.

Association of SNP of cytochrome B with general health conditions and exercise intolerance

The sheep with a mutated form of cytochrome B (variants B, C and D) were observed to show exercise intolerance (Table 4). The individuals with variants B, C and D exhibited subnormal body temperature, tachycardia, weak and feeble pulse, tachypnoea and animals were extremely debilitated with muscle weakness. The sheep were unable to rise and stand of its own. Cases of cardiomyopathy in the sheep with variant D were confirmed by ECG.

Association of SNP of cytochrome B with haematology

The sheep containing mutated variant of cytochrome B (variants B, C and D) were observed to possess decreased haemoglobin concentration, PCV and total erythrocytic count. Similarly, variants B, C and D also possess reduced mean corpuscular volume (MCV), mean corpuscular hemoglobin (MCH) and mean corpuscular hemoglobin concentration (MCHC) compared to that of the wild type variant A.

Total leucocytic count and differential leucocytic count were found to be lower in variants B, C and D. Platelet count was also found to be depressed.

Association of cytochrome B polymorphism with kidney function

In extreme cases of a few animals, multiorgan failure was diagnosed. Few individuals with variants B, C and D were detected with depressed kidney function, diagnosed through clinical symptoms coupled with biochemical tests of kidney function tests for urea level, blood urea nitrogen (BUN), and creatinine level of serum.

Association of Cyt B SNP with liver function test

A few individuals for variant B, C and D were observed to have depressed liver function as revealed by biochemical tests of liver function, revealed through serum levels of total protein, bilirubin, serum glutamic oxaloacetic transaminase (SGOT), serum glutamic pyruvic transaminase (SGPT) and alkaline phosphatase.

Association of *cytochrome* B SNP with serum mineral status

In Garole sheep, the presence of variants B, C and D of the *cytochrome B* gene was associated with a reduction in serum calcium concentrations.

Comparative genetic analyses revealed that ovine *cytochrome B* shares a high degree of sequence similarity with that of other mammalian species. Phylogenetic and genetic distance analyses further indicated a close evolutionary relationship between sheep and humans, underscoring the potential relevance of sheep as a model organism for studying *cytochrome B*-associated metabolic disorders in mammals.

Human cytochrome B was observed to be 77.3% identical to sheep cytochrome B. Hence, sheep may be considered an excellent animal for studying humans as an animal model (See Fig. 5.3 and Fig. 5.4).

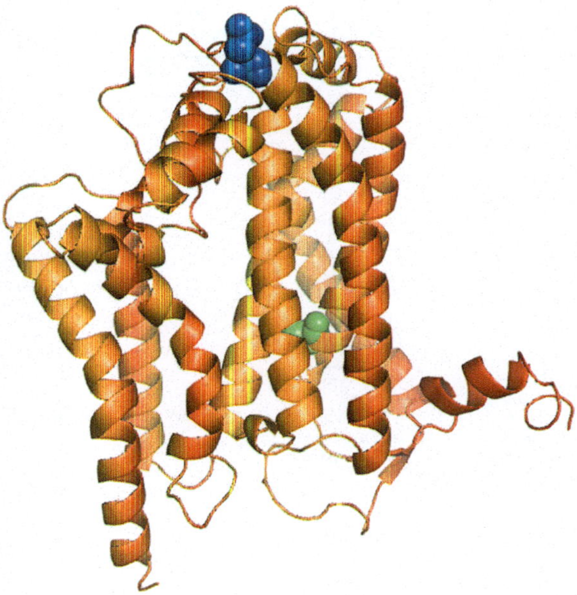

FIGURE 5.3 3D structure of wild type variant A (red) for cytochrome B with SNP at amino acid position 33, green sphere and amino acid position 171, blue sphere.

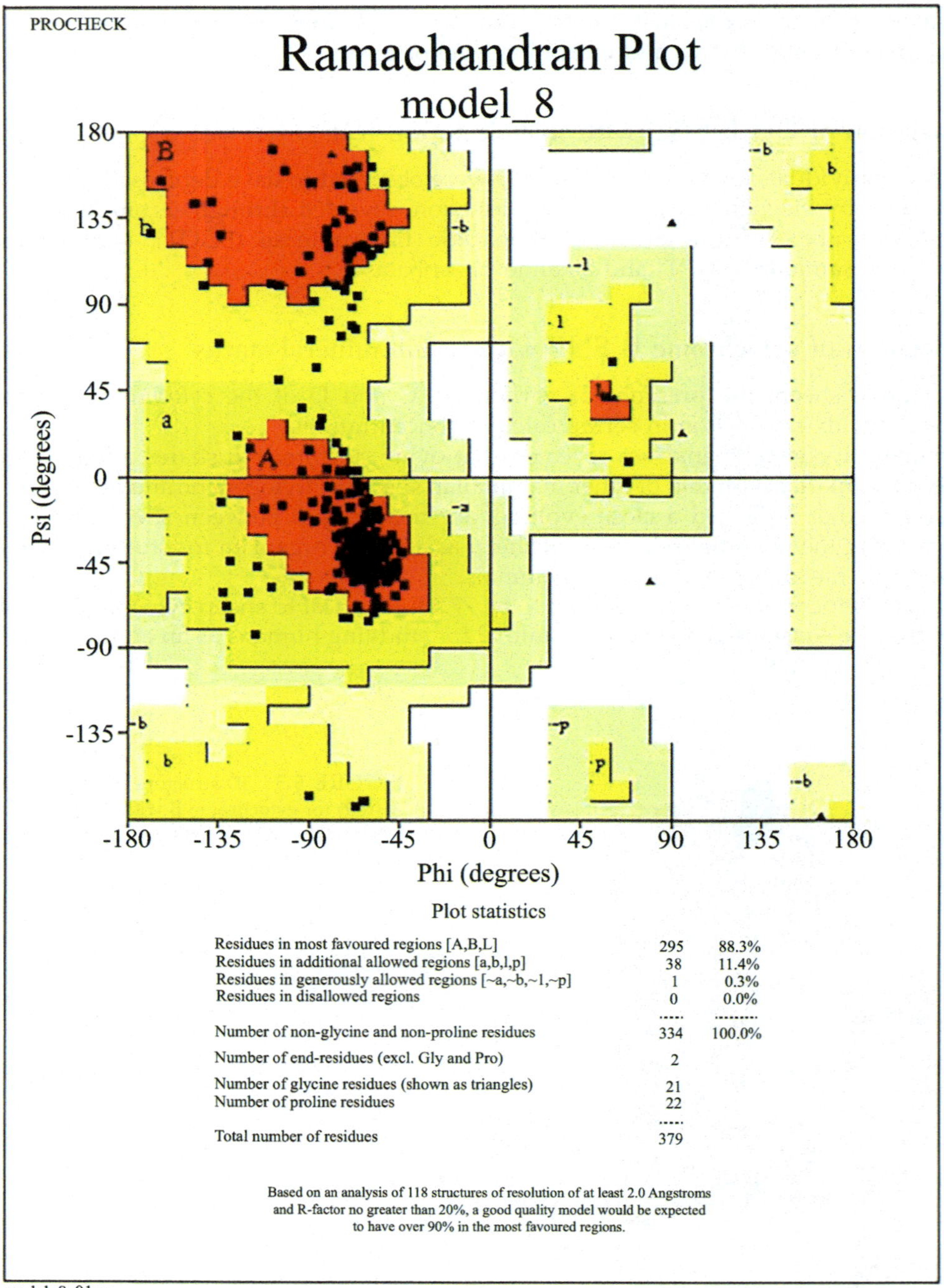

model_8_01.pe

FIGURE 5.4 Ramachandran plot depicting protein stability of cytochrome B gene of sheep.

Further Reading

McInnes, J. (2013). Mitochondrial-associated metabolic disorders: foundations, pathologies and recent progress. *McInnes Nutrition & Metabolism 2013, 10*, 63.

Pal, A., Pal, A. B., Banerjee, S., Batobyal, S., & Chatterjee, P. N. (2019). Mutation in ***Cytochrome B gene*** causes debility and adverse effects on ***health***of sheep. *Mitochondrion, 46*, 393–404. https://doi.org/10.1016/j.mito.2018.10.003.

Prasun, P. (2020). Mitochondrial dysfunction in metabolic syndrome. *BBA - Molecular Basis of Disease, 1866*(2020), 165838.

CHAPTER

6

Mitogenomics and viral disease

Introduction

In our earlier chapters, we have discussed the basis of mitogenomics, its structure and functional significance.

Disease resistance or immunity against viral infection is quite interesting to study. To date, no effective antivirals have been developed that act against viruses and destroy them. Vaccination is an effective method for the control of viral infection. However, due to rapid mutation and emergence of newly evolving viruses, vaccination becomes less effective, particularly for single-stranded RNA (ssRNA) viruses.

Types of pathogenic viruses

Viruses exist in the following forms as per their genetic material.

Class I: dsDNA virus

Class II: ssDNA virus

Class III: double-stranded RNA (dsRNA) virus

Class IV: Positive-sense ssRNA virus

Class V: Negative-sense ssRNA virus

Class VI: RNA viruses that reverse transcribe

Class VII: DNA viruses that reverse transcribe

A lot of studies have been undertaken for viral immunity, mostly involving nuclear genes (Pal and Chakravarty, 2019). In this current chapter, we will discuss how mitochondrial genes are involved in antiviral immunity. Even some work in conjunction with nuclear genes.

In this chapter, we will attempt to address certain key issues about how viruses attack the host? What are the changes occurring in mitochondria? How does mitochondrial DNA affect immunity against viral infection? Is there any interaction between mitochondrial genes and nuclear genes in innate immunity?

Classification of the viruses has been conducted based on their genetic material. Scientists believe that viruses are at the transition between living and nonliving. They hijack the living host genome and replicate themselves, then kill the cell. They interact with the host nuclear as well as mitochondrial genome in different fashion.

Mitochondrial Genomics
DOI: https://doi.org/10.1016/B978-0-443-26655-3.00010-X

In the coming section, we will discuss the role of various kinds of viruses and their possible mode of interaction with host mitochondria. We will discuss the viral pathogens infecting both humans and animals, along with the zoonotic diseases.

Herpesvirus

They are a group of DNA viruses belonging to Alphaherpesviridae family, causing diseases to both humans and animals. In this current chapter, we will discuss the involvement of mitochondria and mitochondrial genetics pertaining to such diseases (see Fig. 6.1).

Infection commences when a viral particle interacts with a cell possessing certain receptor molecules on its surface. Following the attachment of viral envelope glycoproteins to cell membrane receptors, the virion is internalised and disassembled, allowing the translocation of viral DNA to the cell nucleus. So, the nucleus is the site for viral DNA replication and transcription of viral genes. In the course of symptomatic infection, infected cells transcribe lytic viral genes. In certain host cells, a limited quantity of viral genes known as latency-associated transcripts accumulates instead. In this manner, the virus can endure within the cell (and hence the host) indefinitely. Although initial infection frequently presents with a transient clinical illness, long-term latency remains asymptomatic.

Zoonotic herpesviruses: Apart from the herpesviruses considered endemic in humans, certain viruses primarily affecting animals may infect humans, referred to as zoonotic diseases.

We are citing two examples of zoonotic viruses in the following table: Table 6.1

Animal herpesviruses

In animal virology, the best-known herpesvirus belongs to the subfamily Alphaherpesvirinae. Research on pseudorabies virus (Prv), the causative agent of Aujeszky's disease in pigs, has

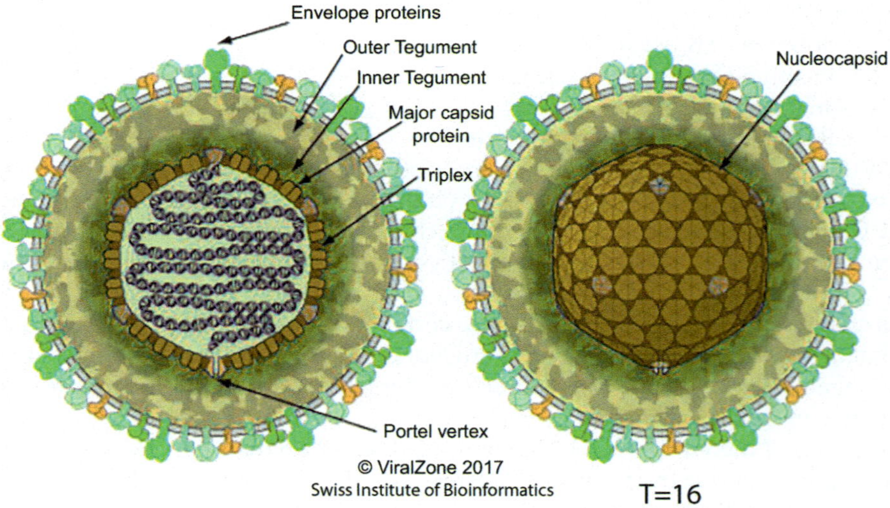

FIGURE 6.1 Replication for all herpesviruses takes place in the nucleus. The herpes viral DNA is transcribed into mRNA within the nucleus of the infected cell.

TABLE 6.1 Viral proteins involved in modulating host mitochondrial dynamics and function.

Protein type and function	Proteins, virus	Affected proteins	Viral infection/ overexpression of viral proteins	Consequences on cell physiology
Structural proteins: Essential for viral infection (spike), attachment (envelope) and assembly of viral components, virion budding (nucleocapsid membrane and core)	**M protein:** SARS-CoV-2 SARS-CoV **N protein:** SARS-CoV **N/E/** **C protein:** HCV	Activation of caspase-3 Upregulation of Drp1, Parkin activation	Protein overexpression Protein overexpression Viral infection	Mitochondrial metabolism (predicted) induces apoptosis. Mitochondria-mediated apoptosis caused by to increase in ROS and cytochrome C release. Enhances mitochondrial fission, inhibits apoptosis, facilitates persistent infection, promotes mitophagy.
Nonstructural proteins: Viral RNA replication	**NSP3:** HCV **NSP4:** SARS-CoV-2 **NSP4a:** HCV **NSP4b:** DENV/ ZIKA DENV HCV **NSP5:** JEV **NSP5a:** HCV **NSP8:** SARS-CoV-2 **NSP13:** SARS-CoV-2	Upregulation of Drp1, targets MAVS in MAMs TIM Complex Upregulation of Drp1, targets MAVS in MAMs Downregulation of Drp 1 MAMs proteins- MFN2, FACL4 MAMs proteins Mitochondrial trifunctional protein Activation of NF- kappaB and STAT-3 TBK1	Protein overexpression Viral infection Viral infection Viral infection Viral infection Both Both Protein overexpression Protein overexpression Protein overexpression	Enhances mitochondrial fission, inhibits innate immune responses, and prevents pro-apoptotic effects of NS4a. Affects protein import machinery or CM formation(predicted). Enhances mitochondrial fission, inhibits innate immune response. Enhances mitochondrial fission, facilitates replication, increases OXPHOS, and triggers mitophagy. Affects metabolism and induces MAVS. Downregulates OXPHOS and metabolism. Impairs β oxidation. Upregulates mitochondrial calcium, increases cellular ROS and expression of survival genes. Interact with mitochondrial ribosome (predicted). Regulates MAVS signalling (predicted).

(*Continued*)

TABLE 6.1 (Continued)

Protein type and function	Proteins, virus	Affected proteins	Viral infection/ overexpression of viral proteins	Consequences on cell physiology
Accessory proteins: Enhance the selective mode of infection	**ORF 3:** SARS-CoV-2	Oligomerization of VDAC1	Protein overexpression	Preserves mitochondrial membrane potential and prevents apoptosis.
	ORF3a: SARS-CoV-2	BAX Oligomerization	Protein overexpression	Promote apoptosis.
		Blocks Type 1 IFN expression	Protein overexpression	Inhibit MAVS signalling.
	ORF3b: SARS-CoV	Drp1, LC3, and MAVS	Protein overexpression	Proteasomal degradation of Drp1 enhances mitochondrial fusion, triggers mitophagy and blocks innate immunity.
	ORF-9b: SARS-CoV SARS-CoV-2.	Interacts with TOM70 Interacts with ETC Components	Protein overexpression	Modulate immune pathway (predicted). OXPHOS (predicted).
	ORF-9C: SARS-CoV-2	NLRX1, NDFIP2		Regulate MAVS and immune response (predicted).

pioneered animal disease control with genetically modified vaccines. Prv is now extensively studied as a model for basic processes during lytic herpesvirus infection, and for unravelling molecular mechanisms for herpesvirus neurotropism, whereas bovine herpesvirus 1, the causative agent of bovine infectious rhinotracheitis and pustular vulvovaginitis, is analysed to elucidate the molecular mechanism of latency. The avian infectious laryngotracheitis and pustular vulvovaginitis virus is phylogenetically distant from these two viruses and serves to underline similarity and diversity within the Alphaherpesviridae.

Nine distinct herpesviruses are recognised as primarily infecting humans. Among these, at least five exhibit high prevalence across global populations and are aetiologically linked to common human diseases. These include: herpes simplex virus types 1 and 2 (also designated human herpesvirus 1 (HHV-1) and HHV-2, respectively), both of which can cause orolabial and genital herpes; Varicella-zoster virus (VZV, HHV-3), the causative agent of varicella (chickenpox) and herpes zoster (shingles); Epstein–Barr virus (EBV, HHV-4), associated with infectious mononucleosis and multiple malignancies; and Human cytomegalovirus (HCMV, HHV-5), a pathogen implicated in a range of clinical conditions, particularly in immunocompromised individuals and during congenital infections.

Over 90% of adults have been infected with at least one of these viruses, and a latent form persists in nearly all individuals who have been infected. Additional HHVs include HHV-6A and HHV-6B, HHV-7, and Kaposi's sarcoma-associated herpesvirus (KSHV, or HHV-8). Human diseases:

1. HSV-1 or HHV-1—orolabial and genital herpes
2. HSV-2 or HHV-2—orolabial and genital herpes
3. VZV or HHV-3—the causative agent of chickenpox and shingles
4. EBV or HHV-4—associated with several diseases such as mononucleosis and some malignancies and
5. HCMV or HHV-5
6. HHV-6A and HHV-6B,
7. HHV-7 and
8. KSHV, also referred to as HHV-8

Animal diseases of zoonotic importance

Other animal diseases:

Duck plague (also known as duck viral enteritis) is a worldwide disease caused by Anatid alphaherpesvirus 1 (AnHV-1) of the family Herpesviridae that causes acute disease with high mortality rates in flocks of ducks, geese and swans.

In the coming subsequent topics, we will be discussing genomics for herpesviruses, interaction of viral genomics with host genomics, role of mitochondria against viral pathogenesis, mitochondrial genes' role in viral infection and interaction of mitochondrial genes with nuclear genes.

It is to be noted that Prv is alphaherpesvirus.

Hepatitis virus

In this current section, we will describe two important life-threatening viruses—hepatitis B and hepatitis C.

Hepatitis C

As the name implies, it refers to inflammation of the liver through infection by the hepatitis C virus (HCV). Hepatitis C infection spreads through the intravenous route, usually through accidental contamination of infected blood or syringes.

The most important criterion of our importance is that there exists individual variation in susceptibility, indicating genetic control (see Fig. 6.2).

Fig. 6.3

Fig. 6.4

In the subsequent topics, we will be discussing genomics for HCVs, interaction of viral genomics with host genomics, role of mitochondria against viral pathogenesis, mitochondrial genes role in viral infection and interaction of mitochondrial genes with nuclear genes.

Other important viruses are hepatitis B, EBV, HCMV. Influenza virus, measles virus and Newcastle disease virus (NDV) and severe acute respiratory syndrome coronavirus.

Other important viruses of animal origin

Rabies
Peste des Petits ruminants (PPR)
Foot and Mouth disease (FMD)
Rinderpest
Avian influenza
Swine fever
Ranikhet disease of poultry

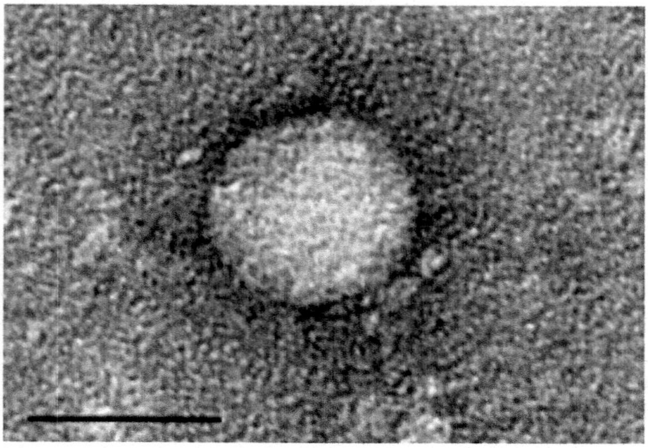

FIGURE 6.2 View of the hepatitis C virus through an electron microscope.

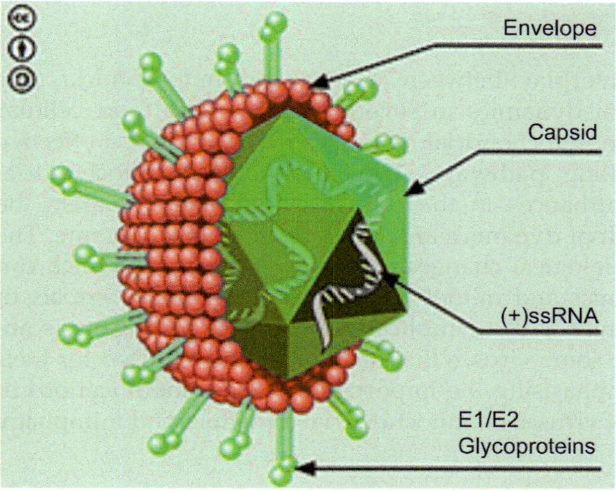

FIGURE 6.3 3D structural model for hepatitis C virus.

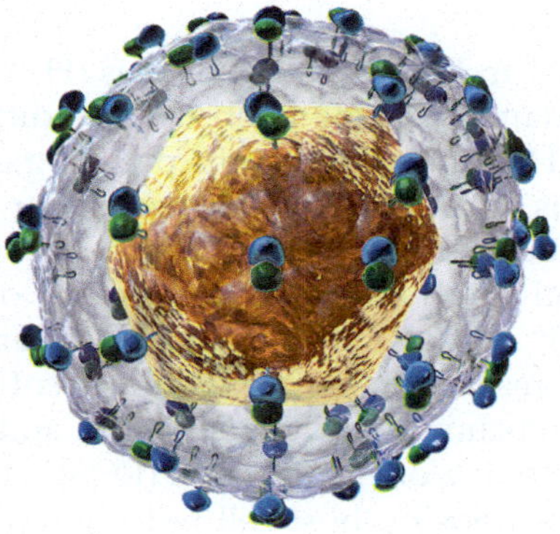

FIGURE 6.4 Detailed structure indicating the surface receptor binding site for hepatitis C virus (HCV).

So far, we are concerned with the genetic aspect of the disease resistance to viruses. In one of our earlier books, we have discussed nuclear genes and their mechanism of action. But in this current context, we will discuss mitochondrial genes and their role against viral infection.

Mitochondria and viral infection

Viral proteins associate to the modulation of host mitochondrial dynamics and function.

Mitochondrial dynamics in response to viral infection

Viruses disrupt mitochondrial pathways and impair mitochondrial functioning to promote their replication. A viral tactic to obstruct the mitochondria-associated antiviral signalling pathway exemplifies the interplay between viruses and mitochondria. The functional significance of mitochondrial dynamics in several mitochondrial and cellular processes indicates that modifications in mitochondrial dynamics may work as an effective viral tactic to disrupt cellular signalling pathways. The sensitivity of mitochondrial dynamics to minor physiological disturbances in the cellular environment renders the impact of viral infections on mitochondrial dynamics an almost unavoidable outcome. The viral infection induces numerous physiological changes in host cells, many of which can directly influence mitochondrial dynamics and mitophagy. The comprehensive impact of viral infection on mitochondrial dynamics and its implications for viral pathogenesis are depicted in Fig. 6.2. Below, we examine some viruses that have been characterised for their impact on mitochondrial dynamics, emphasising the importance of these modifications in the viral infection process. The impact of viruses on mitochondrial dynamics and mitophagy is encapsulated in Table 6.1.

Hepatitis viruses

Chronic infection with HCV and the hepatitis B virus (HBV) has been consistently linked to significant mitochondrial damage in the liver. Despite differences in genomic organisation and life cycle, HBV and HCV have analogous pathologies, leading to chronic hepatitis that advances to liver fibrosis, cirrhosis, and hepatocellular carcinoma. Histopathological characteristics frequently noted in HBV/HCV infection encompass mitochondrial enlargement and a reduction in both the number and structure of mitochondrial cristae [108]. Infections by HBV and HCV induce endoplasmic reticulum (ER) stress, leading to the release of calcium from ER reserves, which is subsequently absorbed by neighbouring mitochondria, resulting in mitochondrial depolarisation and malfunction, along with the formation of mitochondrial reactive oxygen species (ROS). We have recently examined the impact of HBV and HCV infections on host cell mitochondrial dynamics and identified a clear correlation between viral infections and mitochondrial dynamics, mitochondrial equilibrium, and viral endurance. HCV is a positive-sense RNA virus classified within the genus Hepacivirus of the Flaviviridae family. The 9.5 kb RNA genome is translated by an internal ribosome entry site (IRES) element situated in the 5′ untranslated region, yielding a precursor polyprotein of approximately 3000 amino acids, which is then processed by cellular and virus-encoded proteases to generate three structural and seven nonstructural viral proteins with diverse roles. The majority of viral proteins remain bound to the ER membrane, inducing ER stress. HCV generates ER stress and facilitates autophagy, which is essential for HCV RNA translation and replication. HCV infection induces mitochondrial abnormalities, and it is hypothesised that ER-stress-induced depletion of ER Ca^{2+} reserves and their absorption by mitochondria facilitates mitochondrial dysfunctions. HCV proteins have been demonstrated to directly connect with the mitochondria, localising to the outer mitochondrial membrane (OMM) and the mitochondrial-associated membranes (MAM). HCV core protein nonstructural protein 5A (NS5a) proteins have been demonstrated to disrupt complex one activity and facilitate mitochondrial Ca^{2+} absorption, ROS generation, and mitochondrial permeability transition. The HCV NS3/4a protease is localised to the MAM and cleaves MAM-associated mitochondrial antiviral signalling protein (MAVS) to promote immune evasion. HCV infection results in increased ROS in the mitochondria, triggered by calcium signalling. Mitochondrial impairment during HCV infection is associated with the suppression of mitochondrial β-oxidation of fatty acids. This phenomenon is associated with enhanced lipogenic activity and increased lipid uptake during HCV infection, leading to the intracellular accumulation of lipid droplets. Hepatosteatosis is defined by the accumulation of lipid droplets, which expedite the progression to end-stage liver disorders in chronic hepatitis C. Additional impacts on mitochondrial functioning during HCV infection encompass diminished oxidative phosphorylation and ATP production. HCV infection causes perinuclear clustering of mitochondria, a significant feature in cells subjected to oxidative stress. Our research indicates that HCV facilitates mitochondrial fragmentation by promoting dynamin-related protein-1 (Drp1) S616 phosphorylation. Phosphorylation of Drp1 at S616 facilitates its translocation to the mitochondria and induces subsequent mitochondrial fission/fragmentation. HCV-infected cells show better expression of Parkin and PTEN induced kinase 1, in association with mitochondrial recruitment of Parkin. The recruitment of Parkin to mitochondria in HCV-infected cells was linked to a simultaneous enhancement in the autophagic degradation of damaged mitochondria via Parkin-dependent mitophagy, as effectively illustrated by the mitophagy reporter experiment. The depletion of Drp1 and Parkin in HCV-infected cells obstructed HCV-induced mitochondrial fission and mitophagy, respectively, while resulting in a significant enhancement of mitochondrial apoptotic signalling, as seen by substantial cytochrome C release and elevated cysteinyl aspartate-specific proteinase-3 (caspase-3) activity. Our findings indicated that HCV-induced mitochondrial fission, succeeded by full mitophagy, is essential for maintaining mitochondrial homoeostasis during infection and enhancing the survivability of infected cells. The increase of the mitochondrial quality control pathway mitigates the imminent cell death occurring due to mitochondrial damage sustained during infection. Besides enhancing the survival of infected cells, HCV-induced modifications in mitochondrial dynamics significantly undermined mitochondria-associated antiviral signalling, as demonstrated by a significant increase in interferon (IFN) production following the suppression of Drp1 expression in HCV-infected cells. The HCV undermines the innate immune response through NS3/4a protease-mediated cleavage of the MAVS protein, indicating that the impact of modified mitochondrial dynamics on MAVS signalling further enhances this viral approach. These investigations unequivocally demonstrate that HCV-induced modifications of mitochondrial dynamics and mitophagy are significant factors in viral survival, highlighting a novel element of viruses utilising mitochondrial dynamics and quality control mechanisms to enhance viral propagation. Inhibition of mitochondrial fission influences HCV secretion, likely due to modified rates of glycolysis and ATP synthesis. Conversely, ROS generated by accumulating defective mitochondria, can impair lipids and proteins that may be essential for HCV secretion. HBV, in contrast to HCV, the genetic composition of the HBV is comprised of DNA. The replication of HBV takes place through an RNA intermediary and undergoes reverse transcription, a method similar to that of retroviruses. An effective recombinant vaccination for HBV is available, providing a substantial level of protection (about 98%) against infection. HBV encodes a regulatory protein known as the hepatitis B virus X protein (HBx) or HBV X protein. HBx localises to the mitochondria by its interaction with voltage-dependent anion channel (VDAC), modifies membrane potential, and increases calcium and ROS levels, ultimately resulting in mitochondrial damage. Calcium signalling and ROS regulated by HBx are implicated in the activation of latent transcription factors, including STAT-3, nuclear transcription factor-kB, and Nuclear factor of activated T-cells (NFAT). HBV also activates early autophagic mechanisms that subsequently facilitate viral DNA replication. HBV, similar to HCV, also prompted Drp1-dependent mitochondrial fission and Parkin-dependent mitophagy. Disruption of mitochondrial dynamics was noted in cells expressing the full-length HBV genome or solely HBx; however, the HBx-defective (HBV-ΔX) genome exhibited no impact on mitochondrial dynamics, underscoring that HBx is adequate to induce this effect.

Mitochondrial impairment and modification of mitochondrial dynamics

HBV-induced modifications in mitochondrial dynamics and mitophagy significantly enhance mitochondrial quality control, hence maintaining mitochondrial homoeostasis. This facilitates the sustenance of chronic infection by suppression of apoptosis in HBV-infected cells. In cells lacking Parkin, HBV infection induced apoptosis, as demonstrated by the release of cytochrome C, activation of caspase-3, and cleavage of poly (ADP-ribose) polymerase. These data combined suggest that mitochondrial dynamics are crucial for sustaining the persistent phenotype in HBV-infected hepatocytes. The reduction of Parkin also somewhat influenced HBV DNA synthesis, perhaps due to compromised mitochondrial processes or heightened innate immunity.

Epstein–Barr virus

EBV is an oncogenic virus from the herpesvirus family and is associated with many lymphoid and epithelial malignancies. A recent study reveals that the viral latent membrane protein 2A (LMP2A) induces higher levels of mitochondrial fission in gastric and breast cancer cells. Their data indicated that LMP2A-activated Notch pathway induces Drp1 expression, resulting in increased mitochondrial fission. It is important to note that Drp1 overexpression alone did not result in increased fission; rather, the mitochondrial recruitment of Drp1 and the activation of its GTPase activity are crucial for this process.

Advocate for fission: Drp1 has been associated with the invasive capacity and augmented epithelial–mesenchymal transition of breast cancer cells. This study concurrently demonstrates that modified mitochondrial dynamics significantly influence the metastatic behaviour of EBV-associated gastric and breast carcinomas.

Human cytomegalovirus

Cytomegaloviruses (CMVs) are pervasive viruses that present a significant health issue due to the elevated incidence of congenital infections. Severe infections correlate with immunosuppression [129]. McCormick et al. show that fibroblasts infected with HCMV have a distinctive characteristic of punctate and scattered mitochondria, akin to mitochondrial fission, 24 hours postinfection. The mechanism responsible for this modification is the UL37x1 gene product, an early antiapoptotic gene also known as viral mitochondrion-localised Inhibitor of apoptosis (vMIA). vMIA comprises two domains. The first domain spans amino acids 5 to 34 (30 amino acids) and coincides with the mitochondrial signal sequence that directs vMIA to the mitochondria. The secondary domain situated between amino acids 118 and 147 exhibits antiapoptotic effects. A completely functioning recombinant fusion protein generated from vMIA comprises these two domains. vMIA is preserved across all human CMV strains and in other investigated primate CMVs. The vMIA-induced suppression of apoptosis relies on its interaction in conjunction with the pro-apoptotic proteins bcl-2-associated protein X (Bax) and Bcl-2 Antagonist/Killer (Bak). The main sequence of vMIA exhibits minimal resemblance to B-cell lymphoma (Bcl)-xl (antiapoptotic). The ectopic Bax can restore the Mitochondrion-localised Inhibitor of apoptosis (MIA)-induced fragmented phenotype, but the BCL-2 homology 3 (BH3) mutant Bax exhibited no rescuing impact. This study demonstrates the participation of Bax functions. Bak BH3 domains facilitate apoptosis and regulate mitochondrial shape in healthy cells.

Pseudorabies virus (an alphaherpesvirus)

Prv and HSV-1 infections impair mitochondrial motility and morphology in the superior cervical ganglion neurones of mice. During Prv infection, fusion events influenced by glycoprotein B lead to electrical coupling of neurones and heightened action potential firing rates, which lead to an increase in intracellular Ca^{2+} and modified mitochondrial dynamics. The pathway is regulated by the Ca^{2+}-sensitive protein Miro, which obstructs the recruitment of kinesin-1 heavy chain to the mitochondria. The alteration of mitochondrial dynamics induced by Prv is essential for the effective growth and dissemination of Prv, suggesting that modified mitochondrial transport facilitates alphaherpesvirus pathogenesis and infection. This research suggests that kinesin-1 heavy chain (HC) may be usurped and utilised for the transport of viral particles or protein cargo during Prv assembly and/or egress.

Influenza virus

Influenza virus infections result in millions of fatalities globally. A compelling study illustrates that Nucleotide-binding oligomerization domain-containing protein 2 (NOD2) and receptor-interacting protein kinase 2 (RIPK2) react to influenza A virus (IAV) infection by facilitating unc-51 like autophagy activating kinase 1 (ULK1) phosphorylation and initiating mitophagy. RIPK2−/− cells demonstrated impaired mitophagy and elicited heightened inflammasome activation due to the buildup of damaged mitochondria. This reveals that the increase of mitochondrial quality control during IAV infection attenuates inflammasome activation and IL-18 production. The induction of mitophagy occurred independently of Parkin but was linked to ULK1 phosphorylation at serine 555. It is uncertain whether RIPK2 directly phosphorylates ULK1 or if other intermediaries are implicated. This work indicates that NOD2–RIPK2 signalling safeguards against virally induced immunopathology by enhancing ULK1-dependent mitophagy. The IAV protein PB1-F2 interacts with MAVS and influences IFN production. PB1-F2 causes mitochondrial fragmentation, regulates the innate immunological response, and activates the NOD-like receptor pyrin domain-containing protein 3 (NLRP3) inflammasome. PB1-F2 translocates to the mitochondrial inner membrane space through the Translocase of outer membrane 40 (TOM40) channel, resulting in a decrease in mitochondrial membrane potential. The PB1-F2 mutant devoid of the C-terminal region does not influence mitochondrial function.

Measles virus and Newcastle disease virus

Measles is a severe, extremely contagious illness. The Edmonston strain of measles virus (MV-Edm) promotes p62-dependent mitophagy in nonsmall cell lung cancer (NSCLC) cells. Impairment of the autophagic machinery attenuates MV-Edm replication and diminishes virus-induced cytotoxicity in NSCLC cells. Mitophagy induced by MV-Edm leads to a decrease in MAVS protein levels, subsequently attenuating the innate immune response. Silencing of p62 expression inhibited mitophagy and resulted in the preservation of mitochondrial mass in MV-Edm–infected cells. These findings suggest that MV exploits mitophagy to suppress the innate immune response mediated by retinoic acid-inducible gene I (RIG-I)/MAVS signalling. A comparable work by the same group indicates that MV-Edm induces p62-mediated mitophagy, which inhibits apoptosis in infected NSCLC cells by obstructing cytochrome C release, hence facilitating increased viral replication. Autophagy-mediated continuous viral replication ultimately induces necrotic cell death due to ATP depletion, while the suppression of autophagy leads to a significant reduction in MV-Edm's oncolytic efficacy. While both investigations concur that MV-Edm triggers p62-dependent mitophagy, essential for facilitating viral replication, they diverge regarding the physiological consequences of mitophagy. The same group demonstrated that the NDV used a comparable method, utilising p62-mediated mitophagy to enhance viral proliferation. The induction of mitophagy reduced cytochrome C release and intrinsic apoptotic signalling in NDV-infected NSCLC cells, while the suppression of mitophagy amplified the oncolytic effect of NDV. Inhibition of mitophagy likely enhances innate immune signalling, resulting in cell death, or both mitochondrial apoptosis and innate immune signalling may contribute to cell death and oncolysis.

Severe acute respiratory syndrome coronavirus

Severe acute respiratory syndrome coronavirus (SARS-CoV) is a novel coronavirus that has emerged as a fatal pathogen and a significant hazard to human health. The mechanisms by which SARS-CoV evades innate immune signalling remain inadequately comprehended. A recent report reveals that the SARS-CoV protein, open reading frame-9b (ORF-9b), localises to the mitochondria and releases the proteasomal degradation of Drp1, leading to mitochondrial fusion. The expression of 9b in human embryonic kidney 293 cells resulted in approximately a 70% decrease in Drp1 levels, a drop that was responsive to proteasome inhibition but not influenced by autophagy inhibition. Notably, in this work, a decrease in Drp1 expression correlated with compromised MAVS signalling; however, this reduction was less effective than the exogenous expression of ORF-9b protein, indicating the potential involvement of alternative processes. The idea that diminished Drp1 expression or increased mitochondrial fusion inhibits MAVS signalling contradicts the prevailing view that MAVS signalling and IFN production are augmented and suppressed by mitochondrial fusion and fission, respectively. Additionally, ORF-9b facilitates the degradation of MAVS and its interaction partners by recruiting poly(rC) binding protein 2 (PCBP2) and the HECT domain E3 ligase AIP4, resulting in the degradation of MAVS, tumour necrosis factor receptor-associated factor 3 (TRAF3), and TRAF6, hence impairing MAVS signalling and IFN production. Mitochondrial localisation of ORF-9b induces the translocation of PCBP2 from the nucleus to the mitochondria, where it subsequently interacts with AIP4 and facilitates its mitochondrial recruitment. It is yet to be determined whether TRAF3 and TRAF6 are direct targets of AIP4. Suppression of PCBP2 or AIP4 expression restored MAVS-mediated signalling and reinstated IFN production. The results demonstrate that SARS-CoV ORF-9b alters mitochondrial dynamics and targets mitochondrial innate immune signalling to circumvent host innate defence. Nevertheless, additional research is necessary to clarify the physiological importance of increased mitochondrial fusion in viral infection mechanisms or disease development. Despite the efforts undertaken in recent decades to investigate the role of mitochondria in viral infections, numerous topics remain unanswered. Recently, mitochondrial dynamics and mitophagy have garnered significant attention as these processes regulate mitochondrial functioning during viral infections. Studies conducted in recent days have placed mitochondrial dynamics and viral infections at a critical intersection. There is a significant correlation between mitochondrial dynamics and viral infections; nonetheless, comprehensive characterisations and their implications for pathogenesis may enhance the understanding of pathogenic processes. It is crucial to ascertain the differential regulation of mitochondrial dynamics and mitophagy during viral infections and to clarify the many methods by which viruses exploit these modifications for their advantage. The clarification of the functional significance of mitochondrial dynamics and viral pathogenesis will create new opportunities for the therapeutic development of techniques to address viral infections and related disorders. Future investigations must pursue a deeper knowledge of essential mitochondrial processes such as bioenergetics, apoptosis, innate antiviral signalling, and interorganelle communication to clarify the effects of viral infections on mitochondria and mitochondrial dynamics (see Fig. 6.5)

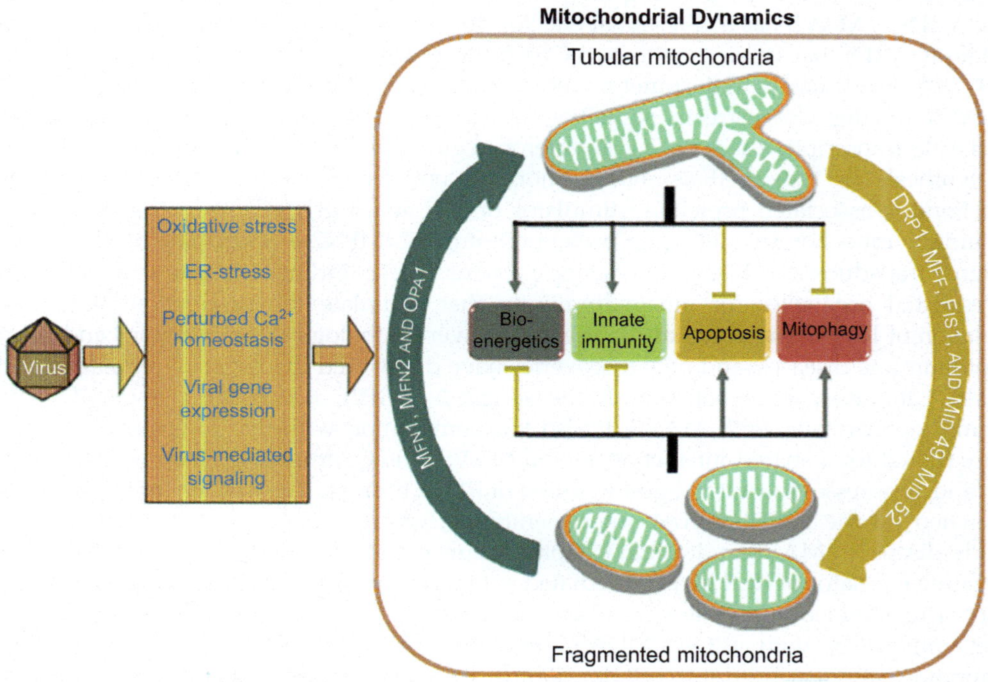

FIGURE 6.5 Virus affecting mitochondria.

Alteration of mitodynamics by viruses and its physiological significance

Viruses leading to the disruption of mitochondrial dynamics and the consequences on cell physiology and viral pathogenesis.

Mitochondria function as a platform for interferon signalling

Retinoic acid-inducible gene I like receptors (RLRs) are cytoplasmic proteins that detect viral RNAs in the cytoplasm through RIG-I and/or melanoma differentiation-associated gene 5 (MDA5). RIG-I is crucial for the synthesis of IFNs in reaction to RNA viruses such as paramyxoviruses, influenza virus, and Japanese encephalitis virus, while MDA5 is vital for immune responses to picornaviruses. RIG-I identifies the uncapped, unmodified 5′-triphosphate end of RNA produced by viral polymerases; however, this molecular signature is lacking in picornaviruses. DNA viruses, such as adenovirus (ADV), HSV-1, and EBV, stimulate IFN-β expression through a RIG-I-dependent mechanism. EBV generates short RNA molecules that stimulate RIG-I-mediated IFN production. Chen and colleagues recently revealed that DNA-dependent RNA polymerase III (Pol-III), which synthesises 5′-triphosphate RNA from cytosolic poly(dA-dT) DNA, is a significant modulator of RIG-I signalling triggered by HSV-1 or ADV infection. These investigations demonstrate that RLRs serve as crucial cytoplasmic sensors of DNA and RNA viruses. In our lab, we observed similar observations with duck plague virus and avian influenza virus. Specific molecules have been observed to be signal transducers downstream of RIG-I and MDA5. IFN-β promoter stimulator 1 (IPS-1; also known as MAVS, Cardif, or virus-induced signalling adaptor (VISA)) has been identified as a RIG-I-binding protein. IPS-1/MAVS has a C-terminal effector domain and an N-terminal caspase recruitment domain (CARD) that engages with two CARDs from RIG-I or MDA5. It has been shown that IPS-1/MAVS is situated on the outer membrane of mitochondria. Upon activation by RLRs, IPS-1/MAVS recruits downstream signalling molecules from many cascades to activate and translocate transcription factors (e.g. interferon regulatory factors [IRF3] or NF-κB), leading to the synthesis of type I IFNs. Furthermore, recent reports indicate that two additional mitochondria-associated proteins—stimulator of IFN genes (STING), primarily located in the ER, and the mitochondrial tethering protein mitofusin 2 (MFN2)—participate in RLR signalling as signal transducers or MAVS-interacting proteins. These findings indicate that mitochondria are essential organelles for innate immunity that stimulate the production of IFNs. Self-regulation of RLR-induced signals is essential to avert pathological overreactions and the onset of autoimmune diseases, and new discoveries have discovered the negative regulators of these signals. Laboratory of genetics and physiology 2 (LGP2), which lacks a CARD domain, coimmunoprecipitates with MAVS. It also has competition with the downstream kinase IkB kinase (IKKε) for a shared interaction site on MAVS. The C-terminal domain of LGP2 interacts with and suppresses the auto-oligomerisation of RIG-I. NLRX1, a constituent of the NLR family, is situated on the mitochondrial outer membrane, where it engages with MAVS. We have already discussed MAVS in detail in Chapter 4. Functional studies report that NLRX1 serves as a negative regulator of RLR MAVS-mediated antiviral signalling. The Atg5–Atg12 conjugate, an essential regulator of autophagy, has been demonstrated to suppress RLR signalling through direct interaction with RIG-I, MDA5, and IPS-1. Autophagy is a physiologically and immunologically regulated intracellular homoeostatic mechanism that sequesters and destroys macromolecular aggregates, cellular organelles like mitochondria, and whole microorganisms or their byproducts. Elevated mitochondrial mass and the buildup of impaired mitochondria in Atg5-deficient cells, indicating that the lack of autophagy led to a ROS-mediated enhancement in RLR signalling. Figure illustrates mitochondrial associated proteins that participate in the activation and suppression of innate immune responses , bioenergetics, apoptosis, and mitophagy in response to viral infection

Mitochondrial adaptor proteins mitochondrial antiviral signalling protein and interferon-beta production

MAVS has 540 amino acid residues and encompasses three functional domains. This consists of an N-terminal CARD (amino acids 10–77) and a C-terminal transmembrane domain (amino acids 514–535). The two domains are linked by a proline-rich region (aa 103–173), which serves as the third domain. The CARD of MAVS engages with the CARDs of RLRs, including RIG-I and MDA5. MAVS is tethered to the mitochondrial outer membrane through its C-terminal transmembrane domain, which has structural similarities to the transmembrane regions of several mitochondrial proteins, including Bcl-xL, Bcl-2, and TOM20. A C-terminal-truncated variant of MAVS, devoid of the transmembrane domain, was identified in the cytosol and was ineffective in facilitating RIG-I/MDA5 signalling. Recent investigations have shown that the homodimerization of MAVS through its transmembrane domains is essential for innate immunological signalling. The proline-rich region in MAVS facilitates the interaction with TRAF family proteins TRAF2, TRAF3, and TRAF6. Although MAVS lacks the proline-rich region, it is still proficient in stimulating IFN-β production. The tumour necrosis factor (TNF) receptor-associated death domain (TRADD) interacts with MAVS through its core region. Studies demonstrated that vesicular stomatitis virus (VSV)–induced IFN-β production was markedly reduced in TRADD-deficient mice, indicating that TRADD functions as a scaffold to recruit TRAF3, TRAF family member- associated NF-κβ activator (TANK), and Fas-associated protein with death domain (FADD)/Receptor-interacting Protein 1 (RIP1) into a higher-order signalling complex, thereby promoting the activation of IRF3 and NF-κB.

The examination of shortened MAVS resulting from splicing variations revealed an additional role of MAVS. Li and colleagues indicated that MAVS splicing variants 1a (exon 2 deletion) and 1b (exon 3 deletion), resulting from a frameshift caused by exon deletion, comprise 131 and 124 amino acid residues, respectively. Both MAVS 1a and 1b possess the putative CARD, facilitating their reception of RLR signals. MAVS 1a and 1b exhibit contrasting effects on RLR MAVS-mediated IFN production. MAVS 1a is devoid of the signalling domain necessary for downstream signal transmission and functions as an antagonist in MAVS full-length-mediated signalling. MAVS 1b can engage with FADD and RIP1 (downstream of RLR signalling) and specifically induce IFN-β production. The results indicate that MAVS signalling may be negatively modulated by its own splice variations.

Stimulator of interferon genes

STING, also known as mediator of IRF3 activation (MITA), MPYS, and ERIS, was recognised as an adaptor in the antiviral pathway that induces IFN-β expression. It was demonstrated that MITA/STING was affixed to the mitochondrial outer membrane and that it stimulated IFN-β production through interaction with MAVS. STING was situated on the ER and translocated from the ER to Sec-5-containing vesicles in response to HSV-1 infection, indicating that STING plays a role in DNA-mediated innate immunity. STING has been shown to translocate from the ER to the Golgi apparatus and to colocalise with TANK binding kinase I (TBK1). The dimerisation of ERIS/STING is essential for self-activation and IFN production. Consequently, STING probably functions as a molecular adaptor to facilitate signalling at the contact points between mitochondria and the ER.

Mitofusin 1 and mitofusin 2

MAVS also associates with MFN2, a dynamin-related GTPase localised to the OMM. Both MFN1 and MFN2 are classified as members of the dynamin-related GTPase family. MFN1 is important to mitochondrial fusion, while MFN2 connects the ER to mitochondria. MFN2 is concentrated at the ER-mitochondria interface, and the disruption of ER-mitochondria interactions is evident in mouse embryonic fibroblasts when MFN2 expression has been suppressed. The Ras-binding domain of MFN2 is essential for its tethering function. The coexpression of MFN2 suppressed the MAVS-dependent activation of the IFN-β promoter. Silencing of endogenous MFN2 expression by siRNA augmented IFN-β promoter activation triggered by measles virus infection and poly(I:C). The heptad region 1 of MFN2 engages with the central region of MAVS, which is essential for interaction with downstream effectors involved in IFN production, indicating that MFN2 functions as an inhibitor of RLR signalling. In cells infected with Sendai virus H4, RLR activation facilitates mitochondrial elongation through MFN1, leading to MAVS-STING interaction at the mitochondria-ER interface. Consequently, MFN1 and MFN2 seem to assume contrasting functions in antiviral responses. These data strongly indicate that the contact areas between the ER and mitochondria, as well as the MAM, serve essential functions as signal mediators in host cell antiviral responses.

Viral evasion of host immune responses

Viruses have evolved mechanisms to circumvent host immune responses, including RLR-mediated IFN production. Several viral gene products have been recognised as antagonists of RLR signalling. These proteins interact with viral RNA, disrupt transcriptional factors downstream of RLRs, or modulate other mitochondria-associated signal transducers. Given that RLR signalling converges in mitochondria, it is logical that viruses would exploit mitochondrial mechanisms to circumvent this aspect of immune responses. The HCV contains NS3/4A, a serine protease that selectively obstructs RIG-I-mediated signalling by cleaving MAVS at Cys508. Cleaved MAVS is incapable of anchoring to mitochondria or facilitating RIG-I signalling. A comparable process has been documented in Flaviviridae GB virus B and hepatitis A virus (HAV) of the Picornaviridae family. HAV encodes serine protease 3C, that cleaves MAVS at Gln428, leading to inhibition of antiviral signalling. Non-structural protein 1 (NS1), produced by the IAV, interacts with RIG-I and suppresses RIG-I-mediated stimulation of IFN-β synthesis. NS1 may disrupt signals from the RIG-I-STING complex by direct interactions. The V proteins expressed by paramyxoviruses interact with MDA5 and particularly suppress MDA5-mediated IFN expression. Viral vMIA, encoded by

UL37x1 from HCMV, is a recognised antiapoptotic and mitochondria-associated protein. vMIA overexpression suppresses MAVS-mediated activation of the IFN-β promoter and mitochondrial fragmentation. Consequently, vMIA may modulate mitochondrial functions, leading to the suppression of IFN production.

Mitochondrial homoeostasis and viral infection

Three models have been proposed to elucidate the mechanism by which mitochondrial membrane permeabilisation (MMP) induces apoptosis:

1. selective enhancement of mitochondrial outer membrane permeabilisation
2. destabilisation of the mitochondrial lipid bilayer
3. permeabilisation of the mitochondrial inner membrane through activation of the permeability transition pore complex (PTPC).

Significantly, proapoptotic or antiapoptotic proteins from the Bcl-2 family have been included in all three models. Bcl-2 was the inaugural recognised mammalian homologue of ced-3, a crucial gene for apoptosis in C elegans. Bcl-2 comprises four unique Bcl-2 homology domains (BH1–4). The Bcl-2 family consists of three subfamilies: the antiapoptotic Bcl-2 subfamily (e.g. Bcl-2 and Bcl-xL), the proapoptotic Bax subfamily incorporating BH1–3 domains (e.g. Bax and Bak), and the proapoptotic BH3-only subfamily (e.g. Bid, Bad, and Bim). Bax and Bak congregate on the mitochondrial and ER membranes, serving as channels or pores that facilitate MMP, leading to the release of cytochrome C and calcium, respectively. Bcl-2/Bcl-xL establishes an inhibitory antiapoptotic complex with activated Bax/Bak. A pro-apoptotic pathway entails BH3-only proteins binding to activated Bax/Bak rather than the repressor Bcl-2/Bcl-xL. In the second concept, Bax is crucial for the fast transbilayer reorganisation of lipids, and the resultant curvature stress in the mitochondrial outer membrane triggers the creation of large holes. The activation of the PTPC leads to the permeabilisation of the inner mitochondrial membrane, triggered by particular stimuli like calcium excess and ROS. In addition, it has been suggested that PTPC is modulated by various elements, such as mitochondrial matrix proteins (e.g. cyclophilin D), intermembrane space proteins (e.g. creatinine kinase), and outer membrane-associated proteins (e.g. peripheral benzodiazepine receptor (PBR)). Bax and Bcl-2/Bcl-xL have also been reported to be responsible for the regulation of PTPC activity. Multiple investigations have found viral antiapoptotic (Table 6.2) and proapoptotic proteins (Table 6.1), some of which demonstrate activities or structures analogous to the Bcl-2 family. We succinctly outline the roles of the viral proteins in the subsequent section (see Fig. 6.6

Cytosolic viral RNA and DNA are recognised by RIG-I, MDA5, Z-DNA-binding protein 1 (DAI) and other unknown DNA sensors. Cytosolic double-strand DNA is recognised by cytosolic DNA receptors such as DAI, or in another pathway, cytosolic DNA-dependant RNA Pol-III synthesises 5′ triphosphate single-stranded RNA, which is recognised by RIG-I. RIG-I and MDA5 recognise long or short viral dsRNA or 5′ triphosphate ssRNA via their C-terminal RNA helicase domain. The two N-terminal CARDs then interact with an N-terminal CARD from MAVS; also referred to as IPS-1, Cardif, and VISA. Anchored to mitochondria via C-terminal transmembrane domain, MAVS receives signals from RLRs via a CARD and transduces the signal to downstream proteins via other regions, including a proline-rich region that interacts with TRAF2 and TRAF3 and a central region that interacts TRADD and TRAF6.

TABLE 6.2 Examples of zoonotic viruses.

Species	Type	Synonym	Sub family	Human pathophysiology
Macaque monkey	CeHV	Cercopithecine herpesvirus 1	α	Quite unusual, 25 human cases reported. Untreated infection is sometimes deadly. Sixteen of twenty-five cases cause fatal encephalomyelitis. At least four cases resulted in survival with severe neurological impairment. Awareness and early treatment are important for laboratory workers facing exposure.
Mouse	MuHV	Murid herpesvirus 68	γ	Zoonotic infections are common in laboratory workers handled infected mice. ELISA tests show a factor of four positive results, due to antibody cross-reaction with other herpesviruses.

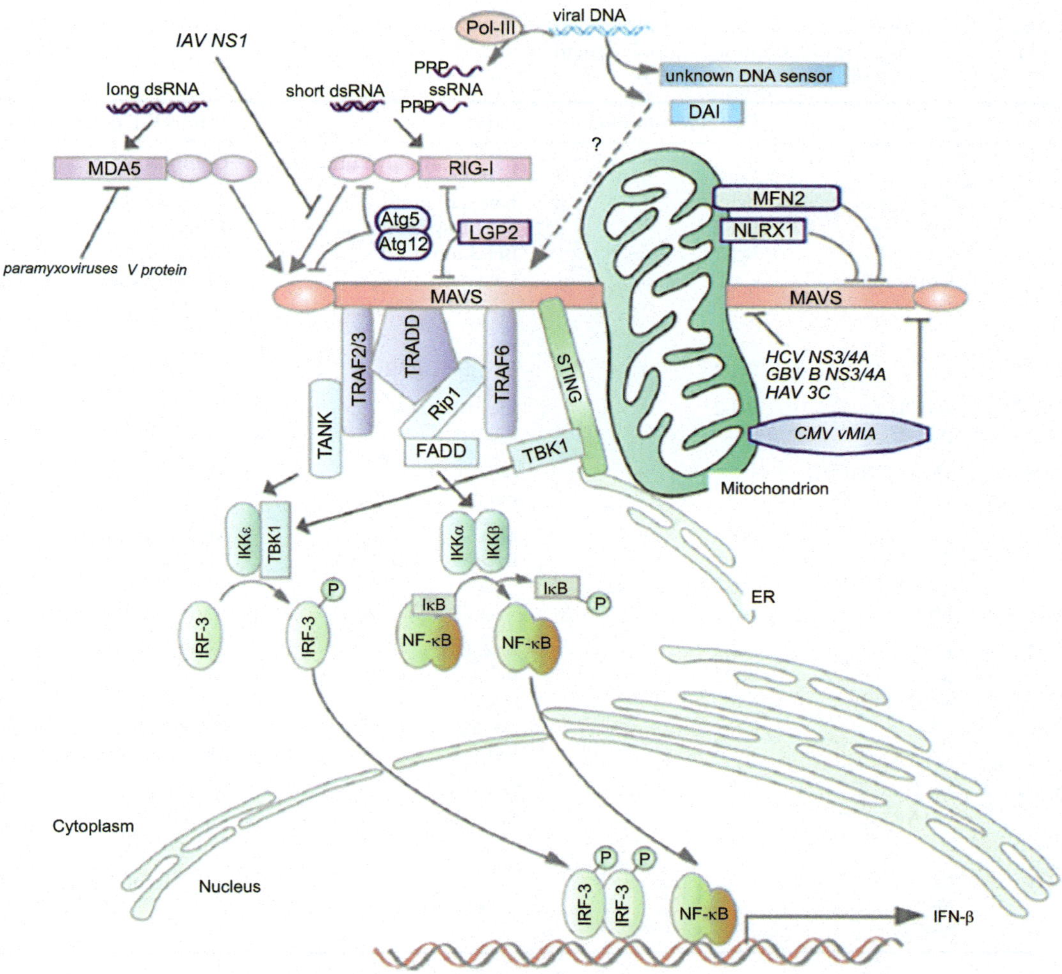

FIGURE 6.6 RLR mediated viral recognition and pathways leading to interferon (IFN-β) production. *RLR*, retinoic acid-inducible gene I like receptors.

TRAF2 and TRAF3 interact with TANK. TANK recruits a kinase complex composed of TBK1 and IKKε, which phosphorylated IRF3 and activation of INF-β transcription. TRADD interacts with TRAF3 and Rip1, whereas TRAF6 interacts with Rip1. Rip1 mediates signals between FADD and IKKα and IKKβ, leading to activation of the transcription factor NF-κβ (see Table 6.3).

Negative regulators of MAVS-mediated signals are also shown. Interactions between LGP2 and MAVS or RIG-I antagonise RIG-I/MAVS-mediated signalling. NLRX1 localises to mitochondria and interacts with MAVS, leading to inhibition of signalling MFN2, a mitochondrial tethering GTPase, interacts with MAVS via the heptad region of MFN2, which interferes with interactions between MAVS and various downstream molecules. The autophagic regulator protein conjugate Atg5/Atg12 also inhibits MAVS signalling. Several viral proteins inhibit

TABLE 6.3 Alteration of mitochondrial dynamics by viruses and its physiological significance. Viruses leading to the disruption of mitochondrial dynamics and the consequences on cell physiology and viral pathogenesis are listed below.

Virus	Protein	Cellular homologue	Partners	Intracellular localisation
ADV	E1A	-	p53, BH-3 only subfamily	Cytoplasm
VACV	E1B-19K F1L N1 CmA	Bcl-2 Bcl-2 Bcl-2 -	Bax subfamily Bax, caspase-9 BH-3 motif Caspase	Mitochondria Mitochondria Cytoplasm Mitochondria Cytoplasm Nucleus
MXV	M11L	Bcl-2	Bax, Bak, PBR	Mitochondria
ASFV	A179L A224L	Bcl-2 -	Bid, Bax, Bak Caspase	Cytoplasm Cytoplasm
CMV	vMIA (UL37x1) vICA (UL36)	- -	Bax, ANT Bak, caspase-8	Mitochondria Cytoplasm
EBV	BHRF1 EBNA3A, EBNA3C EBNA-LP	Bcl-2 -	Bim Expression of Bim	Mitochondria Nucleus
KSHV	KSBcl-2 K7 K13/vFLIP K15	Bcl-2 Bcl-2 - -	Bax, Bak Bcl-2, caspase-3, CAML Caspase-8 HAX-1	Cytoplasm Mitochondria, ER, cytoplasm Cytoplasm Mitochondria, ER
γHV-68	M11 vMAP	Bcl-2 -	Bax, Bak Bcl-2, Bax, Bak	Cytoplasm, nucleus Mitochondria
HCV	E2 NS2 NS5A	- - -	- CIDE-B -	Cytoplasm ER ER

MAVS-mediated signals. NS1 from IAV, for instance, may interfere with the interaction between RIG-I and MAVS. MAVS is cleaved by viral proteases, including NS3/4A from HCV and GB virus B and 3C protein from HAV. Cleaved MAVS does not transmit signals downstream or localise to mitochondrial membranes.

Viral antiapoptotic and proapoptotic proteins that target mitochondria

Deoxyribonucleic acid viruses (ADV), possess two distinctive oncogenes, E1A and E1B, which encode proteins that demonstrate proapoptotic or antiapoptotic functions. E1A promotes apoptosis by activating p53 or enhancing cellular sensitivity to death receptor ligands (e.g. TNF-α, Fas ligand (FasL) and TNF-related apoptosis-inducing ligand (TRAIL)). Bik, a constituent of the BH3-only subfamily, is important to ADV-induced apoptosis. Bik is indeed activated by ADV infection, and the reduction of Bik using siRNA decreases apoptosis in cells infected with ADV. The E1B-19K protein inhibits E1A-induced cell death and was originally recognised as a viral Bcl-2 homologue. Similar to Bcl-2, E1B-19K is predominantly situated near mitochondria, enabling E1B-19K to obstruct multimeric pore formation by members of the Bax subfamily.

The vaccinia virus, classified as an orthopoxvirus, encodes three antiapoptotic proteins: cytokine response modifier A (CrmA), F1L and N1. CrmA is a universal caspase inhibitor, while the other two are viral Bcl-2-like proteins. F1L accounts for the residual antiapoptotic action in viruses devoid of CrmA. F1L targets the mitochondria through its C-terminal domain and disrupts apoptosis by preventing the decline of the inner mitochondrial membrane potential (Δψm) and the release of cytochrome C. Recent findings indicate that F1L possesses significantly different BH domains, which likely facilitate constitutive interaction with Bak and inhibit its activation. During infection, F1L substitutes for the roles of the cellular antiapoptotic Bcl-2 family protein, myeloid leukaemia cell differentiation protein-1. Furthermore, F1L binds to and selectively inhibits caspase-9. The data indicate that F1L acts as a suppressor of pro-apoptotic Bcl-2 family proteins and an inhibitor of caspase-9, hence neutralising two consecutive steps in the mitochondrial cell death process. Numerous F1L orthologs have been discovered in different poxviruses, including variola virus, monkeypox virus and ectromelia virus. The ectromelia virus ortholog's antiapoptotic action has been established. The N1 protein, produced by the N1L gene, exhibits structural similarity to Bcl-xL and other constituents of the Bcl-2 subfamily. N1 decreases apoptosis through the process of binding to the BH-3 motifs of pro-apoptotic proteins. The Myxoma virus, is a leporipoxvirus affecting rabbits, causing a fatal disease marked by widespread tumour-like lesions and extensive immunosuppression. M11L, a protein located in the mitochondrial outer membrane, has been demonstrated to suppress apoptosis by binding to the MMP-inducible, pore-forming proteins Bax and Bak. Structural study indicates that M11L is a Bcl-2 homologue that interacts with Bax and Bak with affinities comparable to those of cellular Bcl-2 and Bcl-xL. M11L further inhibits the release of cytochrome C and the dissipation of Δψm. Furthermore, M11L physically binds with PBR, which is situated on the OMM and interacts with the mitochondrial PTPC, VDAC and adenine nucleotide translocator (ANT). M11L seems to modulate MMP through its action on PBR, a PTPC modulator. African swine fever virus (ASFV) is a substantial dsDNA virus that causes an acute sickness characterised by a pivotal role of apoptosis in its pathogenesis. A179L is a Bcl-2 homologue derived from ASFV that demonstrates antiapoptotic properties. Furthermore, the transient expression of A179L in Vero cells inhibits apoptosis triggered by active forms of Bid (truncated Bid (tBid)). A179L targets both upstream BH3 activators and the essential proapoptotic components Bax and Bak. The expression of A224L is observed throughout the later stages and belongs to the inhibitor of apoptosis protein family. The overexpression of this factor suppresses caspase activity and cell death triggered by diverse stimuli. A224L further stimulates the transcription factor NF-κB, potentially inhibiting apoptosis during the later stages of infection. CMV, a β-herpesvirus, employs various intricate techniques to influence and avoid host defence mechanisms. The highly conserved UL37 gene encodes vMIA, that is specific and essential for viral replication. vMIA prevents apoptosis triggered by many triggers, including death receptor ligation, infection with E1B-19K-deleted ADV and exposure to cytotoxic agents. vMIA localises to the mitochondrial membrane via a mitochondrial transmembrane domain and forms a complex with ANT, a constituent of the PTPC. vMIA is believed to impede the opening of the PTPC and subsequent MMP by creating a complex with ANT. vMIA interacts with Bax, leading to the transport of the complex to the mitochondria and the suppression of hole formation by multimeric Bax. The antiapoptotic effects of vMIA are absent in Bax-deficient cells, suggesting that vMIA may only neutralise Bax to perform its antiapoptotic actions. The viral inhibitor of caspase-8 activation (vICA or UL36 protein) inhibits Fas/CD95-induced apoptosis. The activation of caspase-8 induces crosstalk and activates the intrinsic route through the production and translocation of tBid. UL36/vICA may delay MMP due to the crosstalk effect. Murine cytomegalovirus (mCMV) encodes an antiapoptotic protein known as viral inhibitor of Bak oligomerisation (vIBO). vIBO inhibits Bak-mediated cytochrome C release and apoptosis, indicating that vIBO and vMIA (m38.5 protein in mCMV) have complementary actions. CMV infection confers protection against apoptosis induced by rotenone, a potent inhibitor of mitochondrial respiratory Complex I. This protection is facilitated by a virally generated 2.7-kb RNA fragment (β2.7) that interacts with Complex I and inhibits the localisation of key component genes linked to retinoid/interferon-induced mortality-19. Accordingly, β2.7 RNA stabilises mitochondrial respiratory chain Complex I, thereby maintaining efficient ATP production, supporting cell survival and facilitating completion of the viral replication cycle (see Tables 6.4 and 6.5).

EBV, a γ-herpesvirus, establishes latent infection in B lymphocytes and induces their transformation. The viral genome encodes two Bcl-2 homologues, BHRF1 and BALF1. BHRF1 and Bcl-2 exhibit analogous localisation profiles, three-dimensional conformations and beneficial impacts on cellular viability. BHRF1 preserves mitochondrial membrane potential and suppresses apoptosis induced by diverse stimuli, including c-Myc activation, heterologous viral infection, γ-irradiation and anticancer chemotherapeutic agents. BHRF1 significantly suppresses TRAIL-induced apoptosis, subsequently leading to caspase-8 activation and a reduction in $\Delta\psi m$. In contrast to Bcl-2 and Bcl-xL, BHRF1 does not possess an accessible, preexisting BH-3 binding groove and so is incapable of sequestering and inhibiting proapoptotic BH-3-only proteins. BHRF1's capacity to bind the pro-apoptotic BH-3-only subfamily member Bim, but excluding Bak, offers a protective effect. Notably, a comparatively minor quantity of Bim is associated with BHRF1 in contrast to the extent of Bim expression elicited by apoptosis. Despite the ambiguity surrounding the mechanism, BHRF1 may operate via Bim to inhibit apoptosis. BALF1, a viral Bcl-2 homologue, interacts with the pro-apoptotic proteins Bax and Bak and may antagonise the antiapoptotic activity of BHRF1. However, its precise role in the EBV life cycle remains unresolved. The function of BALF1 in the viral life cycle remains contentious. Moreover, the nuclear antigens EBV nuclear antigen (EBNA3A) and EBNA3C, essential for lymphoblastoid cell lines, have antiapoptotic properties. EBNA3A and EBNA3C function collaboratively as primary regulators in the suppression of the BH-3-only subfamily member Bim (Anderton et al., 2008). The reduction of Bim, a crucial regulator of B-cell survival and a tumour suppressor in B cells, by the nuclear EBNA3 proteins may promote EBV-mediated lymphomagenesis. The EBV nuclear antigen leader protein (EBNA-LP) perturbs host transcriptional processes and

TABLE 6.4 List of proapoptotic proteins located in mitochondria and cytoplasm against pathogenic viruses.

Virus	Protein	Partners	Intracellular Localisation
EBV	BALF1	Bax, Bak	Cytoplasm
HPV	E1^E4	Cytokeratins	Mitochondria, microtubule-associated
	E6	p53	Cytoplasm
	E7	-	Cytoplasm
HCV	NS3	Caspase-8	Cytoplasm
	NS4A	-	Mitochondria
VSV	M	Transcription of Bcl-2 family members	Cytoplasm, nucleus
	P	-	Cytoplasm
IAV	PB1-F2	VDAC1, ANT3	Mitochondria (inner membrane)
AEV	VP3	-	Mitochondria
	2C	Caspase-9, caspase-3	Mitochondria
SARS-CoV	7A	Bcl-xL	Golgi apparatus
	NSP15	-	Cytoplasm
HIV-1	Vpr	ANT, VDAC	Mitochondria, cytoplasm, nucleus
	Tat	-	Cytoplasm, nucleus
	Nef	-	Cytoplasm
	Protease	-	Cytoplasm
HTLV-1	p13II	-	Mitochondria (inner membrane)
HBV	HBx	VDAC3, Hsp60	Mitochondria

interferes with cytoplasmic apoptotic signalling. EBNA-LP interacts with Bcl-2 indirectly via HS1-associated protein X-1 (HAX-1), a mitochondrial apoptosis regulator whose activity is modulated by the serine protease Omi/HtrA2.

KSHV or HHV-8, the causative agent of Kaposi's sarcoma, encodes the viral Bcl-2 homologue ORF16 (KSBcl-2). Overexpression of KSBcl-2 inhibits apoptosis with comparable efficacy to cellular Bcl-2, Bcl-xL, or EBV BHRF1. However, the antiapoptotic mechanism of KSBcl-2 diverges from that of Bcl-2 and Bcl-xL, as KSBcl-2 does not associate with other Bcl-2 family members. KSHV K7 is structurally analogous to survivin-DeltaEx3, a splice variation of human origin.

Survivin, which safeguards cells against apoptosis by an unexplained mechanism. K7 contains a mitochondrial targeting sequence, an N-terminal baculovirus inhibitor of apoptosis repeat (BIR) domain and a putative BH2-like domain. The BIR domain mediates interaction with the effector caspase-3, while the antiapoptotic function of K7 requires the BH2-like domain, which is essential for its association with Bcl-2. Consequently, K7 may function as an adaptor for Bcl-2. K7 can also regulate the intracellular calcium concentration.

TABLE 6.5 List of anti-apoptotic proteins located in mitochondria and cytoplasm against pathogenic viruses.

Virus	Protein	Cellular homologue	Partners	Intracellular localisation
ADV	E1A	-	p53, BH-3 only subfamily	Cytoplasm
VACV	E1B-19K F1L N1 CmA	Bcl-2 Bcl-2 Bcl-2 -	Bax subfamily Bax, caspase-9 BH-3 motif Caspase	Mitochondria Mitochondria Cytoplasm Mitochondria Cytoplasm Nucleus
MXV	M11L	Bcl-2	Bax, Bak, PBR	Mitochondria
ASFV	A179L A224L	Bcl-2 -	Bid, Bax, Bak Caspase	Cytoplasm Cytoplasm
CMV	vMIA (UL37x1) vICA (UL36)	- -	Bax, ANT Bak, caspase-8	Mitochondria Cytoplasm
EBV	BHRF1 EBNA3A, EBNA3C EBNA-LP	Bcl-2 -	Bim Expression of Bim	Mitochondria Nucleus
KSHV	KSBcl-2 K7 K13/vFLIP K15	Bcl-2 Bcl-2 - -	Bax, Bak Bcl-2, caspase-3, CAML Caspase-8 HAX-1	Cytoplasm Mitochondria, ER, cytoplasm Cytoplasm Mitochondria, ER
γHV-68	M11 vMAP	Bcl-2 -	Bax, Bak Bcl-2, Bax, Bak	Cytoplasm, nucleus Mitochondria
HCV	E2 NS2 NS5A	- - -	- CIDE-B -	Cytoplasm ER ER

It interacts with the cellular calcium-modulating cyclophilin ligand and with protein linking integrin-associated protein to the cytoskeleton 1, a regulator of the ubiquitin–proteasome system. The membrane-associated protein K7 may serve as a foundation for antiapoptotic mechanisms. KSHV also encodes additional antiapoptotic factors, including K13, a viral FLICE-inhibitory protein (vFLIP) with homology to Fas-associated death domain-like proteins, and K15, a mitochondrially localised protein. The antiapoptotic function of vFLIP/K13 is ascribed to the suppression of caspase-8 activation and the upregulation of antiapoptotic proteins through NF-κB activation. K15 is localised to the mitochondria and ER and interacts with HAX-1, which can connect Bcl-2 and EBNA-LP to enhance cell survival. Murine γ-herpesvirus 68 expresses M11, a viral Bcl-2 homologue that inhibits apoptosis triggered by Fas and TNF-α in both murine and human cell lines. Despite the low homology between M11 and Bcl-2, structural studies demonstrated that M11 possesses a BH3 binding groove capable of sequestering Bax and Bak. Furthermore, viruses possessing a mutation that resulted in faulty BH3 binding demonstrated compromised sustained replication and reactivation from latency in vivo. A new viral mitochondrial antiapoptotic protein (vMAP) encoded by the M8 gene was recently reported. The N-terminal domain of vMAP engages with Bcl-2, enhancing Bcl-2's localisation to mitochondria and inhibiting Bax activation. vMAP is essential for viral replication in normal cells, but not in Bax/Bak-deficient cells, indicating that the antiapoptotic impact may mitigate host innate immune responses. The human papillomavirus (HPV) E1^E4 protein is generated from spliced transcripts that fuse the early E1 and E4 ORFs and represents the most abundantly expressed viral protein in HPV-infected epithelial tissue. In differentiated human keratinocytes, E1^E4 associates with the cytokeratin network and promotes its destabilisation. E1^E4 subsequently translocates to the mitochondria through an N-terminal leucine-rich domain, prompting the disassociation of mitochondria from microtubules. The isolated mitochondria subsequently cluster near the nucleus, leading to the dissipation of $\Delta\psi m$ and apoptosis. The pro-apoptotic proteins encoded by E6 and E7 exhibit multifunctionality, occasionally inhibiting p53-mediated apoptosis (E6) (Scheffner et al., 1990). E6 E7 apparently sensitises cells to various apoptotic stimuli, perhaps dependent or independent of p53. Ribonucleic acid viruses HCV encodes NS4A and NS3, two pro-apoptotic proteins that assemble a mitochondrial complex with NS3/4A and suppress host innate immune responses by cleaving MAVS. NS4A localises to mitochondria and modifies their intracellular distribution, resulting in the dissipation of $\Delta\psi m$ and the release of cytochrome C. NS3, a viral protease, has been demonstrated to trigger caspase-8-mediated apoptosis independently of its proteolytic function. Recently, the HCV core protein has been recognised as a third pro-apoptotic protein that promotes effects akin to caspase-independent apoptosis, including membrane blebbing and nuclear condensation. In addition, HCV encodes proteins that inhibit apoptosis. NS2 is a transmembrane protein associated with the ER that functions as an inhibitor of apoptosis triggered by cell death-inducing DFFA-like effector B. NS5A is also identified in the ER of HCV-infected cells. This protein activates NF-κB to inhibit cell death caused by ER stress. Moreover, structural protein E2 has been demonstrated to impede TRAIL-induced apoptosis by an unidentified mechanism. VSV is a rhabdovirus that induces neuronal death and the onset of neurological diseases. VSV-infected cells demonstrate early activation of the mitochondrial apoptotic pathway in the absence of viral protein production or other replication processes. The viral matrix protein M and phosphoprotein P, a crucial component of the RNA-dependent RNA polymerase complex, participate in the activation of apoptosis through several mechanisms. The M protein transcriptionally regulates Bcl-2 family proteins, while the mechanism of action for phosphoprotein P remains unidentified. The IAV encodes the PB1-F2 protein, which is integrated into the mitochondrial inner membrane through a C-terminal transmembrane domain. Proposed interacting partners for PB1-F2 are VDAC1 and ANT3, with indications that the pro-apoptotic function of PB1-F2 arises from increased mitochondrial membrane permeability through direct interactions with other mitochondrial membrane proteins. The avian encephalomyelitis virus (AEV) is classified under the *Picornaviridae* family. The transient expression of the AEV structural protein VP3 and the nonstructural protein 2C was demonstrated to induce cell death. VP3 was identified on mitochondria, where it triggered apoptosis through the caspase-3-like protease pathway. The 2C localises to mitochondria and facilitates the outflow of cytochrome C into the cytoplasm, subsequently activating upstream caspase-9 and downstream caspase-3, resulting in death. The functions of 2C during viral replication, albeit conserved among picornaviruses, remain ambiguous. The 7Aprotein of SARS-CoV is not essential for viral replication in cultured cells; however, it facilitates virus-induced apoptosis by suppressing Bcl-xL. The NSP15 protein, encoded by SARS-CoV, suppresses apoptosis generated by overexpressed MAVS in a dose-dependent manner; however, NSP15 does not inhibit staurosporine-induced apoptosis.

Reverse transcription of viruses

Human immunodeficiency virus type 1 (HIV-1) possesses six auxiliary proteins absent in other retroviral types. Viral protein R (Vpr) induces apoptosis. Synthetic Vpr induces rapid dissipation of $\Delta\psi m$ and the release of intermembrane space proteins in intact cells or purified mitochondria by directly interacting with ANT or VDAC, which are components of the mitochondrial PTPC. In lymphoid cells, Vpr-induced MMP is promoted by Bax's interaction with ANT, whereas it is inhibited by Bcl-2 overexpression and PTPC inhibitors. Vpr is not necessary for HIV replication in T cells, although it is requisite in macrophages. The R77Q mutation in Vpr negates its proapoptotic function, indicating that Vpr may be an essential virulence component in HIV infection. The HIV envelope glycoprotein complex leads to activation of MMP via a complicated signal transduction mechanism. Tat induces apoptosis in both infected and exclusively expressed cells, potentially leading to dementia. The expression of Tat induces a reduction in $\Delta\psi m$, an overproduction of ROS, and the activation of caspases. Tat modifies microtubule dynamics, facilitating the proteasomal degradation of microtubule-associated protein 2 and triggering apoptosis through a mitochondrial pathway. Bim, a pro-apoptotic member of the Bcl-2 family that mediates cell death in response to disrupted microtubule dynamics, enhances Tat-induced apoptosis. The Nef protein causes lysosomal membrane permeabilisation, resulting in the cytosolic release of the Bax activator. HIV-1 protease cleaves caspase-8, which subsequently activates Bid to its form tBid. Human T lymphotropic virus type 1 induces adult T-cell leukaemia and is additionally linked to degenerative neuromuscular disorders. The pX region of the genome encodes many proteins that can disrupt functional and transcriptional cellular processes. The p13II protein localises to the mitochondrial inner membrane, where it induces a membrane potential-dependent influx of potassium, resulting in mitochondrial swelling, fragmentation and altered calcium uptake and retention. p13II enhances cellular sensitivity to apoptosis triggered by Fas ligand and ceramide, while inhibiting cell proliferation and tumorigenesis in cells coexpressing myc and ras. The hepatisis B virus X protein (HBx) facilitates the development of hepatocellular carcinomas by functioning as a transcriptional activator of viral or cellular genes. HBx, potentially linked to mitochondria through a proposed transmembrane domain (aa 54–70), enhances the susceptibility of infected and transfected cells to TNF-α-induced apoptosis. HBx has been shown to accumulate in the mitochondria of transfected hepatoma cells, where it interacts with VDAC3, leading to dissipation of the mitochondrial membrane potential ($\Delta\psi m$) and perinuclear clustering of mitochondria. Recent findings indicate that HBx triggers the cyclophilin D-mediated mitochondrial permeability transition, resulting in calcium influx into the cytoplasm. HBx further associates with the mitochondrial matrix chaperone Hsp60 and engages with Hsp70. Additional research is necessary to clarify the mechanisms involved in HBx-related mitochondrial control.

Herpes simplex virus and mitochondria

HSV-1 and HSV-2 are substantial enveloped DNA viruses containing a minimum of 74 unique genes. About 50% of them are nonessential for viral replication in cell culture; however, they may be necessary during the viral life cycle in humans. Herpesviruses, including HSV-1 and HSV-2, have evolved several techniques that facilitate efficient replication or enhance latency. We are focused on the molecular underpinnings of HSV pathogenesis and thus we encapsulate the findings of current investigations regarding the functions of mitochondria during HSV replication.

Herpes simplex virus influences the intracellular compartmentalisation of mitochondria

We have demonstrated that mitochondria react to HSV infection by moving with viral tegument proteins via microtubules to a location near the nucleus where virion maturation and/or tegument assembly takes place. Notably, concentrations of cellular ATP and lactate, as well as $\Delta\psi m$ is sustained for a minimum of 6 hours postinfection before declining in subsequent stages. Previous research indicated that mitochondrial protein production in HSV-infected cells diminishes progressively, reaching approximately 60% of normal levels. Thereafter, mitochondria reveal less calcium absorption, only 65% of the control rate at the stage of advanced stages of infection. These observations suggest that HSV infection negatively regulates mitochondrial function. The viral assembly site of ASFV is cytoplasmic and encircled by many organelles, including mitochondria. The presence of cristae condensation at this location signifies active mitochondrial respiration. Consequently, the envelopment and capsid formation of ASFV are contingent upon calcium gradients and ATP. Consequently, mitochondria next to the viral assembly site presumably provide energy for viral morphogenesis. In cells infected with HSV, however, there is an absence of the evidence for increased mitochondrial activity and the specific impacts of mitochondrial clustering remain ambiguous. The juxtanuclear aggregation of teguments, nucleocapsids and mitochondria in HSV-2-infected cells exhibits similarities to aggresomes. We noted that the disruption of these aggresome-like structures via nocodazole resulted in a modest yet consistent inhibitory effect (10-fold reduction) on the generation of infectious particles (Nozawa et al., 2004) and suggested that the aggresome-like structures do not serve a critical role but rather an auxiliary function in HSV-2 replication. In HSV-1–infected human precursor oligodendroglial KG-1C cells, prominent mitochondrial aggregates have been observed in proximity to intracellular viral particle clusters, without evidence of apoptosis. These findings suggest that HSV-1 may differentially modulate mitochondrial positioning and function during both early and late stages of infection.

Herpes simplex virus UL7 is localised to mitochondria

The HSV UL7 protein is expressed during the late stage of infection and is linked to intracellular capsids and virions in cells infected with HSV. The UL7 gene is nonessential for HSV replication in cell culture, although the mutant virus devoid of UL7 was generated. Reduced plaques in comparison to the wild-type virus. UL7 is found in both mitochondria and cytoplasm and engages with ANT2 and ANT4 in cells infected with HSV-1. ANT proteins

(ANT1 to ANT4) are present in the mitochondrial inner membrane, where they are integrated into PTPC and VDAC. Additional investigation is necessary to clarify the roles of UL7, particularly its potential role as an apoptotic regulator.

Herpes simplex virus UL12.5 destroys mitochondrial deoxyribonucleic acid

The HSV-1 UL12 gene codes for two separate proteins: UL12 and UL12.5. The former is an alkaline nuclease, while the latter is a 500-amino acid polypeptide truncated at the N-terminus, missing the initial 126 residues of UL12. UL12 is recognised for its significant function in viral DNA replication and processing. UL12.5 possesses nuclease and strand-exchange capabilities; however, in contrast to UL12, it does not accumulate in the nucleus. UL12.5 is mostly localised to mitochondria, where it initiates extensive destruction of mitochondrial DNA at the onset of HSV replication. Furthermore, it has been proposed that UL12.5 functions directly within the mitochondrial matrix to destroy mitochondrial DNA via its nuclease capabilities. The biological significance of mitochondrial DNA loss during infection is, however, ambiguous.

US3 suppresses apoptosis

The genomes of HSV-1 and HSV-2 encode a viral serine/threonine protein kinase, US3, which is conserved across α-herpesviruses. US3 is not critical for viral replication in cell culture, although it plays a significant role in multiple facets of HSV replication and pathogenesis. The expression of US3 significantly influences many signal transduction pathways, including c-Jun N-terminal kinase (JNK) and protein kinase B (PKB)/Akt pathways and is responsible for the suppression of mitochondrial electron transport in cells infected with HSV-1. The US3 protein kinase causes phosphorylation of numerous cellular and viral proteins in infected cells. US3 exhibits significant antiapoptotic action in both transfected and infected cells. For instance, US3 effectively inhibits apoptosis triggered by the overexpression of Bad, Bid and Bax. The findings indicate that US3 demonstrates antiapoptotic effects through direct and/or indirect interactions with Bcl-2 family proteins. The exact mechanism through which US3 exerts its antiapoptotic effects is not yet completely understood. We recently reported that a US3-inactivated HSV mutant exhibited selective and robust oncolytic activity against pancreatic tumour cells, indicating that the deletion of US3 is a viable strategy for the creation of oncolytic HSV.

Viral proapoptotic protein

Viral antiapoptotic protein

Mitochondrial DNA and antiviral immune responses

Mitochondria are important organelles providing energy for cellular life. In addition, to supplying energy, they have been demonstrated to regulate programmed cell death pathways and govern innate immunity. Numerous studies indicate that mitochondrial machinery activates cellular inflammation, particularly by triggering antiviral signalling pathways.

Consequent to infection, viruses trigger pathogen-associated molecular patterns, such as dsRNA. Numerous investigations have clarified the biological function of I_B kinase-I and TBK1 as essential kinases for the phosphorylation and activation of IRF7 and IRF3, both in vitro and in vivo. Subsequently, it was demonstrated that cytosolic RNA sensors, such as MDA5 and RIG-I, are crucial for the induction of antiviral immunity. These RNA sensors can identify RNA from viral species and activate mitochondrial-based antiviral signalling, leading to the production of cytokines (IFNs: types I and III) to inhibit viral replication. Increasing data indicate that mitochondrial DNA, alongside mitochondria, has a role in antiviral immunity. Mitochondrial DNA serves as an intrinsic trigger for antimicrobial (antiviral) signalling, and its regulation of homoeostasis works in conjunction with conventional virus detection mechanisms to fully activate innate antiviral immunity. This section elucidates the role of mitochondrial DNA in triggering antiviral immunity. The innate immune system of eukaryotes employs pattern recognition receptors to identify pathogen-associated molecular patterns, such as viral nucleic acids. RIG-I and toll like receptor 7 (TLR7) detect viral RNA, such as that from the influenza virus, whereas NLRP3 identifies intracellular ionic fluctuations following influenza virus infection. The ion channel activity of viroporins, such as EMCV 2B or the M2 protein of the influenza virus, is crucial for the activation of the NLRP3 inflammasome. Viral infection causes mitochondrial DNA to leak into the cytoplasm, hence activating antiviral immunity. A recent study delineated the mechanism by which encephalomyocarditis virus and influenza virus promote mitochondrial DNA release into the cytosol through the viroporin activity of their transmembrane pore-forming proteins. The resulting cytosolic accumulation of mitochondrial DNA activates innate immune signalling pathways mediated by cyclic GMP-AMP synthase (cGAS) and DDX41.

Furthermore, the STING pathway-dependent antiviral signalling is augmented in adjacent cells through gap junctions. The perturbation in the intracellular ionic environment induced by viroporin activity is associated with the efflux of Mn^{2+} from membrane-bound organelles, such as mitochondria and the Golgi apparatus; thus the function of viroporins may be essential for increasing cGAS sensitivity to dsDNA. Upon infection by DNA viruses, the cytosolic ion concentration in host cells is elevated, which immediately activates the cGAS/STING signalling pathway. A separate study proposed that mitochondrial DNA stress serves as a cell-intrinsic inducer of antiviral signalling and claimed that cellular regulation of mitochondrial DNA homoeostasis works in conjunction with established virus detection mechanisms to comprehensively activate innate antiviral immunity. Mitochondrial DNA stress produced by viruses, such as herpesviruses, leads to robust antiviral immunity. The authors illustrated that mild stress resulting from a deficit in mitochondrial DNA-binding protein of transcription factor A mitochondrial, activates cytosolic antiviral signalling to augment the expression of certain IFN-induced genes. They elaborated that abnormal mitochondrial DNA packaging facilitates the release of mitochondrial DNA into the cytosol, where it engages with cGAS and enhances STING-IRF3-dependent signalling, thereby augmenting IFN-induced gene transcription, amplifying type I IFN responses and providing extensive viral resistance. This work indicated that herpesviruses induce mitochondrial DNA stress, which enhances type I IFN responses and antiviral signalling during infection. Numerous studies have elucidated this phenomenon, indicating a substantial increase in mitochondrial DNA in the cytoplasm of transcription factor IID (TFD) cells under stress conditions (e.g. microbial infection). Confocal and electron microscopy of these cells revealed interconnected, elongated networks of mitochondria, indicating mitochondrial hyperfusion, as mitochondrial fission promotes the leakage of mitochondrial DNA into the cytosol. The transduction of bone marrow-derived macrophages and mice embryonic fibroblasts with replication-incompetent retroviruses encoding the mitochondria-targeted HSV-1 UL12 M185 gene product demonstrated mitochondrial hyperfusion, nucleoid expansion and mitochondrial DNA depletion. Moreover, in these cells, herpesvirus infection activated antiviral immunological signalling by promoting the expression of IFN-stimulated genes, type I IFN and phospho-TBK1. Moreover, in recent decades, the global prevalence of the dengue virus has significantly escalated, and it is now endemic in over 100 nations. The prevailing model regarding cellular antiviral molecular processes in dengue virus infection predominantly emphasises RNA sensors, such as RIG-I and MDA5. In this framework, RNA viruses directly engage RIG-I and activate downstream IRF3 signalling, therefore augmenting IFN-I production. Recent investigations indicate that cGAS possesses significant antiviral effects against many positive-strand RNA viruses, including members of the *Flavivirus* family. Nonetheless, negative-strand RNA viruses are excluded, indicating that the distinctive characteristics noted in the life cycle of positive-strand RNA viruses, such as significant internal membrane rearrangements, may provoke a mislocalization of DNA that could serve as a substrate to activate the cGAS/STING signalling pathway. For instance, cGAS-deficient mice exhibited increased rates of infection and mortality relative to wild-type mice infected with the +ssRNA West Nile virus. This infectious agent likely prompts the release of mitochondrial DNA into the cytosol, which is subsequently detected by cytoplasmic nucleic acid sensors, so activating the immune system to eliminate the infection.

Upon entrance into the target cell, the dengue virus translates its polypeptide as one of the early biochemical processes, rapidly colonising the ER membrane to establish the scaffold necessary for reproduction. The aggregation of these viral proteins also extends to the mitochondrial membrane, which is adjacent to calcium exchange with lipid production and the ER. Consequently, dengue virus infection significantly affects this essential cellular organelle, which serves as a crucial centre for innate immune signalling. Research indicates that the C-terminus of the dengue virus M protein directs itself to the mitochondrial membrane, resulting in matrix swelling, permeabilization and possible loss of mitochondrial membrane integrity. Moreover, it has been proposed that the NS2B3 protease, located in the OMM, cleaves MFN1 and MFN2, hence affecting mitochondrial function and structure during viral infection. A recent study demonstrated that cGAS can detect mitochondrial damage during dengue virus infection to initiate type I IFN responses. Mitochondrial DNA can cause activation of the IFN— promoter via cGAS, acting as a danger-associated molecular pattern, which is released during dengue virus infection. Moreover, the dengue virus employs active biological processes to inhibit the cGAS/STING signalling system by targeting two essential host proteins involved in the detection of cytoplasmic RNA and DNA, namely, cGAS and STING. Inhibiting cGAS diminishes STING activation in infected cells and decreases the transfer of cyclic GMP-AMPs (cGAMPs) to neighbouring cells via tight junctions. Hepatocytes from patients with nonalcoholic steatohepatitis exhibit leakage of mitochondrial DNA as microparticles, suggesting that the release of mitochondrial DNA via extracellular vesicles is a viable mechanism for maintaining cellular homoeostasis. Certain virus-infected cells have demonstrated the capacity to produce extracellular vesicles in substantial quantities, which have been associated with antiviral immune responses. Numerous biological mechanisms have been documented concerning the production of these extracellular vesicles following viral infection. For instance, the tiny Rab GTPase is demonstrated to govern the generation of extracellular vesicles, and certain proteins seem to be significant modulators of extracellular vesicle formation in virus-infected cells. The CMV infection elevated the levels of Rab27a, which correlated with CMV production. The HSV-1 utilises the Rab27a protein for its transport, intracellular movement and exocytosis. The interplay between Rab GTPase and the virus governs both viral generation and the assembly of extracellular virus particles. The alternative mechanism posits that the extracellular development in virus-infected cells relies on tetraspanin-dependent signalling pathways.

A recent study revealed that HSV-1 can stimulate the production of CD63 positive extracellular vesicles without altering TSG101 exocytosis or Alix, suggesting that the viral infection activates Endosomal sorting complex required for transport (ESCRT)-independent signalling pathways for extracellular vesicle formation. It was proposed that the cGAS/STING signalling pathway is associated with the IFN-stimulated gene expressions mediated by extracellular vesicles of KSHV, with mitochondrial DNA present on the surface of these vesicles serving as a causative factor. The scientists proposed that DNA-containing extracellular vesicles from cells infected with KSHV could initiate the immune response to the virus, which may be essential for comprehending the milieu of virus-infected cells. This study could not fully elucidate the processes by which Kaposi's sarcoma-related herpesvirus extracellular vesicles mediate responses of IFN-stimulated genes, nor the associated components and biological pathways.

Numerous enquiries persisted, primarily on whether the DNA within the extracellular vesicles exerts the same effect as the DNA located on the extracellular vesicles. Secondly, the precise mechanisms of extracellular vesicle-mediated antiviral responses and their physiological importance in vivo require more investigation. The interaction between T cells and antigen-presenting dendritic cells leads to T cell activation and appears to influence the functional capacity of antigen-presenting dendritic cells.

A recent study showed that T cell antigen-presenting dendritic cells are facilitated by exosomal DNA transfer, underscoring a specific role for antigen-dependent interactions in safeguarding antigen-presenting dendritic cells from pathogen infection. This study additionally offered evidence concerning the method of information transfer from T cells to antigen-presenting dendritic cells. This indicates that following the interaction of these cells, signals are conveyed from the T cell to the antigen-presenting dendritic cells via extracellular vesicles containing mitochondrial and genomic DNA, thereby activating antiviral immune responses through the cGAS/STING cytosolic DNA-recognition signalling pathway and the transcription of IRF3-dependent IFN-regulated genes. This work demonstrated that antigen-bearing dendritic cells treated with extracellular vesicles exhibit increased resistance to future viral infections, indicating that this biological interaction may enhance antiviral immunity in these dendritic cells.

Interleukin-1

A proinflammatory cytokine plays a crucial biological role in numerous physiological processes, including immunological responses, apoptosis, inflammatory responses, cell cycle regulation and central nervous system function. Various triggers trigger the expression of interleukin-1; for instance, its synthesis is particularly responsive to microbial infection. A recent study proposed a new role for interleukin-1 in augmenting cell-intrinsic immunity, highlighting the crucial participation of the cGAS/STING pathway in assimilating microbial and inflammatory signals for host defence. Exogenous interleukin-1 increases the activation of IFN regulatory factor 3 in human fibroblasts, myeloid cells and epithelial cells.

The activation of IFN regulatory factor 3 by interleukin-1 is contingent upon the adaptor of the DNA-recognising signalling pathway, which induces IFN genes, via the detection of cytosolic mitochondrial DNA by cGAS. To ascertain if interleukin-1 therapy can augment mitochondrial DNA leakage, the authors conducted immunoblast analysis utilising the cytosolic contents. Researchers have analysed the concentration of nuclear and cytosolic DNA fractions through quantification of mitochondrial genes: MT-CO2, D-loop, MT-ATP6 and MT-ND1, as well as the nuclear gene: RPL13A. The analyses validated the release of mitochondrial DNA subsequent to interleukin-1 treatment. The authors suggested that the augmentation of mitochondrial bulk and the decline in mitochondrial membrane potential are particular mechanisms contributing to mitochondrial DNA leaking. This study shows that the administration of interleukin-1 can augment IFN synthesis and its activation signalling, hence facilitating an effective innate immune response that restricts dengue virus infection.

Here we discuss some of the studies we conducted in our lab, that explore the innate immune profile for mitochondrial gene and other associated genes regulating innate immunity against two important viral diseases in animal models, such as duck plague (caused by *Alphaherpesviridae*) and avian influenza (H5N1 strain).

Role of mitochondrial genes against viral infection (avian influenza)—a case study

We have conducted trials on chicken and duck models with challenge studies with avian influenza virus (H5N1) in an ABSL3 lab.

Copy number variation for mitochondrial gene

We observed more copy number for mitochondrial DNA in healthy ducks compared to that of infected ones, as revealed through increased expression of mitochondrial cytochrome B gene (see Fig. 6.7).

Fig. 6.8

Better expression of cytochrome B indicates more mitochondrial copy number in healthy (resistant) ducks compared to infected ones.

Cytochrome C is an important component of mitochondria involved in mitochondrial function, including oxidative phosphorylation and others, detailed in Chapter 2. Similarly, we have also observed better expression of the cytochrome C gene (a mitochondrial protein, nuclear encoding gene) in healthy individuals in contrast to avian influenza-infected ducks.

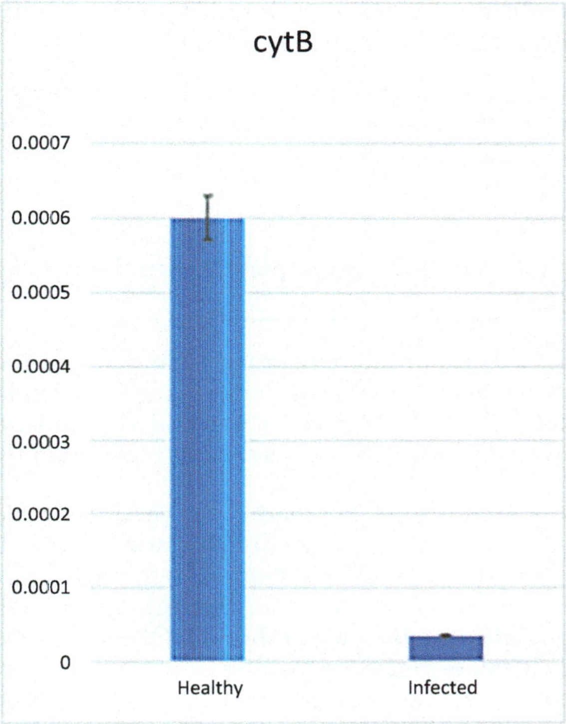

FIGURE 6.7 Expression profile for cytochrome B in healthy control versus avian influenza (H5N1 strain) infected ducks.

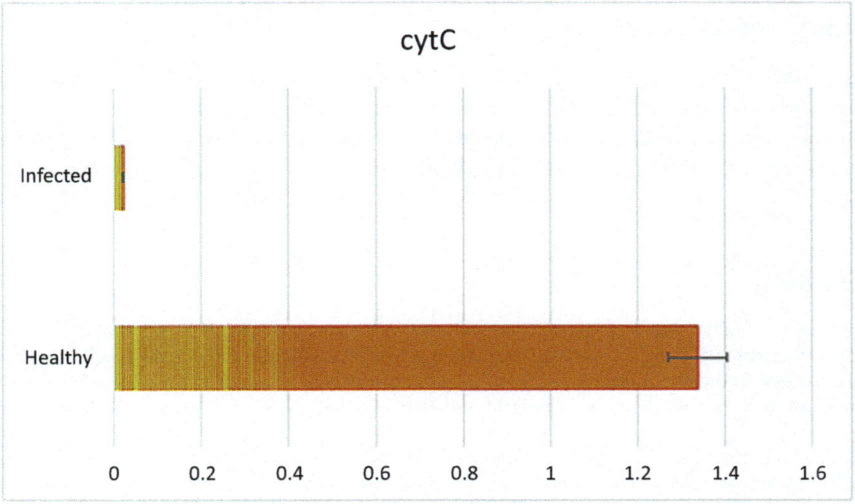

FIGURE 6.8 Expression profile for cytochrome C in healthy control versus avian influenza (H5N1 strain) infected ducks.

Single nucleotide polymorphism identification for mitochondrial genes with reference to avian influenza susceptibility versus resistance

Whole mitochondrial genome sequencing revealed single nucleotide polymorphisms (SNPs) at the cytochrome B gene and D-loop, leading to the ducks being more susceptible to avian influenza. The details have been discussed in Chapter 16.

Role of mitochondrial genes against viral infection (duck plague virus)—a case study

In another similar experiment, we conducted in our lab, where we have detected the potential role of mitochondrial genes conferring immunity against duck plague infection in Bengal ducks. Duck plague (also known as duck viral enteritis) is a worldwide disease caused by AnHV-1 of the family *Herpesviridae* that causes acute disease with high mortality rates in flocks of ducks, geese and swans.

In the rural sector, there was an outbreak of duck plague. But there exists variation in the susceptibility of the duck, some of the ducks were infected and mortality was observed, while some others were healthy. We confirmed the cases of infected ducks through PCR-based detection of duck plague virus and estimation of viral load through QPCR. We observed a significant role of copy number variation as determined through QPCR with the cytochrome B gene, where cytochrome B was better expressed in healthy (resistant) ducks in comparison to infected ones. Similarly, better expression of cytochrome C was observed in healthy (resistant) ducks.

SNP identification for mitochondrial genes with reference to duck plague susceptibility versus resistance

We also identified SNPs in mitochondrial genes as revealed through both Sanger sequencing and whole mitochondrial genome sequencing. We find an association of these SNPs with the susceptibility to duck plague infection in an animal model (duck).

The details regarding whole mitochondrial genome sequencing have been discussed in Chapter 16.

Further reading

Debnath, M., Chakraborty, A., Pal, A. 2025a. Whole mitochondrial genome sequencing reveals the role mitochondrial genes affecting host resistance against avian influenza in Bengal duck. In Proceedings on National symposium on "Reinforcing BioSecurity: Paving the way for a safer One Health Environment". Jan, 2025. venue: West Bengal University of animal and fishery sciences, 37, K.B. Sarani, Kolkata-37. West Bengal, India.

Debnath, M., Chakraborty, A., Pal, A. 2025b. The role of mitochondrial genes affecting host resistance against avian influenza in Haringhata Black chicken has been revealed through Whole mitochondrial genome sequencing. In Proceedings on National symposium on "Reinforcing BioSecurity: Paving the way for a safer One Health Environment". Jan, 2025. venue: West Bengal University of animal and fishery sciences, 37, K.B. Sarani, Kolkata-37. West Bengal, India.

Pal, A., Debnath, M., Chakraborty, A., Hassan, Q. 2025. Copy number variation of mitochondrial DNA has a potent role in conferring immunity against avian influenza. In Proceedings on National symposium on "Reinforcing BioSecurity: Paving the way for a safer One Health Environment". Jan, 2025. venue: West Bengal University of animal and fishery sciences, 37, K.B. Sarani, Kolkata-37. West Bengal, India.

Pal, A., Pal, S. 2025. Mitochondrial genes have role in innate immunity against duck plague infection in duck model. In Proceedings on National symposium on "Reinforcing BioSecurity: Paving the way for a safer One Health Environment". Jan, 2025. venue: West Bengal University of animal and fishery sciences, 37, K.B. Sarani, Kolkata-37. West Bengal, India.

CHAPTER

7

Mitogenomics and bacterial disease

Introduction

Bacterial diseases are quite prevalent in both humans and animals. AMR (antimicrobial resistance) ceatess a major threat in current days for prevention of diseases. Although a lot of studies have been undertaken with host immunogenomic studies against bacterial diseases. However, so far we were mostly concentrated on nuclear genes, particularly immune response genes of the host. In this current chapter, we will focus on mitochondria and mitochondrial genomics regulating bacterial infections.

We will attempt to address the following open questions. How do bacteria attack the host? What are the changes occurs in mitochondria? What is the mechanism of action of mitochondrial DNA (mtDNA) affecting immunity against bacterial infection? Is there any interaction between mitochondrial gene with nuclear genes in innate immunity?

As we have mentioned, bacteria have been classified based on their shape. They interact with host nuclear as well as mitochondrial genomes in different ways.

In the coming section, we will discuss the role of various kinds of bacteria and their possible mode of interaction with host mitochondria. We will discuss the bacterial pathogens infecting both humans and animals, along with the zoonotic diseases.

Classification of bacteria based on their basic shapes:

spherical (cocci),
rod (bacilli),
spiral (spirilla),
comma (vibrios) or
corkscrew (spirochaetes).

Their occurrence may be as single cells, in pairs, chains or clusters. Bacteria are abundant in every habitat on Earth: soil, rock, oceans and even Arctic snow.

In living organisms (in humans and other animals), bacteria mostly exist in two different forms:

Probiotics-beneficial microbes

Pathogenic microbes

In recent days, a lot of research is being undertaken on the role of probiotics against pathogenic microbes and their use in disease control. Even individual variability of host confering natural resistance or species-wise host resistance has been observed against

Mitochondrial Genomics
DOI: https://doi.org/10.1016/B978-0-443-26655-3.00017-2

pathogenic microbes (Pal et al., 2025). Ducks were observed to have more probiotics in comparison to chickens; likewise, ducks have a lower count of pathogenic microbes in comparison to chickens in the study conducted through metagenomic studies (Pal et al., 2025).

Mitochondrial damage-associated molecular patterns (DAMPs) and reactive oxygen species can stimulate antibacterial and antiviral innate immune responses in mammals. In addition to several subcellular organelles, the mitochondrion contains its genetic material: mtDNA. mtDNA is a double-stranded molecule encoding tRNAs, rRNAs and subunits of the respiratory chain. The mtDNA possess a significant degree of similarity. Similar to bacterial ancestors, it features a circular loop and contains a significant quantity of unmethylated DNA in the form of cytidine-phosphate-guanosine (CpG) islands. Numerous lines of evidence suggest that mtDNA plays a role in the control of innate immunity by activating immune signalling pathways or triggering diseases. Recently, the role of mtDNA in antimicrobial immunity has garnered global research interest. Mitochondria typically exist as interconnected networks, located even surrounding cardiac myofibrils and skeletal muscle. Nevertheless, the tiny spheroids that break the mitochondrial network undergo mitophagy, perhaps facilitating the exposure of mtDNA into the cytosol. The liberation of mtDNA from mitochondria under stress conditions, such as microbial infection and its interactions with DNA sensors, including toll-like receptor 9 (TLR9), cGAS and the inflammasome within a cell, are critical events. The influence of bacterial infection on mtDNA release and its effects on subsequent signalling pathways restrict microbial infection. Understanding the correlation between microbial infection and the release of mtDNA, along with the ensuing immunological responses, is crucial for mitigating infection.

Mitochondria and innate immune signalling postmicrobial infection:

Mitochondria are a vital component of eukaryotic cells, serving as the cell's powerhouse by supplying energy for regular physiological operations. Mitochondria are essential for various interconnected anabolic and catabolic processes, including glycolysis, oxidative phosphorylation, fatty acid β-oxidation and the tricarboxylic acid cycle. Moreover, mitochondria have been demonstrated to participate in cellular signalling networks that regulate innate immunity, inflammatory responses, cell survival and apoptosis, as well as calcium signalling. Mitochondria participate in redox signalling by generating H_2O_2 during cellular oxygen deficiency, which stimulates the activation of transcription factors, such as hypoxia-inducible factor, essential for metabolic adaptation in low oxygen conditions. Subsequently, numerous investigations indicated that mitochondrial reactive oxygen species also modulate other signalling pathways (e.g. NF-κB) and this mechanism seems to be a means of the interaction between mitochondrial function and other cellular processes for the preservation of homoeostasis. Increasing evidence indicates that mtDNA modulates immunological processes through its interaction with and regulation of TLR9, inflammasome and stimulator of interferon genes (STING) pathways. This discussion focuses on the existing understanding of mtDNA interactions with the aforementioned components and the resulting signalling cascade. mtDNA and TLR9 is a protein receptor encoded by the TLR9 gene. This protein receptor is expressed in various varieties of immune cells, namely, macrophages and dendritic cells. Moreover, it has been demonstrated that TLR9 resides in the endoplasmic reticulum (ER) of immune cells and is translocated to the endosome following the detection of unmethylated 20-deoxyribose CpG DNA, a characteristic of microbial (e.g. bacterial) DNA. mtDNA, similar to the microbial genome, consists of unmethylated CpG dinucleotides. This DNA functions as a damage-associated molecular pattern, representing endogenous molecules generated by cells experiencing abnormal cell death and has the capacity to activate the host's innate immune response. Class A, B and C ODNs are synthetic DNA nucleotides containing unmethylated CpG sequences akin to bacterial and mtDNA; nonetheless, they may exert varying stimulatory effects on TLR9 activation. Numerous studies indicate that TLR9 detects several types of ODNs: class A ODNs activate plasmacytoid dendritic cells, whereas class B CpG ODNs stimulate B cells. The majority of research illustrated the interaction between mtDNA and TLR9 utilising class A ODNs. In response to bacterial infection or mtDNA leakage, TLR9 detects CpG DNA and initiates immune responses by activating downstream signalling pathways, including NLRP3 inflammasomes, interferon regulatory factors (IRF)-dependent type 1 interferon (IFN) and pro-inflammatory nuclear transcription factor-kB. Multiple lines of evidence indicate that genetic deletion of TLR9 confers protective effects on animal health across diverse tissue injury models. These findings consistently demonstrate a strong association between TLR9 signalling and mtDNA–mediated inflammatory responses. For instance, TLR9-deficient mice exhibited enhanced survival rates in a necrotic lung model characterised by cationic nanocarrier-induced necrosis and mtDNA release. The authors additionally indicated that mtDNA-induced pulmonary inflammation is significantly diminished in MyD88 and TLR mice, highlighting the importance of the TLR9–MyD88 signalling pathway. A further investigation demonstrated that intravenous delivery of mitochondrial debris rich in mtDNA elicited a systemic inflammatory response in wild-type mice, which was significantly diminished in TLR9-deficient animals. Moreover, other independent investigations conducted under diverse experimental conditions revealed that reduced levels of circulating mtDNA positively influence survival in TLR9-deficient mice. A study utilising a mouse model of nonalcoholic steatohepatitis (NASH) indicated that mtDNA from NASH hepatocytes elicited heightened TLR9 activation in comparison to normal livers. This information signifies the correlation between mtDNA and TLR9. This connection triggers the innate immune system and enhances the subsequent inflammatory response. Moreover, the expression level of TLR9 appears to correlate with the concentration of cytosolic mtDNA, indicating that elevated mtDNA release stimulates greater TLR9 expression. The relationship between mtDNA–TLR9 and the subsequent activation of signalling pathways has been extensively investigated in neutrophils. mtDNA stimulates p38 mitogen-activated protein kinase (MAPK) via TLR9, accompanied by the release of matrix metalloproteinases 8 and 9 in neutrophils. A further study found the elevation of mitochondrial membrane potential-9 (MMP-9) and phosphorylated p38 following mtDNA treatment of neutrophils. Furthermore, it was observed that pretreatment with oligodeoxynucleotide (ODN) 2088, a TLR9 inhibitor, inhibited the p38 MAPK pathway.

Activation and MMP-8 efflux:

Another study reveals that intratracheal administration of mtDNA can induce lung inflammation via the TLR9–p38 MAPK pathway. Simultaneously, reports indicate that hip fractures in rats lead to the release of mtDNA. Enhanced activation of TLR9 and NF-κB p65 leads to pulmonary damage. A recent study reveals that the leakage of mtDNA during lung ischaemia-reperfusion results in TLR9-dependent development of neutrophil extracellular traps (NETs), leading to lung damage. Moreover, ascites mtDNA was associated with poorer progression-free survival and an increased likelihood of disease progression. mtDNA in ascites stimulates platelet and neutrophil responses that promote metastasis and inhibit antitumour immunity. The functional roles of additional MAPKs, such as extracellular signal-regulated kinases and c-Jun N-terminal kinases, are not well understood. These results clearly demonstrate that mtDNA activates neutrophils through TLR9 binding, which induces the MAPK signalling pathway, leading to the synthesis of MMP-8 and MMP-9.

Mechanism of mitochondrial DNA internalisation and its interaction with TLR9:

Mitochondrial DNA (mtDNA), normally sequestered within the double-membrane structure of mitochondria, must be released into the cytosol or, when present extracellularly, internalised by host cells to access endosomal compartments. Within endosomes, mtDNA engages TLR9, thereby initiating downstream signalling.

Currently, the mechanisms by which extracellular mtDNA is internalised remain ambiguous; nonetheless, phagocytosis, endocytosis, receptor-mediated endocytosis, and transmembrane (TM) diffusion are considered potential processes for mtDNA uptake. The TM diffusion mechanism appears unsuitable for this internalisation process, as DNA is extremely negatively charged, hindering its passage through the charged pores of the cell membrane. A further study indicated that phagocytosis may serve as an appropriate method, as macrophages originating from monocytes can transport intact mitochondria released during necroptosis. Moreover, numerous studies have demonstrated that the interaction of mtDNA with various cofactors, including high-mobility group box 1 and the receptor for advanced glycation end products, enhances the internalisation process within immune cells. High-mobility group box 1–CpG (class A) complexes induce TLR9/RAGE interaction and facilitate MyD88 recruitment in B cells. This potential biological mechanism requires additional evaluation. Viglianti and his colleagues also suggested an alternative route for mtDNA. Internalisation. It was proposed that the stimulation of autoreactive B cells by mtDNA (CpG DNA) happens subsequent to B cell receptor activation, resulting in the transference of mtDNA to endosomal TLR9.

mtDNA as a catalyst for IFN signalling,

The stimulator of interferon signalling- STING is a cytoplasmic DNA-sensing protein present in the ER. The STING pathway has been demonstrated to facilitate protective immune responses against infections caused by several DNA-containing microbial pathogens. mtDNA resembles microbial DNA and has garnered significant interest as a modulator of innate immunity through the STING pathway [49, 50]. The STING protein is triggered through direct contact with cytosolic DNA, typically originating from intracellular viruses, bacteria, or cyclic GMP–AMP synthetase. Upon activation, it prompts the translocation of IFN regulatory factor 3 to the cell nucleus, thereby stimulating the transcription of type I IFN genes and activating the NF-κB signalling pathway. Growing data reflect that mtDNA, exerts a major role in initiating the STING-mediated IFN response. The absence of apoptotic cysteinyl aspartate specific proteinases (caspases) (e.g. 3, 7 and 9) is responsible for the upregulation of type I IFN gene transcription. This biological phenomenon is linked to Bak-Bax, pro-apoptotic proteins that are part of the BCL2 family. These proteins facilitate the permeabilisation of the mitochondrial outer membrane and enhance the efflux of proteins linked to the intermembrane gap. Moreover, they participate in the release of mtDNA and cytochrome C, which activates the intrinsic apoptotic pathway. Typically, apoptosis does not elicit an inflammatory response. Nonetheless, it has been demonstrated that the suppression of caspases associated with the apoptotic process allows mtDNA to activate cGAS–STING-mediated type I IFN signalling.

mtDNA and the inflammasome:

The inflammasome is a cytoplasmic complex of multiprotein oligomers within the innate immune system that has a role in activating inflammatory responses. mtDNA directly targets them, resulting in the activation of caspase-1 and the maturation of downstream factors, including interleukin-18 (IL-18) and IL-1. To date, six different types of inflammasomes have been identified: NLRP1, NLRP3, NLRP6, NLRP12, AIM-2, and NLRC4-IPAF. Among these, the inflammasome NLRP3 has been studied extensively. Scientists proposed that the inhibition of autophagic proteins elevates the concentration of cytosolic mtDNA in macrophages in response to ATP and LPS. Moreover, the release of mtDNA is facilitated by reactive oxygen species originating from mitochondria, potentially linked to the levels of the NLRP3 inflammasome (also referred to as NALP3). The release of mtDNA is accountable for the production of IL-18 and IL-1β in response to LPS and ATP. The activation of mitochondrial reactive oxygen species has been demonstrated to correlate with increased generation of active IL-1β in a caspase-1 and NLRP3-dependent manner and treatment with mitochondrial reactive oxygen species scavengers mitigates this effect. Multiple investigations have independently established the necessity of mitochondrial-derived reactive oxygen species for the activation of NLRP3, potentially elucidated by their oxidative effect on mtDNA. Shimada and his colleagues have proposed that the oxidised variant of mtDNA imparts inflammatory potential to mtDNA. Mitochondrial reactive oxygen species initiate the oxidative process and facilitate the cytosolic transfer of oxidised mtDNA, which subsequently binds directly to NLRP3. Moreover, it has been demonstrated that nonoxidised mtDNA cannot activate the NLRP3 inflammasome, although it may stimulate IL-1β production via other inflammasomes. The genetic reduction of caspase-1 and NLRP3 results in reduced mtDNA leakage. A prior study investigated the molecular mechanism that enhances mtDNA translocation under conditions of mitochondrial stress, such as the presence of mitochondrial reactive oxygen species. Mitochondrial stress triggers apoptosis, wherein Bcl2 (BAK) and BAX promote mitochondrial outer membrane permeabilisation, resulting in the release of cytochrome c and the activation of apoptotic caspases. During this biological process, the mitochondrial network undergoes coordinated activity, wherein newly assembled BAK/BAX macropores facilitate the efflux of mitochondrial matrix constituents, including mtDNA, into the cytosolic compartment of the cell. The translocation of mtDNA depends on the activation of the NALP3 inflammasome, which ultimately induces the release of IL-18 and IL-1. Subsequently, scientists suggested that the inherent properties of NLRP3 may be responsible for the significant biological function of mitochondrial reactive oxygen species in the NLRP3-dependent cytosolic release of mtDNA. The NLRP3 mostly associates with oxidised mtDNA, subsequently stabilising cytosolic mtDNA after leakage, suggesting a positive feedback loop that activates the NLRP3 inflammasome. Oxidised mtDNA further promotes the leakage of mtDNA, hence stimulating innate immunity. Moreover, in murine macrophages, the liberation of mtDNA after cellular demise serves as a secondary signal that synergises with an additional inflammatory signal. LPS is utilised to trigger the synthesis of IL-1 and activate the NLRP3 inflammasome. Despite the involvement of many molecular mechanisms, including base excision repair and the mismatch repair pathway, in the repair of oxidised mtDNA, our understanding of the repair mechanisms within this organelle remains restricted. Moreover, in certain clinical circumstances (e.g. dementia), oxidative damage to mtDNA is significant. The repair mechanisms may be inadequate and mitochondria likely lack the advanced repair systems seen in the nucleus.

mtDNA and antimicrobial immunity:

In animals, the immune system is equipped with a variety of complementary molecular mechanisms to identify diverse molecules associated with microbial pathogens or their effects on host cells, thereby minimising the likelihood of infectious agents evading immune detection. mtDNA shows a great similarity to bacterial DNA and has been demonstrated to elicit antimicrobial immunity. This section delineates the functional significance of mtDNA as a catalyst for antiviral and antibacterial immunity.

Molecular mechanism for mitochondrial DNA to stimulate innate immune mechanism

Fig. 7.1

mtDNA and antibacterial immunity: Numerous lines of evidence indicate the role of mitochondria in antibacterial immunity through the generation of reactive oxygen species. Mitochondrial-derived reactive oxygen species play a vital role in macrophage-mediated antibacterial immune responses. Apart from the role of mitochondrial reactive oxygen species, as discussed earlier, increasing evidence indicates that mtDNA plays a role in antibacterial immunity. mtDNA, owing to similar ancestry, has characteristics akin to the bacterial genome, as it comprises unmethylated DNA and CpG islands. Accordingly, mtDNA acts as a pathogen-associated molecular pattern with similar functionality and may stimulate antibacterial immunity. This section examines the functional significance of mtDNA in the enhancement of antibacterial immunity (Fig. 7.2).

Following substantial activation signals, diverse immune cells secrete granular proteins and chromatin into the extracellular milieu, resulting in the formation of DNA-based extracellular traps. Extracellular trap generation is especially prevalent in neutrophils, while it also occurs in other immune cells such as basophils, eosinophils, mast cells and macrophages. Increasing data indicate that extracellular traps play a significant role in antibacterial and antifungal immune responses. Eosinophils in human host immune cells generate DNA-based extracellular traps, consisting of antimicrobial peptides, histones and DNA, to ensnare and eliminate microbial invaders. Another study indicated that eosinophils can release mtDNA that functions as an extracellular trap. The mitochondrial source of the spilt DNA was identified by a combination of molecular biology and microscopy approaches. The extracellular trap identifies a microbial pathogen, limits its movement and facilitates its demise. The enzymatic destruction of extracellular DNA renders eosinophils incapable of eliminating bacterial infections in the extracellular environment. Consequently, whereas DNA in extracellular traps is unlikely to be very harmful to bacterial pathogens, it is essential for the antibacterial effects facilitated by eosinophils. The formation

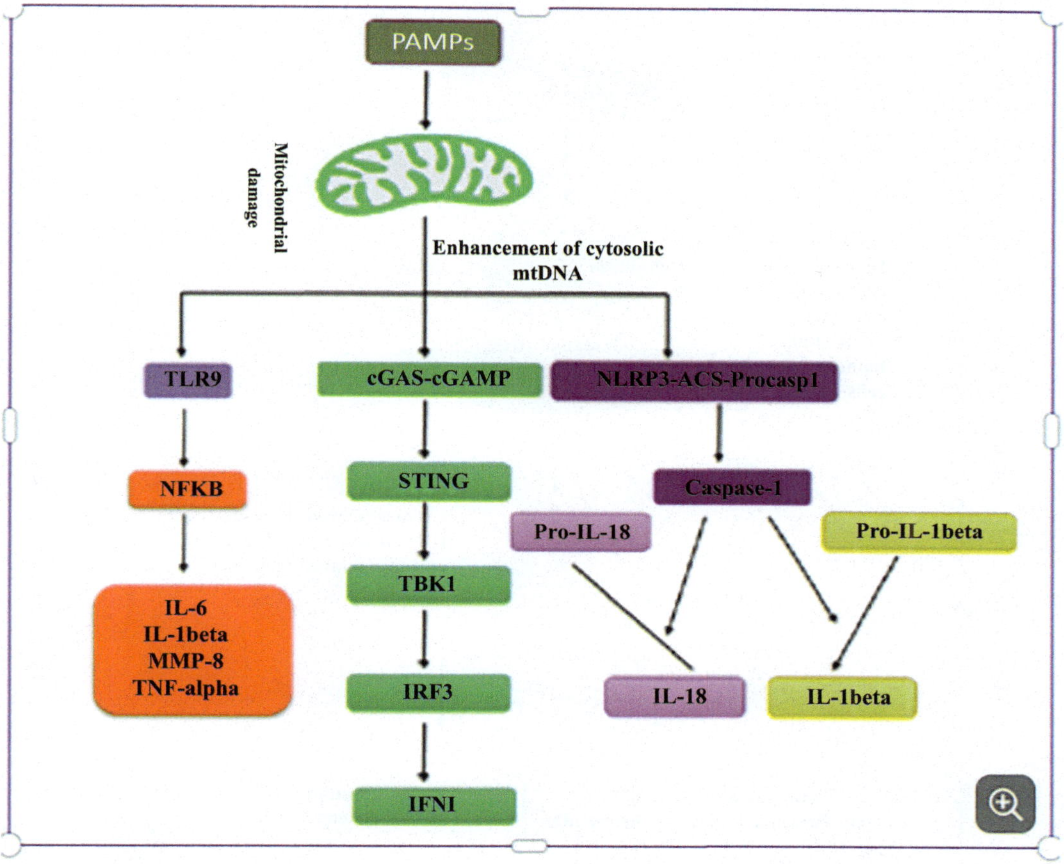

FIGURE 7.1 The schematic diagram represents the molecular mechanism through which mitochondrial DNA stimulates innate antimicrobial immunity. Under stress conditions, mitochondrial DNA is released from mitochondria, which stimulates the production of pro-inflammatory cytokines (e.g. IL6, IL-1β, mitochondrial membrane potential-8 (MMP-8) and TNF-α) and also enhances their release. The escaped mitochondrial DNA following the microbial infection is detected by the cGAS/stimulator of interferon genes (STING) signalling pathway that causes TANK binding kinase 1 (TBK1)/interferon regulatory factors 3 (IRF3)-dependent production of type I interferon (IFN) and also limits replication of microbes. However, the cysteinyl aspartate specific proteinases (caspases), for example caspase-3–7 and 9 activation regulate the intrinsic apoptotic signalling pathway. Furthermore, the escaped mitochondrial DNA after stress also stimulates the NLRP3 inflammasome to recruit and activate caspase-1, which is involved in the cleavage of pro-IL-18 and pro-IL-1β into their active forms.

of eosinophil extracellular traps is contingent upon the generation of reactive oxygen, similar to the extracellular traps produced by neutrophils and mast cells. In contrast to mast cells and neutrophils, eosinophils do not undergo apoptosis upon the release of mtDNA. Subsequent research on mouse and human basophils indicated the production of reactive oxygen species and extracellular traps composed of mtDNA following IL-3 priming in a NADPH oxidase-independent manner. Consequently, the activation of NADPH oxidase is considered a crucial intracellular signalling mechanism for the generation of extracellular

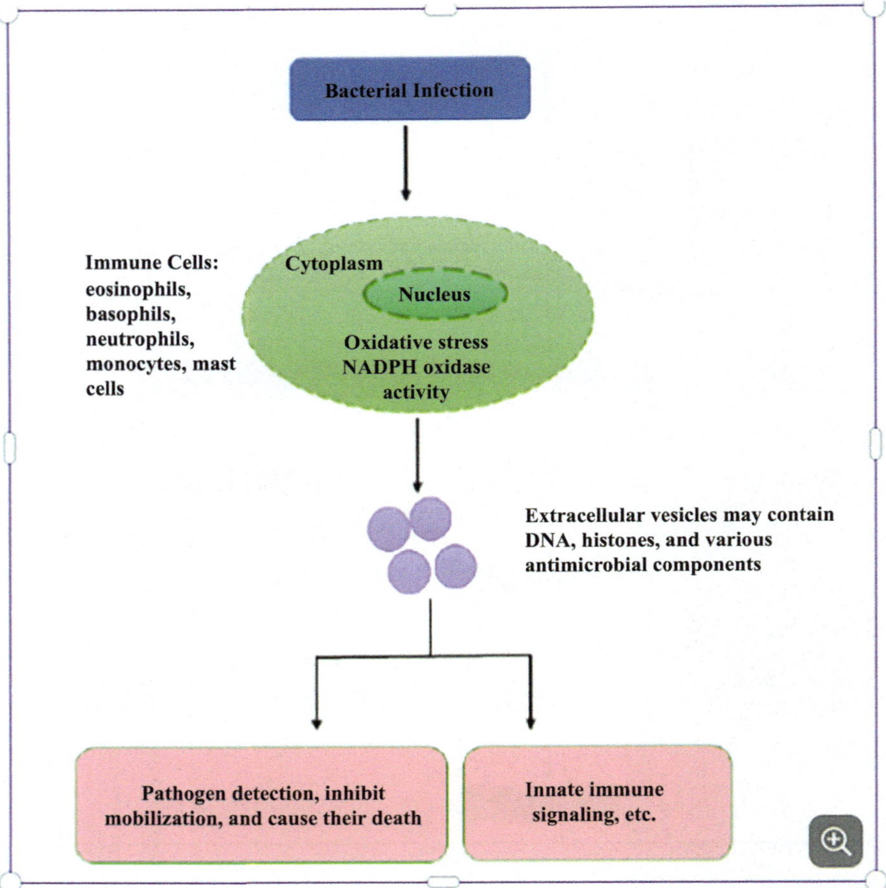

FIGURE 7.2 The schematic diagram represents the effect of bacterial infection on immune cells that causes the formation of mitochondrial DNA-based extracellular vesicles to limit bacterial infection. The extracellular vesicles directly bind to the pathogens and causetheir death. Furthermore, they are involved in the transportation of signals to the other immune cells, etc.

traps. Genetic abnormalities and pharmacological suppression of this NADPH oxidase are linked to diminished production of reactive oxygen species, hence impairing eosinophils' capacity to generate extracellular traps. Although basophils lack phagocytic capacity, they are capable of exerting antibacterial activity by releasing extracellular traps consisting of granular proteins and mtDNA.

The pathogen-induced mtDNA-derived extracellular vesicles in basophils and eosinophils play essential roles in antibacterial immunity. Moreover, a recent study indicated that eosinophils discharge extracellular DNA traps, which serve as essential structural components of eosinophil-dense secretions. This work demonstrated that eosinophil DNA traps captured both bacteria and fungal pathogens. In addition to eosinophils, neutrophils, monocytes and mast cells have been shown to create extracellular traps that contain DNA.

Panton–Valentine toxin from *Staphylococcus aureus* has been demonstrated to induce NADPH oxidase-independent development of extracellular traps in neutrophils, resulting in the production of nuclear blebs that are discharged as extracellular traps. Nonetheless, there is still debate concerning the origin of the DNA and the process of its leakage. Most researchers assert that the DNA in extracellular traps originates from the nucleus and requires cellular apoptosis. In neutrophils, the cell death that results in the creation of neutrophil extracellular DNA traps is referred to as netosis.

Nonetheless, many reasons warrant caution:

Research indicates that NETs can be generated by live neutrophils both in vivo and in vitro. The creation of NETs occurs significantly more swiftly than apoptosis or any other kind of programmed cell death that employs triggers recognised to impede apoptosis.

In numerous investigations, researchers generated neutrophils using phorbol 12-myristate 13-acetate, which is widely recognised for stimulating rapid necrosis akin to cell death. The DNA released into the milieu (extracellular space) of phorbol 12-myristate 13-acetate-induced neutrophils does not display the characteristic structure of extracellular DNA trap fibres. It is essential to note that the structures of normal neutrophils were formed prior to the onset of cell death. Cell death associated with the creation of NETs does not inherently indicate a unique kind of cell death. It is challenging to ascertain how a deceased cell can extrude DNA, given that this procedure seemingly necessitates energy. In addition to the aforementioned controversy regarding the origin of DNA and its leakage mechanism in neutrophils, the antibacterial function of extracellular DNA structures has primarily been investigated through NETs, which comprise histones, DNA and various antimicrobial components, including cathepsin G, proteinase 3, neutrophil elastase, defensins, bacterial permeability-increasing protein and myeloperoxidase. NETs, similar to eosinophil extracellular traps, contain a high concentration of granule proteins that can eliminate invading microbial infections. Additional data concerning the pathogenic importance of extracellular DNA traps have also been identified in numerous microbial diseases, including *Clostridium perfringens, Streptococcus pyogenes* and Streptococcus *aureus*. Invertebrates, akin to their vertebrate counterparts, also demonstrate a comparable phenomenon to DNA-based extracellular traps. A recent investigation in the invertebrate oyster *Crassostrea gigas* demonstrated that haemocytes produce and secrete H5-like and H1-like histones in response to tissue damage and microbial challenge and these histone variants exhibit antimicrobial activity. These antimicrobial histones are linked to extracellular DNA networks released by haemocytes and facilitate the elimination of microbial infections. The DNA released by haemocytes was observed to encircle and entangle vibrios. This immune defence mechanism resembles the extracellular traps of neutrophils recently documented in oysters. Extracellular traps in oysters have been observed in vivo among hemocyte-infiltrated interstitial tissues next to lesions, but unchallenged tissues lack these traps. The extracellular creation in oysters is significantly reliant on the production of reactive oxygen species by haemocytes, indicating that this process is governed by molecular and cellular pathways across both invertebrates and vertebrates. Moreover, extracellular development is an efficient method for controlling microbial infections in invertebrates. *Streptococcus pneumoniae* is the predominant bacterial pathogen responsible for invasive pneumococcal infections and can generate substantial quantities of H_2O_2. Scientists revealed that H_2O_2 could inhibit cell migration and hinder airway epithelial cell healing. A549 cells produced lactate dehydrogenase upon exposure to H_2O_2 and displayed a necrotic phenotype instead of undergoing programmed cell death. A recent study indicated that *S. pneumoniae* released H_2O_2, resulting in mitochondrial damage and significant histopathological injury in the lung tissue of mice. Mitochondrial injury induces the release of mtDNA into the cytosol, thereby activating the STING signalling pathway and promoting the transcription of type I IFNs, which function to restrict bacterial infection. This study presented various evidence concerning the activation of the STING signalling pathway and the production of type I IFNs, suggesting that this cascade may play a role in bacterial clearance; however, questions persist regarding whether this process facilitates bacterial clearance or exacerbates host cell apoptosis. Pneumolysin, a pore-forming toxin, is the principal virulence factor of Gram-positive bacteria (*S. pneumoniae*) and is believed to affect alveolar epithelial cells, thereby stimulating the immune system through the production of danger-associated molecular patterns. Researchers found that the pore-forming toxin pneumolysin, released by bacteria, may significantly disrupt ATP homoeostasis in cells and induce morphological alterations in the mitochondria of human alveolar epithelial cells in vitro. This toxin also induces an inflow of mitochondrial calcium and a reduction in MMP, leading to the opening of the permeability transition pore.

Mitochondria and the release of mitochondrial DNA through microvesicles without the activation of intrinsic apoptosis

mtDNA released in response to *S. pneumoniae* infection robustly activates the innate immune system via the STING signalling pathway, thereby facilitating the clearance of bacterial infection. Under stress conditions, especially during microbial infections, physiological alterations in mitochondrial membranes result in the leakage of mtDNA, which serves as an evolutionarily advantageous biological mechanism in the host, enhancing antiviral and antibacterial signalling in response to pathogen invasions. Nonetheless, the atypical accumulation of impaired mitochondria and the liberation of mtDNA into the cytosol may also contribute to certain inflammatory disorders.

MtDNA transfer between cells:

Recent studies have demonstrated that mtDNA is not confined to individual cells; rather, it can be transferred between cells, as evidenced by research conducted on mouse models and tumour models of tissue injury. A recent study indicated that mouse tumour cells devoid of mitochondria develop tumours solely after acquiring mtDNA from tumour cells within their microenvironment. Mitochondrial transport between cells has been documented during ischaemia-induced damage in the brain, heart and lung epithelium. Mitochondrial transfer between cells appears to be an evolutionarily conserved phenomenon associated with diseases characterised by impaired mitochondrial function, including neuromuscular, neurodegenerative and cardiovascular disorders, as well as ageing and cancer. Numerous biological pathways have been suggested concerning the transmission of mtDNA. Microvesicles and tunnelling membrane nanotubes have been demonstrated to facilitate mitochondrial transport between cells. An in vitro investigation indicated that cell fusion may possibly be a process involved in mitochondrial transfer. Moreover, the transfer of mitochondria between cells, which establishes junctions without necessitating total cellular fusion, cannot be disregarded. In dense tissue architecture, nanotubes or direct cell-to-cell connections, which involve loose junctions centred on components such as cytoskeletal and hemi-connexins, are probable. Connexin 43 has been documented to be present at the junction with nanotubes in several in vitro systems. Moreover, substantial microvesicles containing intact mitochondria with standard respiratory functions have been shown to be generated by astrocytes in cell culture. When introduced into ischaemic brain-damaged mice, these microvesicles were associated with injured neurones, suggesting their potential role in facilitating mitochondrial uptake in the murine brain. Different scenarios could arise based on whether these vesicles undergo endocytosis, subsequently releasing mitochondria into the cytoplasm instead of entering the lysosomal pathway, or whether they fuse with the cell membrane, so discharging their mitochondrial contents into the cytoplasm. In both conditions, the level of cytosolic mtDNA is increased, detectable by the nucleic acid sensors in the cytoplasm. The augmentation of mtDNA, in the cytoplasm may activate the innate immune system. The intercellular transfer of mtDNA and its consequences for innate antimicrobial immunity have been infrequently examined; future studies should prioritise this critical physiological phenomenon to enhance our understanding of the innate immune system. Significant advancements have been achieved in recent years in elucidating the basic mechanisms by which the innate immune system identifies microbial pathogens and integrates signals from pathogen-associated molecular patterns (PAMPs) to initiate a targeted immune response. Recent findings indicate that in conditions of cellular damage and stress, such as microbial infections, mtDNA is typically released into the cytoplasm rather than genomic DNA. This cytosolic mtDNA is essential in restricting microbial infections and contributes to the pathogenesis of different inflammatory disorders by activating signalling pathways. Numerous studies have demonstrated the activation of NLRP3, TLR9 and STING by mtDNA, highlighting its role in the control of innate immune signalling. Numerous intriguing enquiries persist regarding mtDNA and its biological function in innate immunity. Is there an undiscovered mtDNA signalling route that exists? What are the precise biological mechanisms involved in the leakage of mtDNA from the mitochondria? A study indicated that the release of mtDNA is facilitated by mitochondrial permeability transition, as evidenced by sensitivity to cyclosporine A, which is reasonable if mitochondrial permeability transition precedes osmotic swelling and rupture of the mitochondrial membrane. A recent study indicated that the breakdown of the mitochondrial membrane by secreted phospholipase A2 IIA (sPLA2-IIA) generates inflammatory mediators (such as mtDNA and lysophospholipids) that trigger leucocyte activation, resulting in various inflammatory responses. Moreover, many investigations indicate that certain viral proteins facilitate the release of mtDNA into the cytosol. This creates a necessity for a deeper understanding of mtDNA. During *M. tuberculosis* infection, mitochondrial damage results in the release of DNA from mitochondria, which is regarded as a significant source for binding with cGAS. Moreover, the DNA of *M. tuberculosis* is a primary source for the activation of this signalling cascade. Does mtDNA serve as a catalyst for the activation of signalling pathways during microbial infection and if so, does it influence both type I IFN production and selective autophagy? The role of mtDNA in bacterial autophagy is particularly intriguing due to the resemblance of this signalling route to mitophagy, as well as the historical truth that the mitochondrion was originally a bacterium. Additional research and testing are necessary to enhance comprehension of cellular damage outcomes and the role of host mtDNA in microbial infection. Additionally, an intriguing enquiry is to elucidate the process via which DNA-sensing molecules can enhance diverse results of microbial infection. For instance, regarding *M. tuberculosis*, cGAS, STING, and TANK binding kinase 1 (TBK1) are essential for autophagosomal targeting and type I IFN expression. How can these same molecules be concurrently involved in such diverse signalling pathways? One possibility is that posttranslational modifications facilitate protein–protein interactions that promote associations within complexes or at the phagosomal membrane, thereby enhancing the expression of type I IFN. Another possibility is that TBK1 assumes a critical biological role in polarising DNA, discerning outcomes based on its capacity to phosphorylate diverse substrates, including IRF3. The identification and characterisation of these substrates may be essential for comprehending how mtDNA recognition enhances both type I IFN expression and autophagy.

Role of mitochondrial DNA in the PAMP, DAMP and the inflammatory response

Fig. 7.3

During sepsis, bacterial DNA and during severe trauma, mtDNA, function as potent immunostimulatory ligands that signal through TLR9 to activate neutrophils via p38 MAPK signalling. In parallel, formylated peptides released from both bacteria and damaged mitochondria act as chemoattractants, directing neutrophil migration to sites of inflammation and tissue injury through formyl peptide receptor-1. Activation of these pathways contributes to excessive neutrophil recruitment and systemic inflammation, which can culminate in acute lung injury as a manifestation of the systemic inflammatory response syndrome. In this context, mitochondrial components function as DAMPs, whereas bacterial products represent PAMPs.

Pro-inflammatory signalling pathways engaged by mitochondrial DNA

Fig. 7.4

mtDNA can trigger various pro-inflammatory signalling pathways by endosomal localised TLR9 or via cytosolic cGAS–STING or via cytosolic inflammasome (AIM-2 or NLRP3).

Top: TLR9 binds mtDNA in the endosome, eliciting an NF-κβ dependant pro-inflammatory signalling programme.

Middle: cGAS recognises mtDNA in the cytosol and activatesER localised STING, triggering an IFN response.

Bottom: mtDNA-dependant inflammasome activity leads to caspase-1-dependant maturation or pro-inflammatory IL-1 and IL8.

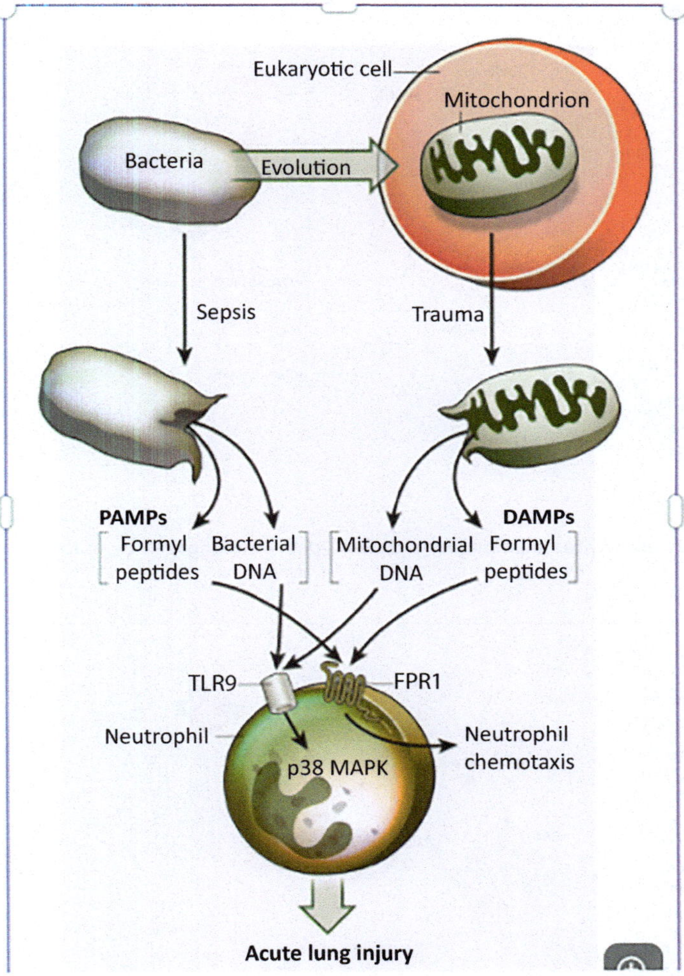

FIGURE 7.3 Pathogen-associated molecular patterns (PAMPs), damage-associated molecular patterns (DAMPs) and the inflammatory response.

mtDNA-dependant activation of cGAS–STING signalling

Fig. 7.5

Various mitochondrial stresses, including bacterial and viral infection, can lead to mtDNA release. Alternatively, activation of BAX and BAK leads to outer mitochondrial membrane permeabilisation and mDNA release. Once cytoplasmic, mtDNA can bind the DNA-sensing protein cGAS that catalyses the production of the secondary messenger 2′3′ cyclic GMP-AMP (2′3′cGAMP) from ATP and GTP. cGAMP binds the adaptor molecule STING on the ER, leading to activation of TBk1 kinase. Active TBK1 phosphorylates the transcription factor IRF3 initiating type I IFN response.

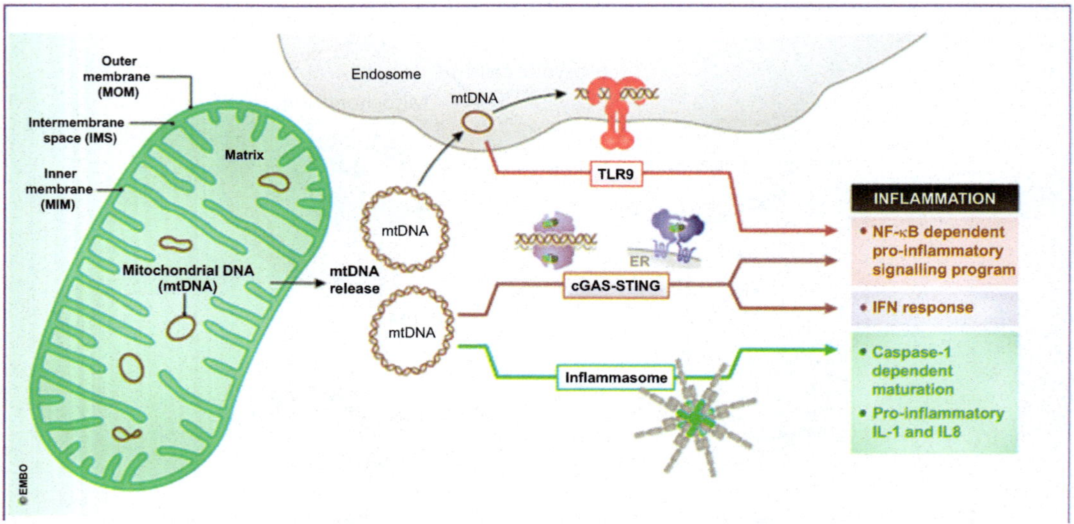

FIGURE 7.4 Overview of pro-inflammatory signalling pathways engaged by mitochondrial DNA.

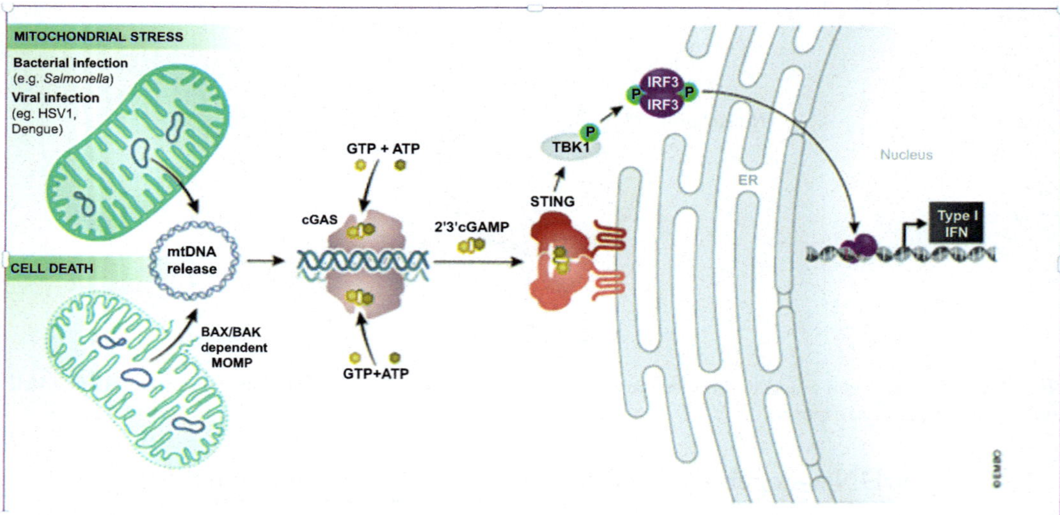

FIGURE 7.5 mtDNA-dependant activation of cGAS–STING signalling.

BAX/BAK-dependant initiation of inflammation

Fig. 7.6

Following a pro-apoptotic stress, BAX and BAK are activated leading to mitochondrial outer membrane permeabilisation. This enables the release of caspase- activating proteins

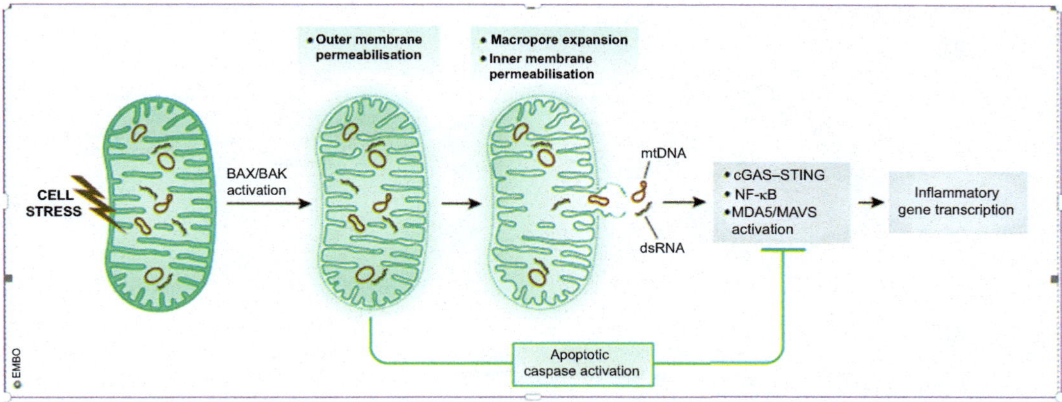

FIGURE 7.6 BAX/BAK-dependant initiation of inflammation. Following a pro-apoptotic stress, BAX and BAK are activated, leading to mitochondrial outer membrane permeabilisation. This enables the release of cysteinyl aspartate specific proteinases (caspase)- activating proteins from the mitochondrial intermembrane space. Following this, macropores form on the mitochondrial outer membrane causing extrusion and permeablization of the inner membrane. This enables the release of mtDNA. Mitochondrial double-stranded RNA (dsRNA) can also be released. Collective release of these molecule triggers inflammation via mitochondrial antiviral signalling protein (MAVS). cGAS–STING and NF-κβ. Caspase activity is antiinflammatory, in part, through direct cleavage and inactivation of inflammatory signalling molecules.

from the mitochondrial intermembrane space. Following this, macropores form on the mitochondrial outer membrane causing extrusion and permeablization of the inner membrane. This enables release of mtDNA. Mitochondrial double-stranded RNA can also be released. Collective release of these molecules triggers inflammation via mitochondrial antiviral signalling protein. cGAS–STING and NF-κβ. Caspase activity is antiinflammatory, in part, through direct cleavage and inactivation of inflammatory signalling molecules.

Noncell autonomous effects of mitochondrialDNA

Fig. 7.7.

Role of mitochondrial DNA against antibacterial immunity (against *Pasteurella multocida*) in duck model—a case study

Mitochondria are more than just the powerhouse of the cell—they are dynamic, semiautonomous organelles with critical roles in energy metabolism, calcium signalling, iron homoeostasis, apoptosis and innate immunity. While the mitochondrion contains 37 of its own genes, it depends on over 200 nuclear-encoded genes for its structural integrity, enzymatic activity and dynamic behaviour. This intimate cross-talk between mtDNA and nuclear genes forms a tightly regulated genomic axis whose disruption is increasingly

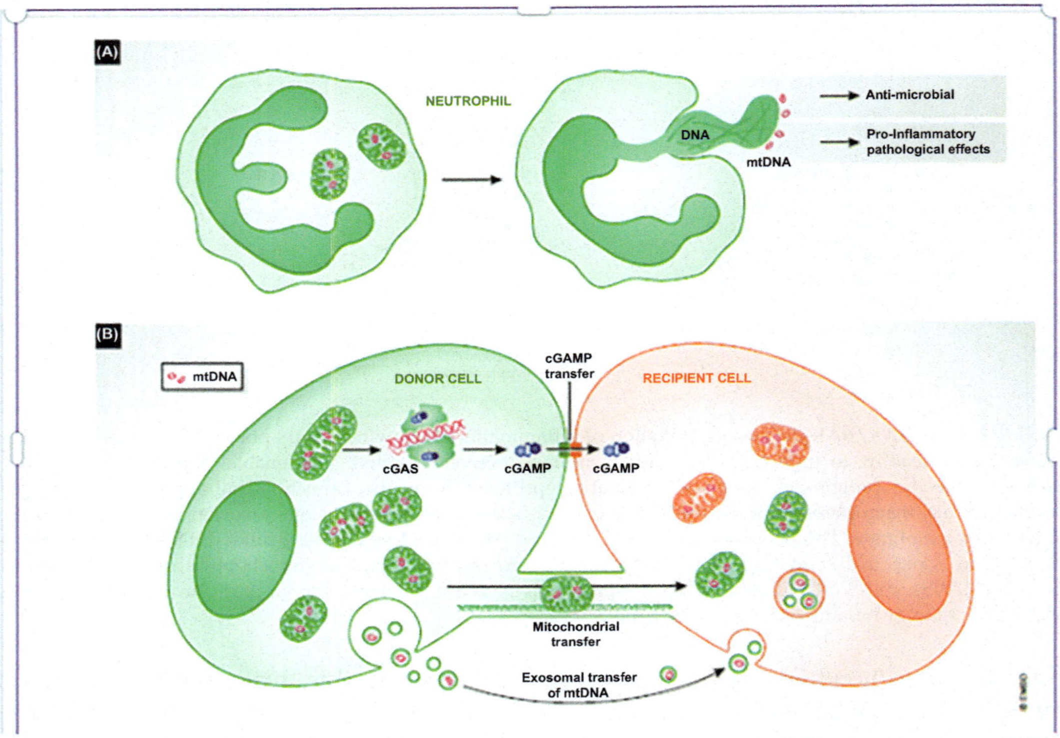

FIGURE 7.7 Noncell autonomous effects of mtDNA. Upon pathogen encounter, neutrophils can extrude DNA (both nuclear and mitochondrial) that forms an extracellular trap for extracellular microbes. Due to pro-inflammatory properties, these DNA neutrophil extracellular traps can also have pathological effects in diseases such as lupus. Mitochondrial DNA can transfer via exosomes or in intact mitochondria to neighbouring cells, impacting on the metabolism and survival of the recipient cell, inflammatory responses to mtDNA can also have noncell autonomous effects. The cGAS-induced secondary messenger cGAMP has been shown to transfer via gap junction, eliciting antiviral interferon responses in neighbouring cells.

recognised as a driver of numerous diseases across species. Despite its importance, the mitochondrial-nuclear dialogue remains underexplored. In this current study, we explored the cross-talk with complex interaction of mitochondria and nucleus regulates the immune response in duck Pasteurellosis (bacterial disease) in Anas platyrynchos as the model organism. We have an *in-depth* study of mitochondrial associated cytochrome B and cytochrome C genes, which have a crosstalk with nuclear NRLP3, IL-18 and STING genes through NOD-like receptor signalling pathway, Phagocytic pathway and Cytosolic DNA sensing pathway. Active binding sites for IL-18 with the bacterial pathogen, histological section and immunohistochemistry further confirm the finding. This is the first report of such cross-talk in an animal model through retrograde signalling of mitochondrial genes on the nucleus.

As we discussed in our earlier chapters, apart from direct mutation in mtDNA and mitochondrial copy number variations, mutations or expression profiling for genes coding other factors/proteins involved in mitochondrial replication, transcription and functions, such as mitophagy, mitochondrial fission and mitochondrial fusion, are equally important and responsible. Accordingly, we have studied three such genes, namely thymidine phosphorylase (TP), transcription factor A, mitochondrial (TFAM), thymidine kinase 2 (TK2), mitochondrial and identified their important role in response to innate immune response against bacterial infection (duck Pasteurellosis) in a duck model.

Case study 1: mitochondrial copy number variation affecting innate resistance against duck *Pasteurellosis*

Host immunity is important to understand host-pathogen interaction, which is a dynamic process between diverse pathogens and the host in all stages of pathogenic infection. Innate immunity is governed by a complex network of gene-encoded mediators and signalling proteins that function sequentially and cooperatively to initiate, amplify and regulate host defence responses. We consider them as immune response genes, regulating proline-rich region, pro-inflammatory and antiinflammatory cytokines. In our lab, we have reported the role of a certain set of nuclear genes in controlling disease resistance or animal health in sheep, cattle, buffalo, goat, chicken and duck models. Simultaneously, we have also reported the role of mitochondrial genes in controlling health and disease resistance, and even reproduction. Mitochondrial genes not only act through oxidative phosphorylation, but they also help in other series of metabolic processes, including calcium and iron metabolic pathways. We have already reported certain preliminary works involving all the 37 mitochondrial genes in duck through a next-generation sequencing approach of whole mitochondrial genome sequencing. Apart from it, duck genomics was also studied. Now, recent evidences indicate the role of mtDNA in the regulation of nuclear immune response genes and thereafter the pathogen is destroyed.

Pasteurellosis has zoonotic importance and can affect humans and other livestock. Duck cholera, caused by *P. multocida*, is known to result in high morbidities and fatalities as well as significant economic losses. Chronic duck cholera results in the localisation of the organism in specific organs and is distinguished from acute duck cholera by severe watery diarrhoea, anorexia, respiratory symptoms and septicaemia. Fowl cholera causes great economic loss, which includes mortality, morbidity, treatment and vaccination costs. An outbreak in a flock of 50,000 birds had an average cost of $0.40 (€0.33) for fowl cholera vaccine per bird, while nonoutbreak flocks that had received the fowl cholera vaccine had an average cost of $0.12 (€0.10) per bird. The treatment cost following an outbreak of fowl cholera in 1985–86 was estimated to be $0.51 (€0.452) per bird. This cost included productivity losses as well. Fifty per cent of the total expenditure was incurred due to immunisation against poultry cholera following the outbreak. In all, California lost 0.5% of its 1985–86 turkey meat supply due to fowl cholera.

Ducks have been considered to be the second most important poultry species after chickens. Indigenous ducks have been observed to have certain unique characteristics in terms of their disease resistance ability against common avian diseases. Ducks can be well maintained under organised farming system with a small water container that allows them to dip their head to wash their eyes without any natural water bodies. It is one of the least explored poultry species with certain unique characteristics in response to the innate immune response. Duck cholera is a limiting disease affecting duck production.

Hence, in this current study, we attempt to characterise certain nuclear innate immune response genes and polypeptide coding mitochondrial genes and to explore their role and certain interactions among them in providing immune response against duck *Pasteurellosis* (bacterial disease) in Anas platyrynchos (duck) model.

Birds and sample collection

The samples were collected from Bengal ducks reared at the livestock farm, West Bengal University of Animal and Fishery Sciences.

Sample collection and RNA isolation

Duck liver tissue (1 g) was collected from slaughtered birds, which were clinically healthy and maintained in the duck house of the LFC Dept, WBUAFS. Samples were collected from adult males (n = 6) randomly and transported to the laboratory immersed in TRIzol, maintaining cold chain for RNA isolation. Total RNA was isolated using the TRIzol extraction method (Life Technologies, USA), as per the standard procedure and was further utilised for cDNA synthesis. cDNA concentration was estimated and samples above 1200 micrograms per ml were considered for further study. Tissue samples for gut-associated lymphoid tissue (GALT), such as the ileocaecal junction, liver, gizzard, caecum, jejunum, spleen and pancreas, were collected (n = 6) each from both healthy and infected with *P. multocida*. Infected samples were confirmed through molecular detection of *P. multocida*.

cDNA synthesis and PCR amplification of the mitochondrial genes cytochrome B and nuclear gene coding for mitochondrial protein-cytochrome C and nuclear gene (NRLP3, IL-18 and STING) gene of Bengal duck

A 20 µL of the reaction mixture containing 5 µg of total RNA, 0.5 µg of oligo dT primer (16–18 mer), 40U of Ribonuclease inhibitor, 1000 M of dNTP mix, 10 mM of DTT and 5U of MuMLV reverse transcriptase was taken with appropriate buffer with thorough mixing followed by incubation at 37°C for 1 hour. The next step is 10 minutes by heating the mixture at 70°C unliganded and then chilled on ice. Afterwards, the integrity of the cDNA was checked by performing polymerase chain reaction (PCR). cDNA concentration was measured with NanoDrop. NRLP3, IL-18, STING, cytochrome B and cytochrome C primer pairs were designed based on the published sequences of chicken and duck origin using DNASTAR software (Hitachi Miraibio Inc., USA) to amplify the full-length open reading frame (ORF) of these genes.

25 µL of reaction mixture was comprised of 80–100 ng cDNA, 3.0 µL 10X PCR assay buffer, 0.5 µL of 10 mM dNTP, 1U Taq DNA polymerase, 60 ng of each primer and 2 mM MgCl2. PCR reactions were conducted in a thermocycler (PTC-200, MJ Research, USA), cycling conditions being initial denaturation at 94°C for 3 minutes, further denaturation at 94°C for 30 seconds, annealing at 61°C for 35 seconds and extension at 72°C for 3 minutes were conducted for 35 cycles followed by a final extension at 72°C for 10 min.

Study of mitochondrial (cytochrome B and cytochrome C) in silico

Predicted peptide sequences of duck mitochondrial (cytochrome B and cytochrome C) were analysed. Proteins were then aligned with those of other species using MAFFT. The analysis was conducted for a sequence-based comparative study. The signal peptide is essential to prompt a cell to translocate the protein, usually to the cellular membrane and ultimately, the signal peptide is cleaved to give a mature protein. Prediction of the presence and location of the signal peptide of these genes was conducted using the software (Signal P 3.0 Sewer-prediction results, Technical University of Denmark). The leucine percentage was calculated manually from the predicted peptide sequence. Disulfide bonds are necessary for protein folding and stability. It is the 3D structure of the protein which is biologically active.

Protein sequence level analysis was conducted (http://www.expasy.org./tools/blast/) for the assessment of leucine-rich repeats (LRR), leucine zipper, detection of leucine-rich nuclear export signals (NES) and detection of the position of GPI anchor, N-linked glycosylation sites. Being receptors, they are rich in LRR, necessary for pathogen recognition and binding. A leucine zipper is essential to assess the dimerisation of IR molecules. N-linked glycosylation is important for the molecule to assess its membranous or soluble form. The leucine-rich nuclear export signal is essential for the export of this protein from the nucleus to the cytoplasm, whereas the GPI anchor is responsible for anchoring in the case of a membranous protein. Prosite was employed for LRR site detection.

Leucine-rich NES were analysed with NetNES 1.1 server, Technical University of Denmark. O-linked glycosylation sites were detected using NetOGlyc 3.1 server (http://www.expassy.org/), whereas N-linked glycosylation sites were assessed through NetNGlyc 1.0 software (http://www.expassy.org/). Sites for leucine-zipper were detected through Expasy software, Technical University of Denmark. Sites for alpha-helix and beta sheet were detected using NetSurfP-protein surface accessibility and secondary structure predictions, Technical University of Denmark. Domain linker sites were predicted. LPS-binding and LPS-signalling sites were predicted based on homology studies with other species of the respective polypeptide. These sites are important for pathogen recognition and binding.

Three-dimensional structure prediction and model quality assessment

3D models of mitochondrial protein (cytochrome B and cytochrome C) polypeptides were modelled through the SWISS-MODEL Repository. Templates containing the greatest identity of sequences with our target template were identified with PSI-BLAST (http://blast.ncbi.nlm.nih.gov/Blast). The PHYRE2 server, based on the 'Homology modelling approach', was used to build a three-dimensional model of these proteins. Molecular visualisation tool as PyMOL (http://www.pymol.org/) was employed for model generation and visualisation of the three-dimensional structure of these proteins under study for duck origin. The structure of duck molecules was evaluated and assessed for its stereochemical quality (through structural analysis and verification server (SAVES), http://nihserver.mbi.ucla.edu/SAVES/); then refined and validated through Protein Structure Analysis (ProSA) web server (https://prosa.services.came.sbg.ac.at/prosa). NetSurfP server (http://www.cbs.dtu.dk/services/NetSurfP/ Peterson et al., 2009) was employed for assessing the surface area of these proteins with relative surface accessibility, Z-fit score and probability of alpha-helix, beta-strand and coil score.

The alignment of 3D structure of these proteins was checked with RMSD estimation to assess the structural differentiation by TM-align software.

Protein-protein interaction network depiction

To understand the protein interaction network of these proteins, we performed the search in the STRING 9.1 database. The functional interaction was determined with a confidence score. Interactions with scores <0.3, scores ranging from 0.3 to 0.7 and scores >0.7 are designated as low, medium and high confidence, respectively. Also, we executed a KEGG analysis, which depicted the functional association of these proteins with other related proteins.

Sample collection

For the present study, six tissue samples were collected from the infected indigenous duck samples, followed by a challenge study and six tissue samples from six healthy ducks following infection. We assign the ducks with mortality as infected or diseased (less immune) and the surviving ducks as healthy (better immune).

Natural challenge of ducks with *Pasteurella multocida*, symptomatic diagnosis, PCR-based detection

Bengal ducks in the experimental duck Farm were subjected to natural infection with *P. multocida* following a natural disaster. We have grouped the ducks into two groups—Infected (with symptoms that ultimately die) and healthy (survivors without any symptoms).

Challenge study with *Pasteurella multocida* in vivo

A challenge study was conducted with *P. multocida* followed by the detection of infective bacterial load at different stages of infection through PCR-based detection. For the challenge study, initially, we incubated the fertile duck eggs (n = 12) in the incubator at 37°C along with water in a Petri dish to maintain the proper humidity, which is essential for the incubation of duck eggs. After one week, we rechecked and screened the fertile embryo through candling. In order to harvest the *P. multocida* bacterial culture, at nine days of incubated fertilised egg, we made a minute hole through needle puncture of the air sac with sterilised needle at the chorioallantoic membrane and inoculated 200 ul of the infective agent in a sterile atmosphere. The needle puncture was rapidly sealed with glue (feviquick). The eggs were again incubated in a sterile environment at 37°C and humid conditions as earlier. The entire process was undertaken at the BSL3 lab. The duck *Pasteurella* bacterial strain employed was the available infection in the state from the clinically infected cases. After 3 days of inoculation, the bacteria were isolated from embryonic fibroblast cells through a viral DNA isolation kit. Quantification of bacterial load was assessed through quantitative PCR. For the challenge study, 0.3 ml of bacterial stock at egg infective dose 50 (EID50) was administered for each embryonated egg. Infective amnio-allantoic fluid was used to determine 50% EID. Initial verification was conducted to ensure that the ducks were free from duck *Pasteurella* infection.

Sample collection

In the present study, six tissue samples were collected from the infected indigenous duck samples following a challenge study. We assign the ducks with mortality as infected or diseased (less immune) and the ducks that survived as healthy (better immune).

Diagnosis and confirmation of *Pasteurella multocida*

The ducks were primarily diagnosed with duck pasteurellosis clinically with the help of initial symptoms. Later on, the confirmatory diagnosis was conducted through molecular detection of the duck *Pasteurella multocida*-specific primers in the infected samples.

Gross anatomical view of the liver in *Pasteurella*-infected duck

We analysed different body organs from infected ducks for clinical diagnosis of duck pasteurellosis.

Confirmation through molecular detection

In the next step, we follow the confirmation of the samples with a molecular PCR-based technique using the following primers as recommended by OIE, Paris. (Table 7.1).

TABLE 7.1 Primers used for the molecular detection of ***Pasteurella multocida***.

Target gene	Primer sequence	Product size
KMT1 (partial)	F5/GCTGTAAACGAACTCGCCAC3′- R5/ATCCGCTATTTACCCAGTGG-3′	460 bp

It involves bacterial DNA isolation through an isolation kit (Himedia), followed by PCR-based amplification.

Assessment of dose of infection

We assessed the dose of bacterial infection standardisation with quantitative PCR. We take the standard vaccine (against duck cholera) as a control for its amplification and quantification.

Differential mRNA expression profiling for cytochrome B and cytochrome C in Bengal duck

Real time-polymerase chain reaction

RNA estimation

The tissue samples were homogenised and submerged in liquid TRIzol. 1 ml of triturated tissue was taken into a 2 ml Eppendorf tube, and 1 ml of chloroform was added. The next step is the centrifugation at the rate of 10,000 rpm for 10 minutes at 4°C in an automated temperature-controlled refrigerated centrifuge machine (Thermo fisher). Three phases of differentiation were identified. The uppermost aqueous phase was collected for RNA isolation into a new Eppendorf tube and an equal volume of 100% isopropanol was added. It was left for a minute at 20°C–25°C temperature and centrifuged at the rate of 10,000 rpm for 10 minutes. After that supernatant was discarded and an equal volume of 70% ethanol was added to the pellet. Again the mixture was centrifuged at the rate of 10,000 rpm for 10 minutes and then the supernatant was discarded. Then the Eppendorf tube containing the pellet was kept at room temperature for air drying. After drying the remaining pellet dissolved with nuclease-free water (Applied Bio-systems, Cat. No. AM9930) and stored at -20°C for future uses.

Qualitative analysis of total RNA

The total RNA was quantified with NanoDrop-8000 (Thermo Scientific, Model No. 8000 Spectrophotometer), taking 1 µl of each sample to determine the concentration and A260/280 ratio.

First strand cDNA synthesis (reverse transcriptase-PCR)

A 5 µl estimated RNA was taken into a PCR micro tube and added 1 µl Oligo dT primer, 1 µl 10 mM dNTP and 13 µl distilled water. After well mixing, it was heated at 65°C for 5 minutes. Then those tubes were quickly chilled on ice for 5–7 minutes and centrifuged. Then 4 µl 5x first strand buffer and 2 µl DTT (100 mM) were added to the tube. It was well mixed and incubated at 37°C for 2 minutes. Then 1 µl M-MLVRT (200 u/µl) was mixed with the reaction mixture and treated at 37°C for 50 minutes and at 70°C for 15 minutes for inactivation. Adjusted automatically in the thermal cycler machine (Applied Biosystems by Life Technologies) by setting the whole programme.

Real-time PCR experiment (SYBR Green-based)

The primers were standardised with respective cDNA samples before being subjected to real-time PCR. The reactions were conducted in triplicate (as per MIQE guidelines) with the total volume of the reaction mixture made up to 20 µl. The experiment was conducted with primers and GAPDH (housekeeping gene) primers calculated as per concentration. A volume of 1 µl of cDNA was mixed with 10 µl of Hi-SYBr Master Mix (HIMEDIA MBT074), and rest volume made upto 20 µl with nuclease-free water. Then the reaction was performed into the ABI 7500 system and run the reaction programme. The delta-delta-Ct (ΔΔCt) method $2^{-\Delta\Delta Ct}$ was employed for the analysis of the result. The primers used for the reaction are as followed:

Histological section

The liver samples were prepared for stereology and histological analysis after being embedded in paraffin and fixed in 10% formalin. The liver tissues were prepared and vertically embedded in paraffin wax after being immersed in Lillie fixative for a week at room temperature. Then, a fully automated rotary microtome (Leica RM2255, Germany) meticulously divided each liver sample into slices that were 4 µm thick. Hematoxylin and eosin were used to stain each of these sections before they were mounted. The systematic uniform random sampling approach was used to choose 10 to 15 sections from each liver sample.

Immunohistochemistry

Paraffin tissue blocks were prepared by the standard manual alcohol-acetone protocol. The 5–6 µm thick paraffin sections obtained from the liver of both the infected and healthy control ducks were taken on Millennia 2.0 adhesion slides (Cat. No. 71863-01, Abcam). The tissue sections were de-paraffinized, followed by hydration in distilled water. The tissue slides were poured with trypsin enzymatic antigen retrieval solution (Cat. No. ab970, Abcam) and kept in an incubator in a humid environment at 37°C for 5–10 minutes. The sections were then incubated for 60 minutes in peroxidase blocking solution (Lot. No. 00065614, Dako) at room temperature to block nonspecific antibody binding activity. After subsequent washing with phosphate buffer saline (PBS), the sections were incubated at 37°C for 2 hours in a humid environment with mouse monoclonal anti-IL-18 antibody in a 1:200 dilution. Immunoreactivity was detected after one-hour incubation at 37°C with a secondary antibody, Rabbit antimouse IgG H&L (HRP conjugated, Cat. No. ab6728; Abcam) in dilution

1:200. Slides were then rinsed 3 times in PBS for 5 minutes each, followed by treatment with freshly prepared DAB solution for 3 minutes (DAB substrate, Cat. No. 34001, Thermo Fisher Scientific). The sections were counterstained with Mayer's haematoxylin, dehydrated in ethanol, cleared in xylene and then mounted in DPX.

Statistical analysis

Microsoft Excel was employed for the descriptive statistical analysis. SYSTAT 13.1 software (SYSTAT software Inc.) was used for statistical analysis and analysis of variance was used to test differences between parameters.

Result

Characterisation of the cytochrome B and cytochrome C derived peptide from duck liver

We have sequenced cytochrome B gene for Bengal duck and predicted its amino acid (aa) (QVX28267.1) sequence through *in silico* methods. In the next step, we have predicted the 3D model for cytochrome B and cytochrome C for Bengal duck. We have equally sequenced cytochrome C and submitted it to gene bank with accession number bankit number 2970653 Fig. 7.8, Fig. 7.9, Fig. 7.10, Fig. 7.11, Fig. 7.12

Differential mRNA expression profiling for mitochondrial genes

Differential mRNA expression profiling for cytochrome B and cytochrome C has been depicted in Figs. 7.6, 7.7, respectively. Gene expression of both cytochrome B and cytochrome C was observed to be better in healthy ducks in comparison to infected ducks Fig. 7.13, Fig. 7.14, Fig. 7.15, Fig. 7.16

In this current research, we attempt to explore the molecular control of host resistance against duck *Pasteurellosis* in the duck model. We attempt to identify the role of mitochondrial genes and their interaction of mitochondrial gene with nuclear genes against bacterial infection (duck *Pasteurellosis*). This is a novel finding and the first report of this kind.

We have characterised, predicted 3D structure modelling and identified the important domains for polypeptide coding mitochondrial genes, cytochrome B and nuclear gene coding for mitochondria-Cytochrome C for the indigenous Bengal Duck.

We have also carried out the experiment in vivo with the isolated pathogenic *Pasteurella spp*. We collected liver samples from both groups and subjected them to differential mRNA expression profiling. Better expression profiling is an indication of better mitochondrial copy number in the individual. We observed better expression of cytochrome B in healthy individuals in comparison to infected ones. Better expression of cytochrome C (mitochondrial protein) was also observed, indicative of more mitochondrial copy number in the liver cells of a healthy individual.

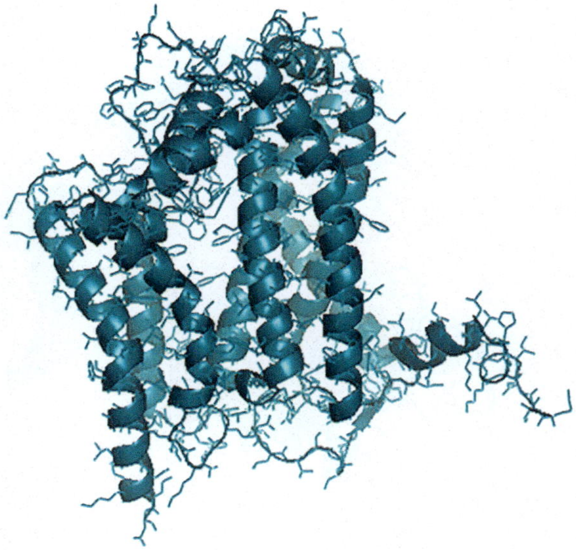

FIGURE 7.8 3D model for duck cytochrome B.

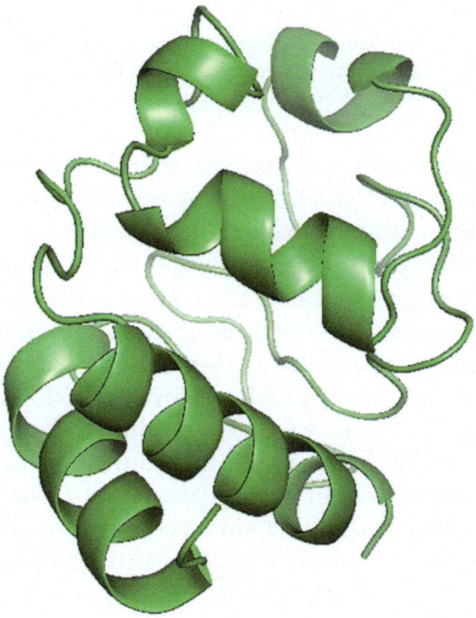

FIGURE 7.9 3D model for duck cytochrome C.

It is to be noted that we have isolated mRNA for mitochondrial genes without isolating the mitochondria. We speculate that the mRNA isolated was from cytoplasm, that is mRNA transcribed from released or escaped mtDNA from the mitochondria.

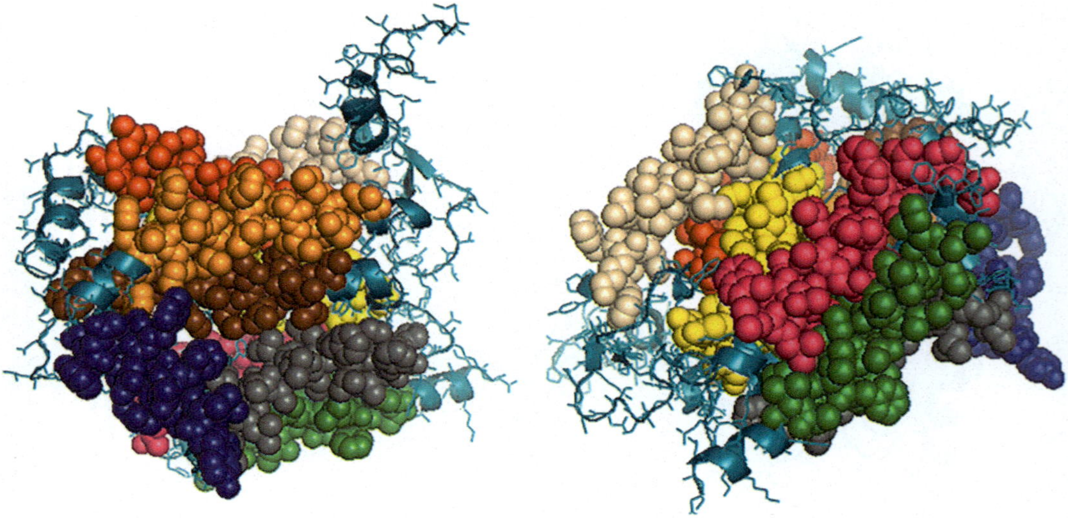

FIGURE 7.10 Important binding sites for duck cytochrome B. Red sphere: amino acid (aa) 34–54, haem binding site, transmembrane (TM) site, helical topology yellow sphere: aa 84, axial binding residue, ligand, Fe (iron binding site) purple sphere: aa 87–107, haem binding site, TM site, helical topology chocolate sphere: aa 116–136, TM site, helical topology blue: 151–171, TM site, helical topology orange 179–199, TM site, helical topology and haem binding site wheat 225–245, TM site, helical topology gray 289–309 TM site, helical topology hot pink 324–344 TM site, helical topology forest green 350–370 TM site, helical topology.

We observed better expression for cytochrome B and cytochrome C in the healthy ducks in comparison to those of infected ducks. We have also observed better expression of IL-18, NRLP3 and STING genes in healthy ducks in comparison to the infected group. Hence, we observed a positive correlation of mitochondrial gene expression along with the identified nuclear genes. KEGG analysis has revealed the role of mtDNA escape from the mitochondria, which triggers various pathways, including cytosolic DNA sensing pathway, NOD-like signalling receptor pathway, the pathway depicting how the body's immune response is triggered with bacterial infection, necroptosis pathway. STRING analysis has revealed how various immune response genes interact with each other and act with CASCADE formation.

Scientists observed that mtDNA escapes during infection and triggers nuclear-derived immune response genes and thereafter acts against and destroys the invading pathogen through various modes. Healthy ducks that survived exposure are expected to have a better mtDNA-mediated and triggered immune response. mtDNA released from mitochondria is able to activate cGAS–STING signalling. Researchers have examined mtDNA release in the context of cell death, while West et al provided evidence that TFAM deficiency promotes mitochondrial stress and mispackaged mtDNA, leading to their expulsion into the cytoplasm, where they bind and activate cGAS and initiate the type I IFN response.

Infection with the bacterial pathogen *Mycobacterium tuberculosis* triggers cGAS activation and subsequent IRF3-dependent type I IFN response. This was presumed to be due solely to the detection of mycobacterial DNA, but other studies have detected a role for mitochondrial stress

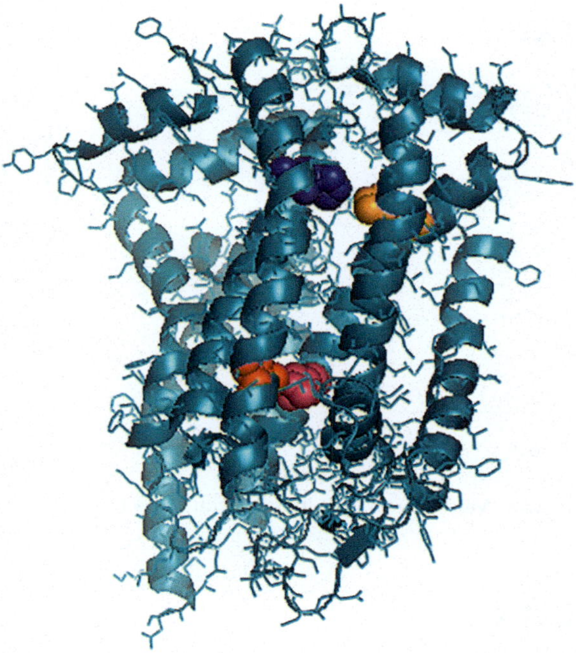

FIGURE 7.11 Important binding sites for duck cytochrome B orange sphere: amino acids (aa) 84 iron binding, axial binding residue, warm pink sphere: aa 98 iron binding, axial binding residue, blue sphere: aa 183, iron binding, axial binding residue, haem binding red sphere: aa 197 iron binding, axial binding residue.

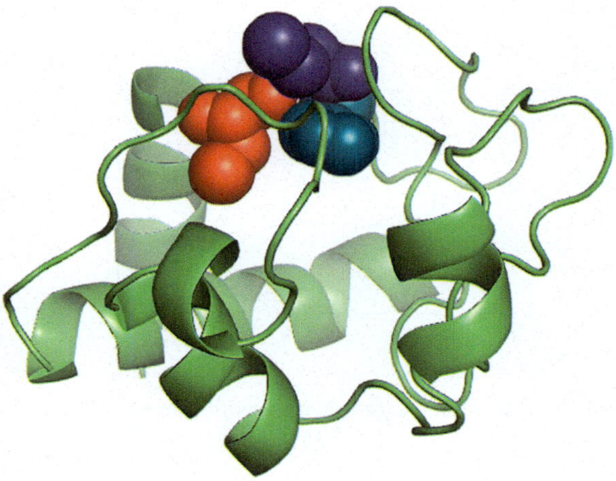

FIGURE 7.12 Important binding site for cytochrome C for duck. Red sphere: amino acid (aa) 15: haem C binding site, blue sphere: aa 18: haem C binding site, deep teal sphere: iron binding site: axial binding residue.

and the subsequent release of mtDNA into the cytoplasm. This observation is strain-dependent, but suggests a role for mitochondrial stress and dynamics in *M. tuberculosis* -induced mtDNA release. Previous work observed cytochrome C release from mitochondria in *M. tuberculosis*-infected cells, suggesting that there may be a possible role for BAX/BAK51-dependent mitochondrial permeabilisation in infection-associated mtDNA release. Several researchers have

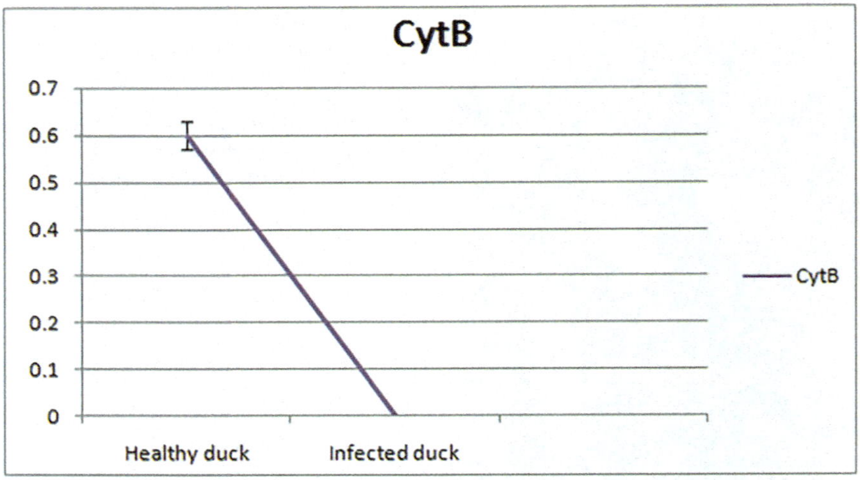

FIGURE 7.13 Differential mRNA expression profiling for cytochrome B gene in duck with respect to *Pasteurella multocida* infection.

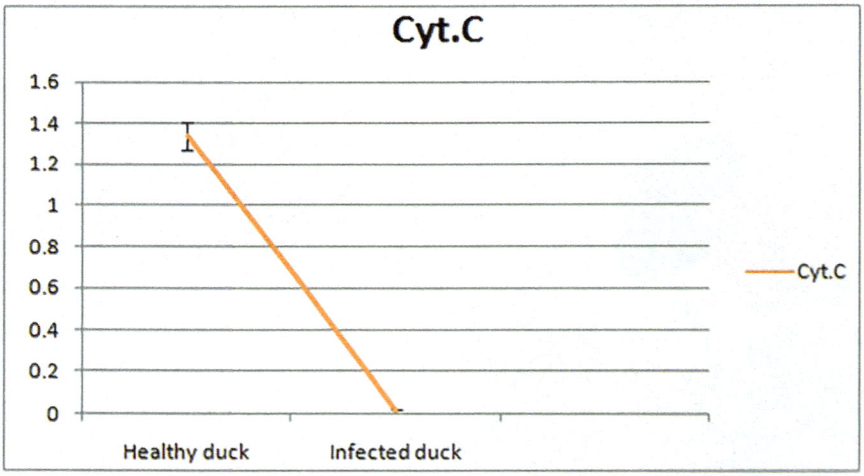

FIGURE 7.14 Differential mRNA expression profiling for cytochrome C gene in duck with respect to *Pasteurella multocida* infection.

depicted the role of mtDNA in cell death, apoptosis and their possible influence on immunity or disease resistance. Mitochondrial permeability is affected by nuclear factor NF-κβ-dependent antitumour activity under caspase deficiency. The action of cytochrome C in maintaining the integrity of mitochondria and its association with disease has been observed by several workers. The nuclear DNA is ready to be expelled as a NET through a highly regulated process involving chromatin decondensation and citrullination of histones. Furthermore, permeabilisation of the plasma membrane is also regulated, which inevitably leads to cell death. The complex cross-talk

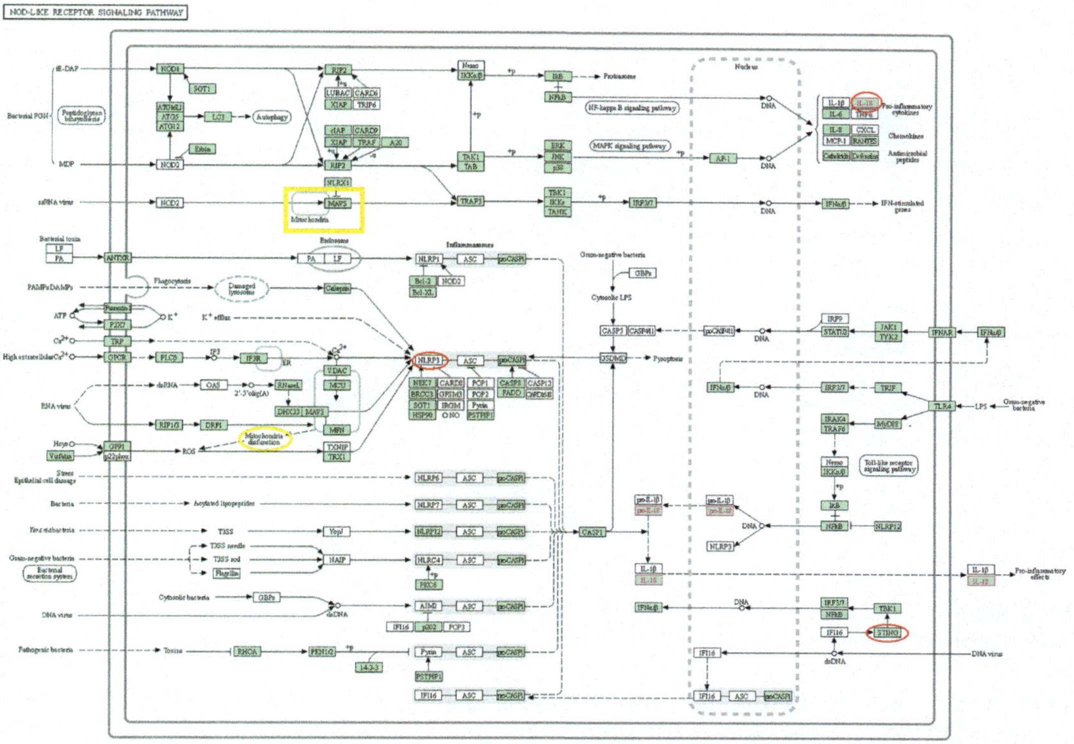

FIGURE 7.15 Molecular pathway analysis for IL-18/NLRP3/STING gene for antibacterial activity in duck.

between mitochondria and NRLP3 and NALP3 has been studied. The complex interaction of mitochondrial permeability and influx of mtDNA has been reported with a possible role in disease incidences, like myocardial infarction. The role of mitochondrial genes in controlling innate immunity against bacterial infection has been reported.

There exists a certain complex interaction among the mitochondria, referred to as mitochondrial dynamics, including mitochondrial fission, mitochondrial fusion, mitophagy and transport. Interestingly, these processes are governed by the factors coded by nuclear genes. Nuclear genes involved in mitochondrial fusion are mitofusin (Mfn1), Mfn2, ocular atrophy factor 1, whereas those involved in mitochondrial fission are Mff, Drp1, with interaction with ER. However, mitophagy involves PTEN induced kinase 1 (PINK1), Parkin, NIX and LC3. Alterations in these normal processes are observed in diseases, equally holds good for bacterial infection like *Pasteurellosis*. Mitochondria are considered more than the motherboard of a cell (https://www.scientificamerican.com). Apart from the influx of mtDNA from mitochondria to cytoplasm, recent reports indicate transfer of intact mitochondria from a cell, transfer to other cells and even presence in blood plasma with free existence and understanding of mitochondrial dysfunction in ageing and its possible role in immune cell and platelet senescence. We identified certain important domains as multiple haem binding sites in cytochrome B and cytochrome C, which are helpful for the

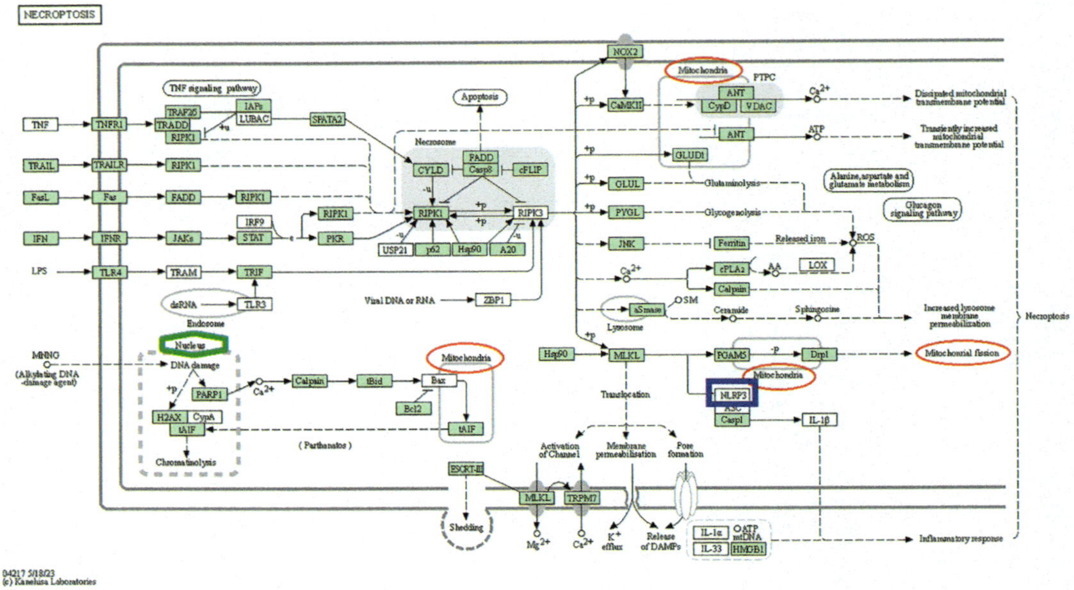

FIGURE 7.16 Molecular pathway analysis for NRLP3 gene and the interaction of mitochondrial DNA with nuclear genes in the duck.

immune response. Similarly, important domains identified in NRLP3, such as NACHT and Pyrin or DAPIN, are promising for immune functions.

Hence, it is evident that there exists a complex interaction between mitochondrial and nuclear genes and their association with diseases. Thus, both mitochondrial and nuclear genes may be employed as biomarkers for certain diseases.

Characterisation of other nuclear origin genes involved in mitochondrial function (replication and transcription) in association to bacterial disease: case study 2

Mitochondrial structural and functional components are controlled by both mitochondrial and nuclear genes. Earlier, we studied whole mitochondrial genome sequencing and its association with bacterial infection (duck *Pasteurellosis*). In this current study, we screened certain nuclear mitochondrial genes, which are basically nuclear genes, but instead of having a direct role in the immunity of the host, they act directly on mitochondrial function. Mitochondria, in turn, have a direct effect on providing antibacterial immunity. Such identified genes were TP/ endothelial cell growth factor 1 (ECGF1) gene, TFAM, TK2. In the current research, we have studied the duck as a model organism with antibacterial host immunity against duck *Pasteurellosis*, a bacterial infection caused by gram-negative bacteria. A better mRNA expression profile for these nuclear mitochondrial genes was observed in healthy ducks in comparison to infected ducks. In the next step, we observed a better expression profile for cytochrome B and cytochrome C in healthy ducks compared to infected ones. Cytochrome B expression may be used as an indicator of mtDNA copy number. Molecular pathway analysis reveals certain important roles for cytochrome C in antibacterial immunity. Certain important domains were identified for these nuclear mtDNA (NUMT) genes with a distinctive effect on mitochondria. Molecular interaction analysis reveals that these NUMT genes interact with each other for effective biological function related to mitochondria. Mitochondria, in turn, have a distinctive action against bacterial infection. This is the first report of NUMT genes in animals. Exploitation of the NUMT genes as a molecular marker or through gene editing may lead to the development of ducks resistant to duck *Pasteurellosis* or duck cholera. NUMT genes may explore a new era for understanding the complex interaction of host immunity against bacterial infection.

Immune response genes have been observed to have innate immune response against various types of bacterial infection, which are of nuclear origin (Chakraborty et al., 2025; Pal & Chakravarty, 2019). Several other studies have also established the role of mitochondrial genes in antibacterial immunity (Chakraborty et al., 2025) and also antiviral immunity (Pal et al., 2022a, 2022b; Debnath, M, 2023). Mitochondrial function is very vital not only in oxidative phosphorylation, but in calcium metabolism, iron metabolism (Pal & Banerjee, 2018; Horowitz & Greenamyre, 2010) and immunity. Mitochondria possess 37 genes of its own, thirteen polypeptide coding genes, rRNA gene, tRNA genes and D loop, aiding mtDNA replication. Apart from these 37 genes, there are numerous genes around 200 which are involved in either mitochondrial integral component or involved in mitochondrial function. Some of the most important genes are TP, TFAM, TK2, mitochondrial and MPV17.

Ducks can readily be used as a model organism, since easy to manage, economically important, have a short generation interval and most importantly with a robust, developed immune system with lymph nodes, which chickens lack. Ducks have been considered to be the second most important poultry species after chickens. Indigenous ducks have been observed to have certain unique characteristics in terms of their disease resistance ability against common avian diseases (Pal, 2021).

We have considered duck *Pasteurellosis* caused by *P.multocida*, gram-negative facultative commensal bacteria, as the target bacterial infection with great economic importance. It has been classified as serotypes 1 to 16 as per their lipopolysaccharide antigen and serogroups A, B, D, E and F based on its capsular component. Serogroup A are generally responsible for fowl cholera. In the early stages of an illness, mortality rates might range from 5% to 20%. Even far higher rates of mortality could occur. Mortality may decrease to 2%–5% each month as the illness progresses into chronicity. Birds with a chronic infection may pass away, suffer from infection for a long time, or recover. An outbreak flock of 50,000 birds had an average cost of $0.40 (€0.33) per bird, while nonoutbreak flocks that had received the fowl cholera vaccine had an average cost of $0.12 (€0.10) per bird. Furthermore, the cost of treating an outbreak of fowl cholera in 1985–86 was estimated to be $0.51 (€0.452) per bird. This cost included productivity losses as well. About half as much was spent on the immunisation against poultry cholera overall as there was on the outbreak. In all, California lost 0.5% of its whole 1985–86 turkey meat supply due to fowl cholera.

In our lab, we had been studying on disease genetics in sheep (Banerjee et al., 2021; Rawat et al., 2023; Banerjee et al., 2023; Rawat et al., 2021) and duck model (Arhga, Manti, Subhomoy). We have studied immunogenetics against parasites (Banerjee et al., 2021; Rawat et al., 2023; Banerjee et al., 2023; Rawat et al., 2021), viral infection, such as duck plague and avian influenza. In this study, we are working on host immunogenetics against bacterial infection. We had studied immunogenetics based on nuclear genes and mitochondrial genes. But in this current study, we aim to study a novel class of genes, which are basically nuclear genes, but act through mitochondria, known as NUMTs. Promising NUMT genes are TP, TFAM, TK2, mitochondrial.

TYMP/ECGF1 encodes for the protein TP, also known as TdRPase, or ECGF1, gliostatin and platelet-derived (PD)-ECGF. Nishino et al. (1999) and other researchers revealed its *involvement in mitochondrial genome maintenance*‾. TFAM codes 246 aa, encodes for protein TFAM. Celestini et al. (2018) and Hao et al. (2020) reported this gene binds to the mitochondrial light strand promoter and functions in mitochondrial transcription regulation. This gene was observed to be the component of the mitochondrial transcription initiation complex, composed at least of TFB2M, TFAM and POLRMT, which is required for basal transcription of mtDNA. TK2 encodes for the protein TK2, mitochondrial. It is composed of 270 aa in *Mus musculus* and 265 aa in *Homo sapiens*. It was reported that TK2 phosphorylates thymidine, deoxycytidine and deoxyuridine in the mitochondrial matrix. They also reported in nonreplicating cells, where cytosolic dNTP synthesis is downregulated, mtDNA synthesis depends solely on TK2 and DGUOK. The TK2 gene is also active in the mitochondrion, enabling TK activity .

The previous work on NUMT genes has mainly been done in human or insect models. So the objectives are to explore the effects of these genes in indigenous ducks based on the following considerations: Characterisation of NUMT genes in indigenous ducks affecting disease resistance against duck *Pasteurellosis* and differential mRNA expression profiling for the identified NUMT genes affecting disease resistance against duck *Pasteurellosis*.

Methodology in brief

Birds and sample collection

The samples were collected from Bengal ducks maintained at the West Bengal University of Animal and Fishery Sciences. Tissue samples were also collected from the slaughterhouse for Bengal ducks.

Sample collection and RNA isolation

Duck liver and kidney tissue (1 g) were collected from slaughtered birds, which were clinically healthy and maintained in the duck house of the LFC Dept, WBUAFS. Adult males (n = 6) were selected for the collection of samples. The tissue was immersed in TRIzol in the vial and transported on ice to the laboratory for RNA isolation. Total RNA was isolated using the TRIzol extraction method (Life Technologies, USA), as per the standard procedure and was further utilised for cDNA synthesis (Pal & Chatterjee, 2009; Pal et al., 2011; Banerjee et al., 2021; Rawat et al., 2021). cDNA concentration was estimated and samples above 1200 micrograms per ml were considered for further study.

cDNA synthesis and PCR amplification of TYMP/ECGF1, TFAM, TK2 gene of Bengal duck

A 20 μL of the reaction mixture was composed of 5 μg of total RNA, 0.5 μg of oligo dT primer (16–18 mer), 40U of ribonuclease inhibitor, 1000 M of dNTP mix, 10 mM of DTT and 5U of

MuMLV reverse transcriptase in an appropriate buffer. The reaction mixture was mixed thoroughly, followed by incubation at 37°C for 1 hour. The reaction was allowed up to 10 minutes by heating the mixture at 70°C unliganded and then chilled on ice. Afterwards, the integrity of the cDNA was checked by performing PCR (Pal et al., 2014a, Pal et al., 2014b; Pal, 2020; Pal, 2021; Pal et al., 2019; Pal et al., 2011). The concentration of cDNA was estimated through NanoDrop. TYMP/ECGF1, TFAM and TK2 gene primer pairs were designed based on the published sequences of chicken and duck origin using Primer 3 plus software (https://www.primer3plus.com) to amplify the full-length ORF of these genes.

A 25 µL of reaction mixture was comprised of 80–100 ng cDNA, 3.0 µL 10X PCR assay buffer, 0.5 µL of 10 mM dNTP, 1U Taq DNA polymerase, 60 ng of each primer and 2 mM $MgCl_2$. PCR reactions were performed in a thermocycler (PTC-200, MJ research, USA) with cycling conditions as, initial denaturation at 94°C for 3 min, further denaturation at 94°C for 30 seconds, annealing at 61°C for 35 seconds and extension at 72°C for 3 minutes were conducted for 35 cycles followed by a final extension at 72°C for 10 min.

cDNA cloning and sequencing

Amplified products of TK, TFAM and TYMP were checked by 1% agarose gel electrophoresis. The products were purified from gel using a gel extraction kit (Qiagen GmbH, Hilden, Germany). pGEM-T easy cloning vector (Promega, Madison, WI, USA) was used for cloning. Then, 10 µL of the ligated product was mixed thoroughly with 200 µL competent cells and heat shock was given at 42°C for 45 seconds in a water bath. Subsequently, the cells were immediately transferred on chilled ice for 5 minutes. and SOC medium was added to it. The bacterial culture was centrifuged to obtain the pellet and plated on an LB agar plate containing ampicillin (100 mg/mL) added to the agar plate @ 1:1000, IPTG (200 mg/mL) and X-Gal (20 mg/mL) for blue-white screening. Plasmid isolation from overnight-grown culture was carried out by a small-scale alkaline lysis method as described (Pal, 2021). Recombinant plasmids were characterised by PCR using primers as reported earlier and restriction enzyme digestion. The gene fragments released by the enzyme EcoRI (MBI Fermentas, USA) were inserted into a recombinant plasmid, which was sequenced by the dideoxy chain termination method with T7 and SP6 primers in an automated sequencer (ABI prism, Chromous Biotech, Bangalore).

Sequence analysis

DNASTAR Version 4.0, Inc., USA software was employed for the nucleotide sequence analysis for protein translation, sequence alignments and contigs comparisons. Novel sequences were submitted to the NCBI Genbank and accession numbers were obtained, which are available in the public domain.

Study of predicted duck TYMP/ECGF1, TFAM, TK2 protein using bioinformatics tools

Predicted peptide sequences of duck TYMP/ECGF1, TFAM and TK2 proteins were then aligned with those of other species using MAFFT (Katoh et al., 2013). The analysis was conducted for a sequence-based comparative study. The signal peptide is essential to prompt a cell to translocate the protein, usually to the cellular membrane and ultimately the signal peptide is cleaved to give a mature protein. Prediction of the presence and location of the signal peptide of these genes was conducted using the software (SignalP3.0 server-prediction results, Technical University of Denmark). The leucine percentage was calculated manually from the predicted peptide sequence. Disulfide bonds are essential for protein folding and stability, ultimately. It is the 3D structure of the protein which is biologically active.

Protein sequence level analysis was employed (http://www.expasy.org./tools/blast/) for the assessment of LRR, leucine zipper, detection of leucine-rich NES and detection of the position of the GPI anchor, N-linked glycosylation sites. Since these are receptors, they are rich in LRR, which are essential for pathogen recognition and binding. A leucine zipper is essential to assess the dimerisation of IR molecules. N-linked glycosylation is important for the molecule to determine its membranous or soluble form. The leucine-rich nuclear export signal is essential for the export of this protein from the nucleus to the cytoplasm, whereas the GPI anchor is responsible for anchoring in the case of a membranous protein. Prosite was used for LRR site detection.

Leucine-rich NES were analysed with NetNES 1.1 server, Technical University of Denmark. O-linked glycosylation sites were detected using NetOGlyc 3.1 server (http://www.expassy.org/), whereas N-linked glycosylation sites were assessed through NetNGlyc 1.0 software (http://www.expassy.org/). Sites for leucine-zipper were detected through Expasy software, Technical University of Denmark (Glick, 1977). Sites for alpha-helix and beta sheet were detected using NetSurfP-protein surface accessibility and secondary structure predictions, Technical University of Denmark (Petersen et al., 2009). Domain linker sites were predicted (Ebina et al., 2009). LPS-binding and LPS-signalling sites (Muroi et al., 2002) were predicted based on homology studies with other species of the respective polypeptide. These sites are important for pathogen recognition and binding.

Three-dimensional structure prediction and model quality assessment

Three-dimensional models of TYMP/ECGF1, TFAM and TK2 polypeptides were predicted through the SWISS-MODEL Repository (Kiefer et al., 2009). Templates possessing the greatest identity of sequences with our target template were identified with PSI-BLAST (http://blast.ncbi.nlm.nih.gov/Blast). PHYRE2 server, based on the 'Homology modelling approach', was used to build a three-dimensional model of these proteins (Kelley, 2015). Molecular visualisation tool as PyMOL (http://www.pymol.org/) was employed for model generation and visualisation of the three-dimensional structure of these proteins under study for duck origin. The structure of duck molecules was evaluated and assessed for its stereochemical quality (through SAVES, http://nihserver.mbi.ucla.edu/SAVES/); then refined and validated through ProSA web server (https://prosa.services.came.sbg.ac.at/prosa) (Wiederstein & Sippl, 2007). NetSurfP server (http://www.cbs.dtu.dk/services/NetSurfP/ Peterson et al., 2009) was used for assessing the surface area of these proteins through relative surface accessibility, Z-fit score and probability of alpha-helix, beta-strand and coil score.

The alignment of 3D structure of these proteins was analysed with RMSD estimation to evaluate the structural differentiation by TM-align software (Zhang et al., 2005).

Protein-protein interaction network depiction

To understand the protein interaction network of these proteins, we performed the search in the STRING 9.1 database (Franceschini et al., 2012). The functional interaction was assessed with a confidence score. Interactions with scores <0.3, scores ranging from 0.3 to 0.7 and scores >0.7 are classified as low, medium and high confidence, respectively. Also, we executed a KEGG analysis, which depicted the functional association of mitochondrial proteins with other related proteins in the antibacterial immunity pathway.

Challenge study with **Pasteurella multocida**, *symptomatic diagnosis, estimation of infective viral load at different stages of infection, PCR-based detection*

For the challenge study, initially, we incubated the fertile duck eggs (n = 12) in the incubator at 37°C along with water in a Petri dish to maintain the proper humidity, which is essential for the incubation of duck eggs. After one week, we rechecked and screened the fertile embryo through candling. To harvest the *P. multocida* bacterial culture, at nine days of incubation of the fertilised egg, we puncture the air sac with the help of a sterilised needle through the chorioallantoic membrane and inoculate 200 µL of the infective agent in a sterile environment. The needle puncture was quickly sealed with glue (Feviquick). The eggs were again incubated in a sterile environment at 37C and humid conditions as earlier. The entire procedure was undertaken at the BSL3 lab. After 3 days of inoculation, the viruses were isolated from embryonic fibroblast cells through a viral DNA isolation kit. Quantification of viral load was estimated through quantitative PCR with the duck *Pasteurellosis* vaccine as a control. To challenge, 0.3 ml of viral stock at EID50 was employed for each bird I/M. Infective amnio-allantoic fluid was used to determine 50% EID. Initial verification was conducted to ensure that the ducks were free from *Pasteurellosis* infection.

Sample collection

For the present study, six tissue samples were collected from the infected indigenous duck samples, followed by a natural challenge study, that is natural outbreak of infection. We designate the ducks with mortality as infected or diseased (less immune) and the ducks that survived as healthy (better immune).

Diagnosis and confirmation of *Pasteurella multocida*

Similar to as discussed.

Challenge study with *Pasteurella multocida* and study of expression profiling in healthy and infected groups

Differential mRNA expression profiling for GALT tissues w.r.t. TYMP/ECGF1, TFAM and TK2 gene.

Real time-polymerase chain reaction

RNA estimation

The tissue samples were cut into small pieces and submerged in liquid TRIzol, followed by thorough grinding with a mortar and pestle. 1 ml of triturated tissue was taken into a 2 ml Eppendorf tube and 1 ml of chloroform was added. Then it was centrifuged at the rate of 10,000 rpm for 10 minutes at 4°C in an automated temperature-controlled refrigerated centrifuge machine (BR Biochem, Life Sciences and REMI C24 Plus). Three phases of differentiation were identified. The uppermost aqueous phase was collected for RNA isolation into a new Eppendorf tube and an equal volume of 100% isopropanol was added. It was left for a minute at 20–25°C temperature and centrifuged at the rate of 10,000 rpm for 10 minutes. After that supernatant was discarded and an equal volume of 70% ethanol was added to the pellet. Again, the mixture was centrifuged at a rate of 10,000 rpm for 10 minutes and then the supernatant was discarded. Then the Eppendorf tube containing the pellet was kept at room temperature for air drying. After drying, the remaining pellet dissolved with nuclease-free water (Applied Bio-systems, Cat. No. AM9930) and stored at -20°C for future use (Pal, 2009, Pal et al., 2011, Pal et al., 2014a).

Qualitative analysis of total RNA

The total RNA was quantified using NanoDrop-8000 (Thermo Scientific, Model No. 8000 spectrophotometer), taking 1 µl of each sample to determine the concentration and A260/280 ratio (Rawat et al., 2021; Banerjee et al., 2021; Pradhan et al., 2018).

First strand cDNA synthesis (reverse transcriptase -PCR)

5 µl estimated RNA was taken into a PCR micro tube and added 1 µl Oligo dT primer, 1 µl 10 mM dNTP and 13 µl distilled water were added. After well mixing, it was heated at 65°C for 5 minutes. Then those tubes were quickly chilled on ice for 5–7 minutes and centrifuged. Then 4 µl 5x first strand buffer and 2 µl DTT (100 mM) were added to the tube. It was well mixed and incubated at 37°C for 2 minutes. Then 1 µl M-MLVRT (200U/µl) was added to the reaction mixture and treated at 37°C for 50 minutes and at 70°C for 15 minutes for inactivation. All the temperature treatments were maintained automatically in the thermal cycler machine (Applied Biosystems by Life Technologies) by setting the whole programme.

Primer designing and synthesis

All the primers were designed using Primer 3 software (v. 0. 4.0) as per the recommended criteria. The primers for the genes used in the experiment were designed (DNASTAR software).

Real-time PCR experiment (SYBR Green-based)

The primers were standardised with respective cDNA samples before being subjected to real-time PCR. The entire reactions were performed in triplicate (as per MIQE guidelines) and the total volume of the reaction mixture was set up to 20 μl. The reaction mixture is set up with TYMP/ECGF1, TFAM, TK2 gene and GAPDH (housekeeping gene) primers calculated as per concentration. A volume of 1 μl of cDNA, 10 μl of Hi-SYBr Master Mix (HIMEDIA MBT074), and the rest of the volume adjusted with nuclease-free water to achieve the total reaction mixture volume. Then the reaction plate was placed into the ABI 7500 system and run the reaction programme. The delta-delta-Ct (ΔΔCt) method was used for the analysis of the result. The primers used for the reaction are as follows:

Result

Characterisation of TYMP/ECGF1, TFAM, TK2 gene in *Anas platyrhynchos*

In the current study, a number of nuclear mitochondrial genes were studied, namely TP/ECGF1, TFAM, TK2 and MPV17 and others. These genes are basically nuclear genes, but code for mitochondrial proteins. This is the first report of such NUMT genes in any livestock or poultry. However, a significant role in bacterial immunity through mitochondria was observed for TP/ECGF1, TFAM, TK2.

Characterisation of thymidine phosphorylase/endothelial cell growth factor1 gene of duck

TP/ECGF1 (Acc no. PV833397), TFAM (Acc no. PV833398), TK2 (PV853950) from duck were amplified and sequenced. The aa sequence for TP in *Anas platyrhynchos* is comprised of 464 aa. It is derived from the gene sequence. The 3D structure of TP is depicted in orange colour. The TP (thymidine and pyrimidine-nucleoside phosphorylases) domain has been depicted at aa position 116–131. Go annotations of TP reveal a domain for glycosyl transferase family 3N-terminal at aa 54–81 and glycosyl transferase family 3 at aa position 92–198. Two disordered regions have been predicted at aa positions 1–20 and 201–239. Compositional bias with glycine residues has been observed in 1–12 aa positions. Secondary structure has been predicted for TP protein Fig. 7.17, Fig. 7.18, Fig. 7.19, Fig. 7.20

Characterisation of TFAM.

The aa sequence for TFAM in *Anas platyrhynchos* is comprised of 272 aa. It is derived from the gene sequence. The 3D structure of TFAM is depicted in Fig. 7.3 a and b in red colour. High mobility group box (HMG-box) A DNA binding domain (green) and HMG-box B DNA binding domain (blue) of TFM of *Anas platyrhynchos* have been depicted at aa positions 45–113 and 157–221 aa positions respectively Fig. 7.21, Fig. 7.22, Fig. 7.23

Characterisation of TK2, mitochondrial

The aa sequence for TK2 in *Anas platyrhynchos* is comprised of 289 aa. It is derived from the gene sequence. The 3D structure of TP is depicted in pink colour Fig. 7.24, Fig. 7.25, Fig. 7.26

Molecular interaction among the genes-STRING analysis

Physiological functions are being regulated through a series of molecular pathways involving various proteins coded by both nuclear and mitochondrial genes. The molecular

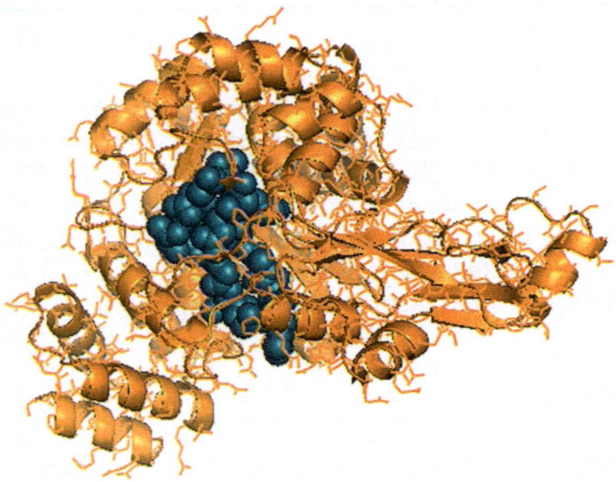

FIGURE 7.17 3D structure of thymidine phosphorylase with THYMID_PHOSPHORYLASE *thymidine and pyrimidine-nucleoside phosphorylases signature*: at amino acid position 116–131 of *Anas platyrhynchos*.

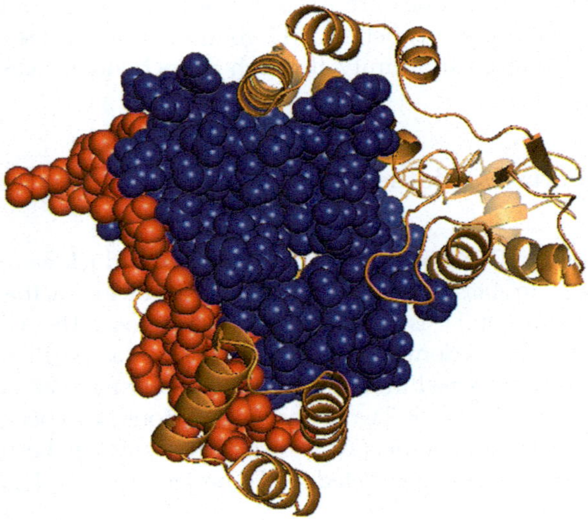

FIGURE 7.18 Important domains identified for thymidine phosphorylase. Domains for glycosyl transferase family 3 N-terminal at amino acid (aa) 54–81 (red sphere) and glycosyl transferase family 3 at aa position 92–198 (blue sphere).

FIGURE 7.19 Two disordered regions predicted at amino acid (aa) positions 1–20 (green sphere), 201–239 aa positions (purple sphere).

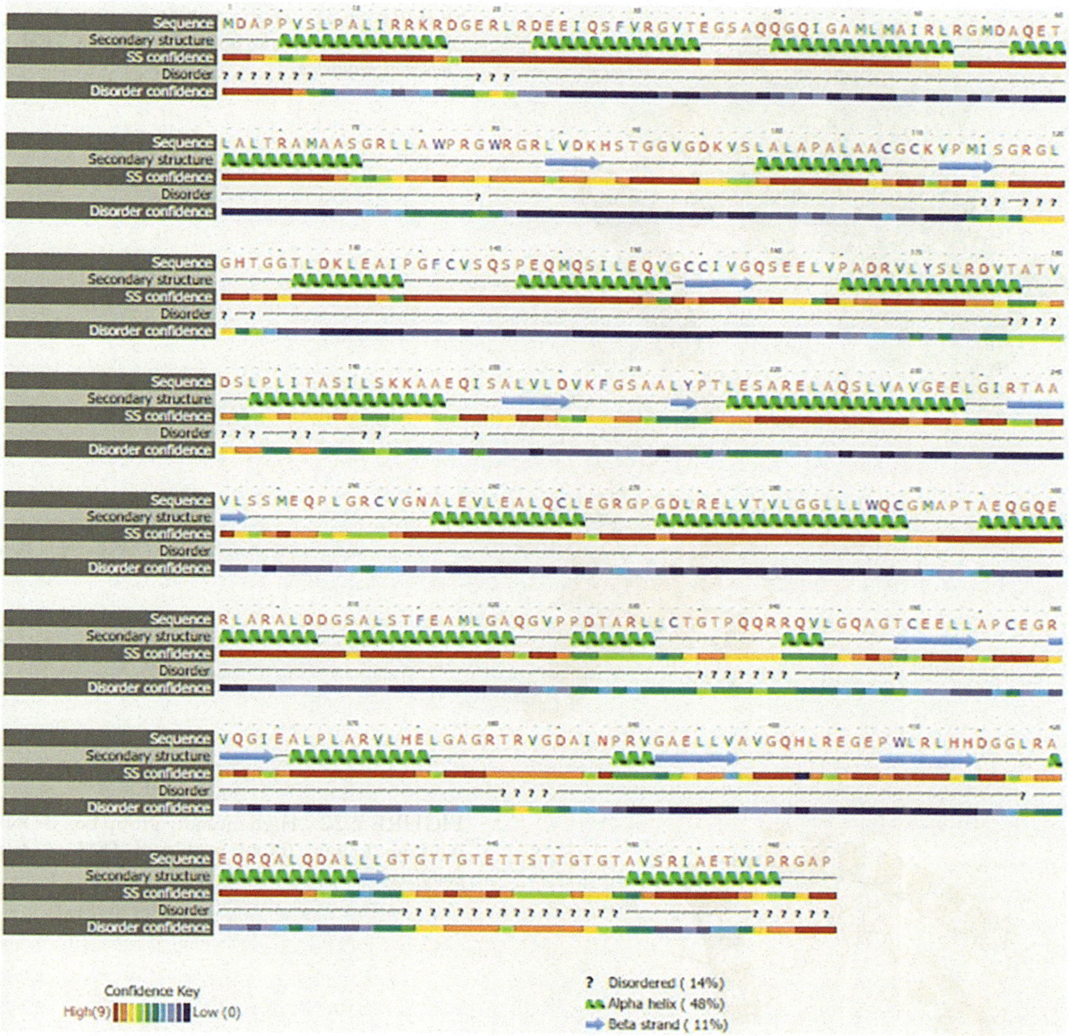

FIGURE 7.20 Secondary structure prediction for thymidine phosphorylase of duck.

interaction analysis among such proteins through STRING analysis is being depicted for TP, TFAM and TK Fig. 7.27, Fig. 7.28, Fig. 7.29, Fig. 7.30

Differential mRNA expression profiling for NUMT genes with respect to disease resistance against duck *Pasteurellosis* in *Anas platyrhynchos*

Unvaccinated ducks were naturally challenged with duck *Pasteurellosis*. The diagnosis was initially conducted through symptomatic diagnosis, followed by final molecular diagnosis of *P. multocida* with the following primer pairs. Individual differences in

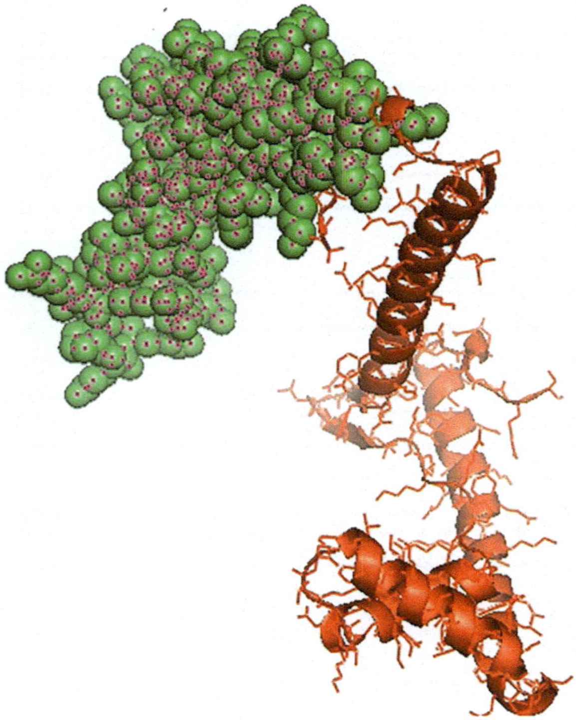

FIGURE 7.21 High mobility group box A DNA binding domain (green) profile of TFM of *Anas platyrhynchos*.

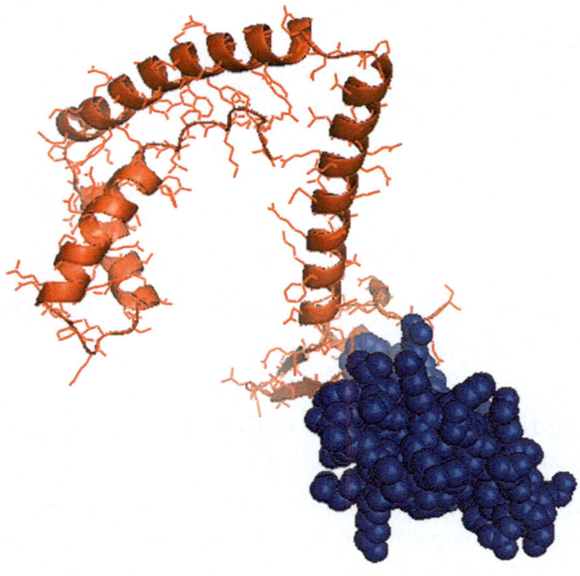

FIGURE 7.22 High mobility group box B DNA binding domain (blue) profile of TFM of *Anas platyrhynchos*.

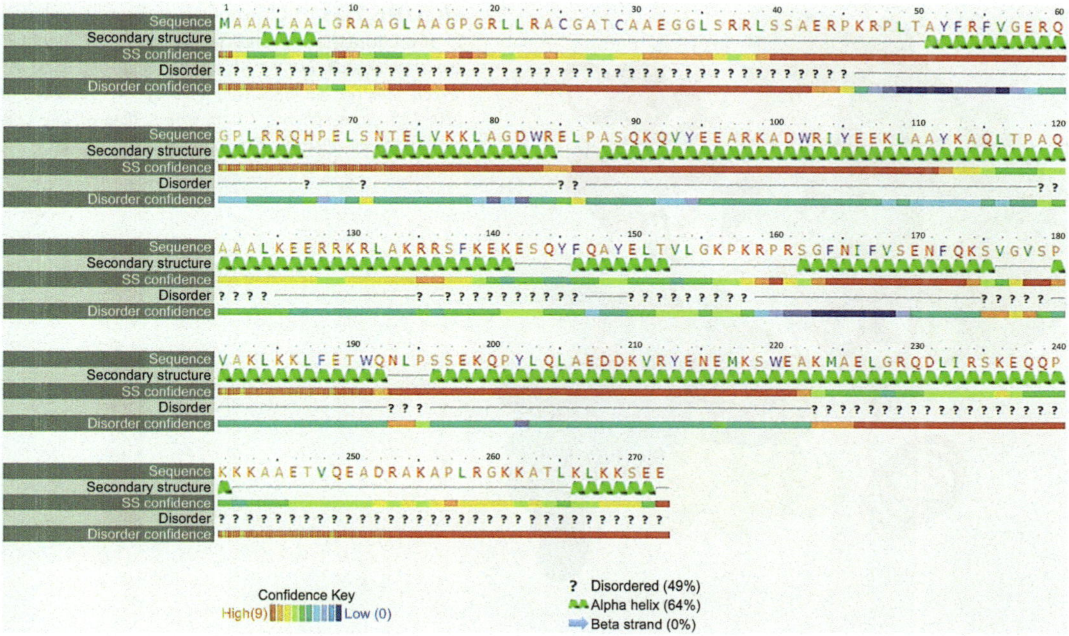

FIGURE 7.23 Secondary structure prediction for TFM of *Anas platyrhynchos*.

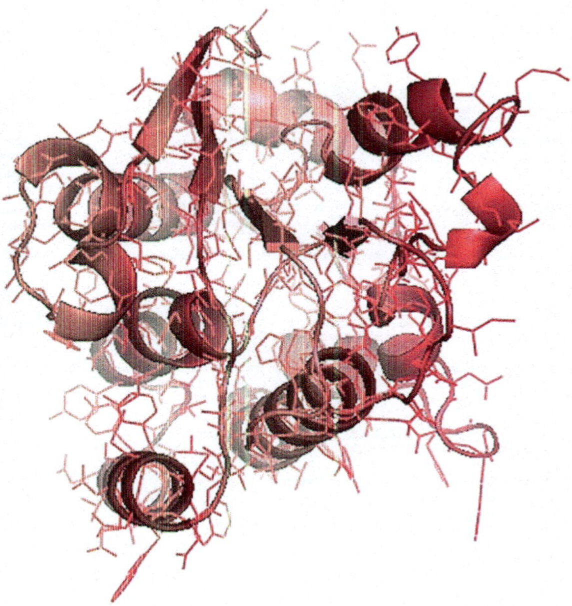

FIGURE 7.24 3D structure of thymidine kinase 2, mitochondrial of *Anas platyrhynchos*.

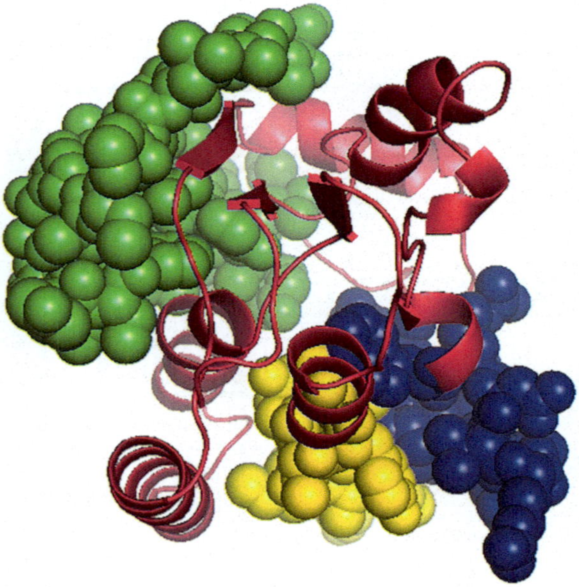

FIGURE 7.25 3D structure of thymidine kinase 2, mitochondrial of *Anas platyrhynchos*. *Domain for deoxynucleoside kinase at amino acid (aa) 230–434 (green sphere). Composition bias of Pro residues (aa 110–125: blue sphere), Composition bias of low complexity (at aa position 126- 137: yellow sphere).*

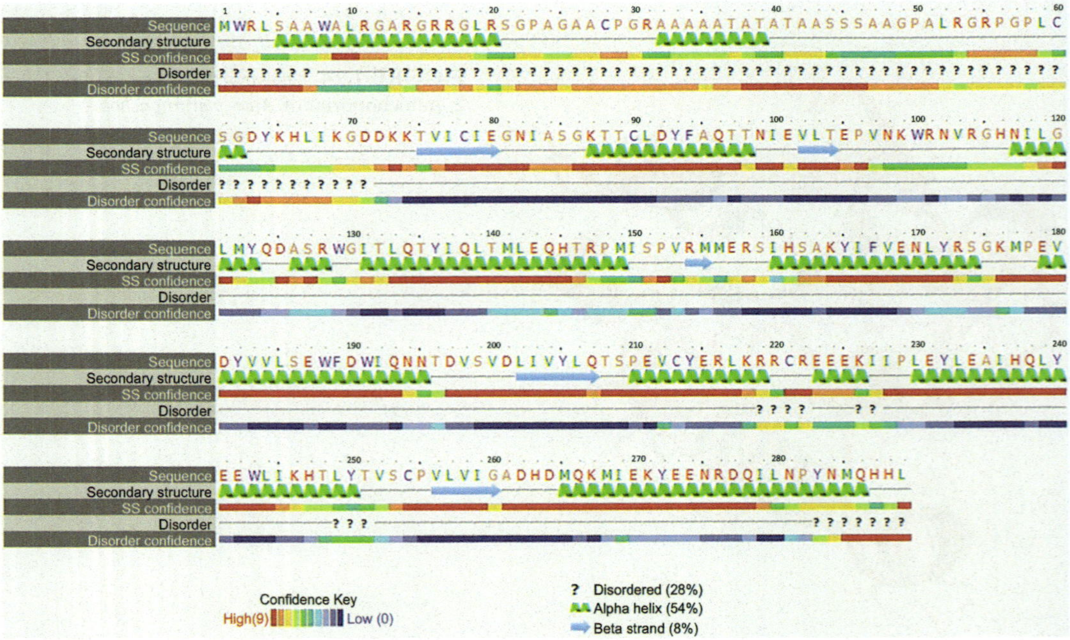

FIGURE 7.26 Secondary structure predicted for thymidine kinase 2, mitochondrial of *Anas platyrhynchos*.

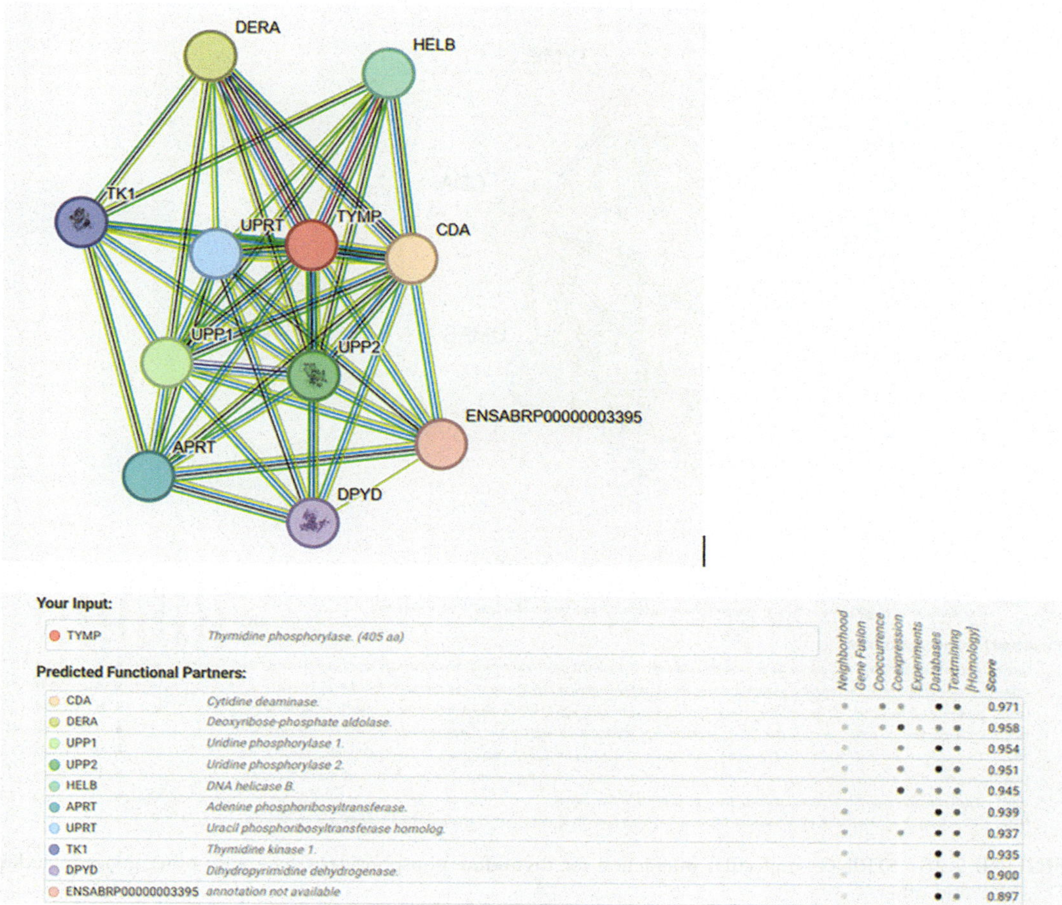

FIGURE 7.27 STRING -molecular interaction for thymidine phosphorylase gene with DERA and other genes.

susceptibility or resistance were observed for Bengal ducks with respect to infection of *P. multocida*.

NUMT genes, such as TP/ECGF1 gene, TFAM, TK2, mitochondrial genes, were upregulated in healthy in comparison to infected ducks: Fig. 7.31, Fig. 7.32, Fig. 7.33

Differential mRNA expression profiling for mitochondrial gene cytochrome B and nuclear origin cytochrome C

Differential mRNA expression profiling for cytochrome B and cytochrome C has been depicted. Gene expression of both cytochrome B and cytochrome C was observed to be better in healthy ducks in comparison to infected ducks.

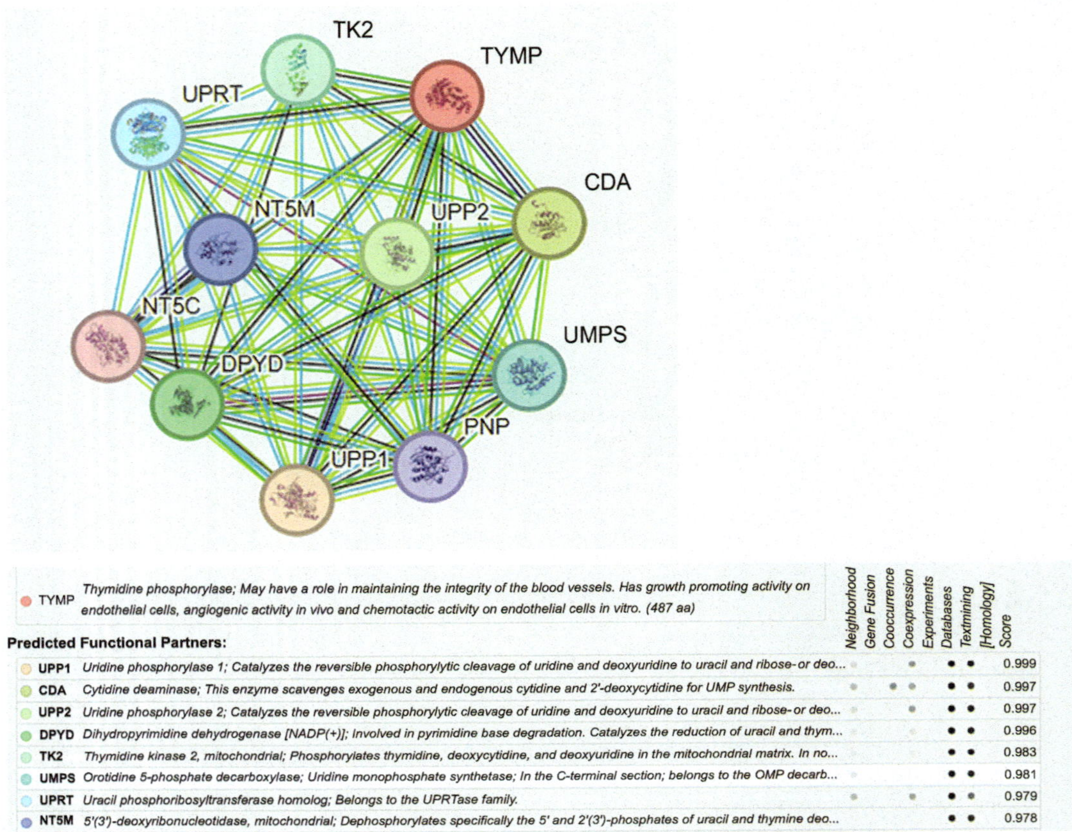

FIGURE 7.28 STRING– molecular interaction for thymidine phosphorylase gene with other genes including Tk1, PNP and others.

KEGG pathway analysis

Mitochondria have a role against bacterial infection. It acts through various pathways-cytosolic DNA sensing pathway, NOD-like receptor signalling pathway, phagocytic pathway (Fig. 7.9 a). Detailed pathways have been depicted for some gram-negative bacteria such as *Pasteurella spp.* namely, *Shigella, E. coli and Salmonella spp.*

NOD-like receptor signalling pathway

Fig. 7.34

Phagocytic pathway

Fig. 7.35

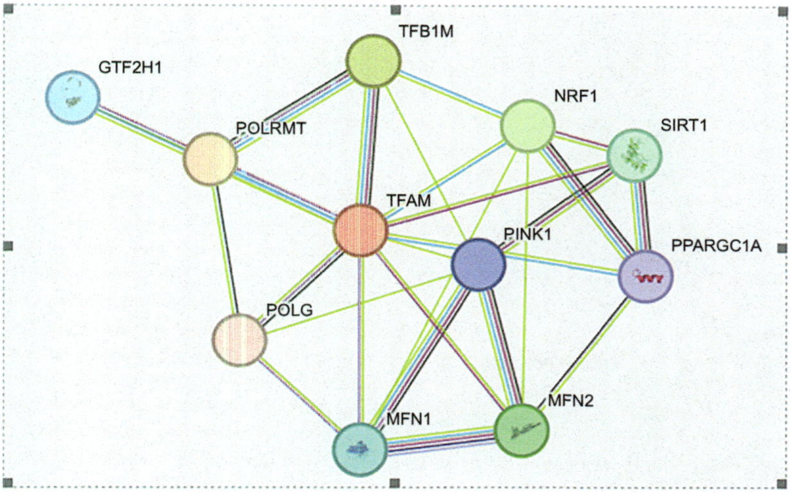

Your Input:

TFAM *Transcription factor A, mitochondrial. (379 aa)*

Predicted Functional Partners:

		Neighborhood	Gene Fusion	Cooccurrence	Coexpression	Experiments	Databases	Textmining	[Homology]	Score
POLRMT	*DNA-directed RNA polymerase; DNA-dependent RNA polymerase catalyzes the transcription of DNA into RNA using the four...*					•	•	•		0.985
TFB1M	*rRNA adenine N(6)-methyltransferase; Belongs to the class I-like SAM-binding methyltransferase superfamily. rRNA adenine...*				•	•	•	•		0.907
NRF1	*Nuclear respiratory factor 1.*						•	•		0.872
MFN2	*Mitofusin 2; Belongs to the TRAFAC class dynamin-like GTPase superfamily. Dynamin/Fzo/YdjA family.*					•		•		0.757
SIRT1	*Sirtuin 1.*					•		•		0.751
MFN1	*Mitofusin 1; Belongs to the TRAFAC class dynamin-like GTPase superfamily. Dynamin/Fzo/YdjA family.*					•		•		0.669
GTF2H1	*General transcription factor IIH subunit 1.*					•	•	•		0.660
PINK1	*PTEN induced kinase 1.*							•		0.644
PPARGC1A	*PPARG coactivator 1 alpha.*						•	•		0.644
POLG	*DNA polymerase gamma, catalytic subunit.*				•	•		•		0.640

Your Current Organism:

Anas platyrhynchos

NCBI taxonomy Id: 8840

Other names: A. platyrhynchos platyrhynchos, Anas platyrhynchos platyrhynchos, common mallard, northern mallard

FIGURE 7.29 STRING-based molecular interaction for TFAM gene with other genes.

Cytosolic DNA-sensing pathway

Fig. 7.36

Role of mitochondria against Gram-negative bacteria

In this section, we attempt to discuss about the functioning of mitochondria through different pathways in immune response against certain economically important gram negative bacteria.

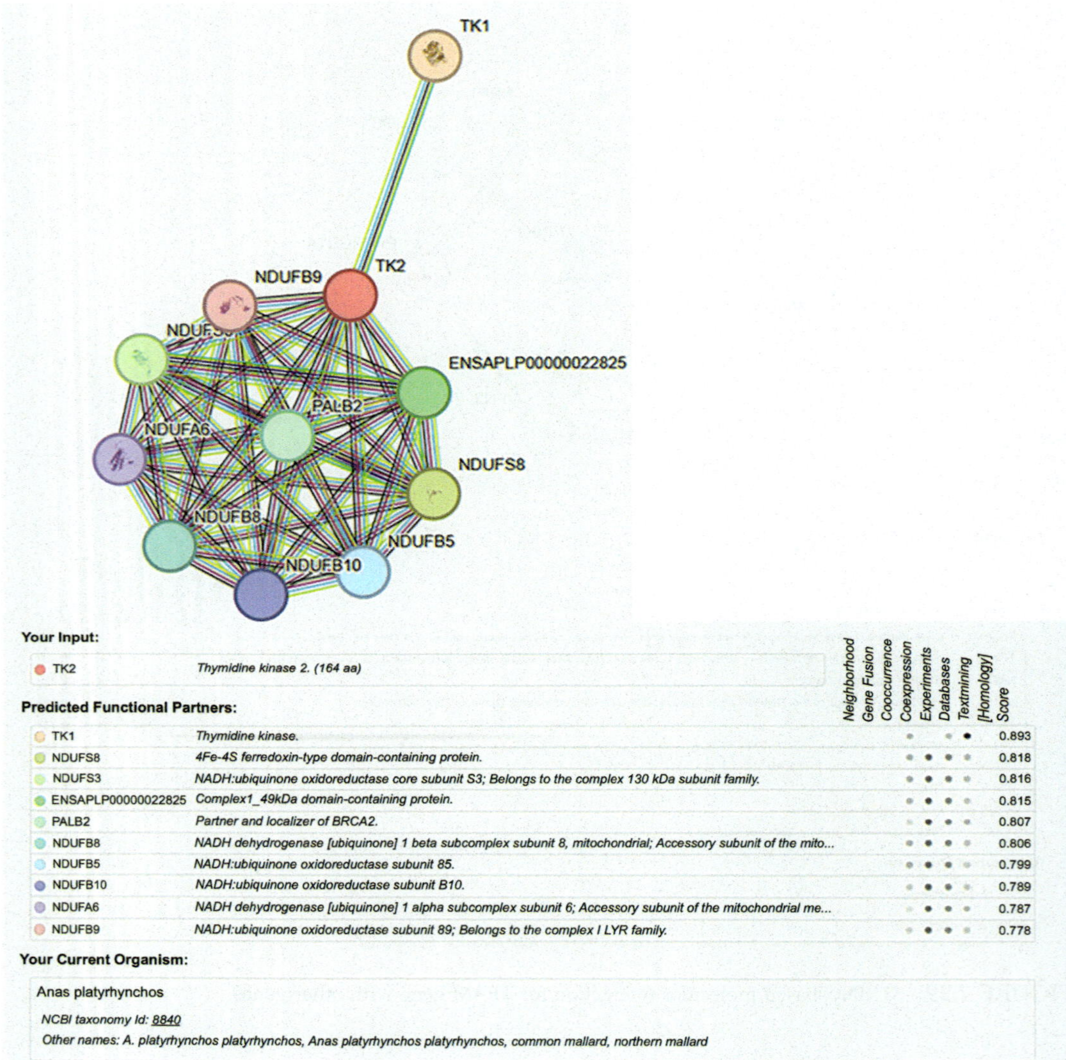

FIGURE 7.30 STRING- based molecular interaction for thymidine kinase (TK) gene with other genes.

Shigellosis

Fig. 7.37

E. coli

Fig. 7.38

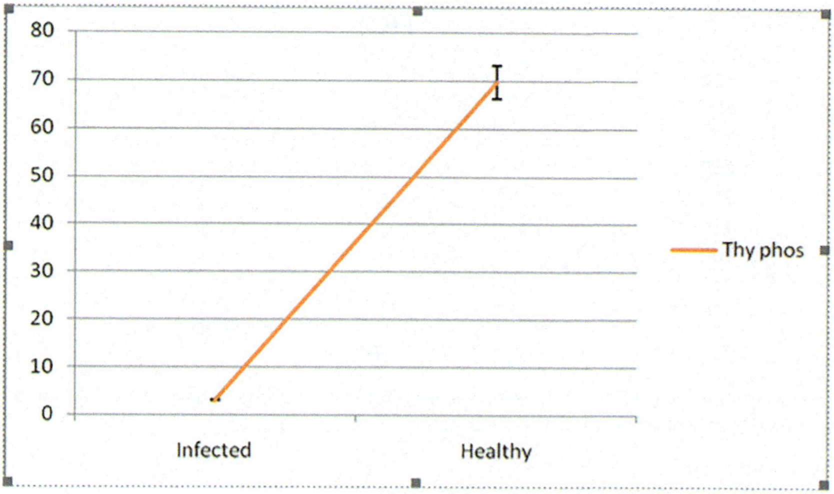

FIGURE 7.31 Differential mRNA expression profiling for as thymidine phosphorylase/endothelial cell growth factor1 gene (nuclear mitochondrial genes) with respect to disease resistance trait in duck.

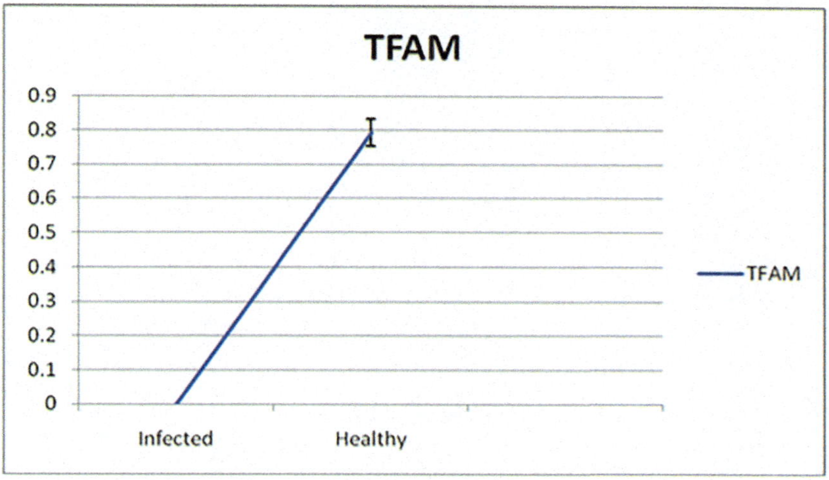

FIGURE 7.32 Differential mRNA expression profiling for transcription factor A, mitochondrial (TFAM) gene (nuclear mitochondrial genes) with respect to disease resistance trait in duck.

Salmonellosis

Fig. 7.39
Fig. 7.40

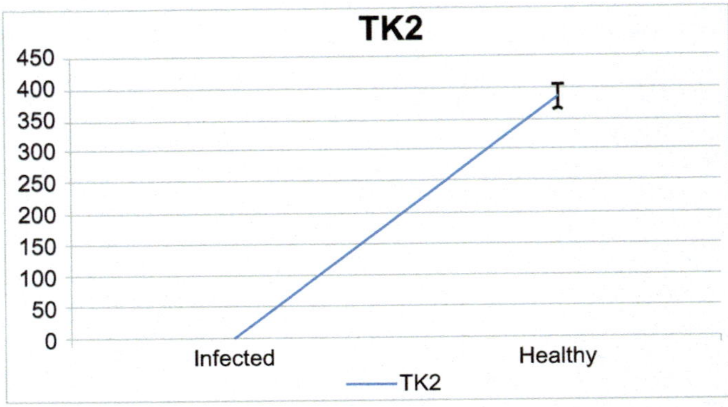

FIGURE 7.33 Differential mRNA expression profiling for thymidine kinase 2 (TK2), mitochondrial gene (nuclear mitochondrial genes) with respect to disease.

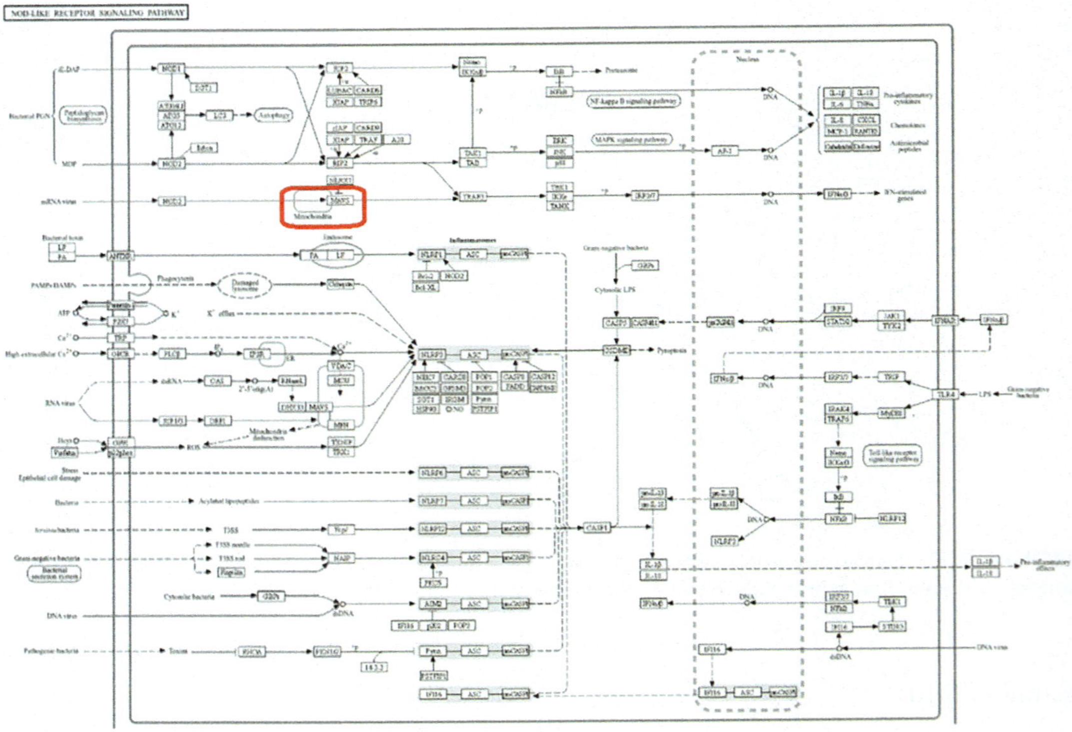

FIGURE 7.34 NOD-like receptor signalling pathway.

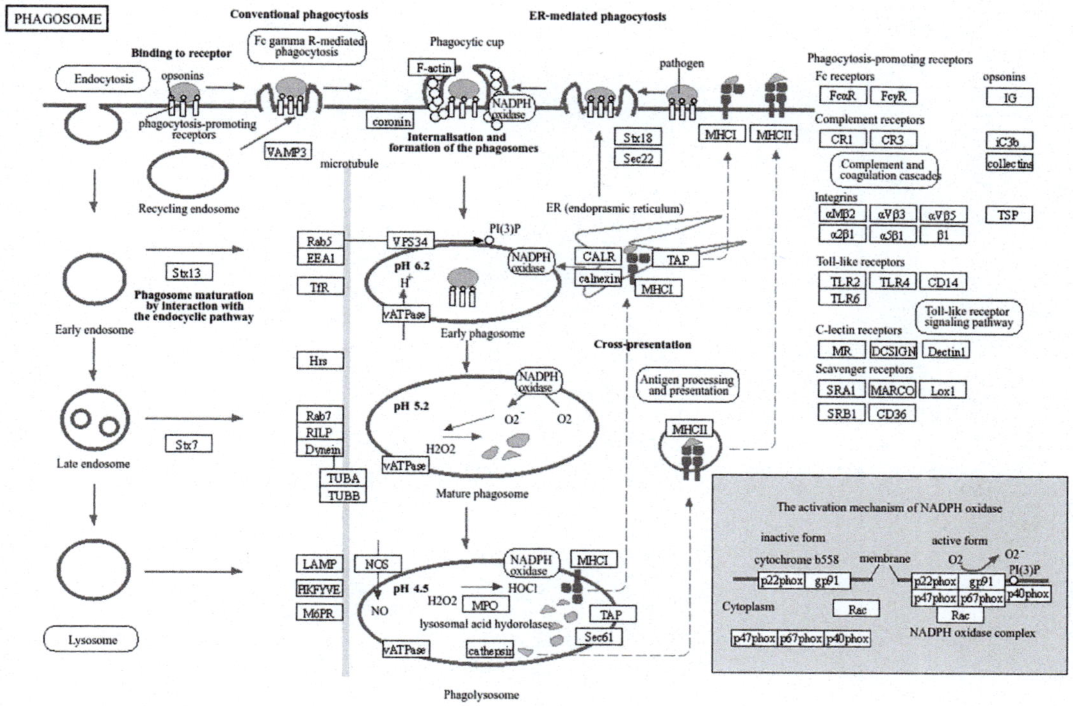

FIGURE 7.35 Involvement of cytochrome B is an indication of involvement of mitochondrial function.

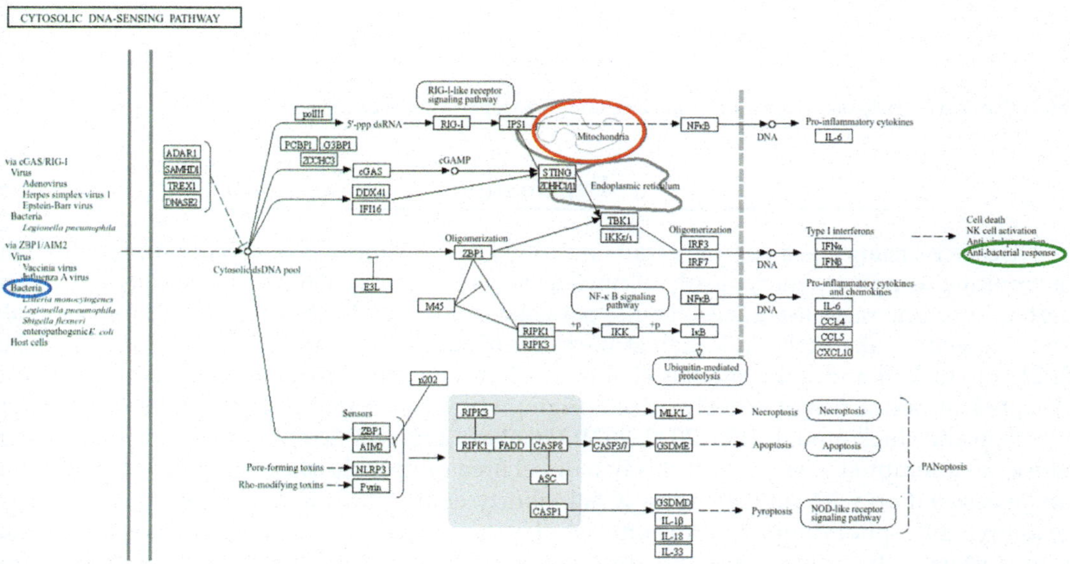

FIGURE 7.36 KEGG pathway analysis-depicting the role of mitochondria against bacterial infection through various molecular pathways.

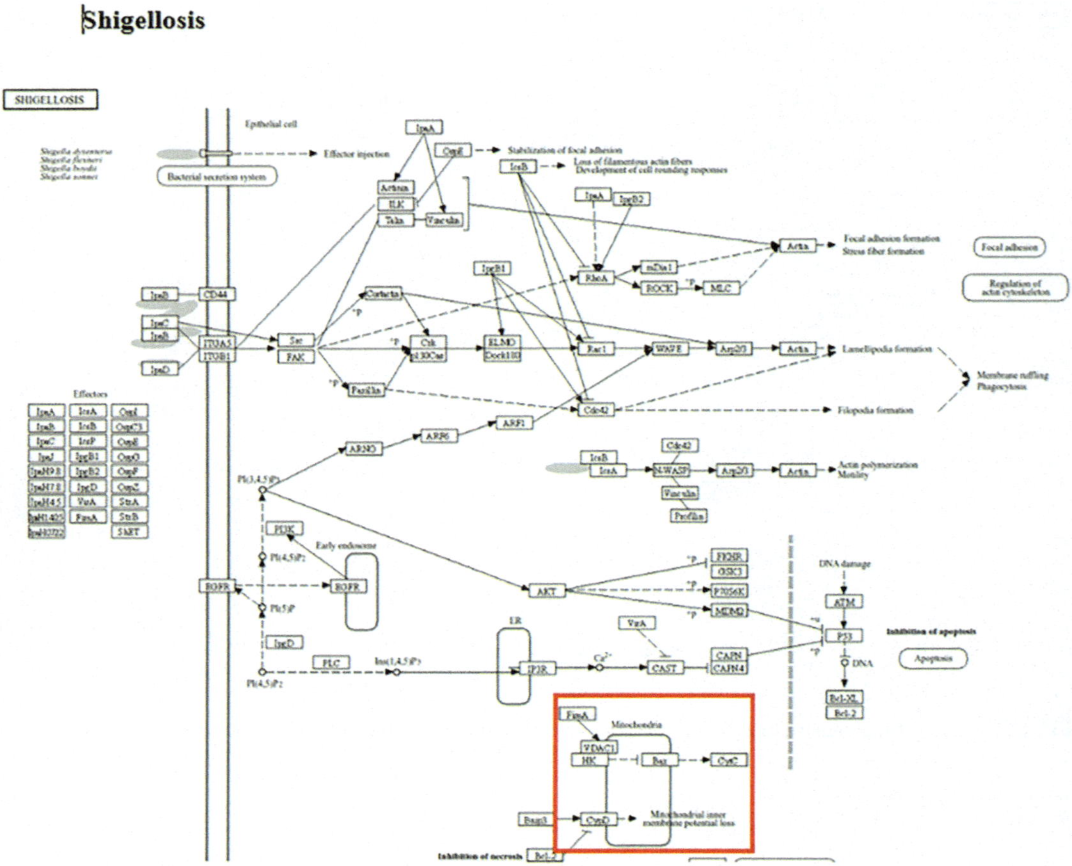

FIGURE 7.37 Role of mitochondria affecting immune response against shigellosis.

Explanation

Disease resistance is a complex trait controlled by polygenes. We usually designate them as immune response genes, mostly of nuclear origin. In our earlier studies, we could explore certain immune response genes against parasitic infestation in sheep model (Banerjee et al., 2021; Rawat et al., 2021) and against bacterial (Chakraborty et al., 2025), viral (Pal, 2020, 2021) in chicken and duck models (Pal et al., 2022a, 2022b). Simultaneously, we have also detected certain genes of mitochondrial origin, which are responsible for disease resistance in sheep and duck models (Pal, 2020, 2021; Pal et al., 2022a, 2022b). In the following research work, an attempted was made to study a novel area of the role of certain nuclear genes that act in association with mitochondria (Chakraborty et al., 2025). It is evident that apart from oxidative phosphorylation, mitochondria play a variety of functions including calcium transporters, cell proliferation and differentiation, free radical production, apoptosis and

2. E.coli

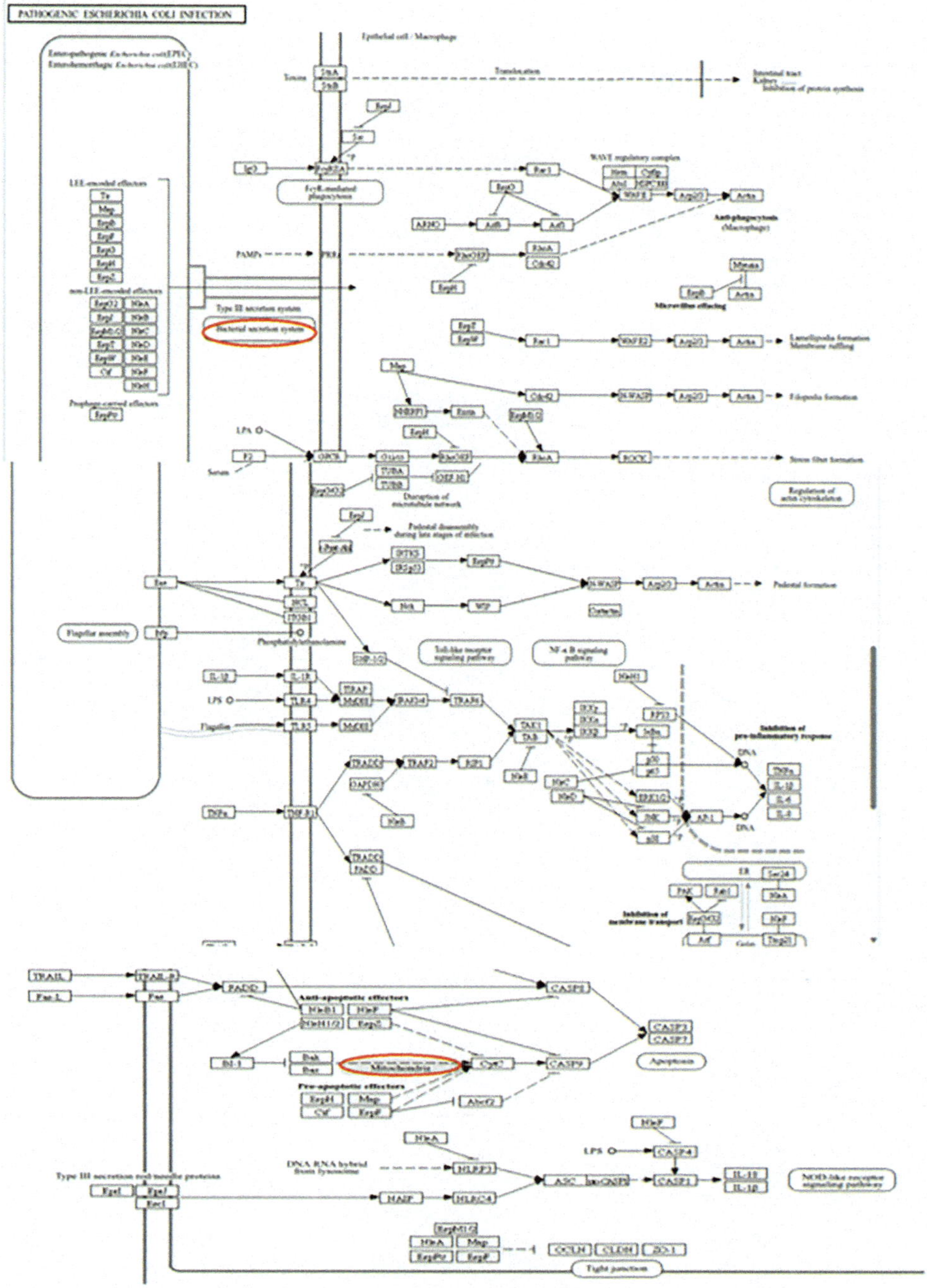

FIGURE 7.38 Role of mitochondria affecting immune response against E. coli.

3. Salmonellosis

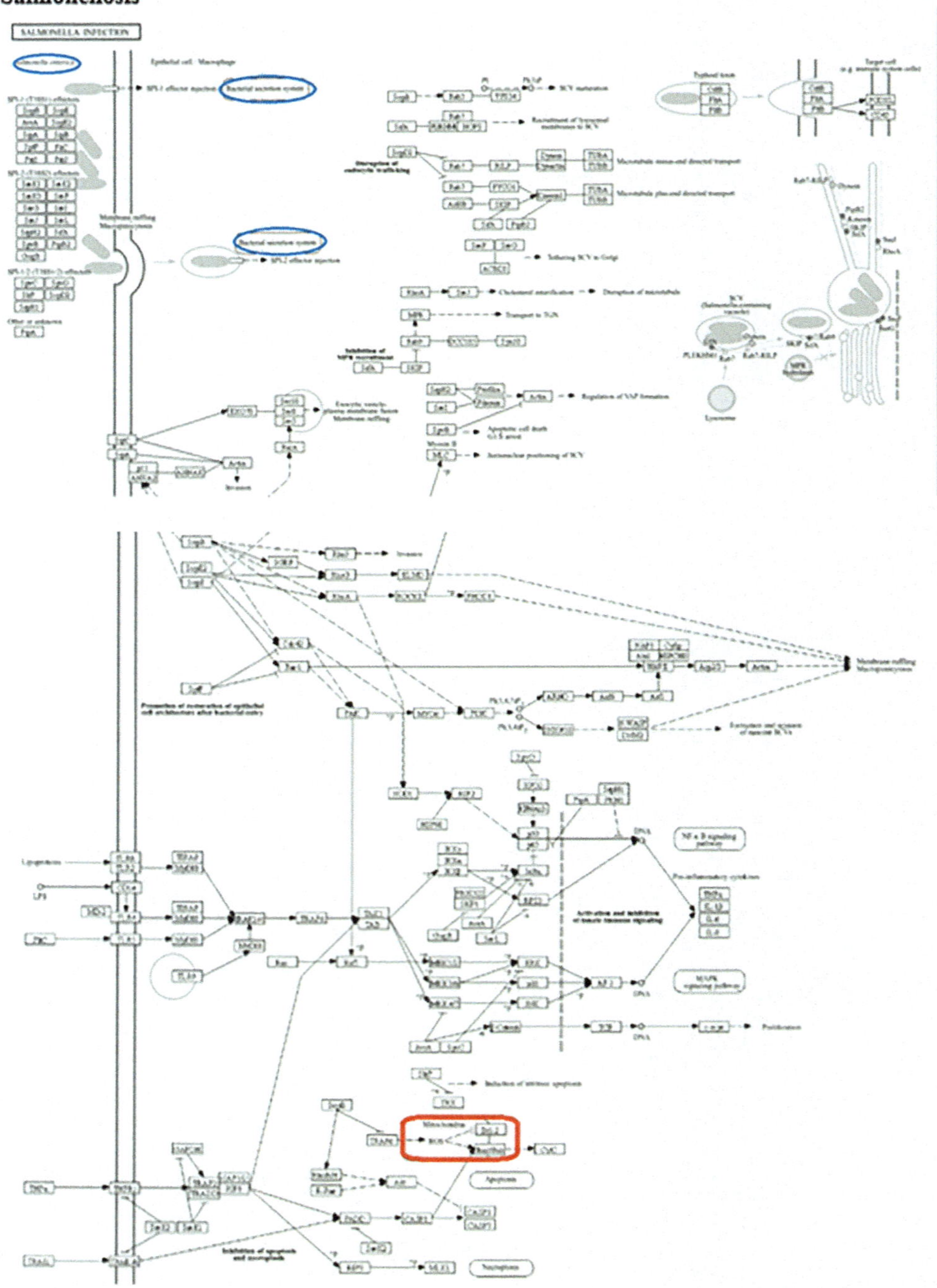

FIGURE 7.39 KEGG pathway Analysis-depicting the role of mitochondria against Gram-negative bacterial infection.

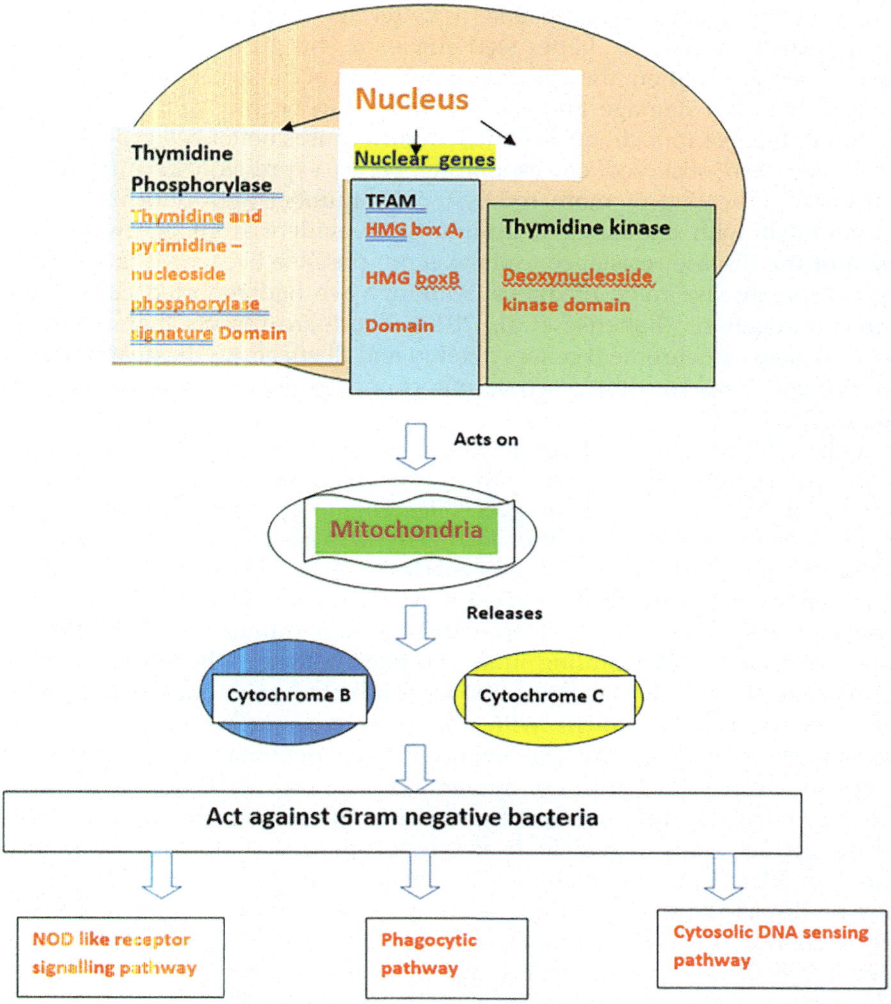

FIGURE 7.40 Mechanism of action: how nuclear mitochondrial (NUMT) genes act against Gram-negative bacterial infection. Better expression of NUMT genes is being manifested in healthy birds upon challenge with pathogenic Gram-negative bacteria. On the other hand, birds with less NUMT gene expression and less mitochondrial- associated gene expression were observed to be infected with *Pasteurella multocida*.

autophagy, ageing, signalling, immunity, growth and reproduction. (Gunter et al., 2000, Maeda & Chida, 2013, Angajala et al., 2018; Pal et al., 2019; Pradhan et al., 2018).

In this current article, we have explored another set of genes which are basically of nuclear origin, but they act through mitochondria, designated as NUMT. We have considered duck as the model organism, considering its hardy nature. In healthy animals, there is upregulation of NUMT genes, which means they are expressed better, which in turn influences mitochondria that act against bacterial infection through various pathways.

Upregulation of these genes is responsible for better nerve impulse transmission, mitochondrial and cellular homoeostasis, better feed efficiency and growth. While in a pathological condition or disease condition, there is downregulation of the genes because there is cellular damage, mitochondrial damage and tissue damage. In healthy animals, upregulation of NUMT genes influences mitochondria, which in turn causes better cell growth, reproduction and metabolism. We observed corresponding better expression of mitochondrial cytochrome B gene, indicative of more mtDNA copy number and cytochrome C, which is directly associated with antibacterial immunity as evident from pathway analysis. The comparison of the disease resistance traits was not possible because it is the first report of NUMT genes on disease traits of ducks. Although we had reported certain preliminary studies on numtogenesis (Banerjee et al., 2020). There are reports that a mutation in the mitochondrial gene cytochrome B causes debility and ill effects on the health of sheep (Pal et al., 2019). We had already studied whole mitochondrial genome sequencing for the duck (Pal et al., 2022a, 2022b).

Singh et al. (2017) reported that numtogenesis or insertion of the NUMTs into the nuclear genome of somatic cells can alter the pathways contributing to tumorigenesis and disrupt the functions of tumour suppressor genes, leading to the development of cancer. The NUMTs can sometimes also directly or indirectly activate oncogenes and thereby further drive development of tumour. Srinivasainagendra et al. (2017) in their studies also revealed that Numtogenesis may lead to the development of cancer. Some study reported that the integration of mtDNA in the IVS 4 acceptor site and human factor VII gene causes a deleterious mutation effect, resulting in severe plasma factor VII deficiency and bleeding diathesis. Turner et al. (2003) in their studies conducted on a patient with Pallister-Hall syndrome characterised a patient with the de novo nucleic acid transfer from the mitochondrial genome to the nuclear genome. They demonstrated that this transfer or numtogenesis is a novel mechanism of human inherited disease. Goldin et al. (2004) in their studies revealed an inherited transfer of mtDNA fragment into the nuclear DNA, causing numtogenesis, which leads to genetic disease mucolipidosis IV. In healthy animals, there is upregulation of NUMT genes, which means they are expressed better, while in pathological conditions or disease conditions, there is downregulation of the genes because there is cellular, mitochondrial and tissue damage. In healthy animals, upregulation of these genes causes better cell growth, reproduction and metabolism. Upregulation of these genes is responsible for better nerve impulse transmission, mitochondrial and cellular homoeostasis, better feed efficiency and growth. Whereas in a pathological condition, there is cellular-level stress and hence the genes are not expressed well.

NUMT genes identified with significant effect on bacterial immunity were TP, TFAM, TK2, revealed through differential mRNA expression profiling. We characterised them and identified important domains, molecular interactions among them in mitochondrial function.

In our current study, we observed better expression of TP in healthy ducks in comparison to *Pasteurella* (bacteria) infected. Nishino et al. (1999) revealed its involvement in mitochondrial genome maintenance. Thus in healthy ducks, due to better expression of TP, mitochondrial genomes were better maintained. This leads to better metabolism in healthy subjects. Important domains were identified as *thymidine and pyrimidine-nucleoside phosphorylases signature*, glycosyl transferase family 3N-terminal, glycosyl transferase family.

Thymidine and pyrimidine-nucleoside phosphorylases act as a catalyst for the reversible phosphorolysis of thymidine to thymine and 2dDR1P, thereby maintaining nucleoside homoeostasis for efficient DNA replication and cell division. Two disordered regions have been predicted along with the compositional bias with glycine residues. Molecular interaction of TP is associated with cytidine deaminase, deoxyribose phosphate aldolase, uridine phosphorylase1, uridine phosphorylase 2, DNA helicase B, adenine phosphoribosyl transferase, uracil phosphoribosyl transferase, TK 1, dihydropyrimidine dehydrogenase, TK2, mitochondrial, Orotidine 5′-phosphate decarboxylase, uridine monophosphate synthetase, 5′(3′) deoxyribonucleotidase, mitochondrial, purine nucleoside phosphorylase, 5′(3′) deoxyribonucleotidase, cytosolic. TYMP/ECGF1encodes for the protein TP, also known as TdRPase, or ECGF1, gliostatin and PD-ECGF. In mice, it is comprised of 471 aa, in rats 476 aa and 482 aa in humans (Uniprot, P19971 • TYPH_HUMAN, Uniprot, Q99N42 • TYPH_-MOUSE Uniprot, Q5FVR2 • TYPH_RAT). Accordingly, this gene was observed to be involved in the disease of Mitochondrial DNA depletion syndrome 1 (MTDPS1), MNGIE type in humans, which is a multisystem disease associated with mitochondrial dysfunction. This disease is clinically characterised by its onset between the second and fifth decades of life; there is ptosis, gastrointestinal dysmotility (often pseudo-obstruction), progressive external ophthalmoplegia, cachexia, diffuse leukoencephalopathy, peripheral neuropathy and myopathy.

Differential mRNA expression profile depicts better expression of TFAM in healthy ducks compared to those of infected ones. TFAM possess very important mitochondrial functions, thus leading to better mitochondrial copy number and better secretion of cytochrome C. Increased amount of mitochondria with increased concentration of cytochrome C triggers antibacterial immunity. We identified two important domains in duck TFAM as HMG-box A DNA binding domain and HMG-box B DNA binding domain and predicted their site. The knowledge of this region will be helpful for future gene cloning, knock-in for gene editing purposes. In molecular biology, the HMG-box is a protein domain which is involved in DNA binding. The domain is composed of approximately 75 aa residues that collectively mediate the DNA-binding of chromatin-associated high-mobility group proteins. TFAM interacts with other molecules for its effective biological function, namely, DNA-directed RNA polymerase, rRNA adenine N(6) methyl transferase (TFB1M), nuclear respiratory factor1, Mfn2, Mfn1, Sirtuin1, General transcription factor IIH subunit 1, PINK1, PPARG coactivator 1 alpha, DNA polymerase gamma indicating its close association to mitochondrial function, while these genes affect vital functional role in mitochondria. TFAM codes 246 aa, encodes for protein TFAM. We identified important domains Celestini et al. (2018) and Hao et al. (2020) reported that this gene binds to the mitochondrial light strand promoter and functions in mitochondrial transcription regulation. This gene was observed to be the component of the mitochondrial transcription initiation complex, composed at least of TFB2M, TFAM and POLRMT, which is required for basal transcription of mtDNA. It was also reported that the involvement of this gene in transcription initiation at the mitochondrial promoter and positively regulates DNA-templated transcription. This gene is required for accurate and efficient promoter recognition by the mitochondrial RNA polymerase. It promotes transcription initiation from the HSP1 and the light strand promoter by binding immediately upstream of transcriptional start sites and unwinding of DNA. TFAM knockdown causes a reduction of mtDNA copy number and abnormalities in the embryos of

Zebrafish. TFAM was observed to bend the mitochondrial light strand promoter DNA into a U-turn shape via its HMG boxes. It is also required for the maintenance of normal levels of mtDNA, as well as playing a role in organising and compacting mtDNA, enabling chromatin binding. TFAM was observed to be a part of a protein-containing complex, which is located in the mitochondrial nucleoid. Gangelhoff et al. (2009) reported that this gene enables mitochondrial promoter sequence-specific DNA binding. TFAM is involved in mitochondrial transcription and enables mitochondrial transcription factor activity. Gaudet et al. (2011) reported that this gene enables transcription, cis-regulatory region binding and DNA binding and bending. He also revealed its involvement in mitochondrial respiratory chain complex assembly. Due to mutation in this gene, a severe disease condition is seen known as MTDPS15, hepatocerebral type, which is characterised by severe intrauterine growth restriction, neonatal-onset hypoglycemia and liver dysfunction. There is also depletion of mtDNA in the liver and skeletal muscle and abnormal mitochondrial morphology is observed in skeletal muscle. Hepatic pathology includes cirrhosis, steatosis and cholestasis. There is rapid progression to liver failure and death with no neurological impairment or other organ involvement.

TK2 gene was observed to be better expressed many-folds in healthy ducks compared to infected ones. TK2 encodes the protein TK2, mitochondrial. Important domains identified include deoxynucleoside kinase and its site was also predicted. The predicted site will be helpful for gene cloning and gene editing through *knock in* for the production of disease-resistant birds. Mammals have four deoxyribonucleoside kinases, the cytoplasmic (TK1) and mitochondrial (TK2) TKs, and the deoxycytidine (dCK) and deoxyguanosine (dGK) kinases, which salvage the precursors for nucleic acid synthesis. It is composed of 270 aa in *Mus musculus* and 265 aa in *Homo sapiens*. It was reported that TK2 phosphorylates thymidine, deoxycytidine and deoxyuridine in the mitochondrial matrix. They also reported in nonreplicating cells, where cytosolic dNTP synthesis is downregulated, mtDNA synthesis depends solely on TK2 and DGUOK. This gene can be widely used as a target of antiviral and chemotherapeutic agents. They also reported that it enables deoxycytidine kinase activity. Gaudet et al. (2011) reported that this gene is active in the cytoplasm and enables deoxynucleoside kinase activity. The TK2 gene is also active in the mitochondrion, enabling thymidine kinase activity. Researchers reported its involvement in the mtDNA metabolic process and mt DNA replication (Akman et al., 2008, Zhou et al., 2008). This gene is also involved in the deoxycytidine metabolic process, DNA biosynthetic process, phosphorylation, thymidine metabolic process, nucleobase-containing compound metabolic process and pyrimidine- nucleoside salvage. A mutation in the TK2 gene causes the disease 'Mitochondrial DNA depletion syndrome 2 (MTDPS2)', which is characterised by the childhood onset of muscle weakness associated with the depletion of mtDNA in skeletal muscle. There is wide clinical variability; some patients have onset in infancy and show a rapidly progressive course with early death due to respiratory failure, whereas others have a later onset of a slowly progressive myopathy in humans. TK2 gene was observed to have molecular interaction with TK1, 4Fe-4S ferredoxin type domain containing protein (NDUFS8), NADH-ubiquinone oxidoreductase core subunit S3 (NDUFS3), Partner and localiser of BRCA2, NADH dehydrogenase (ubiquinone)1 beta subcomplex subunit 8 (NDUFB8), NADH ubiquinone oxidoreductase subunit B5(NDUFB5), NADH ubiquinone oxidoreductase subunit B10 (NDUFB10), NADH ubiquinone oxidoreductase subunit B9(NDUFB9),

NADH dehydrogenase (ubiquinone)1 alpha subcomplex subunit 6 (NDUFB6). These interaction reflects its close association with mitochondrial function.

In order to further explore the role of NUMT genes, how they act through mitochondria, we have further assessed the differential mRNA expression profiling of cytochrome B and cytochrome C genes. Better expression profile for cytochrome B in healthy ducks is an indication of increased copy number of mtDNA, or mitochondria. Better expression of cytochrome C is directly related to antibacterial immunity, as evident from molecular pathway analysis against various bacteria (E. coli, Salmonella, Shigella). Pathway analyses directly depict the role of mitochondria against bacterial infection through cytosolic DNA-sensing pathway, NOD-like receptor signalling pathway and Phagocytic pathway.

Conclusion

The better expression profile for the other two genes, was also observed as TFAM and TK2 affecting reproduction. Nuclear genes involved in mitochondrial structure and functionnamely, TYMP, TFAM and TK2 were upregulated in healthy in comparison to infected ducks, which in turn influences mitochondria for the production of more mtDNA or mitochondria as a whole.

References

Akman, H. O., et al. (2008). Thymidine kinase 2 (H126N) *knock in* mice show the essential role of balanced deoxynucleotide pools for mitochondrial DNA maintenance. *Hum Mol Genet, 17*, 2433–2440.

Banerjee, S. et al. Role of NUMT in disease occurrence and immunity. BIORXIV/2020/326355 (2020).

Banerjee, S., Pal, A., Mandal, S. C., Chatterjee, P. N., & Chatterjee, J. K. (2021). RIG-I has a role in *immunit*y against *Haemonchus contortus*, a gastrointestinal parasite in *Ovis aries*- a novel report. *Frontiers in immunology*. https://doi.org/10.3389/fimmu.2020.534705 08 January 2021.

Banerjee, S., Pal, A., Pal,Ab., Chatterjee, J.K.2023. Ovine MyD88 have active role in host resistance against Haemonchus contortus infection. doi: https://doi.org/10.1101/2023.10.08.561398.

Celestini, V., et al. (2018). Uncoupling FoxO3A mitochondrial and nuclear functions in cancer cells undergoing metabolic stress and chemotherapy. *Cell Death Dis, 9*, 231.

Chakraborty, A., Pal, A., Batabyal, S., Chatterjee, P.N. 2025. Mitochondrial mutations causes susceptibility to bacterial infection (Duck Pasteurellosis) in duck model- a study revealed through *whole mitochondrial genome sequencing (NGS) study, Proceedings on National conference on "Advancing Animal Health and production through Biochemical and Biotechnological innovations"* University of Animal and Fishery Sciences, 37, K.B. Sarani, Kolkata-37. West Bengal. India. Pin-700037.

Debnath, M., Pal,A., Chakraborty, A., Pal, S., Pal,Ab. 2022. Genomics for reproduction in Anas platyrynchos-a novel report. https://doi.org/10.1101/2022.05.29.493861.

Ebina, T., Toh, H., & Kuroda, Y. (2009). Loop-length-dependent SVM prediction of domain linkers for high-throughput structural proteomics. *Biopolymers, 92*, 1–8. https://doi.org/10.1002/bip.21105.

Franceschini, A., Szklarczyk, D., Frankild, S., Kuhn, M., Simonovic, M., Roth, A., Lin, J., Minguez, P., Bork, P., Von Mering, C., & Jensen, L. J. (2012). STRING v9.1: protein-protein interaction networks, with increased coverage and integration. *Nucleic Acids Research, 41*, D808–D815. https://doi.org/10.1093/nar/gks1094.

Gangelhoff, T. A., Mungalachetty, P. S., Nix, J. C., & Churchill, M. E. (2009). Structural analysis and DNA binding of the HMG domains of the human mitochondrial transcription factor A. *Nucleic Acids Res, 37*, 3153–3164.

Gaudet, P., Livstone, M. S., Lewis, S. E., & Thomas, P. D. (2011). Phylogenetic-based propagation of functional annotations within the Gene Ontology consortium. *Brief Bioinform, 12*, 449–462.

Goldin, E., et al. (2004). Transfer of a mitochondrial DNA fragment to MCOLN1 causes an inherited case of mucolipidosis IV. *Hum Mutat, 24*, 460–465.

Glick, D. M. (1977). Revised edition*Glossary of Biochemistry and Molecular Biology*. London, UK: Portland Press, 324.

Gunter, T. E., Buntinas, L., Sparagna, G., Eliseev, R., & Gunter, K. (2000). Mitochondrial calcium transport: mechanisms and functions. *Cell Calcium, 28*, 285–296.

Hao, Z., et al. (2020). N6-Deoxyadenosine Methylation in Mammalian Mitochondrial DNA. *Mol Cell, 78*, 382–395.

Horowitz, M. P., & Greenamyre, J. T. (2010). Mitochondrial iron metabolism and its role in neurodegeneration. *J. Alzheimers Dis. 20*(Suppl 2), S551–S568. https://doi.org/10.3233/JAD-2010-100354.

Katoh, K., & Standley, D. M. (2013). MAFFT multiple sequence alignment software version 7: improvements in performance and usability. *MolBiolEvol, 30*, 772–780.

Kelley, L., et al. (2015). The Phyre2 web portal for protein modeling, prediction and analysis. *Nature Protocols, 10*, 845–854. https://doi.org/10.1038/nprot.2015.053.

Kiefer, F., Arnold, K., Künzli, M., Bordoli, L., & Schwede, T. (2009). The SWISS-MODEL Repository and associated resources. *Nucleic Acids Res, 37*, D387–D392.

Maeda, Y., & Chida, J. (2013). Control of cell differentiation by mitochondria, typically evidenced in dictyostelium development. *Biomolecules, 3*, 943–966.

Muroi, M., Ohnishi, T., & Tanamoto, K. (2002). Regions of the Mouse CD14 Molecule Required for Toll-like Receptor 2- and 4-mediated Activation of NF-κB. *Journal of Biological Chemistry, 277*, 42372–42379. https://doi.org/10.1074/jbc.M205966200.

Nishino, I., Spinazzola, A., & Hirano, M. (1999). Thymidine phosphorylase gene mutations in MNGIE, a human mitochondrial disorder. *Science, 283*, 689–692.

Pal, A., & Chatterjee, P. N. (2009). Molecular cloning and characterization of CD14 gene in goat. *J. of Small Ruminant Research, 82*, 84–87.

Pal, A., Chakravarty, A. K., & Chatterjee, P. N. (2014a). Polymorphism of Growth hormone gene and its association with seminal and sexual behavioral traits in crossbred cattle. *Theriogenology, 81*, 474–480.

Pal, A., Chatterjee, P. N., & Sharma, A. (2014b). Sequence characterization and polymorphism detection in bubaline CD14 gene. *Buffalo Bulletin, 32*(2), 138–156.

Pal, A., Sharma, A., Bhattacharya, T. K., Chatterjee, P. N., & Chakravarty, A. K. (2011). Molecular characterization and SNP detection of CD 14 gene in crossbred cattle. *Molecular Biology International, vol. 2011*(13), 2011. https://doi.org/10.4061/2011/507346 Article ID 507346.

Pal, A. (2020). *Bengal duck-The hardy breed of West Bengal*. West Bengal University of Animal and Fishery Sciences, West Bengal, India: DREF, 55 Published by.

Pal, A. (2021). Protocols in Advanced Genomics and allied techniques. In *Springer Protocols Handbooks*. Springer Publication. https://doi.org/10.1007/978-1-0716-1818-9 ISBN 978-1-0716-1817-2 ISBN 978-1-0716-1818-9 (eBook).

Pal, A., & Banerjee, S. (2018). Mitochondrial replacement therapy-a new remedy for defects in reproduction. *Indian. J. Anim. Sci. 88*(6), 637–644.

Pal, A., & Chakravarty, A. K. (2019). *Genetics & Breeding for the Disease resistance of livestock*. Elsevier Publication. Academic publisher ISBN: 9780128164068.

Pal, A., Chakraborty, A., Debnath, M. 2022a. Molecular evolution and characterization of domestic duck (*Anas platyrynchos*) and Goose (*Anser indicus*) with reference to its wild relatives through whole mitochondrial genome sequencing. https://doi.org/10.1101/2022.07.19.500621.

Pal, A., Chakraborty, A., Debnath, M. 2022b. Whole mitochondrial genome sequencing- a novel approach for studying Phylogenomics for *Anas platyrynchos*. https://doi.org/10.1101/2022.07.11.499548.

Pal, A., Pal, A. B., Banerjee, S., Batobyal, S., & Chatterjee, P. N. (2019). Mutation in ***Cytochrome B*** gene causes debility and adverse effects on ***health*** of sheep. *Mitochondrion, 46*, 393–404. https://doi.org/10.1016/j.mito.2018.10.003.

Pal, A., S. Banerjee, S. Batabyal, P.N. Chatterjee. 2025. Whole mitochondrial genome sequencing reveal certain mutations in cytochrome B and D loop adversely affecting the health of sheep. In the Proceedings of International Conference on "*Precision Animal Breeding through Genomics, artificial intelligence and machine learning*", 13-14th November, 2025. ISAGB conference at West Bengal University of Animal and Fishery Sciences, Kolkata, West Bengal, India.

Pradhan, M., Pal, A., Samanta, A. K., Banerjee, S., & Samanta, R. (2018). Mutations in ***cytochrome B gene*** effects female reproduction of Ghungroo pig. *Theriogenology, 119*, 121–130. https://doi.org/10.1016/j.theriogenology.2018.05.015.

Petersen, B., Petersen, T. N., Andersen, P., Nielsen, M., & Lundegaard, C. (2009). A Generic Method for Assignment of Reliability Scores Applied to Solvent Accessibility Predictions. *BMC Struct. Biol, 9*, 51. https://doi.org/10.1186/1472-6807-9-51.

Rawat, K. Pal, A. et al., 2023. Role of IL-6 and IL10 immune response genes against nematode *Haemonchus contortus in ovis aries- a novel report*. doi: https://doi.org/10.1101/2023.08.30.555538

Rawat, K., Pal, A., Banerjee, S., Pal, Ab, Mandal, S.C., Batabyal, S. 2021. Ovine CD14- an immune response gene has a role against gastrointesinal nematode *Haemonchus contortus* -a novel report. Frontiers in Immunology. https://doi.org/10.3389/fimmu.2021.664877.

Singh, K. K., Choudhury, A. R., & Tiwari, H. K. (2017). Numtogenesis as a mechanism for development of cancer. *Semin Cancer Biol, 47*, 101–109. https://doi.org/10.1016/j.semcancer.2017.05.003.

Srinivasainagendra, V., Sandel, M. W., Singh, B., Sundaresan, A., Mooga, V. P., Bajpai, P., Tiwari, H. K., & Singh, K. K. (2017). Migration of mitochondrial DNA in the nuclear genome of colorectal adenocarcinoma. *Genome Med, 9*(1), 31. https://doi.org/10.1186/s13073-017-0420-6.

Turner, C., et al. (2003). Human genetic disease caused by de novo mitochondrial-nuclear DNA transfer. *Hum. Genet. 112*, 303–309.

Wiederstein, M., & Sippl, M. J. (2007). ProSA-web: interactive web service for the recognition of errors in three-dimensional structures of proteins. *Nucleic Acids Research, 35*, W407–W410. https://doi.org/10.1093/nar/gkm290.

Zhang, Y., & Skolnick, J. (2005). TM-align: a protein structure alignment algorithm based on the TM-score. *Nucleic Acids Research, 33*, 2302–2309. https://doi.org/10.1093/nar/gki524.

Zhou, X., Solaroli, N., Bjerke, M., Stewart, J. B., Rozell, B., Johansson, M., & Karlsson, A. (2008). Progressive loss of mitochondrial DNA in thymidine kinase 2-deficient mice. *Hum Mol Genet, 17*(15), 2329–2335. https://doi.org/10.1093/hmg/ddn133.

Further Reading

Bernardi, P. (1999). Mitochondrial transport of cations: channels, exchangers, and permeability transition. *Physiol. Rev. 79*, 1127–1155.

Bliksøen, M., Mariero, L. H., Ohm, I. K., Haugen, F., Yndestad, A., Solheim, S., Seljeflot, I., Ranheim, T., Andersen, G.Ø., Aukrust, P., et al. (2012). Increased circulating mitochondrial DNA after myocardial infarction. *Int. J. Cardiol. 158*, 132–134.

Chen, W., Zhao, H., & Li, Y. (2023). Mitochondrial dynamics in health and disease: mechanisms and potential targets. *Sig Transduct. Target. Ther. 8*, 333. https://doi.org/10.1038/s41392-023-01547-9.

Ehinger, J. K., Westerlund, E., Frostner, E.Å., et al. (2024). Mitochondrial function in peripheral blood cells across the human lifespan. *npj Aging, 10*, 10. https://doi.org/10.1038/s41514-023-00130-4.

Ferree, A., & Shirihai, O. (2012). Mitochondrial dynamics: the intersection of form and function. *Adv. Exp. Med. Biol. 748*, 13–40. https://doi.org/10.1007/978-1-4614-3573-0_2.

Izzo, V., Bravo-San Pedro, J. M., Sica, V., Kroemer, G., & Galluzzi, L. (2016). Mitochondrial permeability transition: new findings and persisting uncertainties. *Trends Cell Biol. 26*, 655–667.

Kausar, S., Yang, L., Abbas, M. N., Hu, X., Zhao, Y., Zhu, Y., & Cui, H. (2020). Mitochondrial DNA: A Key Regulator of Anti-Microbial Innate Immunity. *Genes. (Basel), 11*(1), 86. https://doi.org/10.3390/genes11010086.

Patrushev, M., Kasymov, V., Patrusheva, V., Ushakova, T., Gogvadze, V., & Gaziev, A. (2004). Mitochondrial permeability transition triggers the release of mtDNA fragments. *Cell Mol. Life Sci. 61*, 3100–3103.

Riley, J. S., & Tait, S. W. (2020). Mitochondrial DNA in inflammation and immunity. *EMBO Rep. 21*(4), e49799. https://doi.org/10.15252/embr.201949799.

Sahu, J. 2022. Role NUMTs (Nuclear mitochondrial DNA in controlling growth, reproduction and disease resistance in ducks. M.V.Sc Thesis submitted to West Bengal University of Animal and Fishery Sciences, Kolkata.

Song, X., Hu, W., Yu, H., Wang, H., Zhao, Y., Korngold, R., & Zhao, Y. (2020). Existence of Circulating Mitochondria in Human and Animal Peripheral Blood. *Int. J. Mol. Sci. 21*(6), 2122. https://doi.org/10.3390/ijms21062122.

Williams, G. S., Boyman, L., & Lederer, W. J. (2015). Mitochondrial calcium and the regulation of metabolism in the heart. *J. Mol. Cell Cardiol. 78*, 35–45. https://doi.org/10.1016/j.yjmcc.2014.10.019 Jan.

CHAPTER

8

Mitogenomics and parasitic disease

Introduction

In our earlier chapters, we have discussed the basis of mitogenomics, its structure and functional significance.

Disease resistance or immunity against parasitic infection is quite interesting to study.

Classification of parasites

Ectoparasite: These parasites inhabit the external body surface of the individual.
Endoparasite: These parasites remain inside the body.
Endoparasites are broadly classified further as:

Nemathelminthes (roundworm)
Platyhelminthes (flatworm)

Antiparasitic agents, popularly known as antihelminthics, are employed to treat parasitic infestation. But in recent days, like bacterial infection, new resistant variants are evolving, which are less effective with the antihelminthics available. Satisfactory vaccines are lacking.

Parasitic infestation is common in children and as occupational hazards such as persons engaged in the medical profession and veterinarians. Animals are most prone as they eat raw food containing cysts of parasites.

A lot of studies have been undertaken for parasitic immunity, mostly involving nuclear genes (Pal & Chakravarty, 2019; Banerjee et al., 2021; Rawat et al., 2021). In this current chapter, we will discuss how mitochondrial genes are involved in parasitic immunity. Even some work in conjunction with nuclear genes.

In this current chapter, we will address key important questions: How do parasites attack the host? What are the changes occurs in mitochondria? How does mitochondrial DNA (mtDNA) affect immunity against parasitic infection? Is there any interaction between mitochondrial gene with nuclear genes in innate immunity?

As we have mentioned, parasites have been classified based on their genetic material. They interact with the host nuclear as well as mitochondrial genomes in different ways.

Mitochondrial Genomics
DOI: https://doi.org/10.1016/B978-0-443-26655-3.00012-3

In the coming section, we will discuss the role of various kinds of parasites and their possible mode of interaction with host mitochondria. We will discuss the parasitic pathogens infecting both humans and animals, along with the zoonotic diseases.

Impact of host mitochondria on nutrient uptake for the parasite

A case study

Obligate intracellular pathogens rely on nutrients supplied from host cells for their survival. Restricting their capacity to obtain particular nutrients can impede their growth or trigger a dormant condition. In contrast to small-molecule metabolites that can readily diffuse within the host cell cytoplasm, the acquisition of host fatty acids (FAs) presents a distinct barrier for intracellular microorganisms due to two primary factors: FAs are predominantly stored as triglycerides within membrane-bound lipid droplets (LDs) and mitochondria, the metabolic centres of eukaryotic cells where β-oxidation occurs, serve as a primary destination for FAs derived from LDs. The mechanisms by which a microbe accesses membrane-bound host FAs and the capacity of host organelles, such as mitochondria, to mitigate nutrition syphoning remain unresolved enquiries pertinent to the examination of most host-pathogen interactions. The functional flexibility of mitochondria is primarily influenced by alterations in their shape. Cytosolic signals can modulate mitochondrial fuel utilisation efficiency by influencing fission, fusion and cristae remodelling. In starvation, autophagy is activated to release nutrients and mobilise FAs from LDs through lipophagy, resulting in mitochondrial elongation and an increase in cristae surface area to avoid autophagic degradation and enhance ATP production efficiency. Moreover, these morphological alterations improve the efficacy of β-oxidation by augmenting mitochondrial interaction with and absorption of FAs from LDs. Notably, host mitochondria migrate to the vacuoles, where various intracellular evolutionarily distinct prokaryotic and eukaryotic pathogens, such as the bacteria *Legionella pneumophila* and *Chlamydia psittaci*, as well as the parasite *Toxoplasma gondii*, proliferate and replicate. *Toxoplasma* infects approximately one-third of the human population and an extensive variety of animal species. Similar to *Chlamydia*, it extensively scavenges host lipid species. The morphology of mitochondria in response to infection varies according to the specific microbe: mitochondria elongate around the *Toxoplasma* vacuole; conversely, they fragment upon *Listeria* entry or *Vibrio cholerae* infection, a process crucial for facilitating their proliferation. The reason for the early elongation of mitochondria around the *Toxoplasma* vacuole remains unclear. We demonstrate that *Toxoplasma* utilises host cell lipophagy to acquire important FAs necessary for its growth, thereby elucidating the link between an intracellular bacterium, availability to vital lipids and mitochondrial morphology. In contrast, host mitochondria undergo fusion with the parasitophorous vacuole, thereby sequestering FAs from *Toxoplasma* and suppressing its replication. Our findings indicate that mitochondrial fusion and FA metabolism serve as a crucial defence mechanism against an intracellular parasite.

Toxoplasma depletes host lipid droplets and accumulates lipid droplet-derived fatty acids: case study

Fig. 8.1

Combined bright-field epifluorescence images of FL C12-pulsed cells at 24 hours postinfection (hpi). Regions of interest (ROIs) designated for quantifying mean fluorescence intensity (MFI) are delineated in black (cytoplasm of uninfected cells), blue (nuclei of infected cells), or grey (*Toxoplasma* vacuoles). Scale bar: 10 mm. Mean ± standard error of the mean from $n = 3$ separate studies showing relative FL C12 accumulation in ROIs of uninfected cell cytoplasm or *Toxoplasma* vacuoles in the experiments conducted in Fahrenheit (F). **$P<.01$, ****$P<.0001$ for *Toxoplasma* vacuole compared to uninfected cytosol; *$P<.001$ for 6 hpi versus 24 hpi, as determined by two-way ANOVA. Histograms of FL C12 MFI, ascertained via flow cytometry for *Toxoplasma*: RFP+ isolated at 24 hpi from unpulsed (black line) and FL C12-pulsed MEFs (grey line). *Toxoplasma* proliferates within a membrane-enclosed vacuole and utilises host FAs. Given that FAs are predominantly deposited in LDs as triglycerides and that LDs are transported to vacuoles where *Toxoplasma* and several intracellular infections proliferate, we hypothesised that *Toxoplasma* obtains FAs from host LDs. LDs are actively restructured when FAs are necessary for energy conversion or the creation of complex compounds. To ascertain the fate of host LDs during *Toxoplasma* infection, we examined adipocytes that possess substantial LDs with well-defined dynamics. Electron microscopy (EM) research demonstrated that LDs over 5 mm^2 in surface area were nearly entirely absent in adipocytes containing *Toxoplasma* vacuoles. We augmented these findings using confocal imaging of LDs labelled with the neutral lipid marker BODIPY 493/503 in *Toxoplasma*-infected human intestinal epithelial Caco-2 cells, murine embryonic fibroblasts (MEFs) and myoblasts. Consequently, *Toxoplasma* infection in mammalian cells leads to the reduction of LD size. We subsequently determined whether the detected loss of host LDs was a result of the trafficking of host LD-derived FAs to *Toxoplasma*. Cells treated with BODIPY-FL-C12 (FL C12), a 12-carbon saturated FA covalently attached to the hydrophobic end of the BODIPY fluorophore (thus analogous to a long-chain FA), esterify it in LDs, facilitating the subsequent tracking of LD-derived FAs. During infection, FL C12 was markedly and progressively redirected to the vacuole of live *Toxoplasma*, while deceased parasites did not absorb the fluorescent FA. Notably, we did not detect an increase in FL C12 fluorescence in the cytoplasm of uninfected cells or in the nucleus of infected cells, indicating that host FAs released from host LDs exclusively accumulates in the parasite vacuole. Additionally, flow cytometry verified that parasites extracted from MEFs treated with FL C12 prior to infection accumulated the fluorescent FA FL C12. Collectively, these data indicate that during infection, host FAs are diminished in LDs and assimilated by *Toxoplasma*.

Lipophagy in host cells infected by *Toxoplasma*.

Intact host LDs have been observed within *Chlamydia* and *Leishmania* vacuoles during infection, whereas *Toxoplasma* internalises host LDs into the parasite vacuole after the replenishment of infected cells with exogenous FAs. Nevertheless, host LDs labelled with FL C12 were absent in the *Toxoplasma* vacuole at the early, middle, or late stages of infection, indicating that LD catabolism occurred prior to host FA absorption by *Toxoplasma*. Lipids can be digested via autophagy to supply FAs for mitochondrial β-oxidation. Given that host autophagy is activated during *Toxoplasma* infection and facilitates parasite multiplication, we hypothesised that *Toxoplasma*

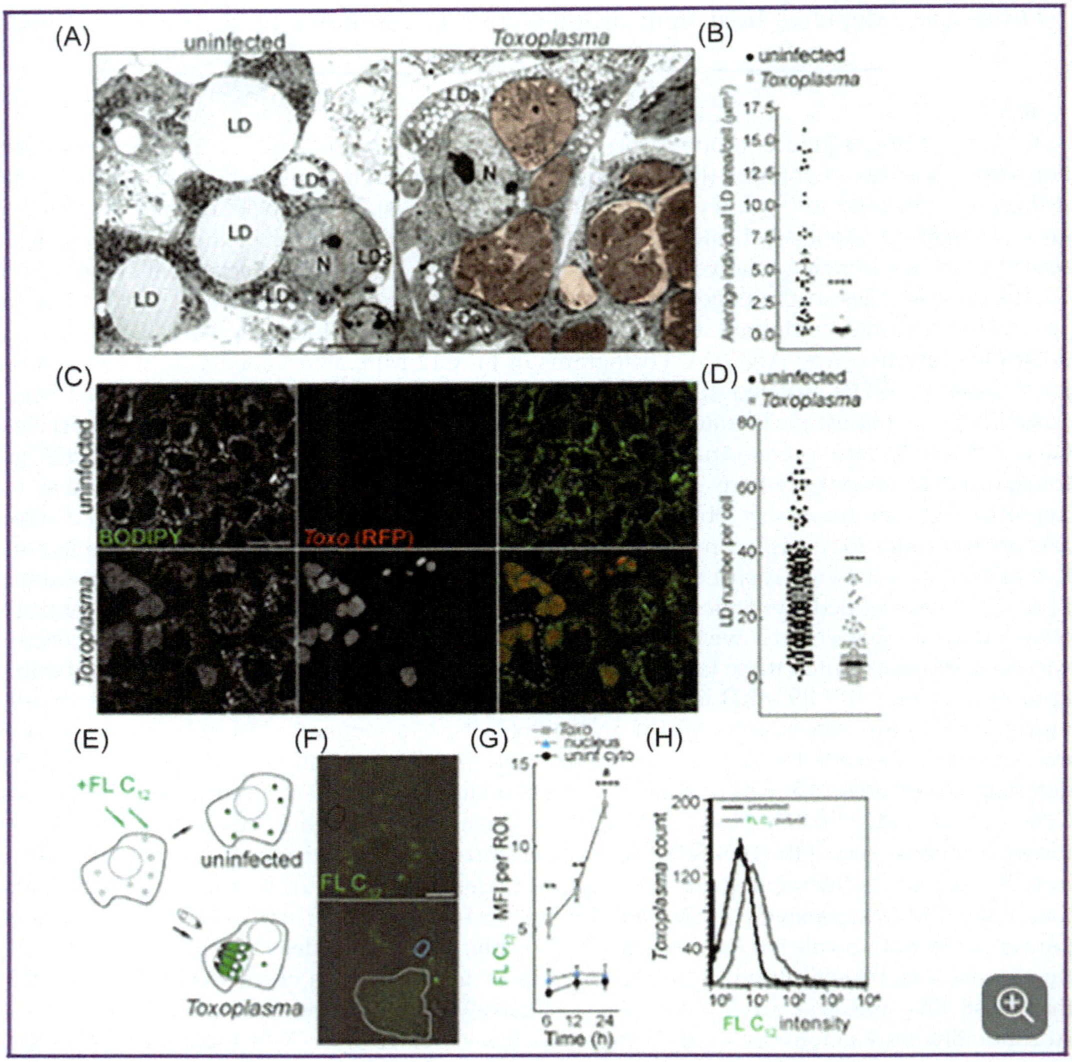

FIGURE 8.1 *Toxoplasma* depletes host lipid droplets (LDs) and accumulates LD-derived fatty acids (FAs). Electron micrographs of mouse primary adipocytes 24 hour post-*Toxoplasma* infection. *Toxoplasma* vacuoles are pseudocoloured red. Scale bar, 5 μm. Dot plot of the average area of an individual LD in n = 40 cells in experiments as in (A). Means ± standard error of the mean (SEM) are plotted. ***$P \leq .0001$ in an unpaired t-test. Confocal images of mock- and *Toxoplasma* RFP+(red)-infected Caco-2 cells labelled with BODIPY 493/503 (green) and imaged 24 hour postinfection. Selected cells are contoured by a dashed line in merged panels. Scale bar, 25 μm. Dot plot of LD number in experiments as in (C) in n = 200 cells. Means ± SEM are plotted. **** $P \leq .0001$ in an unpaired t-test. Schematic representation of fluorescent FA-pulse and infection challenge assay. MEFs were incubated overnight with FL C12, rinsed, mock or *Toxoplasma*-infected and analysed by flow cytometry or fluorescence microscopy.

utilises lipophagy, the autophagic degradation of LDs, to get host FAs from these droplets. Initially, we verified that *Toxoplasma* stimulates autophagy in MEFs, human foreskin fibroblasts and Caco-2 cells, with a reduction in the LD population observed at the later phases of *Toxoplasma* infection. We subsequently enquired if *Toxoplasma* induced lipophagy by examining the association of LDs with autophagosomes during infection. The levels of perilipin-3-RFP+ LD apposition to light chain 3-GFP+ autophagosomes were elevated two- to threefold. Figures depict the early and middle stages of infection. As anticipated, these structures were subsequently lost during infection, coinciding with the disappearance of LDs. Ultimately, EM investigation of *Toxoplasma*-infected MEFs revealed circular electron-transparent structures recognisable as LDs within autophagosomes. Consequently, lipophagy transpires during *Toxoplasma* infection.

Lipophagy enhances *Toxoplasma* acquisition of FAs.

It was attempted to investigate the function of host lipophagy in *Toxoplasma* FA uptake. Considering that lipophagy receptors have not been identified, it was investigated if the impairment of host cell autophagy influenced *Toxoplasma's* acquisition of host FAs. Imaging and flow cytometry demonstrated that parasites derived from Atg7_/_ MEFs, which lack macroautophagy, accumulated around 50% less RC12 (a 12-carbon chain saturated FA) and FL C12 compared to parasites from Atg7+/+ MEFs. To validate the significance of LDs as a lipid source for *Toxoplasma*, we investigated the uptake of FAs by parasites in diglyceride acyltransferases 1 and 2 knockout mouse embryonic fibroblasts (Dgat1_/_,2_/_), which exhibit a deficiency in lipid storage owing to disrupted triglyceride synthesis and LD formation. Parasites exhibited a markedly reduced uptake of FL C12 in Dgat1_/_,2_/_ MEFs. On the other hand, the suppression of cytoplasmic adipose triglyceride lipase by the administration of atglistatin (Mayer et al., 2013) did not diminish FA uptake by *Toxoplasma*, suggesting that ATGL-mediated lipolysis is unnecessary for the parasite acquisition of FAs. These researchers explain that *Toxoplasma* utilises host lipophagy to get FAs.

Inhibition of host cell autophagy and triglyceride accumulation restricts *Toxoplasma* proliferation

Toxoplasma metabolises and assimilates host FAs into storage organelles and dividing membranes, which are crucial mechanisms for parasite proliferation. We hypothesised that reduced FA acquisition could adversely impact *Toxoplasma* proliferation in Atg7_/_ MEFs, regardless of the advantageous role of autophagy against certain intracellular pathogens, including *Toxoplasma*. We developed a flow cytometry-based method utilising *Toxoplasma*-derived RFP as an indicator of parasite proliferation. After infection with RFP+-*Toxoplasma* at a low multiplicity of infection of 2, monolayers were washed 2 hpi and examined at 24 hpi, following multiple replication cycles prior to the commencement of cellular lysis and, thus reinvasion events. We confirmed that under these conditions, the RFP readout indicated parasite proliferation within a single growth cycle, rather than multiple infected cells, asynchronous infection, or reinfection of adjacent cells. A notable decrease in *Toxoplasma* burden was detected via microscopy and flow cytometry at 24 hpi, but not before replication events in cells treated with the parasite growth inhibitor pyrimethamine, indicating that RFP fluorescence serves as a dependable indicator of *Toxoplasma* proliferation. Utilising this methodology, we determined that at 24 hpi, the *Toxoplasma* burden in Atg7_/_ MEFs was approximately 30% lower than in Atg7+/+ MEFs. Supplementation at 2 hpi of BSA-conjugated.

Unsaturated FAs oleate and linoleate, for which *Toxoplasma* is anticipated to be auxotrophic, partially ameliorated the growth deficiency of the parasite in Atg7_/_ MEFs. Accordingly, autophagy enables the mobilisation of LD–derived FAs that are required for *Toxoplasma* proliferation. Parasite growth was significantly reduced in LD–deficient Dgat1^–/–,2^–/– MEFs. The growth of *Toxoplasma* was diminished in LD-deficient Dgat1_/_,2_/_ MEFs. The incorporation of exogenous FAs and cholesterol, which are diminished in Dgat1_/_,2_/_ MEFs and whose esterification is crucial for *Toxoplasma* proliferation, rectified the *Toxoplasma* growth deficiency in these MEFs. The reduced *Toxoplasma* proliferation in Atg7_/_ or Dgat1_/_,2_/_ MEFs was not attributable to variations in invasion, as the parasite load prior to replication events (2 hpi) remained unchanged in both cell lines. These findings endorse a scenario in which host lipophagy provides FAs and cholesteryl esters for *Toxoplasma* proliferation. The proliferation of *Toxoplasma* is contingent upon the lipid reserves of the host cell. To further substantiate the function of host LD-derived FAs in *Toxoplasma* proliferation, we altered host lipid metabolism. We hypothesised that redirecting host FAs towards their accumulation in LDs could mimic the reduced proliferation of *Toxoplasma* seen in Dgat1_/_,2_/_ MEFs lacking FA storage. To prevent potential toxic effects on noninternalised parasites, we infected MEF monolayers and supplemented them with free linoleic acid (LA) at 2 hpi, when the parasites were intracellular, as the addition of free, long-chain unsaturated FAs to cells elevates triglyceride levels and LD accumulation. LA induced the anticipated LD formation and, notably, reduced parasite development by 50%. If the LA therapy diminished *Toxoplasma* proliferation by promoting FA sequestration and esterification into LDs, the activation of lipophagy would counteract this effect. When lipophagy was increased by depleting lipids from the culture conditions, the LD formation and *Toxoplasma* growth impairment caused by LA administration were ameliorated and FA intake by the parasites was enhanced. The elimination of lipids postinfection elevated parasite load in Atg7+/+ MEFs, but not in Atg7-/- MEFs, thereby reinforcing the significance of lipophagy in providing resources for *Toxoplasma*, with lipids essential for its development.

Initial host mitochondrial FA uptake inhibits *Toxoplasma* proliferation.

Modifications in FA metabolism, such as promoting lipid storage or catabolism, have been proposed as a strategy to reduce FA availability in cancer cells. We used this paradigm to study *Toxoplasma* infection and deduced that if our model is accurate, altering FA availability could influence parasite proliferation. The peroxisome proliferator-activated receptor δ agonist GW0742, which facilitates FA metabolism and insulin, which encourages FA storage, reduced *Toxoplasma* load. A hindrance in the entry of acyl-coenzyme A (CoA) into mitochondria. Inhibition of carnitine-palmitoyl CoA transferase I (CPT1) by etomoxir promoted the growth of *Toxoplasma*. We examined whether the suppression of fatty acid oxidation (FAO) fostered an environment conducive to parasite proliferation by diminishing mitochondrial reactive oxygen species (ROS) generation. The mitochondrial ROS scavenger MitoTEMPO marginally promoted parasite proliferation in untreated cells, although it did not influence the etomoxir-induced enhancement of *Toxoplasma* growth. Consequently, we hypothesised that the *Toxoplasma* growth-enhancing effects of etomoxir were associated with its suppression of FAO and investigated the influence of mitochondrial shape, which governs mitochondrial FA absorption, on parasite proliferation. Mitochondria associate with the vacuole of several prokaryotic and eukaryotic microorganisms, including *Toxoplasma*, around which they extend. At 6 hpi, mitochondria appeared elongated and encircled the parasitic vacuole. Notably, etomoxir somewhat reduced mitochondrial size during

Toxoplasma infection. Mitochondrial elongation observed during *Toxoplasma* infection parallels the morphological adaptations required to support mitochondrial FAO under nutrient-deprived conditions. Upon pulsing cells with RC12, we observed enhanced RC12 uptake at 6 hpi in mitochondria surrounding *Toxoplasma* vacuoles, in contrast to mitochondria in noninfected cells. Administration of etomoxir or an autophagy inhibitor inhibited the noted enhancement in mitochondrial RC12 absorption. At 24 hpi, mitochondria in *Toxoplasma*-infected cells were fragmented and displayed reduced overlap with RC12, presumably due to the buildup of RC12 within the parasite vacuoles. Etomoxir treatment, which facilitated parasite proliferation, resulted in a reduction in perimeter length and an enhancement of mitochondrial circularity, indicative of mitochondrial fragmentation. Moreover, etomoxir enhanced RC12 staining of intra-parasite organelles, indicating that parasites assimilate additional FAs when their host cells are unable to perform FAO. These findings support a model in which mitochondrial FA metabolism during *Toxoplasma* infection modulates parasite proliferation by limiting its access to host-derived FAs.

Mitochondrial fusion thereby constrains *Toxoplasma* growth and restricts fatty acid availability

Mitochondrial fusion is essential for the distribution and oxidation of FAs from LDs during food scarcity. To investigate whether host mitochondrial remodelling inhibited parasite growth by restricting FA intake, we initially elucidated the molecular mechanism responsible for the alterations in mitochondrial network shape. Mitochondria in MEFs deficient in the outer membrane pro-fusion proteins mitofusin 1 and 2 (Mfn1_/_,2_/_) did not elongate surrounding parasite vacuoles at 6 hpi, as assessed by confocal microscopy, nor at 12 hpi, as evaluated through morphometric analysis of EM images. At subsequent infection time points, mitochondria in Mfn1+/+,2+/+ MEFs exhibited fragmentation, resembling the length of mitochondria from Mfn1_/_,2_/_ MEFs, thereby indicating that mitochondria utilise their fundamental fusion machinery to elongate surrounding the vacuole during the early infection process.

We subsequently examined the effects of mitochondrial fusion modulation on mitochondrial FA absorption and parasite proliferation. In Mfn1_/_,2_/_ MEFs, mitochondrial FA uptake was undetectable at 6 hpi and thus, *Toxoplasma* acquired more FA exhibited enhanced growth efficiency, as evidenced by fluorescence-activated cell sorting analysis, plaque size assessment and parasite count per vacuole. Due to the potential confounding effects of various mitochondrial abnormalities resulting from the persistent ablation of fusion, we verified that parasite burden was elevated in Mfn1flx/flx,2flx/flxMEFs, where Mfn1 and Mfn2 were present. Precisely ablated by Cre recombinase introduced via viral infection or through parasite-derived secretion. The CPT1 inhibitor etomoxir significantly enhanced *Toxoplasma* proliferation in Mfn1+/+,2+/+ MEFs, but not in Mfn1_/_,2_/_ MEFs. The reintroduction of Mfn1, necessary for mitochondrial FA uptake from LDs, partially inhibited parasite growth in Mfn1_/_,2_/_ MEFs. These data endorse a hypothesis in which mitochondrial fusion augments FAO, hence constraining parasitic activity, limiting access to host FAs and hence hindering its expansion.

Host-pathogen interaction:

An essential element of the host-pathogen interaction is the appropriation of host resources by microorganisms. Pathogens have consequently evolved mechanisms to extract host compounds, while host cells have developed countermeasures. This study illustrates that although a pathogen can use host lipophagy to acquire intracellular lipid reserves, host mitochondria encircle the pathogen vacuole to competitively absorb FAs, restricting the parasite's access to essential host resources, thereby limiting its expansion. While our research concentrated on *Toxoplasma*, a protozoan parasite, the fight between host organelles and pathogens for scarce nutrients serves as a conceptual framework applicable to most intracellular diseases. This competition involves the exploitation of metabolic processes by pathogens, including host proteasomal degradation for amino acid acquisition and host autophagy for access to various metabolites, such as amino acids during *Francisella* infection and FAs, as demonstrated herein. Besides FAs, host autophagy may furnish *Toxoplasma* with a diverse array of nutrients, including nucleosides and sterols, since the supplementation of FAs linoleate and oleate only partially supported parasite proliferation in autophagy-deficient cells. The battle for resources between mitochondria and *Toxoplasma* presumably transcends FAs. Organelles also oxidise pyruvate and amino acids, participating in one-carbon metabolism vital for anabolic pathways. Organelles in the host that may compete for nutrition with infections include the endoplasmic reticulum and LDs (both FA and sterol reserves), which also transport to vacuoles encasing additional pathogens. The revelation that the mitochondrial antiviral signalling (MAVS) protein must be strictly localised on the mitochondrial outer membrane to initiate the antiviral response, resulting in the production of type I interferons and pro-inflammatory cytokines, solidified the function of mitochondria as signalling centres in innate immunity. Our research presents an alternative paradigm for mitochondrial involvement in innate immunity: they initiate a 'metabolic defence' against *Toxoplasma* infection via the sequestration of FAs, necessitating mitochondrial fusion. Consequently, it is unsurprising that microbes have developed various strategies to manipulate mitochondrial dynamics: *V. cholerae* regulates mitochondrial trafficking; mitochondrial fission frequently occurs in infected cells, induced by a secreted toxin, as seen with *Listeria monocytogenes*, or by an effector protein in the case of *L. Pneumophila*. In contrast and unexpectedly, the bacterium *Chlamydia trachomatis* derives advantages from maintaining the host mitochondrial network. It will be intriguing to examine whether the establishment of contact sites between specific microbes, such as *Toxoplasma* and mitochondria, constitutes a universal strategy to impede the metabolic innate immunity delineated herein. Our findings indicate an additional dimension of mitochondrial participation in innate immunity. Mitochondria can protect host cells from infection by fusing and absorbing FAs, thereby sequestering nutrients from intracellular parasites. The competitive structure between bacteria and host cells for nondiffusible metabolites presents opportunities for treatment techniques that adjust host metabolism to restrict pathogen access to nutrients and eventually manage illness.

Toxoplasma infection triggers lipophagy

Studies have been undertaken how toxoplasma (parasite) initiates lipophagy and its link to mitochondria.

Fig. 8.2

Fig. 8.3

Fig. 8.4

Fig. 8.5

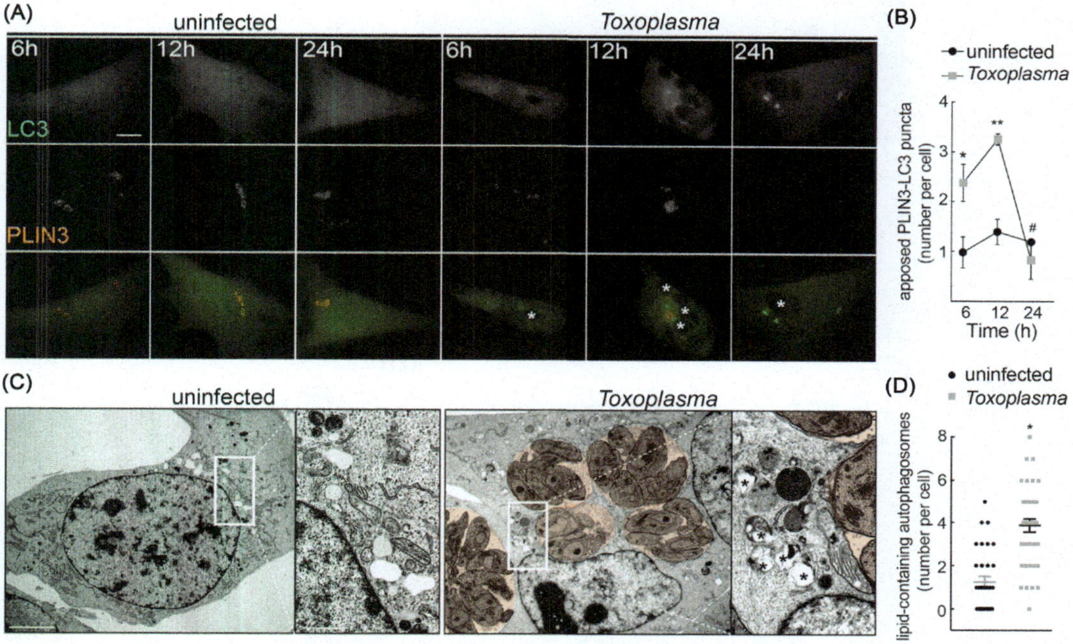

FIGURE 8.2 *Toxoplasma* infection triggers lipophagy. Representative fluorescent images of MEFs expressing light chain 3 (LC3)-GFP (green) and PLIN3-RFP (red) fixed and imaged. Scale bar, 10 μm. Dot plot of the number of apposed LC3-PLIN3 puncta per cell in experiments as in (A). Mean ± standard error of the mean (SEM) is plotted. $P \leq .01$ infected versus noninfected. Electron microscopy (EMs) of uninfected and *Toxoplasma*-infected wild-type MEFs at 24 hpi. *Toxoplasma* vacuoles are pseudocoloured red; lipid-containing autophagosomes are marked with asterisks. Scale bars, 5 μm (main panel) and 2 μm(insert). Dot plot of the number of lipid-containing autophagosomes determined from n = 40. EMs acquired as in (C), Mean ± SEM is indicated, $P \leq .05$, in an unpaired t-test.

Mitochondrial fatty acid uptake during *Toxoplasma* infection limits parasite proliferation

In this section, we will attempt to explain the role of host mitochondria in controlling parasitic infection. Here we discuss Toxoplasma spp as the model parasite infecting host.

Fig. 8.6

Host mitochondrial fusion restricts *Toxoplasma* proliferation

Mitochondrial fusion is a mechanism, where two mitochondria fuses together, and the mechanism is controlled by certain proteins referred to as Mitofusin (Mfn). We have discussed in detail in Chapter 2. Certain genes code for mitofusins, which are of practical relevance in controlling the proliferation of *Toxoplasma* spp.

Fig. 8.7

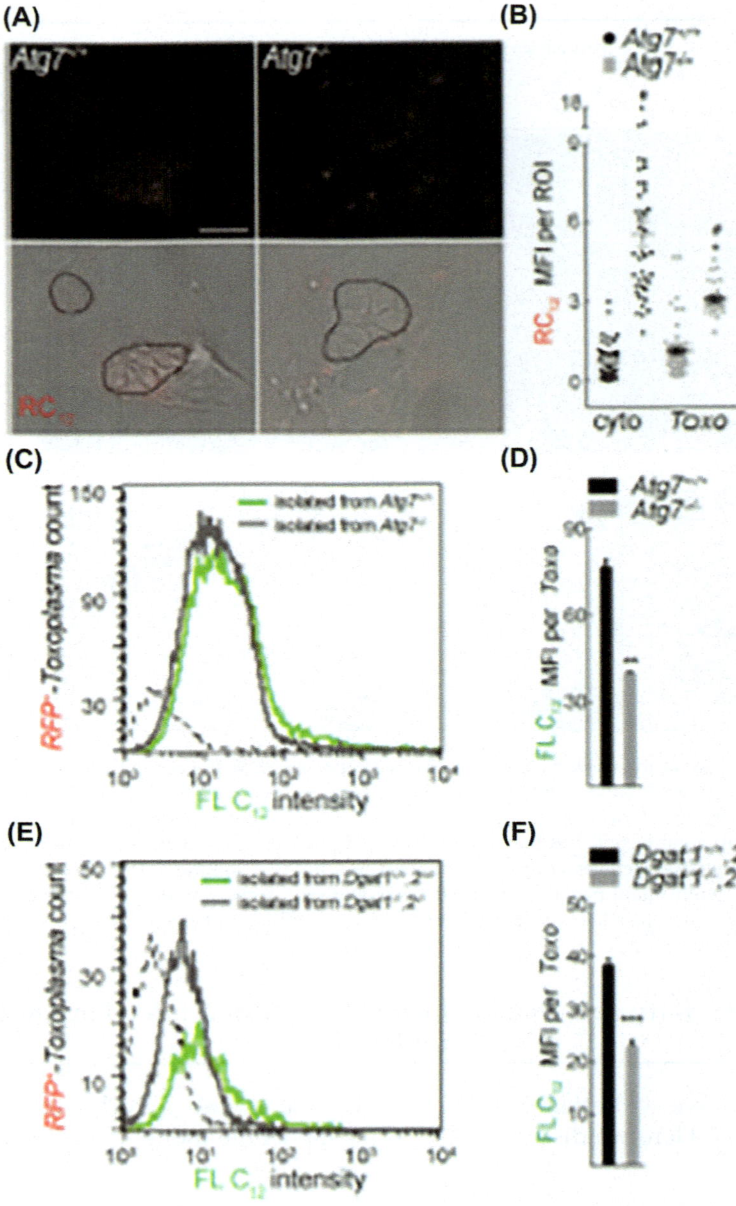

FIGURE 8.3 *Toxoplasma* acquires fatty acids by stimulating host lipophagy (A) RC12 and combined bright-field and fluorescence pictures of mouse embryonic fibroblasts of the specified genotype. Cells were subjected to a pulse with RC12 and either mock- or *Toxoplasma*-infected, subsequently fixed and photographed at 24 hours postinfection (hpi). *Toxoplasma* vacuoles are delineated in black. Scale bar: 10 mm. (B) RC12 accumulation in the cytosol (cyto) of uninfected MEFs of the specified genotype or within *Toxoplasma* vacuoles (Toxo) of infected cells at 24 hpi. The data represent the mean plus the standard error of the mean (SEM) from three studies, as indicated in (A). $P<.05$ for *Toxoplasma* vacuole compared to uninfected cytoplasm; *$P<.05$ for Atg7+/+ versus Atg7_/_ as determined by two-way ANOVA. (C and D) Flow histogram (C) and flow cytometry analysis (D) of FL C12 mean fluorescence intensity (MFI) in RFP+-*Toxoplasma* extracted from FL C12-pulsed MEFs of the specified genotype at 24 hpi. The dotted line in (C) represents the background fluorescence of parasites extracted from unlabelled MEFs. Data represent the mean plus SEM from three independent studies, with **$P<.01$ determined by an unpaired t-test. (E and F) Flow histogram (E) and flow cytometry analysis (F) of FL C12 MFI in RFP+-*Toxoplasma* isolated from FL C12-pulsed Dgat1+/+, Dgat1+/+ and Dgat1−/− MEFs at 24 hpi. The dotted line in (E) expresses the background fluorescence of parasites extracted from unlabelled MEFs. The data represent the mean plus SEM from three independent experiments, with a significance level of ***$P<.001$ determined by an unpaired t-test.

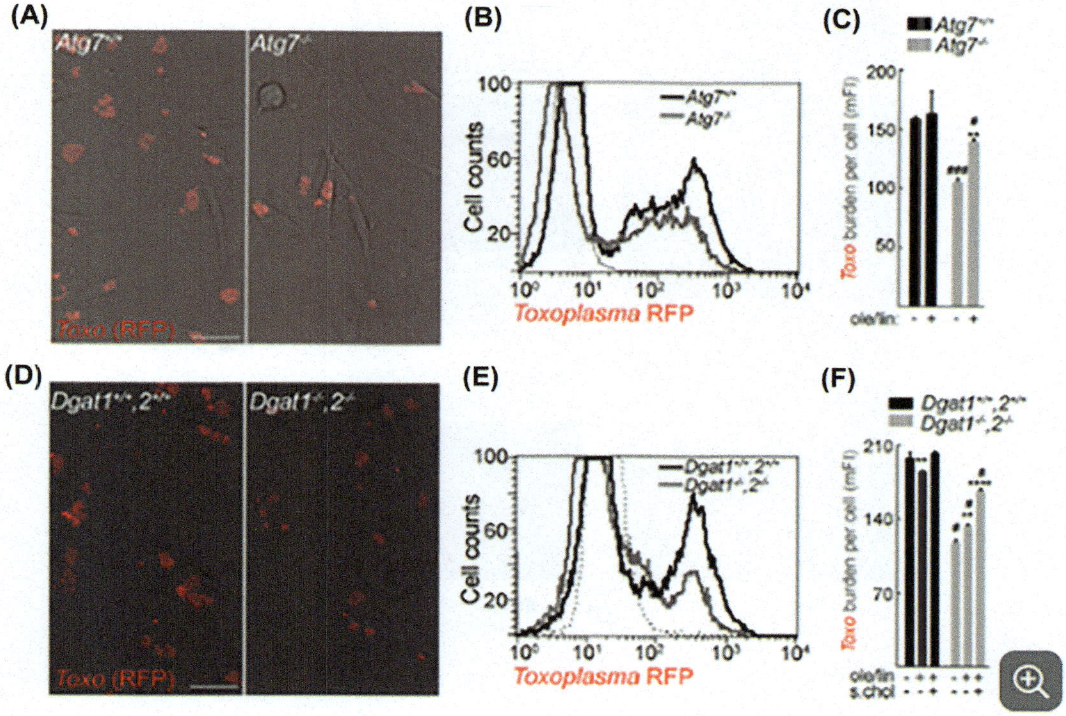

FIGURE 8.4 Optimal *Toxoplasma* proliferation requires host autophagy and triglycerides. Representative combined bright-field epifluorescence pictures of MEFs with the specified genotype 24 hours postinfection (hpi) with RFP +-*Toxoplasma*. Cells were washed at 2 hpi and incubated in full media devoid of parasites. Scale bar measuring 25 mm. Representative flow cytometry histogram of the RFP channel at 24 hpi in RFP+-*Toxoplasma*-infected MEFs of the specified genotype. The dotted line reflects background fluorescence in uninfected MEFs. (C) The experiments were performed as described in (B), with the addition of 1 mg/mL fraction V BSA or 30 mM oleate-linoleate-albumin (ole/lin) to MEFs at 2 hpi. The next step is flow cytometric analysis of median RFP+-*Toxoplasma* burden (mFI). The data represent the mean plus standard error of the mean (SEM) from three independent experiments, with **$P<.01$ indicating significance between treated and untreated groups, *$P<.05$ and ***$P<.001$ denoting significance between wild-type and Atg7_/_ as determined by two-way ANOVA. Representative amalgamated bright-field epifluorescence pictures of mouse embryonic fibroblasts (MEFs) of the specified genotype 24 hpi with RFP+-*Toxoplasma* (multiplicity of infection = 2). Cells were washed at 2 hpi. Scale bar, 25 mm. Flow cytometry histogram shows the RFP channel at 24 hpi in RFP+-*Toxoplasma*-infected MEFs of the specified genotype. The dotted line represents background fluorescence in uninfected MEFs. (F) Experiments were conducted as described in (E), with the addition of 1 mg/mL portion V BSA, 30 mM ole/lin, or a combination of ole/lin and 13 SyntheChol to MEFs at 2 hpi. The analysis of *Toxoplasma* burden (RFP mFI) was performed at 24 hpi using flow cytometry. Data represent mean ± SEM from n = 3 separate experiments, **$P<.01$, ****$P<.0001$ for treated versus untreated and *$P<.0001$ for Dgat1+/+,2+/+ against Dgat1_/_,2_/_ as determined by two-way ANOVA.

Lack of mitochondrial genome in functional mitochondria in eukaryotic parasite: a peculiar case

The anaerobic or microaerophilic protozoan parasites, including the enteric human infections *Entamoeba histolytica, Giardia intestinalis, Cryptosporidium parvum, Blastocystis hominis* and the

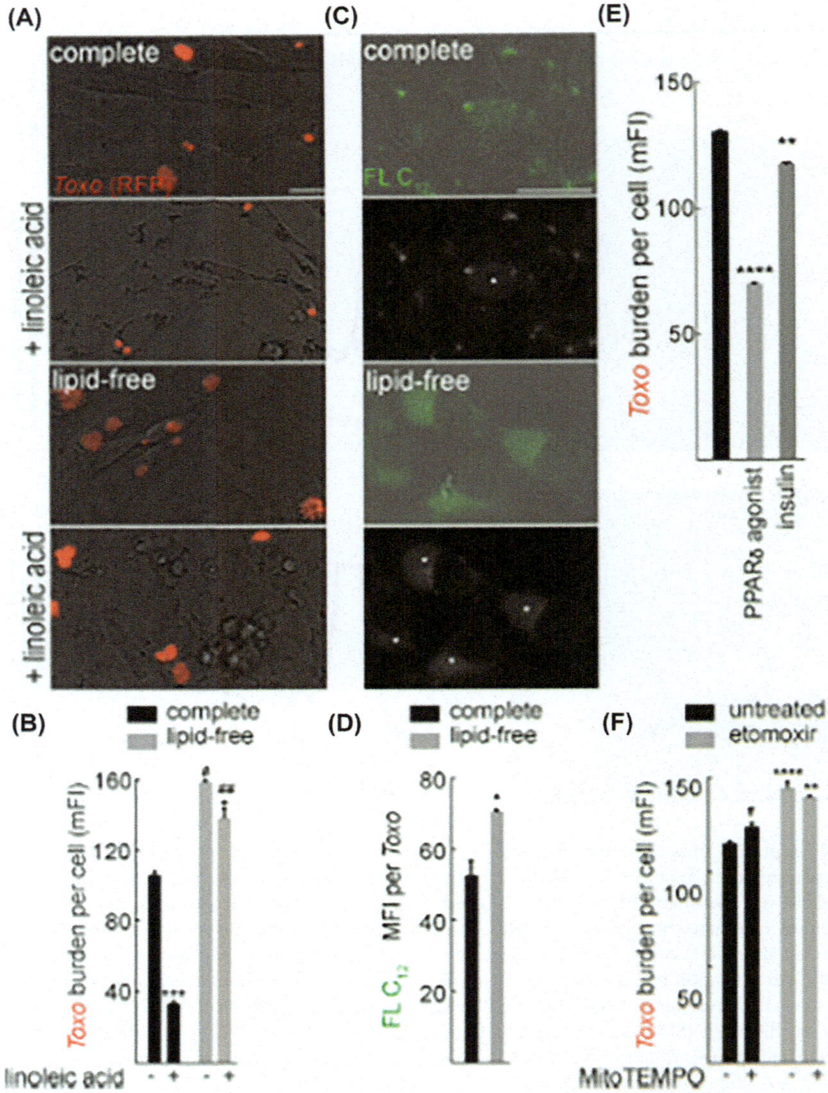

FIGURE 8.5 Lipid storage and mobilisation in host cells characterise *Toxoplasma* proliferation. (A) Illustrative composite bright-field epifluorescence pictures of wild-type MEFs at 24 hours postinfection (hpi) with RFP +-*Toxoplasma* (multiplicity of infection (MOI =2)). MEFs were washed at 2 hpi and incubated in media supplemented, where applicable, with 0.32 mM free linoleic acid. Scale bar measures 25 mm. (B) The experiments were conducted as described in (A), with the exception that MEFs were assessed via flow cytometry for *Toxoplasma* load (RFP mean fluorescence intensity (MFI)). The data represent the mean plus standard error of the mean (SEM) from three studies. Statistical significance is indicated as ***P<.001 for LA-treated versus untreated and *P<.001, **P<.0001 for lipid-free versus full media, assessed using two-way ANOVA. Merged bright-field and fluorescence pictures of FL C12-pulsed MEFs infected with *Toxoplasma*. Infected monolayers were washed at 2 hpi, cultured with the specified media, fixed and photographed at 24 hpi. *Toxoplasma* vacuoles are denoted by asterisks. Scale bar: 20 mm. Fluorescence-activated cell sorting analysis of FL C12 MFI of RFP+-*Toxoplasma* obtained from studies as described in (C). The data represent the mean plus SEM from three studies, with a significance level of *P<.05 determined by an unpaired t-test. Mouse embryonic fibroblasts (MEFs) were infected with RFP+-*Toxoplasma* (MOI = 2), washed at 2 hpi, treated as specified and assessed at 24 hpi using flow cytometry to determine *Toxoplasma* burden (RFP mean fluorescence intensity (MFI)). The data represent the mean plus SEM from three studies, with statistical significance shown as **P<0.01 and ****P<.0001 for treated versus untreated groups, analysed using one-way ANOVA. (F) MEFs were infected with RFP+-*Toxoplasma* (MOI = 2), washed at 2 hpi, treated as specified, and analysed at 24 hpi by flow cytometry for *Toxoplasma* burden (RFP mFI). The data represent the mean plus the SEM from three studies, with statistical significance indicated by *P<.05. MitoTEMPO-treated versus untreated, **P<.01, ****P<.0001. Comparison of etomoxir-treated and untreated groups using two-way ANOVA.

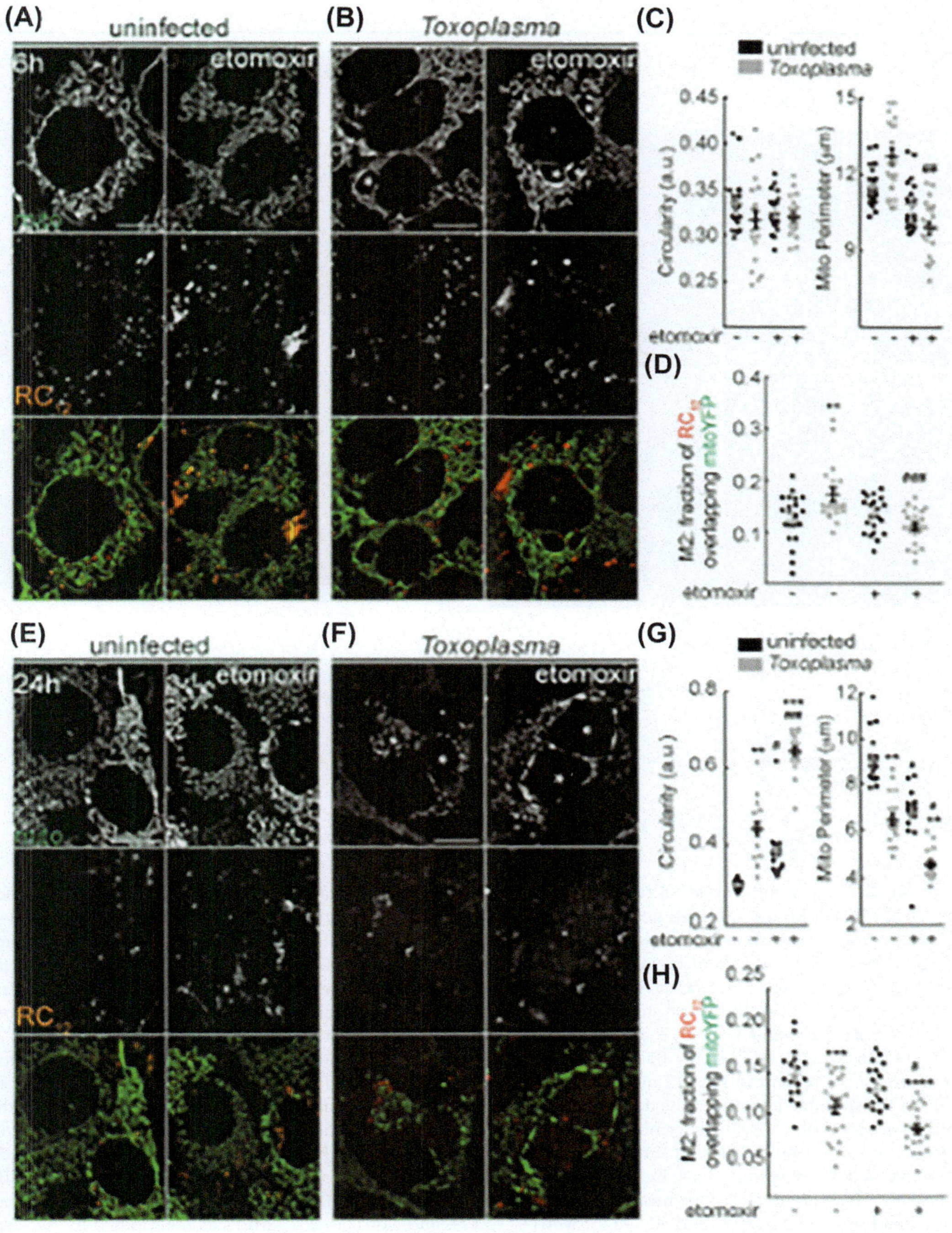

FIGURE 8.6 Initial mitochondrial fatty acid (FA) uptake during *Toxoplasma* infection constraints proliferation of parasites (A and B) MEFs expressing mtYFP, as specified, which were cultured overnight with RC12 and subsequently infected with *Toxoplasma*. Cells were washed and treated with etomoxir at 2 hours postinfection (hpi), as indicated and confocal z-stacks of mtYFP and RC12 were obtained at 6 hpi. Parasite vacuoles are denoted by asterisks (mitoYFP panel) and arrowheads (RC12 panel). Scale bar: 10 mm. (C) Analysis of average perimeter and circularity of mitochondria in uninfected cells (A) and infected cells (B). The data represent the mean ± SEM from n = 20 fields, comprising over 200 cells for uninfected samples and over 30 cells for *Toxoplasma*-infected samples. ***P*< .01, comparison of untreated and etomoxir-treated groups with two-way ANOVA. Fraction of RC12 overlapping with mitoYFP in trials as described in (A) and (B). The data represent the mean ± SEM from n = 20 fields, comprising over 200 cells for uninfected samples and over 30 cells for *Toxoplasma*-infected samples. ***P*<.01 for uninfected compared to *Toxoplasma*-infected, ****P*<.001 for untreated versus chloroquine-treated, determined by two-way ANOVA.

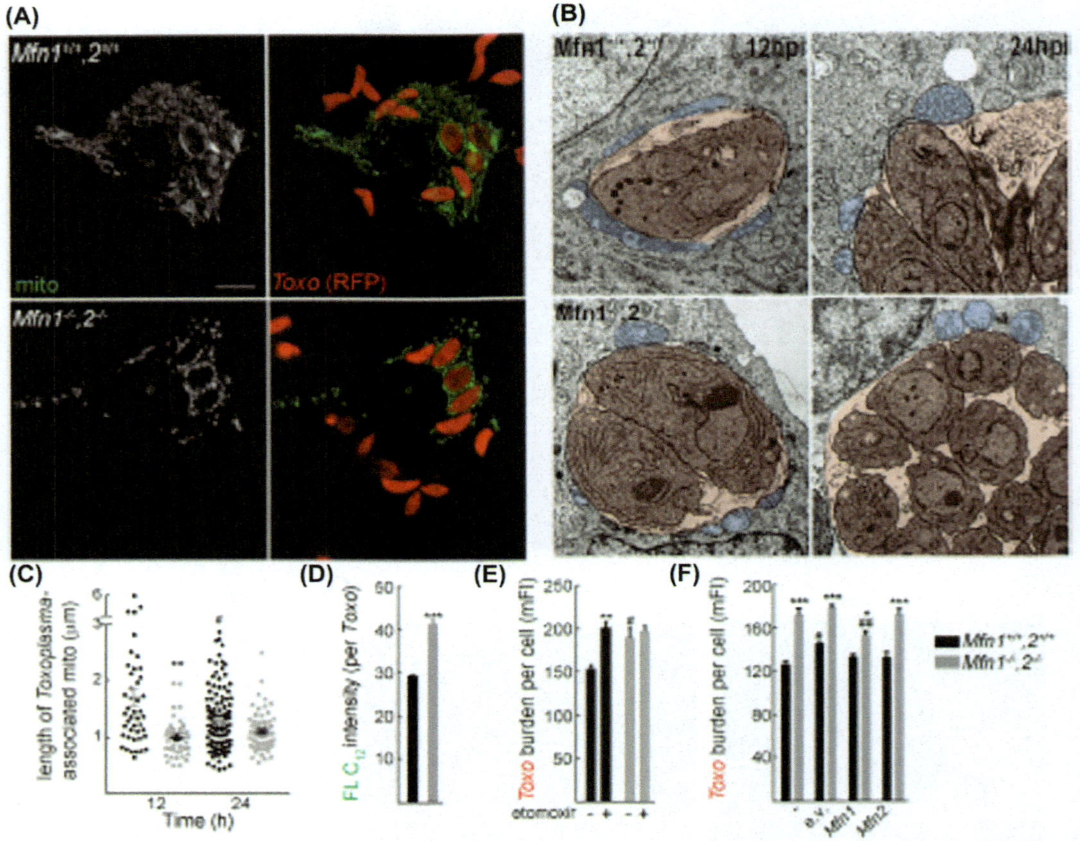

FIGURE 8.7 Host mitochondrial fusion constraints *Toxoplasma* proliferation (A) Z-stack projections of confocal images of MEFs with the specified genotype expressing mtYFP, photographed at 6 hours postinfection (hpi) with RFP+-*Toxoplasma* parasites. Scale bar: 10 mm. (B) Electron microscopy (EM) study of *Toxoplasma* vacuoles and corresponding host mitochondria in mouse embryonic fibroblasts (MEFs) of the specified genotype at the designated time points. *Toxoplasma* vacuoles are depicted in red, while mitochondria are illustrated in blue. Scale bar: 2 mm. (C) Morphometric study of the length of *Toxoplasma*-associated mitochondria in tests as described in (B). Data are expressed as mean ± standard error of the mean (SEM) from n = 40 EMs, $^{**}P<.01$ for Mfn1+/+,2+/+ compared to Mfn1+/+,2+/+, $^{*}P<.01$ for 12 hours against 24 hours as determined by two-way ANOVA. Flow cytometry measurement of FL C12 mean fluorescence intensity of RFP+-*Toxoplasma* isolated from FL C12-pulsed MEFs of the specified genotype. The data represent the mean plus SEM from three independent experiments, with a significance level of $^{***}P<.001$ determined by an unpaired t-test. MEFs of the specified genotype were infected with RFP+-*Toxoplasma* (multiplicity of infection (MOI) = 2), washed at 2 hpi and cultured in complete media supplemented with etomoxir as indicated, followed by analysis of *Toxoplasma* burden (RFP mFI) via flow cytometry. The data represent the mean plus SEM from three studies, with $^{**}P<.01$ indicating significance between treated and untreated groups and $^{*}P<.05$ for Mfn1+/+,2+/+ and Mfn1_/_,2_/_ MEFs as determined by two-way ANOVA. Mouse embryonic fibroblasts (MEFs) of the specified genotype, transfected as indicated, were infected with RFP+-*Toxoplasma* (MOI = 2), subsequently replaced with full media at 2 hpi, and analysed by flow cytometry for *Toxoplasma* burden (RFP mean fluorescence intensity) at 24 hpi. The data represent the mean plus SEM from three studies, with statistical significance indicated as $^{*}P<.01$ and $^{***}P<.0001$ for Mfn1+/+,2+/+ compared to Mfn1_/_,2_/_; and $^{*}P<.05$, $^{**}P<.01$ for Mfn1 or Mfn2 expression relative to the empty vector (e. v.) as determined by two-way ANOVA. In panels (C) to (F), black denotes Mfn1+/+,2+/+, while grey signifies Mfn1_/_,2_/_, as shown by the legend in the bottom right corner. Refer to figure S7.

urogenital tract parasite *Trichomonas vaginalis*, may survive in oxygen-deprived environments. Notwithstanding their existence in adverse conditions, these infections have employed several tactics to endure within their hosts. The release of extracellular vesicles (EVs) has emerged as a significant focus in the investigation of pathogenesis for these parasites. EVs are heterogeneous, membrane-bound entities that significantly contribute to cellular communication by conveying information via cargo and regulating the host's immune system. This review delineates various facets of the recently characterised EVs from anaerobic protozoa, encompassing their function in adhesion, modulation of the immune response and omics analysis to elucidate the potential of these EVs in the pathogenesis of diseases induced by anaerobic parasites.

Role of mitochondrial genes in parasitic immunity: case study with animal model

In our studies in our lab, we have explored the role of mitochondria in innate immune response against parasitic infestation. We are studying the economically important gastrointestinal parasite *Haemonchus contortus* in sheep.

As we have already described, mitochondria possess 37 genes with 13 polypeptide coding genes. Any deviation in the normal mitochondrial genome will lead to defects in disease resistance against parasitic resistance. As we explained in our initial chapters, any structural or functional loss of mitochondria leads to diseases.

1. Deleterious single nucleotide polymorphisms/deletions.
2. Less copy number (CN) for mitochondria.
3. Disruption in the normal functioning of mitochondria – fission, fusion, mitophagy, transcription, replication.

Deleterious single nucleotide polymorphisms/deletions

In our lab, we have detected certain mutations that lead to lowered innate immunity in sheep against deadly gastrointestinal parasite *H. contortus*. We performed whole mitochondrial genome sequencing for sheep from two different groups: healthy versus infected (infested with *H. contortus*). The sheep were classified as infected versus healthy based on faecal egg count (FEC) and molecular detection of *H. contortus* based on the cytochrome oxidase 1 (Cox1) gene of the parasite. Mutations were detected in the cytochrome B and D loop genes of the mitochondria of sheep.

Less copy number for mitochondria

We assess the CN variation among expressed mitochondrial genes through the detection of differential mRNA expression profiling for mitochondrial genes. Cyt B, cyt C and D loop mRNA expression was assessed. Better expression indicates more CNs.

Infected individuals were observed to have less CN for mtDNA, in comparison to healthy individuals. Similar observations were also followed for other mitochondrial genes, ND1, ND2, Cox1, ATP6 and ND6 of the host (sheep).

Characterisation of the mitochondrial genes of sheep involved in antiparasitic immunity: 3D protein modelling along with important domain identification

Characterisation of the cytochrome B gene from sheep abomasum: case study 1

The cytochrome B gene of sheep was characterised (gene bank accession number **2996374**) from abomasum (gut tissue). 3D structure was predicted with important domains identified, revealing four sites for iron binding and one ubiquinone site. Transmembrane helical domains for cytochrome B have been identified. Cytochrome B works in conjunction with multiple other proteins, the majority of which are cytoplasmic, while some are nuclear. Molecular interaction of cytochrome B with other proteins are being depicted through STRING analysis. One of the major roles for cytochrome B is in immunity against infections, which it execute through multiple mechanisms, such as defects in the phagosome pathway and hindrance in the biosynthesis of various nucleotide sugars, making the animal more prone to diseases (see Fig. 8.8).

Fig. 8.9

Fig. 8.10

Fig. 8.11

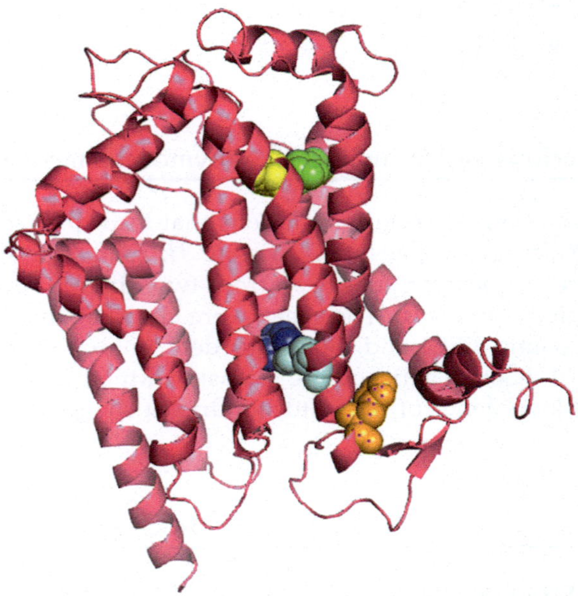

FIGURE 8.8 3D structure of cytochrome B gene from abomasum of sheep (hot pink), binding sites for iron at amino acid positions 83 (green sphere), 97 (blue sphere), 182 (yellow sphere), 196 (cyan sphere) and ubiquinone binding site at amino acid position 201 (orange sphere).

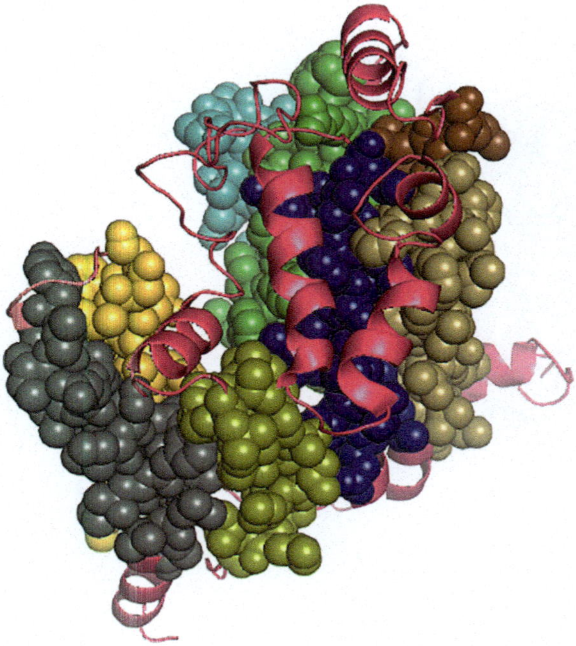

FIGURE 8.9 3D structure of cytochrome B gene from abomasum of sheep (hot pink) with transmembrane helical sites at amino acid positions 33–53 (chocolate sphere), 77–98 (forest green sphere), 113–133 (density blue sphere), 178–198(sand yellow sphere), 226–246 (deep teal sphere), 288–308 (deep olive sphere), 320–340 (bright orange sphere), 347–367 (grey sphere).

Diagnosis for estimation of faecal egg count for *Haemonchus contortus* in sheep

FEC was estimated for strongyle eggs from faeces through the McMaster slide under a microscope fitted with a monitor. Wide variability was observed between noninfected (healthy) and infected sheep. The strongyle eggs were confirmed for *H. contortus* through the visualisation of adult *Haemonchus* from the abomasum (see Table 8.1).

Table 8.2

Confirmation of faecal egg/larvae from *Haemonchus contortus*: molecular detection

Quantitative PCR (qPCR) is one of the most sensitive techniques. The positive samples reflect expression of the Cox1 gene for *Haemonchus cortortus* samples, in contrast to negative samples.

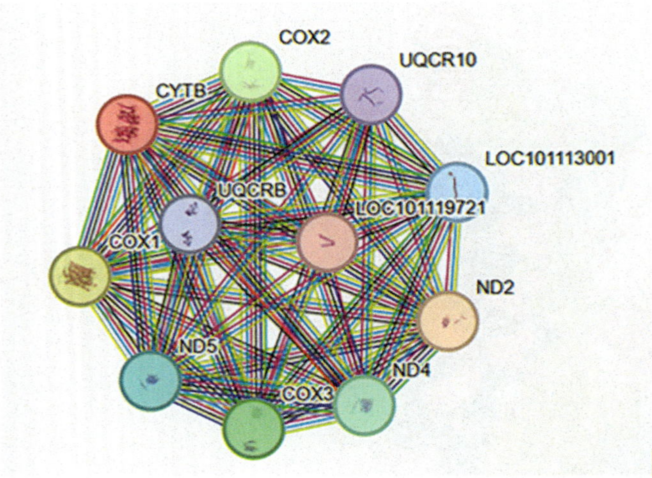

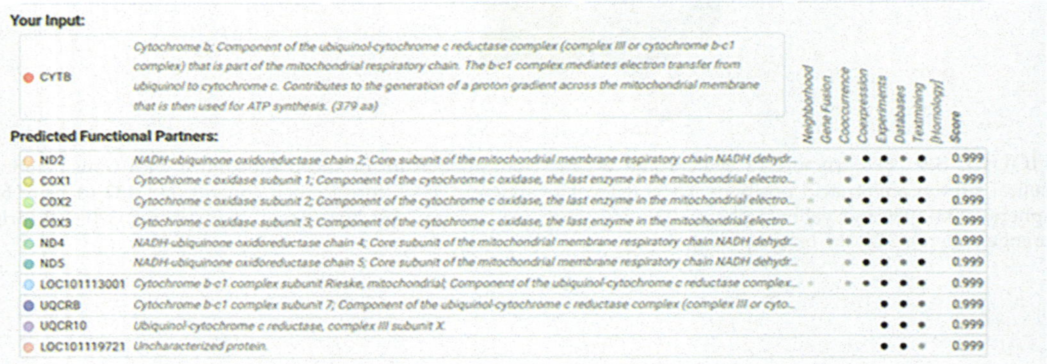

Your Input:

CYTB	*Cytochrome b; Component of the ubiquinol-cytochrome c reductase complex (complex III or cytochrome b-c1 complex) that is part of the mitochondrial respiratory chain. The b-c1 complex mediates electron transfer from ubiquinol to cytochrome c. Contributes to the generation of a proton gradient across the mitochondrial membrane that is then used for ATP synthesis. (379 aa)*

Predicted Functional Partners:

		Neighborhood	Gene Fusion	Cooccurrence	Coexpression	Experiments	Databases	Textmining	[Homology]	Score
ND2	*NADH-ubiquinone oxidoreductase chain 2; Core subunit of the mitochondrial membrane respiratory chain NADH dehydr...*			•	•	•	•	•		0.999
COX1	*Cytochrome c oxidase subunit 1; Component of the cytochrome c oxidase, the last enzyme in the mitochondrial electro...*	•		•	•	•	•	•		0.999
COX2	*Cytochrome c oxidase subunit 2; Component of the cytochrome c oxidase, the last enzyme in the mitochondrial electro...*	•		•	•	•	•	•		0.999
COX3	*Cytochrome c oxidase subunit 3; Component of the cytochrome c oxidase, the last enzyme in the mitochondrial electro...*	•		•	•	•	•	•		0.999
ND4	*NADH-ubiquinone oxidoreductase chain 4; Core subunit of the mitochondrial membrane respiratory chain NADH dehydr...*		•	•	•	•	•	•		0.999
ND5	*NADH-ubiquinone oxidoreductase chain 5; Core subunit of the mitochondrial membrane respiratory chain NADH dehydr...*			•	•	•	•	•		0.999
LOC101113001	*Cytochrome b-c1 complex subunit Rieske, mitochondrial; Component of the ubiquinol-cytochrome c reductase complex...*	•		•	•	•	•	•		0.999
UQCRB	*Cytochrome b-c1 complex subunit 7; Component of the ubiquinol-cytochrome c reductase complex (complex III or cyto...*					•	•	•		0.999
UQCR10	*Ubiquinol-cytochrome c reductase, complex III subunit X.*					•	•	•		0.999
LOC101119721	*Uncharacterized protein.*					•	•	•		0.999

FIGURE 8.10 STRING analysis for molecular interaction studies.

Estimation of haematological and biochemical parameters of infected sheep compared to healthy sheep

We have estimated the haematological and biochemical parameters for the assessment of haematology, as well as liver and kidney function tests. *H. contortus* reveals affecting multifunctional. Infected sheep were observed to have significantly reduced haemoglobin concentration and Total erythrocyte count. Characterisation of cytochrome B revealed important haem binding sites.

Values denoted by different superscripts indicate statistically significant differences among groups. Specifically, the means sharing the superscript 'a' are significantly higher than those with the superscript 'b', with significance determined at $P \leq .05$ (see Table 8.3, Table 8.4).

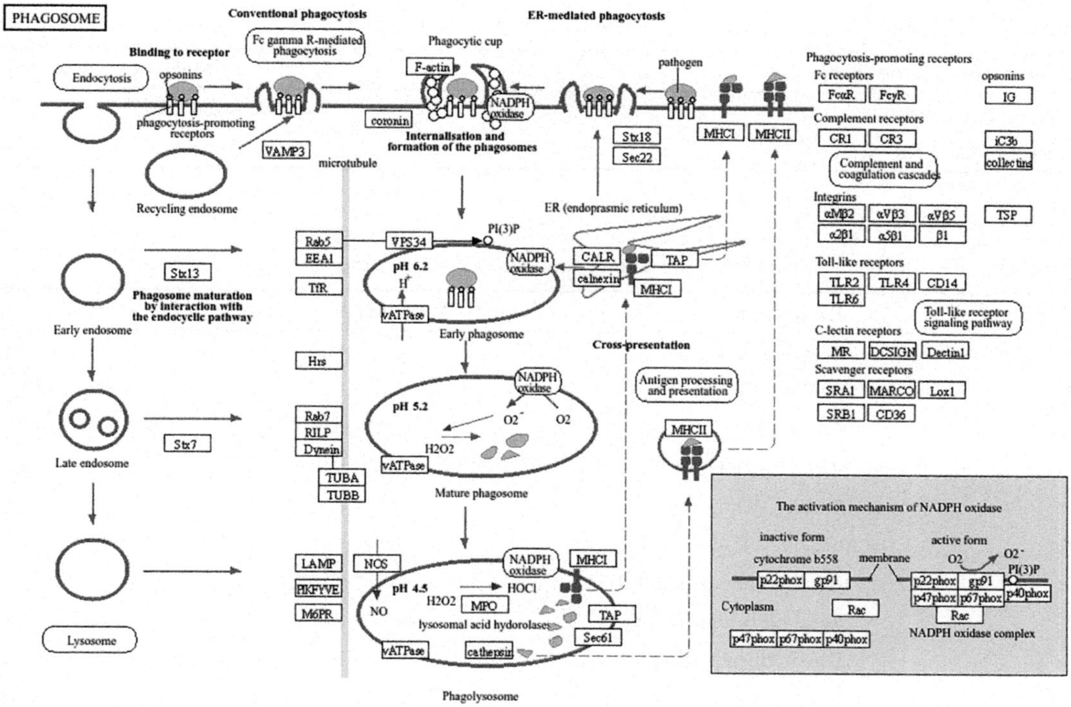

FIGURE 8.11 Kyoto Encyclopedia of Genes and Genomes (KEGG) analysis depicting the role of cytochrome B phagosome pathway.

TABLE 8.1 Estimation of faecal egg count from healthy/noninfected and infected/diseased sheep.

S. no	Parameter	Disease	Healthy
1.	Hb (gm/dl)	$7.96^{b} \pm 0.19$	$10.37^{a} \pm 0.19$
2.	ESR (mm)	1.00 ± 0.00	0.90 ± 0.04
3.	PCV%	$27.84^{b} \pm 0.67$	$36.25^{a} \pm 0.64$
4.	TEC/cumm	$3.98^{b} \pm 0.05 \times 10^{6}$	$4.95^{a} \pm 0.08 \times 10^{6}$
5.	TLC/cumm	$9200.00^{a} \pm 159.23$	$6587.50^{b} \pm 210.81$
5.	Neutrophil%	$53.50^{a} \pm 1.00$	$30.25^{b} \pm 0.59$
6.	Eosonophil%	$16.62^{a} \pm 0.49$	$2.12^{b} \pm 0.44$
7.	Basophil%	0.00	0.00
8.	Lymphocyte%	$29.12^{b} \pm 0.29$	$60.87^{a} \pm 0.29$
9.	Monocyte%	1.37 ± 0.18	1.87 ± 0.29

TABLE 8.2 Haematological parameters for healthy versus infected sheep.

Health status of the sheep	Mean faecal egg count
Noninfected (healthy)	50 ± 5.56^a
Infected (diseased)	550 ± 9.45^b

TABLE 8.3 Biochemical parameters for kidney function test for healthy versus infected sheep.

S. no	Parameters	Disease	Healthy
1.	Total protein (g/dl)	$6.11^b \pm 0.02$	$7.77^a \pm 0.03$
2.	Albumin (g/dl)	$1.92^b \pm 0.04$	$2.80^a \pm 0.03$
3.	Globulin(g/dl)	$4.18^a \pm 0.06$	$4.97^b \pm 0.06$
4.	Albumin:globulin	$0.46^b \pm 0.01$	$0.58^a \pm 0.01$
5.	S.G.P.T (IU/L)	$39.75^a \pm 1.4$	$32.12^b \pm 1.0$
6.	S.G.O.T (IU/L)	$86.87^a \pm 2.6$	$68.87^b \pm 2.7$
7.	Alkaline phosphatase IU/L	$85.25^a \pm 3.89$	$64.87^b \pm 2.09$
8.	Total bilirubin (mg/dl)	$0.67^a \pm 0.07$	$0.37^b \pm 0.03$
9.	Direct bilirubin (mg/dl)	$0.30^a \pm 0.02$	$0.16^b \pm 0.01$
10.	Indirect bilirubin (mg/dl)	$0.37^a \pm 0.09014$	$0.21^b \pm 0.05$

TABLE 8.4 Biochemical parameters for kidney function test for healthy versus infected sheep.

S. no	Parameter	Diseases	Healthy
1.	Glucose (mg/dl)	$55.50^b \pm 1.21$	$68.62^a \pm 2.33$
2.	Creatinine (mg/dl)	$1.71^a \pm 0.05$	$1.20^b \pm 0.02$
3.	Uric acid (mg/dl)	$0.89^a \pm 0.01$	$0.76^b \pm 0.04311$
4.	Urea (mg/dl)	$48.37^a \pm 1.2$	$36.12^b \pm 1.56$
5.	BUN (mg/dl)	$23.62^a \pm 0.73$	$16.87^b \pm 0.95$

Fig. 8.12
Fig. 8.13
The mean values that have superscript were showing a significant difference between groups. The superscript 'a' was showing high mean value than 'b'. Significant at ($P \leq .05$).

Values denoted by different superscripts indicate statistically significant differences among groups. Specifically, the means sharing the superscript 'a' are significantly higher than those with the superscript 'b', with significance determined at $P \leq .05$.

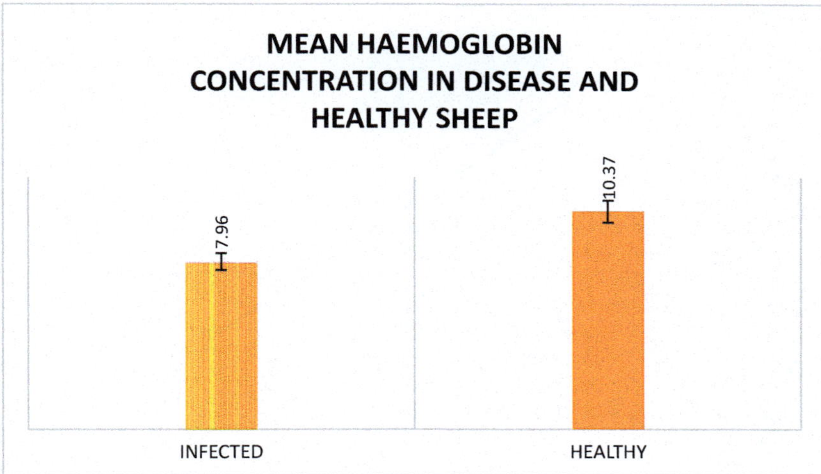

FIGURE 8.12 Mean haemoglobin concentration (gm/dL) in disease and healthy sheep.

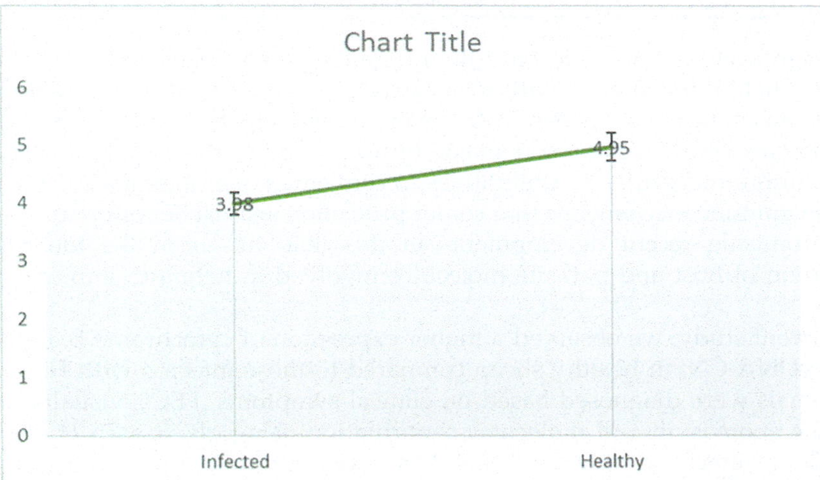

FIGURE 8.13 Mean TEC/cumm in disease and healthy sheep.

Differential mRNA expression profiling for cytochrome B gene from abomasum of sheep

We have analysed the abomasum of sheep from both infected and healthy counterparts. We observed a better expression profile for Cyt B in healthy sheep compared to *Haemonchus*-infected sheep (see Fig.8.14).

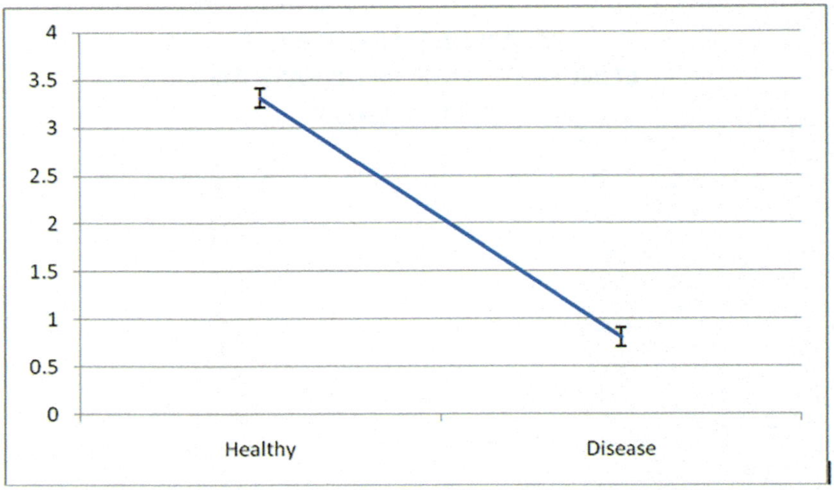

FIGURE 8.14 Differential mRNA expression profiling for sheep cytochrome B with respect to healthy versus infected ones.

Discussion

Despite significant advances in parasite immunology, challenges remain in fully understanding how helminths interact with their hosts, tolerate complex immune responses, and exploit host mechanisms to persist. Moreover, translating this knowledge into effective control strategies continues to be a major hurdle. For the development of prophylactic vaccines or immunotherapies as alternatives to conventional anthelmintics, it is essential to elucidate the immune mechanisms that confer protection against helminth infections. One of the most promising recent developments in this field has been the identification and characterisation of host and parasite molecules involved in helminth-immune cell interactions.

In our current study, we observed a higher expression of cytochrome B, indicative of an increased mtDNA-CN, in healthy sheep compared to those infected with *H. contortus*. The infected animals were diagnosed based on clinical symptoms, FECs, visualisation of adult worms in the abomasum and molecular confirmation using qPCR with *H. contortus* Cox1 gene–specific primers. In our previous laboratory studies, we also observed higher expression levels of the immune response genes retinoic acid-inducible gene I (RIG-I), CD14, IL-6 and IL-10 in healthy sheep compared to infected counterparts. Mitochondria are known to play a crucial role in the regulation of immune responses. They can release mtDNA and other mitochondrial components in response to infection or cellular stress. This extracellular mtDNA functions as a danger-associated molecular pattern, which activates the innate immune system, stimulating the release of pro-inflammatory cytokines and the activation of diverse immune cell populations. Furthermore, mitochondria mediate these immunological effects through a complex network of interactions with nuclear-encoded genes, highlighting their pivotal role in coordinating cellular defence mechanisms. We also observed molecular interactions of the cytochrome B gene with several other genes – most of mitochondrial origin, along with a few nuclear-encoded ubiquinone-related genes. Analysis revealed four iron-binding sites within cytochrome B, consistent with its role in haem biosynthesis. The current research observation depicts three different mechanisms for the association of the mitochondrial CN (revealed through cytochrome B gene expression) with antiparasitic resistance. First mechanism is the enhanced hematopoiesis via iron/haem synthesis, secondly, Innate immune activation through cross-talk with RIG-I/MAVS and other nuclear genes, thirdly, *Phagocytosis* and metabolic support (depicted through KEGG analysis).

H. contortus is a gastrointestinal parasite that primarily infects the abomasum of sheep, causing severe blood loss. Mortality is most commonly observed in young lambs, which possess lower blood volume and smaller body size. Since cytochrome B is directly involved in haem formation and iron binding, it indirectly contributes to hematopoiesis. Consequently, lambs with a lower mtDNA-CN, reflected as reduced cytochrome B expression, are likely to exhibit reduced blood production and volume, rendering them more susceptible to fatal *H. contortus* infections.

Cytochrome B also contributes to apoptotic regulation through its ubiquinone-binding site, which may have an indirect influence on immune responses in sheep. Our KEGG pathway analysis further indicated that cytochrome B participates in phagosome formation, suggesting a potential role in parasite engulfment by immune cells. Additionally, it is involved in the biosynthesis of nucleotide sugars, as revealed through biochemical pathway analysis, thereby supporting DNA and RNA synthesis and ultimately enhancing immune function.

Mitochondria play multifaceted roles in oxidative phosphorylation, calcium metabolism, iron metabolism and immune defence against pathogens – particularly bacteria and viruses. They are crucial for innate immune responses, constituting the first line of defence against parasitic infections. In our earlier studies, we identified mutations in the cytochrome B gene in sheep, which were associated with severe debility involving cardiac and other physiological impairments. Mitochondria are central regulators of cellular metabolism and immune signalling. Their proper functioning is vital for energy production, generation of ROS and maintenance of cellular redox balance – all of which are essential for effective immune responses and parasite elimination. Any defect in mitochondrial metabolic signalling can compromise immune competence and predispose the host to disease susceptibility.

Mitochondria play a central role in apoptosis, a programmed cell death process that serves as a crucial defence mechanism during infection. When host cells are infected with parasites, apoptosis can be triggered to limit parasite proliferation and prevent the spread of infection. This process is partly regulated by mitochondrial factors, including the release of cytochrome C from mitochondria, which subsequently initiates the caspase-dependent apoptotic cascade. In addition to apoptosis, mitochondria play a significant role in regulating inflammation. Mitochondrial dysfunction can result in excessive ROS production, leading to heightened inflammatory responses. Such dysregulated inflammation can be detrimental during parasitic infections, causing collateral tissue damage. Hence, mitochondrial signalling is vital in maintaining the balance between pro-inflammatory and antiinflammatory pathways. The MAVS protein is another key component of the mitochondrial membrane, known for its essential role in host antiviral defence. MAVS activates downstream signalling pathways that lead to interferon production, which is central to antiviral immunity. Although primarily characterised in viral infections, the MAVS pathway underscores the broader immunomodulatory role of mitochondria, which may also extend to antiparasitic defence mechanisms.

In our earlier study, we observed elevated expression of RIG-I in healthy sheep. RIG-I, a pattern recognition receptor, operates through the MAVS pathway, as demonstrated in viral immunity. In the present study, we also report a higher mtDNA-CN in healthy sheep, which may be directly or indirectly associated with RIG-I activity in mediating defence against parasitic infection. Our molecular docking studies further revealed that RIG-I can interact with surface proteins of *H. contortus*. This finding suggests a potential mitochondrial – nuclear cross-talk mechanism that may underlie a coordinated immune response against the pathogenic parasite *H. contortus*. In our earlier work, we reported that the RIG-I gene plays a pivotal role in antiparasitic immunity against *H. contortus*. We demonstrated that RIG-I binds to surface proteins of the parasite, with the predicted binding region spanning histidine (His) 302 to proline (Pro) 763 – corresponding to the parasite's α-tubulin and β-tubulin proteins. Since α- and β-tubulins are the major constituents of microtubules, each binding two molecules of GTP (one exchangeable on the α-chain and one nonexchangeable on the β-chain), this interaction may have important structural and signalling implications. Our analyses identified RIG-I binding sites within the 300–750 amino acid region for most of the parasitic surface proteins studied. The α-tubulin, β-tubulin, lectin, galectin and cysteine protease proteins of *H. contortus* were found to be the most promising RIG-I–binding partners. This represents the first report implicating RIG-I in antiparasitic infection, as prior research had only established its antiviral function. Following its interaction with the parasite, RIG-I acts through the mitochondrial protein MAVS, triggering the secretion of interleukins (IL-6 and IL-10) that ultimately contribute to parasite destruction. In our current study, we observed higher expression of IL-6 and IL-10 in healthy sheep compared with infected animals, supporting their protective immunomodulatory roles. Similarly, we previously identified CD14 as a novel receptor for *H. contortus*. Structural alignment analyses revealed that α-tubulin of *H. contortus* aligns with ovine CD14 between Glu and Ala, while a similar alignment was also observed between β-tubulin and CD14 across the same amino acid region (Glu–Ala). These findings suggest potential host – parasite protein interactions that may be critical for parasite recognition and immune activation.

The present study may represent the first molecular investigation of the mitochondrial genome in relation to antiparasitic resistance. Although limited studies have explored the role of mitochondria in parasite – host interactions, some research has demonstrated mitochondrial involvement in other parasitic infections. For instance, in *T. gondii*, mitochondria have been shown to regulate metabolic efficiency through mitochondrial fission, fusion and cristae remodelling, as well as through the induction of autophagy, lipophagy and mitochondrial elongation, processes that allow the parasite to evade autophagic degradation while enhancing ATP production. Furthermore, *Toxoplasma* increases the efficiency of β-oxidation by promoting mitochondrial association with LDs and enhancing the uptake of FAs. Interestingly, host cell mitochondria are often recruited to the vacuolar membranes that harbour a wide range of intracellular pathogens – including bacteria such as *L. pneumophila* and *C. psittaci*, as well as the protozoan parasite *T. gondii*. Conversely, certain parasitic lineages, such as dinoflagellates and *Giardia*, appear to lack classical mitochondria or contain only rudimentary mitosomes, likely reflecting their adaptation to derive energy directly from the host.

The haemoglobin concentration and total erythrocyte count were observed to be significantly lower in infected sheep compared to healthy controls. A reduced mitochondrial CN, reflected by the decreased expression of cytochrome B, was also evident in the infected group. In our earlier study, we identified four iron-binding sites in cytochrome B that are crucial for haem biosynthesis, which in turn is essential for haemoglobin formation. Enhanced haem synthesis contributes to increased haemoglobin production and a higher erythrocyte count. Consequently, the infected sheep exhibited both lower erythrocyte counts and reduced mtDNA-CNs, indicating a functional relationship between mitochondrial activity and erythropoiesis.

In addition to its role in haem synthesis, mitochondria may contribute to parasite destruction through phagosome formation, as supported by our molecular pathway analyses. These findings collectively suggest that mtDNA exerts a substantial influence on antiparasitic immunity. Recent studies have increasingly focused on the assessment of mtDNA-CN in relation to various pathological conditions. Reports indicate that mtDNA-CN varies across tissues and also exhibits species-specific differences. Multiple methodological approaches have been utilised to estimate mtDNA-CN, including qPCR based on differential expression relative to nuclear DNA, digital PCR and high-throughput DNA sequencing. Alterations in mtDNA-CN have been associated with diverse physiological and pathological traits, such as personality characteristics, low-density lipoprotein cholesterol levels, cardiovascular disease risk, metastatic breast cancer, multiple sclerosis, and metabolic traits.

Interpretation

Mitochondrial CN variation plays a critical role in parasitic immunity against *H. contortus* in sheep. We characterised cytochrome B from the sheep gut and identified four iron-binding sites responsible for haem production, leading to improved haemoglobin synthesis and higher erythrocyte counts in healthy animals possessing greater mtDNA-CNs compared to infected ones. Another potential antiparasitic mechanism of mitochondria involves phagosome formation and biosynthesis of nucleotide sugars. Furthermore, the RIG-I protein was found to bind to the α- and β-tubulin surface proteins of *H. contortus*, subsequently recruiting the MAVS protein and stimulating the release of interleukins IL-6 and IL-10, thereby contributing to the host's immunological defence. Future studies include the inclusion of mitochondrial genome in genomic selection programme, as well as evolvement of future resistant animals through a gene editing approach.

Characterisation of other mitochondrial genes: case study 2

In addition, other mitochondrial genes were systematically characterised and their functionally important domains were identified. Subsequently, CN variation was evaluated in conjunction with differential mRNA expression profiling to assess gene dosage effects and transcriptional regulation (see Fig. 8.15).

Fig. 8.16

Fig. 8.17

Fig. 8.18

Oxygen binding site: 244 (bright orange sphere)

Fig. 8.19

Fig. 8.20

Molecular interaction among mitochondrial genes (STRING analysis)

It is evident that all the biological functions are governed by molecular pathways. Each molecular pathway comprises of a lot of proteins, each in turn are controlled by their respective genes.Here we represent string analysis to depict the role and molecular interaction among various mitochondrial genes.

Fig. 8.21

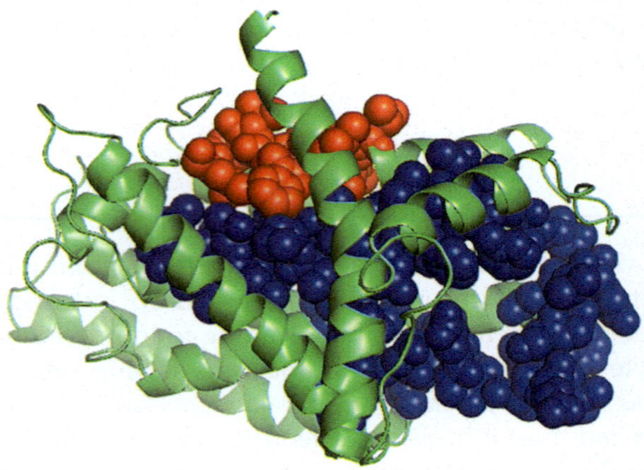

FIGURE 8.15 3D structure for NAD1, where red sphere, (87–96aa) represents NADH binding site, while blue sphere (199–247aa) represents Flavin Mononucleotide (FMN) binding site.

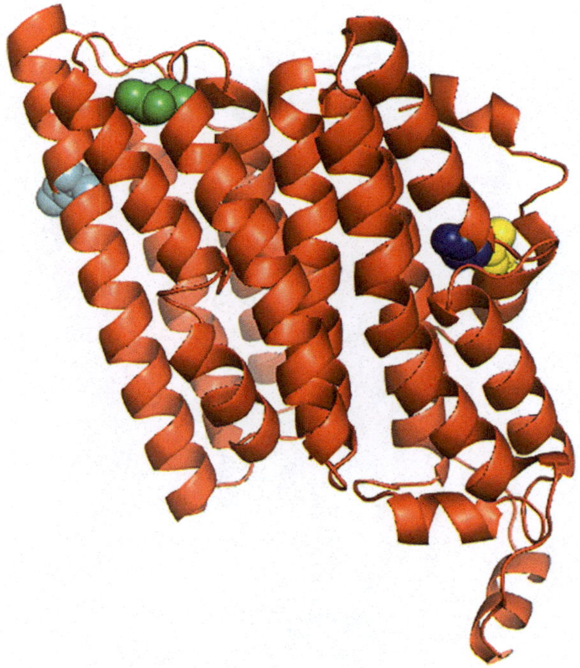

FIGURE 8.16 3D structure for ND2 with the sites for mutation at N150D: green sphere, G259S: blue sphere, A331S: yellow sphere, L71P: cyan sphere.

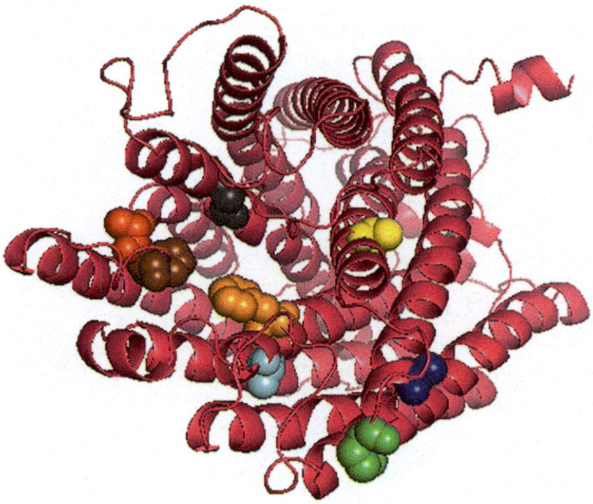

FIGURE 8.17 3D structure for cytochrome oxidase 1, along with the binding sites were predicted. Calcium binding sites: 40 (blue sphere), 45 (green sphere), iron binding site: 61 (yellow sphere), 376 (cyan sphere), 378 (orange sphere). Copper cation binding site: 240 (deep grey sphere), 290 (red sphere), 291 (chocolate sphere).

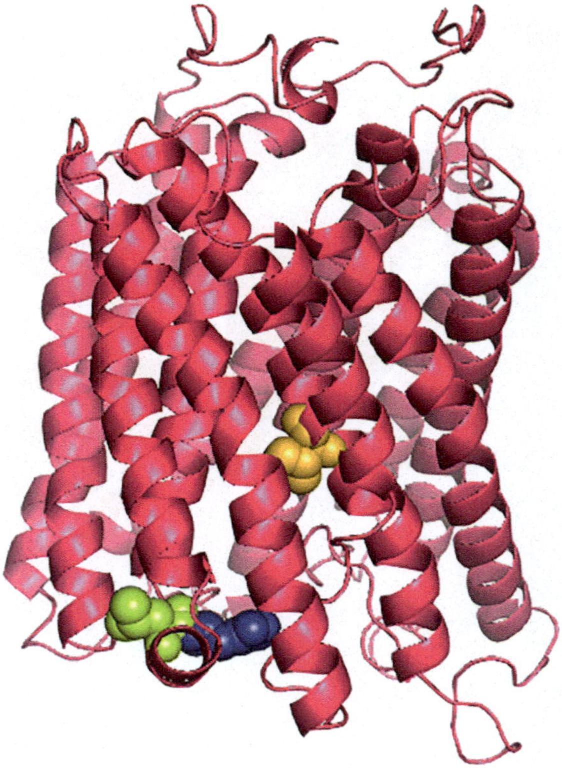

FIGURE 8.18 3D structure for cytochrome oxidase1 with the binding sites was predicted. Magnesium binding site: 368 (limon sphere), 369 (deep blue sphere). Oxygen binding site: 244 (bright orange sphere).

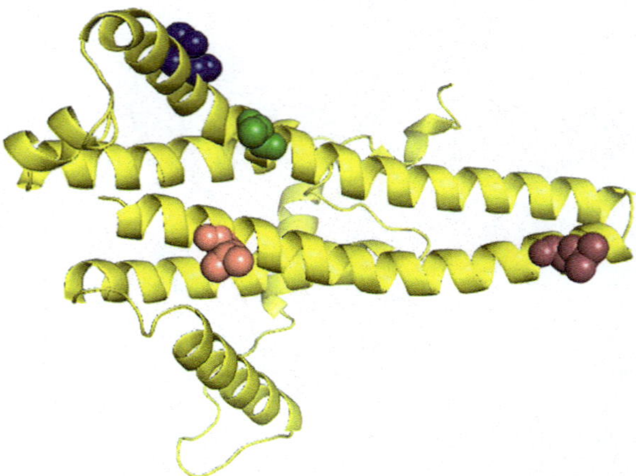

FIGURE 8.19 3D structure for ATP6 with the mutations or single nucleotide polymorphisms (SNPs) detected in ATP6 resulting into diseases: SNPs detected were L156R (forest green sphere), L156P forest green sphere, I192T (raspberry sphere), S148N (density blue), L217P (deep salmon sphere).

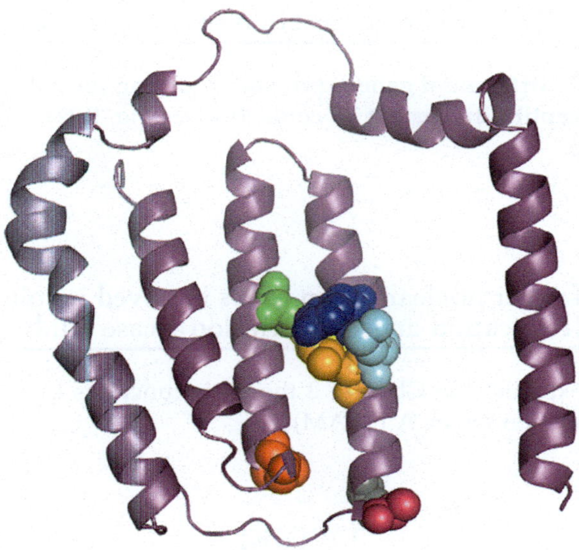

FIGURE 8.20 3D structure for ND6 with sites for mutation predicted as I26M (red sphere), G36S (green sphere), Y59C (blue sphere), L60S (yellow sphere), M63V cyan sphere), M64I (orange sphere), M64V (orange sphere), A72V (grey sphere), A74V warm (pink sphere).

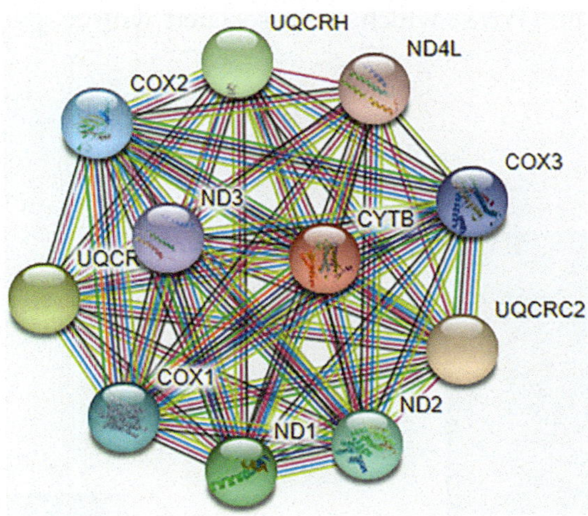

FIGURE 8.21 STRING analysis depicting the molecular interaction among mitochondrial genes in sheep as the host.

Copy number variation of mitogenome

As assessed through **differential gene expression profiling for mitochondrial genes w. r. t. resistance versus susceptibility against *H. contortus*** (see Fig. 8.22).

Fig. 8.23

Fig. 8.24

Characterisation of other nuclear origin genes involved in mitochondrial function (replication and transcription): case study 3

We have already explained in Chapter 2 the basic functions of TWINKLE, MPV17 and mitochondrial transcription factor A (TFAM).

TWINKLE

The mtDNA helicase TWINKLE, encoded by the *TWNK* gene, is a core component of the mtDNA replication apparatus. TWINKLE exhibits 5′→3′ helicase activity and is responsible for unwinding double-stranded mtDNA, a critical step required for accurate replication and maintenance of the mitochondrial genome. Genetic studies have identified nearly 60 pathogenic mutations in *TWNK*, which are associated with a spectrum of mitochondrial

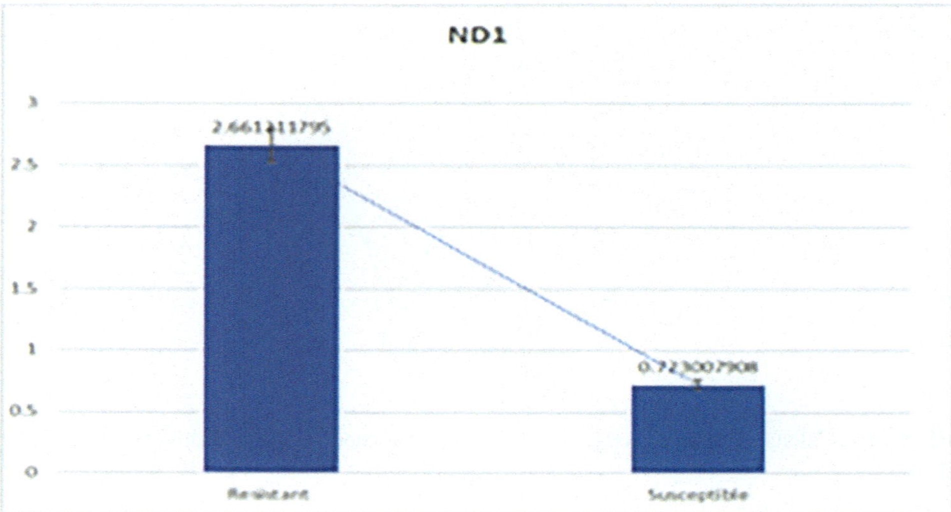

FIGURE 8.22 Differential mRNA expression profile for mitochondrial gene ND1 in sheep w. r. t. healthy and infected (parasitic, H*aemonchus contortus*).

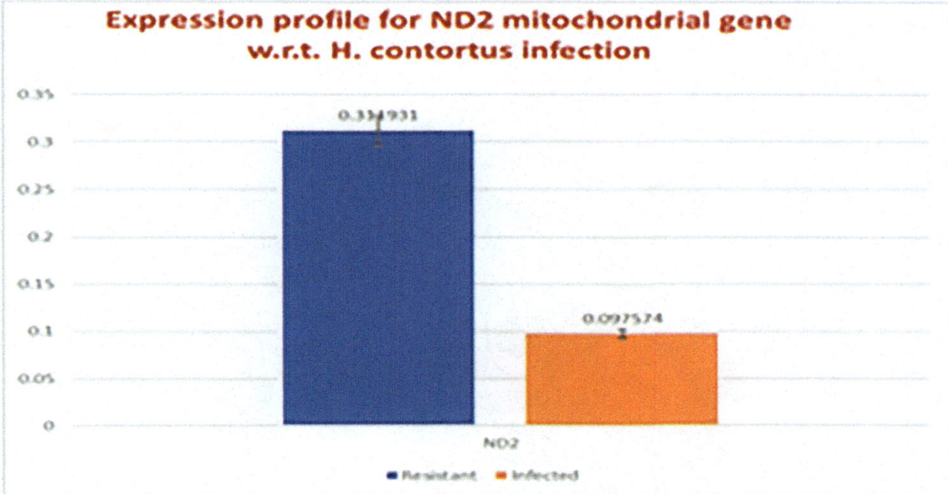

FIGURE 8.23 Differential mRNA expression profile for mitochondrial gene ND2 in sheep w. r. t. healthy and infected (parasitic, *Haemonchus contortus*).

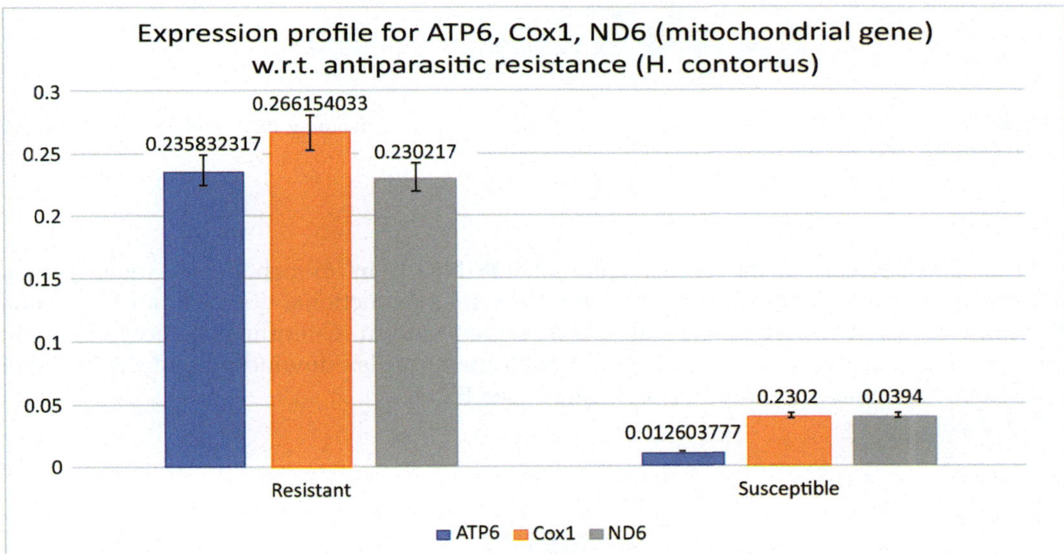

FIGURE 8.24 Differential mRNA expression profile for mitochondrial gene ATP6, cytochrome oxidase 1, ND6 in sheep w. r. t. healthy and infected (parasitic, Haemonchus *contortus*).

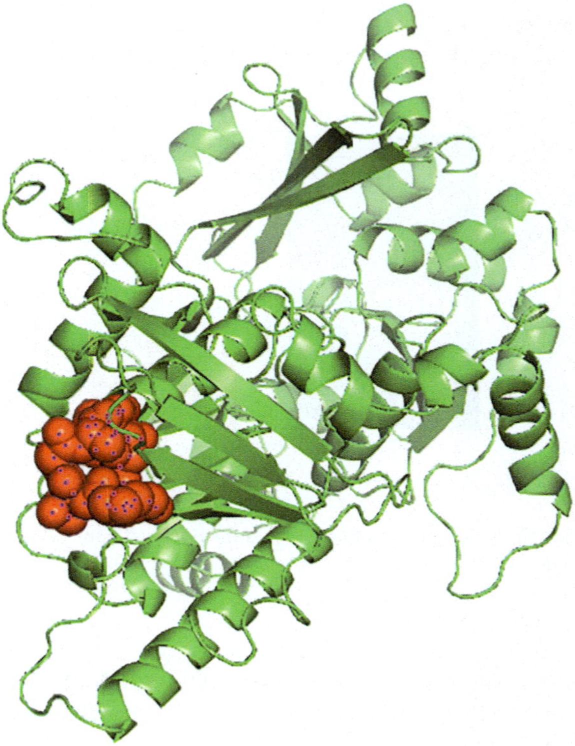

FIGURE 8.25 TWINKLE, mtDNA helicase (green) with red spheres denoting ATP binding site (amino acid position 415–422).

disorders, underscoring the critical role of TWINKLE in mitochondrial integrity and function. Structurally, TWINKLE comprises three major functional domains: an N-terminal primase-such as domain, a central linker region and a C-terminal helicase domain. Collectively, these domains orchestrate DNA binding, duplex unwinding and the initiation and elongation phases of mtDNA replication (see Fig. 8.25).

Fig. 8.26

MPV17

The MPV17 gene encodes an inner mitochondrial membrane protein that is critically involved in regulating mitochondrial deoxynucleotide pool homeostasis and maintaining mtDNA integrity. This protein is essential for preserving mtDNA stability and ensuring proper mitochondrial genome replication and overall mitochondrial homeostasis (see Fig. 8.27).

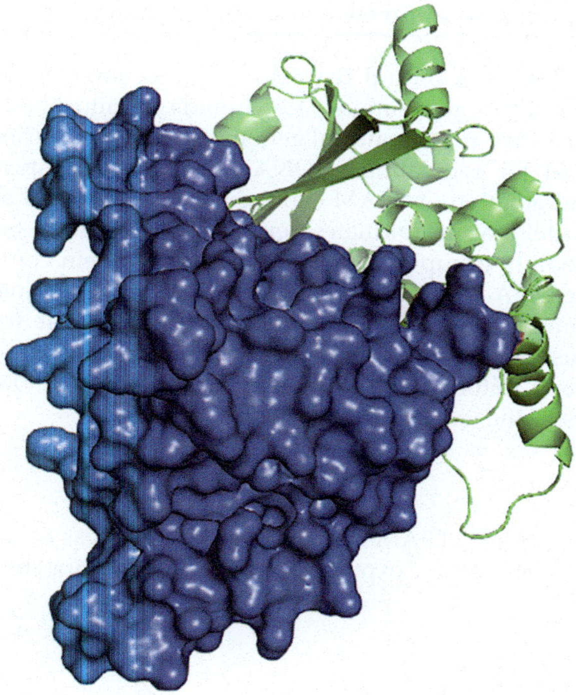

FIGURE 8.26 3D structure prediction for ovine TWINKLE, mtDNA helicase (green) with blue surface denoting SF4 helicase domain (amino acid position 384–635).

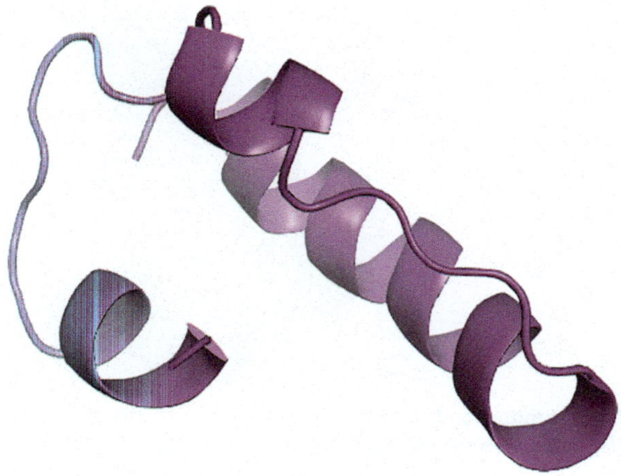

FIGURE 8.27 3D structure for MPV17 for sheep.

Mitochondrial transcription factor A

This gene encodes TFAM, a central regulator of mitochondrial gene expression and a critical contributor to mtDNA replication. TFAM binds to mitochondrial promoter regions to initiate transcription of the mitochondrial genome. Studies in murine models have shown that TFAM is essential for regulating mtDNA-CN and is indispensable for embryonic development. Targeted ablation of TFAM expression in cardiac and skeletal muscle tissues results in a mouse model that recapitulates key features of Kearns–Sayre syndrome.

TFAM is a high-mobility group (HMG) DNA-binding protein containing two HMG-box domains that facilitate DNA binding and bending. This DNA-bending activity is crucial for transcription initiation in mammalian mitochondria, in contrast to yeast mitochondria, where the homologous protein Abf2 does not require such bending. In addition, TFAM is thought to play a role in mitochondrial genome packaging, as its DNA-binding activity is largely nonsequence-specific (see Fig.8.28).

Fig. 8.29

Fig. 8.30

Fig. 8.31

Better expression of MPV17, TFAM and TWINKLE genes was observed in healthy sheep which were resistant upon natural exposure to a pathogen (parasite) in comparison to the existing ones.

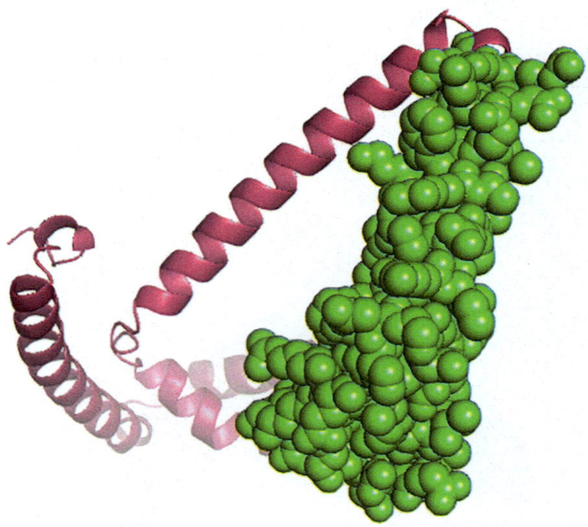

FIGURE 8.28 3D structure for mitochondrial transcription factor A (warm pink), with domain for high-mobility group boxes, ***A DNA-binding domain (green sphere), amino acid position 50–118.***

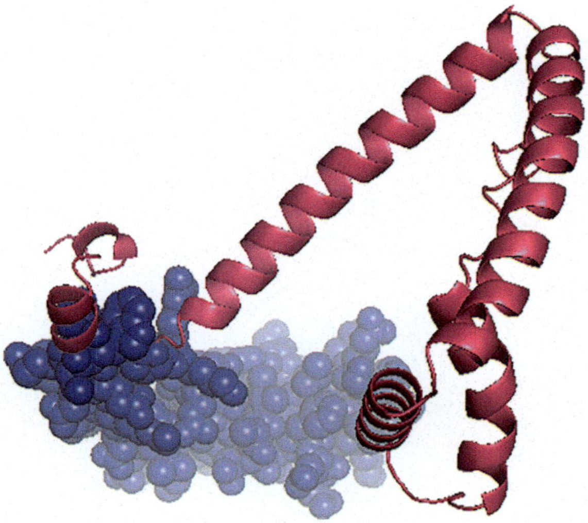

FIGURE 8.29 3D structure for mitochondrial transcription factor A (warm pink), with domain for high-mobility group *boxes B, DNA-binding domains (green sphere), amino acid position 155–219*.

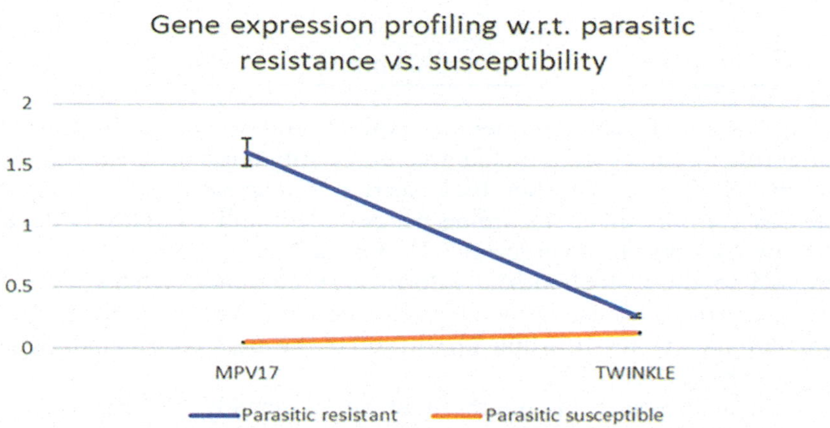

FIGURE 8.30 Gene expression profiling w. r. t. parasitic resistance versus susceptibility.

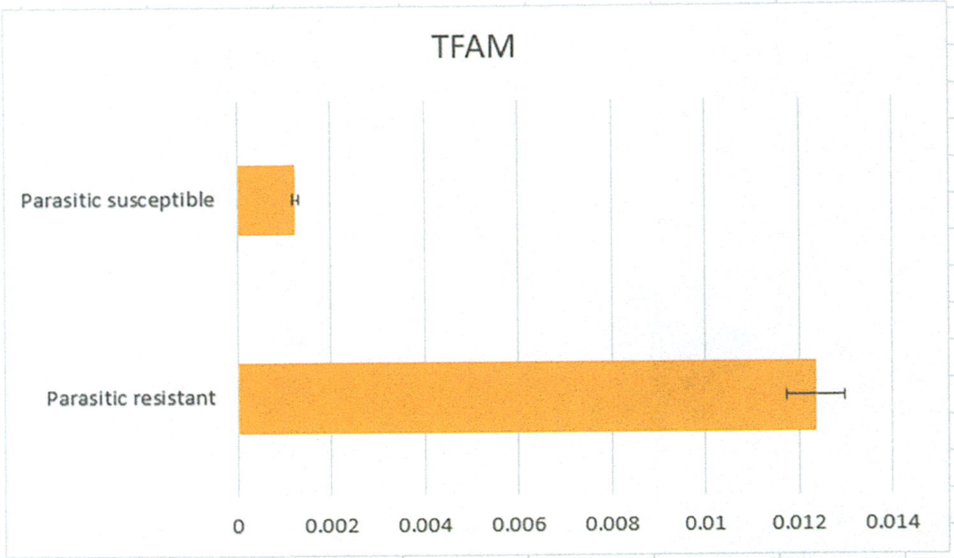

FIGURE 8.31 Differential mRNA expression profiling for mitochondrial transcription factor A (TFAM) w. r. t. parasitic resistance versus susceptibility.

Cytochrome C

Cytochrome C is a small, soluble, haem-containing protein encoded by nuclear DNA that plays a pivotal role in mitochondrial oxidative phosphorylation. It functions as an essential electron carrier within the mitochondrial electron transport chain, facilitating electron transfer between complex III (cytochrome bc_1 complex) and complex IV (cytochrome C oxidase), thereby contributing to efficient ATP production.

It is localised to the mitochondrial intermembrane space, where it functions as a mobile electron carrier, transferring electrons between complex III (cytochrome bc_1 complex) and complex IV (cytochrome C oxidase) of the electron transport chain, thereby supporting ATP generation.

Structurally, cytochrome C is composed of approximately 100–105 amino acids and contains a covalently bound haem prosthetic group with a central iron atom capable of reversible oxidation–reduction. This redox flexibility underlies its efficiency in electron transfer reactions. The protein is peripherally associated with the outer surface of the inner mitochondrial membrane through electrostatic interactions. Cytochrome C is highly conserved among eukaryotes, reflecting its critical and evolutionarily conserved function in cellular energy metabolism. Fig. 8.32

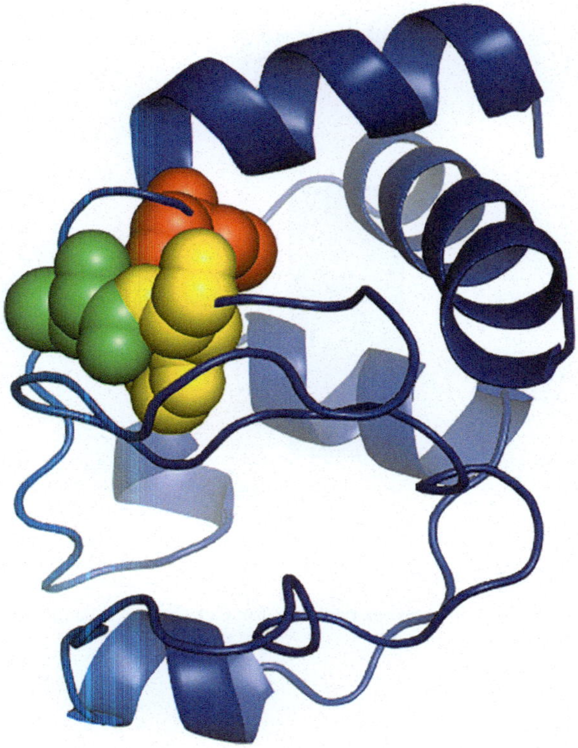

FIGURE 8.32 3D structure for cytochrome C (blue), with the domains for haem C binding domain (amino acid position 15, red sphere), haem C binding domain (aa position 18, green sphere) and axial binding residue (aa position 19, yellow sphere).

Differential mRNA expression profiling for ovine cytochrome C w. r. t. healthy versus infected (gastrointestinal parasite *Haemonchus contortus*)

Fig. 8.33

Cytochrome C was observed to have a profound role in innate antiparasitic immunity, a study explored in sheep model against *H. contortus*. Healthy sheep that were resistant upon natural exposure to pathogen (parasite) exhibited a better expression profile for the cytochrome C gene.

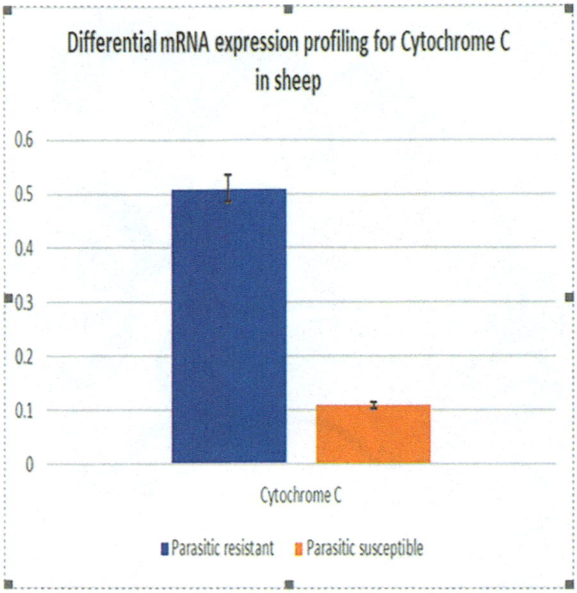

FIGURE 8.33 Differential mRNA expression profiling for Cyt C w. r. t healthy vs. infected (gastrointestinal parasite ***Haemonchus contortus***).

References

Banerjee S., Pal, A.., Mandal,S.C., Chatterjee,P.N., Chatterjee, J.K. 2021. RIG-I has a role in immunity against Haemonchus contortus, a gastrointestinal parasite in Ovis aries- a novel report. Frontiers in immunology. https://doi.org/10.3389/fimmu.2020.534705.

Pal, A. & Chakravarty, A.K. 2019. Genetics & Breeding for the Disease resistance of livestock. Elsevier Publication. Academic publisher. ISBN: 9780128164068.

Rawat, K., Pal, A. , Banerjee,S., Pal, A., Mandal, S.C, Batabyal, S. 2021. Ovine CD14- an immune response gene has a role against gastrointesinal nematode Haemonchus contortus -a novel report. Frontiers in Immunology. https://doi.org/10.3389/fimmu.2021.664877.

Further Reading

Deb, D. 2025. Phenotypic and genomic analysis of the growth, reproduction and parasitic resistance in Garole sheep, PhD Thesis to be submitted to West Bengal University of Animal and Fishery Sciences, Kolkata-37. West Bengal, India.

Deb, D., Pal, A., Rahaman, S., Samanta, R., Chatterjee, P.N. 2025. Exploring the Host immunity against parasites-a molecular approach in sheep model. In Proceedings on Conference on "Reinforcing BioSecurity: Paving the way for a safer One Health Environment." Jan, 2025. venue: West Bengal University of animal and fishery sciences, 37, K. B. Sarani, Kolkata-37. Table 8.2 West Bengal, India.

Deb, D., Pal, A., Samanta, R. et al. 2025. Copy number variations in mitochondrial genome have a potent role in parasitic immunity against Haemonchus contortus-A novel report. In the Proceedings of International Conference on *"Precision animal breeding through Genomics, artificial intelligence and machine learning"*, 13-14th November, 2025. West Bengal University of Animal and Fishery Sciences, Kolkata, West Bengal, India.

Pal, A. et al. 2025. SNPs and indel identified in mitochondrial genome is associated to antiparasitic immunity against *Haemonchus contortus in sheep model*-A novel report. Submitted to Genomics.

Pal, A., Banerjee, S., Rawat, K., Batabyal, S., Deb, D., Chatterjee, P.N. 2025. "Mitochondrial DNA copy number variation is associated with antiparasitic immunity in sheep infected by Haemonchus contortus." Proceedings on National conference on "*Advancing Animal Health and production through Biochemical and Biotechnological innovations*" University of Animal and Fishery Sciences, 37, K. B. Sarani, Kolkata-37. West Bengal. India. Pin-700037.

Pal, A., Banerjee, S., Rawat, K., Deb, D. and Chatterjee, P.N. 2025. Copy number variations in mitochondrial DNA has a potent role in parasitic immunity against *Haemonchus contortus*-A novel report. https://doi.org/10.1101/2025.09.10.675310.

Pernas, L., Bean, C., Boothroyd, J. C., & Scorrano, L. (2018). Mitochondria Restrict Growth of the Intracellular Parasite *Toxoplasma gondii* by Limiting its Uptake of Fatty Acids. *Cell Metab*, *27*, 886–897.

CHAPTER

9

Mitogenomics and reproduction

Introduction

As discussed in earlier chapters, mitochondria play a vital role in basic physiological process, any deviation will cause defects. Similar is the case of the reproductive process. In this chapter, we will discuss the role of mitochondria in both male and female reproduction. Then we will discuss how abnormalities in the mitochondrial genome lead to defects in reproduction.

The male reproduction

Influence of mitochondrial DNA (mtDNA) on male reproduction and fertility: Numerous evidences indicate that mtDNA mutations in males result in detrimental repercussions. mtDNA haplogroup T in humans is linked to diminished sperm motility. Mitochondrial abnormalities significantly contribute to reduced lifespan in numerous taxa. The 'Mother's curse' hypothesis predicts that males induce mitochondrial dysfunctions. Laber's hereditary optic neuropathy (LHON) results from base pair substitutions in germline mtDNA, predominantly affecting male individuals. Recent data has also revealed the paternal transmission of mtDNA. Male infertility is typically linked to asthenozoospermia or oligoasthenozoospermia. This is typically documented in patients with mitochondrial disorders, which are caused by point mutations or numerous deletions of mtDNA. Sperm are demonstrated to be more vulnerable to the loss of mtDNA, which may impact sperm motility and fertility. A correlation has been noted between semen quality and the respiratory chain functions in sperm mitochondria; however, semen quality is also influenced by point mutations, mtDNA single-nucleotide polymorphisms (SNPs) and mtDNA haplogroups. In the semen of infertile men, a significantly low concentration of somatic mtDNA deletions has been detected. This factor induces infertility by affecting sperm motility. To date, few research has been conducted on cattle species. Mutations in cytochrome B were reported to be correlated with libido and seminal traits in male Garole sheep. Subsequent NGS investigations revealed that the entire mitochondrial genome sequencing identified alterations in other mitochondrial genes connected with male reproductive function in Garole sheep as an animal model.

Mitochondrial Genomics
DOI: https://doi.org/10.1016/B978-0-443-26655-3.00006-8

Mechanism via which mutations in mtDNA influence reproduction:

Mitochondria are the primary suppliers of ATP in the body and are crucial for spermatogenesis, differentiation and the appropriate functioning of the germ line. Nonetheless, genetic mutations or anomalies might result in significant repercussions for spermatogenesis and the fertilisation process. Dysfunctions in mtDNA might result in the total cessation of spermatogenesis, thereby causing male infertility. The critical elements in male infertility are the vitality and vigour of the spermatozoa. Spermatozoa necessitate a substantial amount of energy for optimal functionality and survival. The sperm mitochondria are situated in the midpiece, enabling the rapid provision of substantial energy. Sperm motility and propulsion rely on the densely concentrated mtDNA within the flagellum. This energy pack is utilised to propel and drive sperm, a crucial role in male fertility. Reduced sperm motility leads to male infertility, hence impacting the reproductive process. The motility of sperm is significantly reliant on ATP produced through oxidative phosphorylation in the mitochondrial sheath. Consequently, these are the significant characteristics to be noted in male fertility. mtDNA, located near the inner mitochondrial membrane, is subjected to highly carcinogenic O_2 radicals due to the generation of superoxide radicals as byproducts of oxidative phosphorylation occurring in the respiratory chain. The release of superoxide radicals renders mitochondria a principal intracellular source of reactive oxygen species (ROS). Consequently, mtDNA exhibits a propensity for mutations. ROS have a crucial function in male reproduction. During normal physiological processes, spermatozoa generate ROS at low levels. Defective mtDNA results in the formation of aberrant proteins and disrupts the electron transport chain, resulting to elevated ROS levels, which adversely affect spermatogenesis and contribute to male infertility. The existence of free oxygen radicals was noted as early as 1943. The function of these radicals was documented subsequently in 1989. Optimal or reduced levels of ROS are crucial for the proper functioning of spermatozoa, including hyperactivation, acrosome response, oocyte fusion, capacitation, motility and fertilisation. Normal spermatozoa functions are stimulated when incubated with modest concentrations of hydrogen peroxide radicals. Other superoxide radicals, including nitric oxide and superoxide anion, have also been reported to facilitate sperm capacitation and acrosome processes. The production of ROS by spermatozoa is well-documented as essential for sustaining their normal function. Excessive generation may result in dysfunction of sperm under certain clinical situations.

Female reproductive system:

The influence of mtDNA on female reproduction and fertility: mitochondria, as the predominant organelle in the embryo, are crucial to the female reproductive system. All physiological changes occurring in the reproductive system are contingent upon the energy supplied by the mitochondria and mtDNA (Fig. 9.1).

Ovulation in mammals, zygote activation and fertilisation are contingent upon mtDNA and the energy provided by mitochondria. Dysfunction in mtDNA can lead to abnormalities in embryonic development and fertilisation. The mitochondria within the oocyte serve as primary energy sources for the growing embryo; indeed, oocytes predominantly rely on ATP generated by oxidative phosphorylation. The rate of ATP consumption increases with oocyte maturation. mtDNA in mice is essential for oocyte maturation and embryonic development. In pigs, immature oocytes may lead to a reduction in mitochondrial membrane potential. The mtDNA copy number in an oocyte determines oocyte quality

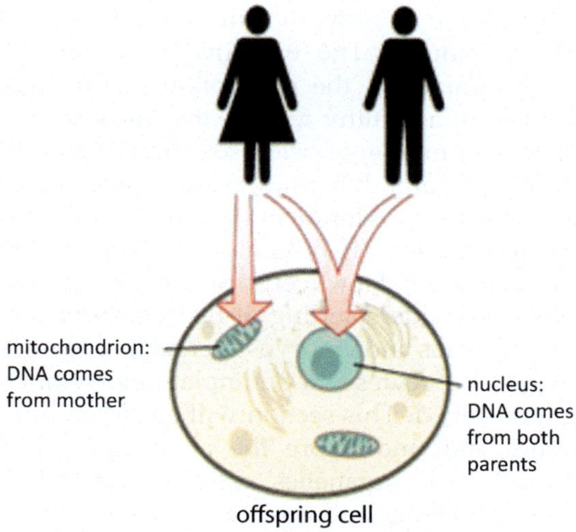

FIGURE 9.1 Mechanism of mitochondrial DNA transmission.

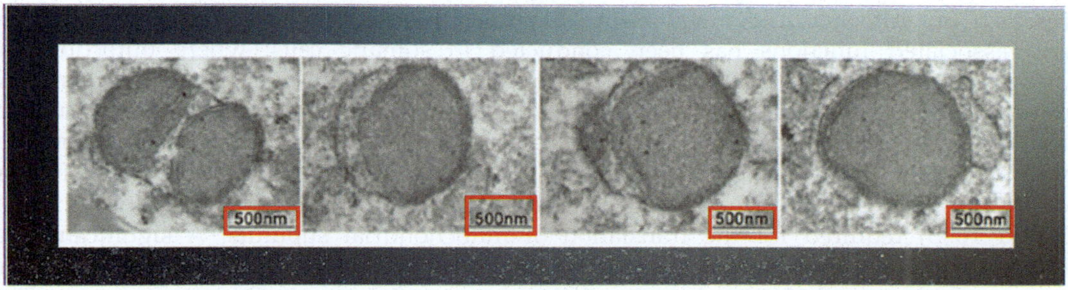

FIGURE 9.2 Transmission electron microscopy of mitochondria. Transmission electron microscopy reveals spherical mitochondria devoid of cristae in a two-cell stage pig oocyte at 48 hours post in vitro fertilisation.

and is associated with many ovarian diseases. An individual with poor-quality oocytes and many ovarian diseases exhibited a reduced quantity of mtDNA molecules compared to a healthy individual, who possessed 256,000 copies of mtDNA. Mammalian eggs and early embryos rely on oxidative phosphorylation since the glycolytic pathway is inhibited by the inhibition of the enzyme phosphofructokinase. Mitochondria facilitate the regulation of calcium wave oscillations, which are ubiquitous signals during fertilisation, as they efficiently transmit intracellular biological information. Research has demonstrated that mutations in cytochrome B are linked to female reproductive functions in Garole sheep. The link of cytochrome B with female reproductive function was identified in the Ghungroo pig model. Subsequently, the entire mitochondrial genome was sequenced, revealing associations between various genes and reproductive functions using sheep as a model.

mtDNA is crucial in the life cycle as it regulates the body's energy generation activities. It is also essential in the processes of animal and human reproduction. Enhancements in mitochondrial functions and the quality of mtDNA result in improved reproductive processes and healthier individuals. The examination of mtDNA facilitates the early identification of genetic abnormalities, the conception of three-parent babies (where the zygote contains cytoplasm from a healthy female, the nucleus from the biological mother and sperm from the father) and marker-assisted selection in animals (Fig. 9.2).

Section 1: mitochondrial structure, dynamics, features and their roles in oocyte maturation, fertilisation and embryonic development: given the pressing necessity to enhance success rates in in vitro fertilisation (IVF) clinics, innovative strategies are required to address mitochondrial dysfunctions. However, prior to the implementation of therapeutic interventions, a comprehensive understanding of the metabolic activities and dynamics of mitochondria in unaffected oocytes and embryos is essential. Numerous facets of mitochondrial activity and malfunction in oocytes and preimplantation embryos remain unexamined, leaving many enquiries unresolved. This section will elucidate our present comprehension of mitochondrial dynamics and underscore the areas that require further elucidation. Mitochondria are multifunctional organelles essential for cellular energy production, calcium homoeostasis, cell signalling, apoptosis and various other cellular processes. The dynamics and motility of mitochondria are associated with their functions, and mitochondria possess a significant capacity to adjust their functions, dynamic behaviour and metabolic activities in response to varying environmental situations. The majority of research on mitochondrial metabolism, dynamics, biochemistry, molecular composition and associated functions has been conducted in somatic cells, where mitochondrial dysfunctions have been linked to particular structural and molecular anomalies related to diseases such as neurological disorders, cardiovascular disease, diabetes and cancer. Mitochondria are semiautonomous organelles possessing a 16.6 kb circular mtDNA genome, which encodes 13 vital protein subunits for complexes I, III, IV and V of the respiratory chain, along with 22 tRNAs essential for mitochondrial protein synthesis and two rRNAs (12 S and 16 S) required for the translation of mitochondrial subunits. A significant and functionally interdependent link exists between the nuclear genome and the mitochondrial genome, with around 1500 nuclear-encoded genes essential for mitochondrial function. The majority of mitochondrial polypeptide transcripts are translated in the cytoplasm and subsequently transported into the mitochondria. Examples include the mitochondrial transcription factor A (TFAM), which is essential for mtDNA replication; additionally, TFAM transcription is associated with the transcription of catalytic DNA polymerase gamma (POLG) and auxiliary POLG2 subunits. Mitochondrial activity relies not only on overall dynamics and motility but also on specific features essential for cell cycle-specific functions, including the mitochondrial transition pore, membrane potential, localisation and distribution patterns and sufficient quantities of mtDNA. Dysfunctions entail elevated levels of ROS and calcium excess. The mitochondrial permeability transition pore (mPT) is a nonselective, high-conductance channel comprised of ANT, the voltage-dependent anion channel and cyclophilin-D. The mPTP is situated in the inner membrane, and an elevation in mitochondrial inner membrane permeability to ions and solutes with molecular weights up to approximately 1500 Da results in matrix swelling. Alterations in the mPTP are crucial for mitochondrial maturation, as demonstrated in differentiating cardiomyocytes. The mitochondrial membrane potential (Ψm) is a consequence of mitochondrial activity and is crucial for oocyte quality; conversely, poorly polarised mitochondria in oocytes lead to malformed embryos. Various probes exist to measure Ψm, including the mitochondrial membrane potential indicator 5,5',6,6'-tetrachloro-1,1'3,3'-tetraethylbenzamidazol-carboncyanine (JC-1). JC-1 is a cationic dye whose mitochondrial absorption is directly correlated with the degree of Ψm. A higher concentration of JC-1 aggregate correlates with increased mitochondrial uptake, producing a red fluorescent emission signal, whereas green fluorescence signifies the presence of JC-1 as a monomer. An elevated red: green ratio signifies a more polarised, or more negative and hyperpolarised mitochondrial inner membrane. JC-1 has been utilised to assess mitochondrial maturation and differentiate between immature and mature mitochondria. In somatic cells, mitochondria establish a very intricate dynamic network and undergo fusion and fission to fulfil the functional requirements of particular cell types. Mitochondria are transported along microtubules to their functional positions in most cell types, utilising the motor proteins dynein and kinesin for efficient intracellular movement. The locations of mitochondria may vary according to their distinct functions during the cell cycle. In certain instances, activity is necessitated in specific localised regions within a cell. Mitochondria localise to the nuclear periphery before the disintegration of the nuclear envelope. The localisation of mitochondria signifies favourable developmental potential, whereas their dispersion throughout the cytoplasm at the anticipated moment of nuclear envelope disintegration suggests reduced developmental potential in cells or embryos. As stated in the introduction, early embryonic cells, such as pluripotent stem cells, possess mitochondria that exhibit a rounded morphology and a limited number of cristae (Fig. 9.1), commonly described as functionally immature mitochondria, which are indicative of undifferentiated cells. Although our understanding of specific mitochondrial functions in embryonic cells remains incomplete, it is established that mitochondria play a crucial role in oocyte maturation, fertilisation and preimplantation development. They may also provide ATP for energy-intensive processes, including nuclear envelope breakdown and the assembly and disassembly of microtubules for meiotic and mitotic spindle formation.

Oocyte maturation: Mitochondria are essential for oocyte maturation and serve as accurate measures of oocyte quality attained throughout this process. Mammalian eggs are halted at the GV stage within the ovaries of the neonate, corresponding to the diakinesis phase of prophase I. The stalled oocytes persist at the GV stage until puberty, when follicle-stimulating hormone prompts the maturation of tiny antral follicles into the preovulatory stage. Oocyte maturation in the ovary progresses with the resumption of meiosis from prophase I (germinal vesicle stage; GV) and the ejection of the first polar body (PBI), subsequently leading to the initiation of meiosis II (MII) and the arrest of the oocyte at the metaphase II stage of the second meiosis. The quantity of mitochondria differs among species; nonetheless, in the majority of mammalian species, mitochondria are situated in proximity to the MII spindle. This localisation indicates their significance in spindle development and the preservation of spindle integrity. Meiotic spindle development initiates in the majority of mammals at the oocyte's centre following the rupture of the nuclear envelope of the GV. The germinal vesicle nuclei undergo breakdown (GVBD) process is initiated by the activation of cumulus cells via luteinizing hormone. The perinuclear increase of mitochondria indicates oocyte quality and facilitates the precise timing of GVBD. The assembled spindle relocates to the oocyte cortex and is secured to the cortex by an actin filament cap. Oocyte maturation is crucial for the development of oocyte characteristics that facilitate good fertilisation and embryo development. During oocyte maturation, the oocyte enlarges and experiences remodelling at both cellular and molecular levels. This remodelling necessitates ATP, presumably provided by mitochondria, facilitating prompt and precise cytoplasmic and nuclear maturation. The ATP provided by mitochondria is crucial for protein phosphorylation and dephosphorylation, which are regulatory processes essential for oocyte maturation. These processes encompass centrosome and microtubule dynamics necessary for the formation of meiotic spindles during meiosis I (MI) and MII. The formation of the MI and MII spindles is essential for precise chromosomal segregation and the two consecutive asymmetric cell divisions that produce small polar bodies and a big polarised oocyte. The cytoskeleton is crucial for cellular and molecular remodelling, and it is vital for the precise redistribution of mitochondria during the maturation process. The oocyte halted at the MII stage represents the ultimate outcome of oocyte maturation. The mature mammalian MII egg typically contains around 100,000 to 200,000 mtDNA copies, with pig oocytes reaching almost 300,000 copies. The quantity of mtDNA in oocytes is positively linked with fertilisation and developmental ability. Additional markers of favourable developmental potential include the arrangement of mitochondria around the MII spindle in most animals and positive staining with brilliant cresyl blue (BCB). The BCB dye is metabolised by glucose-6-phosphate dehydrogenase (G6PD), serving as an indication of G6PD activity. BCB has been consistently utilised to assess oocyte quality. G6PD is produced and stored throughout the oocyte development phase; it is a rate-limiting enzyme essential for NADPH synthesis via the pentose phosphate pathway. The various levels of BCB reflect differences in oocyte quality as assessed by G6PD. Oocytes without adequate G6PD to reduce the dye stain blue and exhibit diminished developmental competence, whereas oocytes that do not stain blue possess G6PD and convert the dye to a colourless solution, suggesting these oocytes may exhibit greater developmental competence. Additional indices of oocyte quality have been employed and examined in references. Functional motor proteins and protein kinases essential for meiotic spindle assembly and chromosomal alignment necessitate ATP, and inadequate ATP synthesis correlates with spindle defects. Both meiotic divisions in humans are susceptible to errors, and chromosomal aneuploidies have been documented in both the first and second meiosis. Malfunctioning motor proteins and kinetochore-associated kinases may lead to chromosomal aneuploidies in females with impaired oocyte quality. Irregularities in polo-like kinase functions have also been associated with aneuploidy. Given that ATP is essential for the operation of protein kinases, diminished kinase activity may lead to mitochondrial malfunction.

Fertilisation: Upon ovulation, the MII oocyte enters the oviduct, where fertilisation transpires. Fertilisation triggers the conclusion of MII, characterised by the expulsion of the PBII, resulting in the haploid maternal contribution to the oocyte. Mitochondria are inherited maternally in mammalian embryos, with the MII oocyte containing all mitochondria necessary for development to the blastocyst stage. The sperm's mitochondria are obliterated during fertilisation, but the oocyte's immature mitochondria are evenly allocated to the dividing daughter cells during the initial and subsequent cell divisions. The unequal distribution results in disproportionate patterns of mitochondrial inheritance in two- to four-cell stage human embryos, causing cell lysis in the blastomere that lacks mitochondria. The loss of cells from the developing embryo may provide detrimental effects on embryonic implantation.

Development of preimplantation embryos: The remodelling of mitochondrial characteristics is crucial for differentiating when heightened energy is required for embryonic development. During preimplantation embryo development, mitochondrial architecture undergoes many transformations, evolving from a simple spherical structure to exhibiting more complicated morphologies, including well-developed forms of cristae, a more compact matrix and an extended or branching morphology. This intricate alteration in shape is associated with a bioenergetic shift from predominantly glycolytic to aerobic oxidative phosphorylation metabolism. This transition entails an augmentation in the density of respiratory chain complexes and ATP synthesis. Numerous proteins facilitate the process of mitochondrial differentiation, some of which have been identified, while many others remain unidentified. Mitochondrial differences become particularly evident during blastocoel formation as cellular differentiation leads to the development of trophectoderm (TE) and inner cell mass (ICM) cells. Mitochondrial elongation and enhanced cristae development are initially detected in TE cells as they differentiate into an epithelial layer, whereas ICM cells maintain spherical mitochondria with few or no cristae. The spherical morphology in ICM cells is also indicative of pluripotent stem cells and many cancer cell types. A significant number of studies on maturation, fertilisation, and embryonic development have been conducted using animal models, including mice, rabbits, bovines and pigs, with fewer investigations focused on humans; however, this research domain is advancing more swiftly due to the demand for improved success rates in IVF clinics.

Mitochondrial dysfunction in quality-compromised oocytes:

As stated earlier, oocyte quality is crucial for fertilisation and the development of viable progeny. Compromised egg quality is associated with infertility, developmental abnormalities, diminished blastocyst cell count and embryo loss; nevertheless, the processes responsible for these effects remain poorly understood. The quality of oocytes is attained during the maturation process, as discussed earlier. Maturation abnormalities may arise from various causes, many of which are linked to mitochondrial dysfunction resulting from inadequate ATP, calcium homoeostasis, hormonal influences, among others. Various metabolic alterations or illnesses, including obesity, diabetes and ageing, contribute to diminished oocyte quality. Metabolic disorders impact not only the oocytes of the adult female but also all oocytes generated in the developing foetus during maternal gestation, as the detrimental intrauterine environment directly influences oogenesis in the foetus and the primordial germ cells of the embryo from which the subsequent generation arises. As stated in Section 1, adequate maturation of oocytes, encompassing both nuclear and cytoplasmic aspects, is crucial for achieving fertilisation competency. Mitochondria play a vital role in cytoplasmic maturation by supplying ATP, which is necessary for essential cytoplasmic and cellular functions. Effective cytoplasmic maturation necessitates a minimum mtDNA copy number, as evidenced in mouse oocytes, bovine oocytes, porcine oocytes and human oocytes. In contrast to the consistent copy numbers observed in mouse, bovine and porcine oocytes, human oocytes exhibit considerable variability, ranging from 20,000 to 598,000, with a mean of 193,000 and additional mean values of 256,000 [85], 314,000 and 795,000 copies. The considerable variation in human oocytes may represent the diversity of IVF patients examined and could signify the broad spectrum of oocyte quality in humans. associated with fertility and reduced fertility/infertility issues. A requisite threshold level of mtDNA is essential for facilitating fertilisation and embryonic development. The threshold quantity of mitochondria in the MII oocyte is crucial, as all mitochondria are inherited maternally. Mao et al. have conducted an in-depth study on the regulation of oocyte mtDNA copy number, demonstrating that follicular fluid, epidermal growth factor and neuroregulin 1 significantly influence in vitro maturation and subsequent embryonic development in pigs. Oocyte ageing is closely linked to mitochondrial dysfunction, with several underlying causes identified, including: an increase in mtDNA damage accompanied by alterations in copy number and mutational load; modifications in mitochondrial gene expression; a reduction in mitochondrial membrane potential; variations in mitochondrial dynamics; heightened density of the mitochondrial matrix; an increased incidence of ruptured mitochondrial membranes; a decline in $\Delta\varphi m$; diminished ATP synthesis and metabolic reactions within the electron transport chain; and elevated production of ROS. The overproduction of ROS is directly linked to oxidative energy metabolism or calcium overload, which may induce the opening of the transition pore and consequent death. ROS are produced during ATP synthesis and can induce oxidative damage to mtDNA if not neutralised, leading to mutations and deletions.

Mitochondrial DNA

Mitochondria are particularly susceptible to oxidative stress-induced damage due to the limited presence of repair enzymes for mtDNA. The mutation rate in mtDNA is 10 to 20 times greater than that of nuclear DNA, perhaps due to its restricted DNA repair capability. Various mitochondrial deletions and mutations have been documented, with the most prevalent deletion detected in the human mitochondrial genome as a 4977 bp deletion occurring inside two 13 bp repeats, commencing at locations 8470 and concluding at 13,459. The presence of the 4977 bp deletion in mtDNA is regarded as an indicator of ageing and is associated with diminished energy output. Accumulations of mutational burdens in mtDNA adversely impair mitochondrial functioning and significantly affect the development of the preimplantation embryo. Furthermore, oocyte mitochondria harbouring abnormalities may transmit these alterations to the progeny, who may acquire a predisposition to metabolic disorders. Chromosomal aneuploidies are recognised consequences of oocyte ageing; nonetheless, the temporal cascades and specific processes resulting in spindle integrity loss and aneuploidy remain under investigation. The specific cause-and-effect mechanism remains unidentified. Nonetheless, various characteristics are distinctly noted during the ageing process, including reductions in motor protein functions as discussed. Diabetes and obesity are primary contributors to mitochondrial dysfunction, which significantly affects oocyte quality in women with these metabolic illnesses. Additional dysfunctions associated with metabolic diseases are likely, numerous dysfunctions remain unaccounted for. Diabetes and obesity have emerged as significant global issues, surpassing the population of the undernourished, with estimated figures of 1.5 billion and 1 billion, respectively. Diabetes and obesity are both strongly linked to infertility and have led to a rise in people pursuing ART procedures in IVF clinics. Mitochondrial dysfunctions associated with ageing oocytes are also observed in oocytes from diabetic and obese females, including diminished membrane potential, decreased blastocyst development and imbalances in trophoblast and ICM populations, leading to embryo loss.

Strategies to mitigate mitochondrial dysfunction in oocytes to enhance IVF success rates:

A significant challenge in attaining effective IVF and embryo development in IVF clinics is the subpar quality of mitochondria in oocytes from obese and diabetic women, women of advanced reproductive age and women with severe metabolic illnesses. This limitation has prompted many strategies to address the deficiencies, including cytoplasmic transfer, which involves augmenting an oocyte with donor cytoplasm that contains healthy mitochondria (mitochondrial supplementation). Although this method had previously elicited concerns, cytoplasmic transfer is now seen more positively as a viable therapy to eradicate mitochondrial abnormalities, facilitating the birth of healthy infants devoid of such mutations. Nonetheless, there are concerns to consider. In ooplasmic transfer (OT), out of 13 pregnancies, two foetuses exhibited a karyotype of 45, XO (Turner syndrome). One of the foetuses was expelled spontaneously, while the other pregnancy was deliberately ended. This information is documented in a Food and Drug Administration (FDA) report (http://www.fda.gov/OHRMS/DOCKETS/ac/02/briefing/3855B1_01.doc). Although various regulations are implemented in different nations and research lacks backing from federal funding organisations, therapeutic applications remain feasible. The option is contingent upon parental consent for the surgery. These new considerations enable novel methods for cytoplasmic transfer into quality-compromised oocytes to address mitochondrial dysfunctions or deficits. Such procedures are particularly significant given the increasing demand for IVF. The frequency of procedures has risen markedly in recent years. The rise in demand is partially attributable to the global escalation of obesity and diabetes, as well as the tendency of delaying childbirth, which results in older oocytes exhibiting mitochondrial deficiencies. Novel therapeutic alternatives for quality-compromised oocytes are required to address inadequate mitochondrial functioning. Research on mitochondria has advanced significantly in recent years due to the association of numerous developmental disorders and diseases with dysfunctional mitochondria, including neurological diseases, diabetes, cancer, immune system disorders, ageing and other conditions that impact a substantial portion of the global population. Given the versatility of mitochondria and their experimental manipulability, it is feasible to devise novel therapeutic techniques aimed at restoring mitochondrial functioning once the specific deficiencies are identified. In oocytes with degraded quality and defective mitochondria, the transfer of mitochondria-rich cytoplasm from donated fresh oocytes has rectified the mitochondrial defects. The advantages and complexities have been examined, revealing that the origin of the donor cytoplasm significantly influences outcomes. The strong functional dependency of mitochondria on the nuclear genome necessitates that the donor cell type be closely linked to the receiving cell to ensure effective coordination of mtDNA and nuclear DNA. Ovarian or oocyte-differentiated cells exhibit the highest success rate in transferring donor cytoplasm with healthy mitochondria into compromised-quality oocytes. Recent studies have focused on cells of oogonial origin in the quest for oogonial stem cells or stem-like cells, amidst ongoing disputes in the area. This research has enhanced the characterisation of oogonial-derived cells. High-quality germline mitochondria are ideally sourced from oocyte precursor cells and should be utilised to mitigate any impairments in oxidative phosphorylation that may arise from mitochondrial heteroplasmy (the presence of multiple mtDNA genotypes). Heteroplasmy may pose challenges and induce adverse effects; nonetheless, somatic cell nuclear transfer (SCNT) tests have demonstrated that heteroplasmy is acceptable to a certain degree. No deleterious consequences are reported in women with 3% mtDNA heteroplasmy. Oocyte progenitor cells have been documented in both mice and humans. These cells may function as superior mitochondrial donors, having differentiated into the oocyte lineage without prolonged residence in a postmitotic state, thereby meeting the ideal conditions for supplying high-quality germ line autologous homoplasmic mtDNA devoid of mutations and deletions. Given that various mitochondrial mutations, deletions and nucleotide variations are present in quality-compromised oocytes and blastocysts from older women and those with metabolic disorders like obesity or diabetes, novel treatments aimed at enhancing the healthy mitochondrial population through cytoplasmic transfer from uncompromised cells could benefit mothers facing fertility issues and their offspring, who might otherwise inherit deficient and suboptimal mitochondrial function. In several instances, a minimal quantity of mitochondrial transfer is adequate to restore normal mitochondrial function and avert the transmission of substandard mitochondria to subsequent generations. Alternative sources for mitochondrial transfer may encompass stem cell cytoplasm, as stem cell mitochondria exhibit numerous characteristics associated with embryonic immature mitochondria, including rounded morphology, reliance on glycolysis for energy production and adaptability to differentiation under varying environmental conditions. In addition to cytoplasmic transfer, dietary factors, including CoQ10 supplementation, have been evaluated for their potential to enhance mitochondrial function in oocytes of reduced quality. CoQ10 facilitates electron transport within the mitochondrial respiratory chain, rendering it an essential enzyme for energy production. Alternative therapeutics are under consideration, and many methodologies have been suggested to formulate methods for mitochondrial pharmacology aimed at identifying druggable mitochondrial targets. Distinct from methods aimed at augmenting mitochondria via cytoplasmic transfer are the strategies presently under consideration to eradicate inherited mtDNA mutations. To prevent the transmission of mtDNA mutations to progeny, two oocytes from distinct females are required, and many methodologies are available. In one method, either the GV or the MII spindle and corresponding chromosomes are extracted from the patient's unfertilised oocyte with faulty mitochondria and transplanted into an enucleated donor oocyte containing healthy mitochondria. The donor egg predominantly comprises the donor's mtDNA, while the nuclear DNA belongs to the patient, although a portion of the patient's mtDNA is also transferred with the nucleus. The reconstructed egg can now be fertilised with the patient's partner via the intracytoplasmic sperm injection (ICSI) method, therefore carrying the future parents' nuclear DNA while retaining the donor's mitochondria. The bisecting embryo, possessing normal mitochondria along with maternal and paternal genomes, may then be implanted into the patient's uterus. This method has been described as 'three-parent in vitro fertilisation'; yet, the nuclear genetic material is solely provided by two parents. It is mostly employed to avert the transmission of hereditary mitochondrial illness. This methodology has been thoroughly examined over the past year and has received approval from the FDA [106]. It was previously assessed in 2012 by the UK Human Fertilisation and Embryology Authority (HFEA). This approach is detailed in multiple recent publications. In an alternative method, the patient's oocyte, which harbours faulty mitochondria, is fertilised with the partner's sperm and permitted to develop to the zygote stage, retaining the abnormal mitochondria. The patient's pronuclei (PN) are subsequently extracted and inserted into an enucleated donor oocyte with viable mitochondria. The rebuilt zygote is subsequently prepared for transfer into the patient's uterus. A contemporary method involves enucleating the recipient oocyte by excising the MII spindle; a donor oocyte is fertilised, and the PBII is extracted and subsequently transferred to the recipient oocyte. This process results in the recipient oocyte containing the nuclear DNA of the donor oocyte while retaining minimal mitochondrial content, as the PB comprises a limited quantity of mitochondria and primarily consists of a single set of chromosomes, which is naturally extruded to restore the haploid state of the oocyte, as elaborated in section 1. The PBII possesses an equivalent quantity of nuclear DNA as the oocyte. The recipient oocyte, possessing healthy mitochondria, now incorporates the nuclear DNA of the donor and can subsequently be utilised for in vitro fertilisation according to standard IVF protocols. Modifications of these existing processes can prevent the transmission of mutant mtDNA to offspring, enabling the birth of infants free from mitochondrial disorders inherited from affected parents. Alternative therapy strategies are being actively explored to enhance mitochondrial dysfunction, including those utilising gene-editing techniques. For instance, targeted eradication of defective mitochondrial genomes in patient-derived cells using mitochondrially-targeted transcription activator-like effector nucleases (mitoTALENs) has been documented, wherein mitoTALEN expression resulted in sustained reductions of deletion or point mutations in mtDNA within these cells. Over the past decade, significant progress has been made in prospective therapy for quality-compromised oocytes, notably benefiting women with metabolic illnesses like obesity, diabetes, or metabolic alterations related to ageing. Dietary elements like CoQ10 and the transfer of mitochondria from cells with intact mitochondrial function have created new opportunities for therapeutic advancements that may enhance the success rates of oocytes in women with diminished oocyte quality. Optimal conditions for such transfers necessitate that mitochondrial supplementation be derived from the patient's own cells to ensure effective communication between mtDNA and nuclear DNA, thereby preserving homoplasmy in offspring; utilising cells of ovarian or oocyte origin; and employing mitochondria from high-quality cells to mitigate the risk of transmitting deletions or mutations to the offspring. Mitochondria from stem cells may serve as an ideal source for enhancing the quality of impaired oocytes.

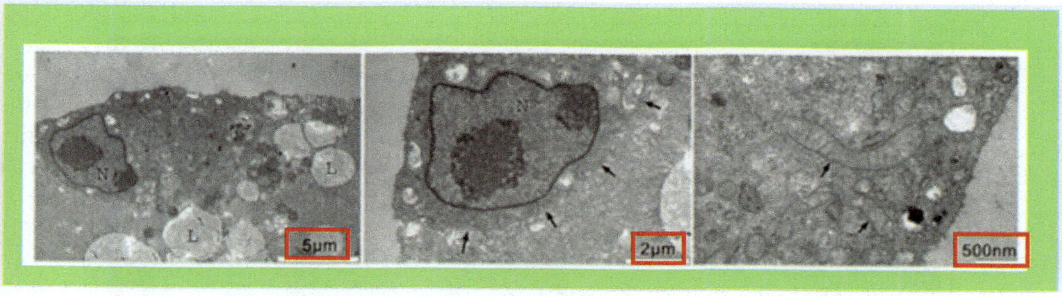

FIGURE 9.3 Transmission electron microscopy of mitochondria following somatic cell nuclear transfer (SCNT). Transmission electron microscopy of a reconstructed somatic nuclear transfer oocyte at 30 minutes postfusion. The donor cell remains distinctly isolated from the enucleated oocyte (shown by arrows) at this stage of SCNT. These cells possess somatic cell mitochondria characterised by an elongated morphology and cristae, which coexist alongside spherical oocyte mitochondria at least until the blastocyst stage, illustrating heteroplasmy in the developing preimplantation embryo.

Electron microscopic view of mitochondria after somatic cell nuclear transfer

It becomes quite important to visualize the mitochondria following somatic cell nuclear transfer. Transmission electron microscopy is an important means to visualize as shown below (Fig. 9.3).

Fluorescent microscopic view of mitochondria after somatic cell nuclear transfer

See Fig. 9.4.

Mitochondrial supplementation

Innovative reproductive technology to avert mtDNA disorders: current standard methods for reducing mtDNA disease transfer include prenatal diagnosis and preimplantation genetic diagnosis (PGD) (Fig. 9.5). Prenatal diagnosis is a method that collects cells at various stages of early gestation and can mitigate the risk of serious mtDNA disorders. The possibility of termination when the mutation burden is elevated. The primary challenge of prenatal diagnosis is the variability of mtDNA mutation burden across different tissues and its dependence on the stage of foetal development at the time of sampling. The complexities of mtDNA inheritance render it exceedingly challenging to establish a secure threshold within which mtDNA disease is improbable to occur in the future offspring. PGD is an IVF-derived technology, established over 20 years ago, that has lately been utilised to mitigate or eradicate the danger of mtDNA illness. This entails genetic testing of one or more cells extracted from an early embryo, facilitating the selection of embryos with little or no mutations for uterine transfer. For reliability, it is essential that the mtDNA mutation load

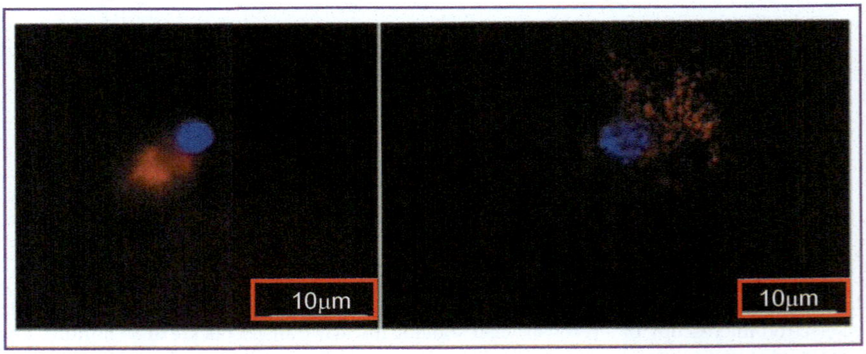

FIGURE 9.4 Fluorescence microscopy of mitochondria following somatic cell nuclear transfer. Fluorescence microscopy of donor cells preloaded with CMXRos, 30 minutes postfusion with enucleated oocytes. Donor-derived mitochondria are situated near the donor nucleus and disseminate into the cytoplasm in developing embryos exhibiting heteroplasmy, as observed up to the blastocyst stage.

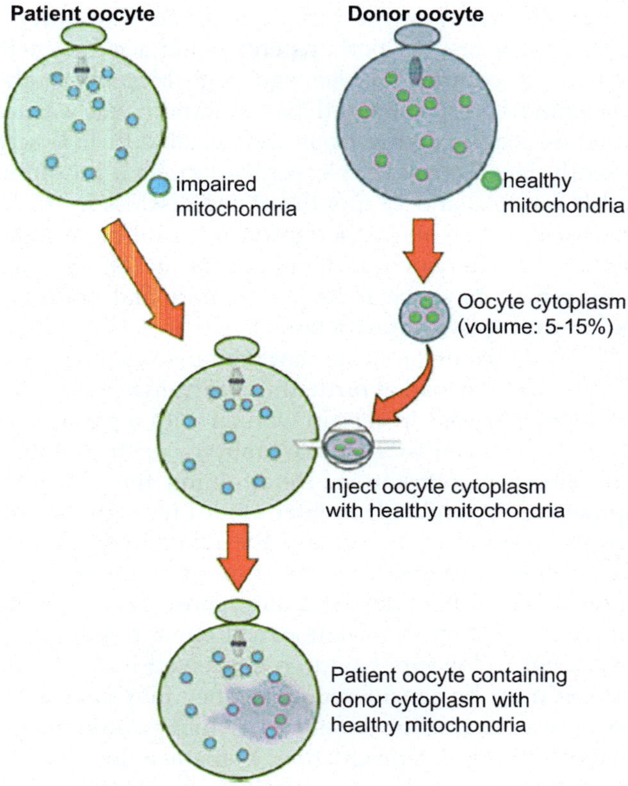

FIGURE 9.5 Schematic illustration of mitochondrial supplementation. Schematic depiction of ooplasmic mitochondrial transport. An oocyte from a patient with deficient mitochondria is injected with donor mitochondria to enhance the mitochondrial population necessary for proper development.

identified in the biopsy accurately reflects the entire embryo, necessitating consistent segregation of the mtDNA mutation throughout preimplantation embryo development. Research utilising heteroplasmic mouse models has demonstrated modest variance among blastomeres, averaging 2%, with sporadic outliers noted. Conversely, research on non-human monkey embryos has demonstrated significant variability in heteroplasmic mtDNA levels among blastomeres, potentially attributable to the experimental methods employed in the generation of heteroplasmic embryos. Furthermore, restricted research on human embryos has demonstrated minimal heteroplasmic variation across the blastomeres of cleavage-stage embryos, with certain outliers. A significant association in heteroplasmy levels has been documented between TE cell biopsies and the residual blastocyst in both murine models and human embryos harbouring mtDNA abnormalities. The inaugural successful instance of PGD to mitigate the risk of mtDNA disease was documented, who conducted a blastomere biopsy on a patient with NARP (m.8993 T>G), ultimately resulting in the birth of a healthy infant. Subsequently, some infants have been born utilising PGD to mitigate the transmission of particular mtDNA mutations. These primarily pertain to blastomere biopsy and TE biopsy at the blastocyst stage. The latter offers the benefit of supplying additional cells for testing, perhaps enhancing accuracy without compromising embryonic development in contrast to cleavage-stage biopsie. In the initial instance of blastocyst biopsy, an embryo with a 12% mtDNA mutation load was transplanted, yielding a newborn with mutation loads varying from undetectable to 15% across different tissues. The results subsequently faced considerable contention, as a reexamination in a different laboratory of the kid at 6 weeks and 18 months of age disclosed mtDNA mutation loads of 42% and 52%, respectively. The child experienced intricate neurological and multisystem issues that were neither characteristic nor definitive for m.3243 A>G-related mitochondrial disease; however, they may have been linked to perinatal complications such as pre-eclampsia and intrauterine growth restriction, both of which are associated with this mtDNA mutation. The PGD results indicate that the mtDNA mutation load assessed in the embryo aligns with the mutation load at birth, thus affirming that PGD is a viable method to mitigate the risk of mtDNA disease in certain women with a pathogenic mtDNA mutation. To achieve success, the woman must generate embryos with mutation levels beneath the crucial threshold for illness manifestation. Determining the percentage threshold below which clinical symptoms are improbable in later adult life might be challenging. The safety level corresponds to the type of mutation and the related clinical characteristics. Certain mtDNA mutations exhibit a comprehensive spectrum of mutation loads, exemplified by the m.3243 A>G mutation, whilst others demonstrate a more biased distribution, as seen with the m.8993 T>G mutation. Further information will emerge with an increased number of PGD cases and subsequent follow-up. Children conceived by PGD for mtDNA disorders. The couple's preferences must be considered, since they may choose to transfer an embryo with a mutation load over a specific threshold, which would nevertheless represent a reduced risk compared to natural reproduction. A metaanalysis of existing pedigree data indicates that a mutation load of 18% correlates with a 95% or greater probability of remaining unaffected, regardless of the type of point mutation. This percentage criterion is frequently utilised in documented PGD investigations, and in the majority of patients, embryos are discovered with a mutation load below 18% after several ICSI cycles. The precise threshold remains arbitrary, and it is hypothesised that additional, yet-to-be-identified elements (including nuclear factors) will be crucial. Furthermore, not all embryos with a mutation burden beneath the threshold will result in a successful pregnancy. Consequently, PGD will not be an appropriate reproductive choice for all women possessing a harmful mtDNA mutation, including those with homoplasmic mtDNA mutations or those who generate only oocytes with elevated mutation loading. This underscores the necessity for other methods to mitigate the danger of mtDNA disease transmission, such as NT.

The application of NT technologies in both animal and human models: The advancement of germline NT techniques, referred to as mitochondrial donation under UK law, offers an alternative reproductive solution enabling women with pathogenic mtDNA mutations to conceive a genetically related child while mitigating the risk of mtDNA disease. This innovative IVF procedure entails the extraction of the nuclear genome from an oocyte or zygote harbouring an mtDNA mutation, thereafter transferring it to an enucleated donor oocyte or zygote possessing wild-type mtDNA. The rebuilt egg or zygote predominantly contains mtDNA from the recipient donor, hence diminishing the likelihood of serious mtDNA disorders in the progeny. The nuclear genome can be transmitted from unfertilised oocytes through methods such as maternal spindle transfer (MST), polar body transfer (PBT) and germinal vesicle transfer (GVT), or from fertilised oocytes (zygotes) using a procedure called pronuclear transfer (PNT). PNT and MST are the most extensively researched approaches to date, with substantial scientific data supporting their capacity to prevent the spread of mtDNA diseases. The data primarily encompass the developmental capacity of rebuilt embryos, which must be evaluated alongside embryo quality and the extent of karyoplast-derived mtDNA (mtDNA carryover) transferred with the nuclear genome. It is essential that this level remains modest to mitigate the danger of mtDNA illness in progeny. The technology for NT between oocytes or zygotes was originally established utilising animal models. Several initial experiments were published before to the detection of dangerous mtDNA mutations, therefore the potential of utilising NT to avert the spread of mtDNA disease had not been contemplated. These studies were significant since they experimentally validated the viability of various transfer procedures, indicating the possibility for further advancement in the development of reconstituted oocytes/zygotes and the birth of healthy progeny. Subsequently, NT experiments have been disseminated across other animal models, corroborating the efficacy of these approaches in preventing mtDNA illness. Recent research has conducted NT using human oocytes and, in certain cases, determined that the techniques employed for animal oocytes were not immediately applicable to human oocytes. This underscores the necessity of conducting such research in humans, since the results are essential for assessing both the safety and efficacy of NT prior to its introduction as a clinical treatment. The most comprehensive preclinical investigation of human oocytes to date concentrated on PNT to mitigate the risk of mtDNA disease transmission, while analogous evidence exists, albeit to a lesser degree, for MST and PBT. Currently, there is no scientific evidence supporting the preference of PNT over MST, or vice versa, for mitigating the risk of mitochondrial illness. Subsequently, the advancement of both PNT and MST in animal and human research will be examined comprehensively, followed by a succinct overview of PBT and GVT.

PNT: Following standard fertilisation, two PNs with distinct membranes become apparent and travel to the middle of the zygote, each harbouring a haploid set of chromosomes. The PNT approach entails excising them and transferring them to a recipient zygote from which its own PN has been eliminated (Fig. 9.3A). A proficient approach for executing this transfer across mouse zygotes was established, as seen in Fig. 9.3, which depicts NT methodologies and the drift/reversion of heteroplasmic mtDNA. The methodologies of PNT, MST, PB1T/PB2T and GVT involve the isolation of the karyoplast from the patient's oocyte or zygote, which is subsequently transferred into an enucleated recipient oocyte or zygote. (B) In the NT procedure, perinuclear mitochondria from the nucleus donor may inadvertently be transferred into the recipient cytoplast, potentially resulting in a low level of karyoplast-derived mtDNA heteroplasmy in the reconstructed embryos, as established 30 years ago by McGrath and Solter, which has since underpinned all contemporary NT techniques. The method entailed the extraction of polar nuclei from a donor zygote utilising cytoskeletal inhibitors (cytochalasin B and colcemid or nocodazole), facilitating their removal within a membrane-bound karyoplast containing minimal cytoplasm, thereby enhancing zygote survival postprocedure. The karyoplast was subsequently transferred to an enucleated zygote using a cell-fusing agent called haemagglutinating virus of Japan envelope. Following culture, 96% of reconstituted zygotes progressed to the morula or blastocyst stage, yielding live progeny upon transfer to pseudopregnant females (live birth rate: 16%). Employing a comparable methodology, Meirelles and Smith conducted PNT between zygotes from two mouse strains with distinct mtDNA genotypes, facilitating the examination of mtDNA segregation throughout preimplantation development. The analysis of the reconstituted zygotes indicated that the average concentration of karyoplast-derived mtDNA was 19% (range: 16%–23%), whereas the concentration in the tissues of the resulting progeny ranged from 0% to 69%. It is essential to acknowledge that the primary objective of the work was to generate heteroplasmic mice, and hence, efforts were not directed towards reducing karyoplast-derived mtDNA. The reported mtDNA carryover levels are significantly elevated compared to a recent study that found an average mtDNA carryover of 0.29% after PNT in mice (Table 9.1). A pertinent study by Sato et al. detailed the application of PNT in a mouse model of mitochondrial illness (mito-mouse) characterised by a significant mtDNA deletion (ΔmtDNA) that caused respiratory deficiencies in the mice. The results validated that PNT can diminish the transfer of mutant mtDNA and indicated that the resulting children may have a longer lifespan compared to those from unmanipulated control zygotes, thus reinforcing the application of PNT for the prevention of mtDNA disease. The inaugural study to do PNT with human zygotes was a proof-of-concept investigation reported by Craven et al. This investigation involved the exchange of PNs of two improperly fertilised human zygotes (1PN or 3PN). This effectively diminished mtDNA carryover levels to an average of less than 2%, with most reconstituted zygotes exhibiting undetectable amounts of karyoplast-derived mtDNA. The incidence of blastocyst formation following PNT was diminished relative to unmanipulated controls (8.3% vs 16.6%, respectively) (Table 9.1), which may be attributed to the lack of a paternal/maternal genome or an erroneous number of centrosomes in the reconstituted zygotes, as abnormally fertilised zygotes were utilised. Furthermore, as improperly fertilised zygotes often experience arrest during preimplantation embryonic development, it is improbable that

TABLE 9.1 Incidences or types of reproductive abnormalities affected by mitochondrial gene.

S. no	Reproductive process
1.	Sperm motility and abnormality
2.	Asthenozoospermia and oligozoospermia
3.	Low level of mtDNA in semen
4.	Defective sperm motility
5.	Ovulation, zygote activation, fertilisation
6.	Oocyte maturation, embryo development
7.	Mitochondrial copy number in oocytes, ovarian ageing
8.	Litter size, percentage of abortion, age at first maturation, age at parturition, conception rate associated with the cytochrome B gene of mitochondria in female sheep and pig as model
9.	Libido, reaction time and seminal characteristics, such as semen concentration, semen volume, semen motility, sperm abnormality, live and dead count, acrosome integrity test and its association with the cytochrome B gene in male sheep
10.	Whole mitochondrial genome sequencing and its association with female reproductive traits of sheep

the blastocyst rate would match that of normal zygotes, even in the absence of intervention. Hyslop et al. just published the inaugural preclinical investigation on PNT utilising normally fertilised (2PN) human zygotes. The study demonstrated that the PNT procedure, originally designed for improperly fertilised human zygotes, was poorly tolerated in normal zygotes, necessitating refinement to enhance PNT efficacy. The principal alteration pertained to the timing of PNT, with PN transfer occurring shortly after the completion of MII (~8 hours post-ICSI, early PNT) instead of prior to the first mitosis (~16 hours post-ICSI, late PNT), resulting in enhanced survival rates of the resultant zygotes (92% for early PNT versus 59% for late PNT) (Table 9.1). This resulted in a markedly enhanced blastocyst rate of 62% in the autologous PNT group and facilitated the production of high-quality blastocysts. The mtDNA carryover was less than 2% in most PNT blastocysts, whereas the examination of aneuploidy and global gene expression showed no significant differences between early PNT blastocysts and unmanipulated controls. Hyslop et al. noted that in one of five human embryonic stem cell (hESC) lines obtained from early PNT blastocysts, the concentration of karyoplast-derived mtDNA gradually escalated during culture, despite an initially low level at the blastocyst stage (Table 9.1) (Fig. 9.3B). This prompted the writers to ascertain that PNT possesses the capacity to significantly mitigate the risk of mtDNA disorders but it may not ensure complete protection and should thus be evaluated alongside prenatal screening.

MST: The maternal spindle, or metaphase II (MII) spindle–chromosome complex, is established in oocytes during the second meiotic division. The MST technique entails excising the spindle and transferring it to an enucleated MII oocyte from which the spindle has been removed, subsequently followed by the fertilisation of the rebuilt oocyte (Fig. 9.3A). The MII spindle is a microtubular structure exhibiting birefringent features, enabling visualisation by polarised microscopy instead of intrusive fluorescent labelling. The initial effective technique for MST was documented in nonhuman primates (*Macaca mulatta*) by Tachibana et al. The spindle–chromosome complex of donor MII oocytes was extracted as a karyoplast and subsequently transferred to an enucleated MII recipient oocyte using HVJ-E (Fig. 9.3A). The resultant oocytes exhibited similar fertilisation rates (by ICSI) and blastocyst formation rates (95% and 61%, respectively) to those of unmanipulated control oocytes. Functional ESC lines were successfully obtained from MST blastocysts with normal euploid karyotypes and negligible mtDNA contamination (detection threshold of 3%). Furthermore, the implantation of MST embryos into surrogate females led to the delivery of three healthy offspring, with undetectable amounts of mtDNA carryover in all examined tissues (Table 9.2). This procedure has since been applied to mouse oocytes with varying mtDNA genotypes, demonstrating a blastocyst formation rate of about 90% and an average mtDNA carryover of less than 1% in reconstituted ST oocytes (Table 9.2). These animal investigations validate that MST is technically viable, with reconstituted oocytes demonstrating normal fertilisation and development to the blastocyst stage at a comparable rate to unmanipulated controls. Moreover, MST blastocysts can provide healthy progeny with negligible amounts of karyoplast-derived mtDNA, so affirming the efficacy of this approach in averting the transmission of mtDNA disorders. The inaugural study doing MST in human MII oocytes was published by Tachibana et al. (2013), which validated the technique's viability while also noting a considerable percentage of human MST zygotes exhibiting aberrant fertilisation following ICSI. The elevated incidence of abnormal

TABLE 9.2 Summary of the included studies on pronuclear transfer (PNT) in animal and human models.

Species(s)	Origins	Efficiency	mtDNA carryover
Anials	Zygotes from C3H/He and C57BL/6J mice	Survival rates: >90%; blastocysts development = Ctr	ND
	Zygotes from C57BL/6 and NZB/BinJ mice	ND	Reconstructed zygotes: 19% (16%–23%) versus 58% (maternal tissues) versus 0%–69% (resultant progeny)
	Zygotes from NZB/BinJ and C57BL/6 mice	ND	Four-cell embryos: 17.3%; blastocysts: 29.6%
	Zygotes from mito mice (ΔmtDNA, large-scale deletions)	ND	ΔmtDNA in newborn mice: 11% (6%–21%) (20–40 days after birth) versus 66% (51%–73%) (Cte); ΔmtDNA in following 300 days: 23% (5%–44%), 12%↑ (300 days), survived >300 days versus 14%↑ (170 days), died at 218–277 days (Ctr)
	Zygotes from NZB/OlaHsd and B6D2/F1 mice	Two cell embryos: 88%; blastocysts development: 100% ~ Ctr	Reconstructed oocytes/embryos: 0.29% ± 0.75%
Humans	Abnormally fertilised donated zygotes (IPN/3PN)	Eight-cell stage: 22.2%-22.7%; blastocyst formation: 8.3% (PNT) versus 16.6% (Ctr)	Reconstructed embryos: <2% (1.68% ± 1.81%)
	Donated MII oocytes; fertilised by ICSI	Zygote survival: 92% (ePNT) versus 59% (LtPNT); blastocyst formation (ePNT): 62% (Altg-PNT), 40% (Het-PNT) versus 56% (Ctr)	Reconstructed blastocysts: <2% (combination of fresh oocytes and vitrified pronuclei achieved the lowest carryover); five ESC lines derived from ePNT blastocysts; <2% in four lines; mtDNA drift in I line (from 0% to 60%)
	Fertilised oocytes from a patient with two failed IVF cycles	Survival rate: 5 out of 7 reconstructed zygotes; triplet pregnancy after transfer of these five embryos (no live birth)	Samples of foetus: 0%

Ctr, nonmanipulated controls; mtDNA, mitochondrial DNA; ND, not defined; ICSI, intracytoplasmic sperm injection; ePNT, early transfer, 8–10 hours post-ICSI; LtPNT, late transfer; 16–20 hours post-ICSI; Atlg-PNT, autologous PNT, which involved removal and replacement of PN in the same zygote; Het-PNT, heterologous PNT between zygotes from different donors involved reciprocal transfer between zygotes originating from fresh and vitrified oocytes; ESC, embryonic stem cell; IVF, in vitro fertilisation; PN, pronuclear.

fertilisation in human MST zygotes compared to controls (52% vs 13%, respectively) was not evident in nonhuman primate MST zygotes and may result from either premature oocyte activation induced by the procedure or spindle damage during enucleation and transfer. The blastocyst formation rate in MST oocytes exhibiting normal fertilisation was similar to that of controls (62% vs 76%, respectively), whereas the concentration of karyoplast-derived mtDNA in reconstituted MST embryos was minimal (0.5% ± 0.4%) and persisted at a low level during the ongoing culture of hESC lines derived from MST blastocysts (0.6% ± 0.9%). A concurrent study conducted MST in human MII oocytes, subsequently activating reconstituted oocytes parthenogenetically. It revealed that premature activation of oocytes could be prevented through partial depolymerisation of the spindle–chromosome complex via reduced temperature or cryopreservation prior to MST execution. The study also verified minimal mtDNA carryover in rebuilt MST embryos, which diminished to undetectable levels over the prolonged culture of hESC lines produced from MST blastocysts. Recent investigations by Yamada et al. (2016) and Kang et al. similarly reported consistently low or undetectable levels of karyoplast-derived mtDNA in the majority of hESC lines produced from MST blastocysts. Nonetheless, a few hESC lines exhibited an elevation in karyoplast-derived mtDNA throughout cultivation. In certain hESC lines, there was a temporary elevation in mtDNA carryover, subsequently followed by a decrease, whereas in others, a total reversion to the karyoplast-derived mtDNA genotype was observed. While this suggests that even minimal mtDNA carryover after NT could elevate to levels surpassing the threshold for mitochondrial illness, the findings should be approached with care, since mtDNA segregation in hESCs may significantly differ from in vivo conditions. Furthermore, the majority of studies on mtDNA segregation in hESCs have utilised stem cells with heteroplasmic mtDNA that lacks a pathogenic mtDNA mutation, which may segregate in a distinct manner. The research conducted by Kang et al. was the inaugural study to implement mitochondrial replacement therapy utilising human metaphase II (MII) oocytes from women with a pathogenic mtDNA mutation, thereby offering the initial evidence that MST can diminish the transmission of mtDNA mutations to minimal levels (1%) in MST blastocysts. The derivation of hESC lines from a restricted number of MST blastocysts showed variable effects on the quantity of karyoplast-derived mtDNA, despite the presence of the mtDNA mutation; one hESC line exhibited an increase in mtDNA carryover, while two lines maintained levels below 1%. Currently, there is a singular report on the therapeutic use of MST utilising human MII oocytes to avert the transmission of mtDNA diseases, leading to live children. The procedure conducted by Zhang et al. (2017) in Mexico, owing to the absence of approval in the United States, culminated in the inaugural live birth of an ostensibly healthy infant. The case pertained to a 36-year-old lady who suffered four miscarriages and two deceased children, attributed to elevated levels of a maternally transmitted mtDNA mutation (>95%) associated with Leigh syndrome (m.8993 T>G). Subsequent to MST, the mean mtDNA mutation load in the MST blastocyst designated for embryo transfer was 5.73%, whereas the mean in various tissues obtained postnatally was 1.60% ± 0.92% (range: 0%–9.23%). This study indicates that MST can significantly diminish the transfer of a harmful mtDNA mutation and that MST blastocysts can lead to a successful pregnancy and live birth. It is significant that the mtDNA carryover level identified in the MST blastocyst transferred to the uterus exceeded that reported in prior research examining

MST's capacity to avert mtDNA disease transmission. Moreover, despite the low amounts of mtDNA carryover detected in neonatal tissues, it remains uncertain if the mtDNA mutation could accumulate in different tissues or revert to the karyoplast-derived mtDNA genotype throughout time. This underscores the necessity of long-term monitoring of infants born after NT, as the data is crucial for validating the safety of this experimental treatment.

Transfer of PBs: PBs are diminutive entities expelled from oocytes and zygotes during the initial and subsequent meiotic divisions. The PB1 is diploid, but the PB2 is haploid, and both possess identical genetic material to the oocyte or zygote. PB1T entails the extraction of PB1 from an MII oocyte, followed by transfer to an enucleated MII oocyte containing the maternal spindle.

Extraction and fertilisation of the reconstructed oocyte PB2T entails the extraction of PB2 from a fertilised PN stage. Zygote and the transfer to a recipient zygote containing the female pronucleus are eliminated. Both strategies have been executed effectively. Conducted in murine oocytes, illustrating the capability of PBT to generate Embryos that are developmentally competent and produce viable progeny. Recently, PBT has been conducted in mice for investigation purposes. The capability of these approaches to inhibit the transmission of mtDNA illness. The research validated the viability of PBT while evaluating the approach in relation to both MST and PNT regarding maternal mtDNA transmission. This disclosed imperceptible or minimal levels. Less than 2% of mtDNA carryover occurs after PBT, in contrast to elevated levels following MST (approximately 6%) and PNT (approximately 24%). The statistics corroborate that PBT diminishes mtDNA carryover, which is unsurprising considering PBs are diminutive and possess minimal cytoplasmic content, but also indicate that PBT leads to reduced levels of mtDNA produced using karyoplasts in comparison to other NT methodologies. The mtDNA carryover documented for MST and PNT in this investigation, however, exceeds those documented in alternative investigations and is not necessarily representative. Furthermore, subsequent to PB1T, the formation of meiotic spindles was allegedly contingent upon the residual meiotic factors. Activity of recipient enucleated metaphase II oocyte. Given that the centrioles originate from sperm, PB1T in unfertilised MII oocytes with a cytoplasmic deficit of PB1 may avoid multipolar spindles or aneuploidy resulting from aberrant centrioles. For PB2T, a potential issue is whether isolated PB2 comprises sufficient centrioles to facilitate the development of mitotic spindles due to Its restricted cytoplasm (Schatten et al., 2014). To data, limited research has examined this matter. However, in mice, PB2T zygotes possess seem to facilitate standard preimplantation development and subsequent transfer of these PB2T embryos into pseudopregnant mice has produced viable progeny. This signifies that a deficiency of PB2 in the cytoplasm may not hinder development. of rebuilt embryos, particularly in murine models. At present, there exists solely one report of PBT in human MII oocytes, indicating that reconstituted PB1T oocytes demonstrated the ability for normal fertilisation with ICSI, an analogous rate to unaltered controls. PB1T zygotes exhibited the potential for further development; nevertheless, blastocyst formation was restricted. with the majority (~58%) ceasing development before to the blastocyst stage. The efficacy of the production of hESC from PB1T blastocysts was minimal, potentially attributable to substandard blastocyst quality; nonetheless, subsequent examination of these hESCs demonstrated no substantial variation in gene expression when contrasted to the controls.

This prompted the authors to deduce that PB1T may offer an alternative method to enhance the application of NT. To address mitochondrial dysfunction, but preclinical data are presently deficient and the restricted embryonic growth necessitates additional enquiry. No literature reports exist for PB2T at present. In human oocytes, probably owing to the challenge in differentiating male and female PN within human zygotes, which renders the procedure technically demanding.

NT of the GV: The GV is the nucleus of an immature egg that is halted in prophase I of the initial meiosis. The GVT technique entails excising the GV and transferring it to an enucleated oocyte from which the GV has been removed, subsequently followed by in vitro maturation (completion of meiosis) and fertilisation of the reconstituted oocyte (Fig. 9.3A). GVT has been employed in both animal and human research to examine the mechanisms underlying oocyte meiosis and maturation, revealing that reconstituted GVT oocytes can develop to the blastocyst stage and produce viable offspring, at least in murine models. The possibility of GVT substituting mutated mtDNA and averting the propagation of mtDNA illness has also been examined in mice. Neupane et al. (2014) conducted germline mitochondrial transfer using mouse oocytes and determined that mtDNA carryover was undetectable in the resulting embryos. Nonetheless, it was reported that, despite reconstituted GVT oocytes possessing the capability to complete maturation and undergo cleavage postparthenogenetic activation, all GVT zygotes experienced arrest and failed to progress to the blastocyst stage, likely attributable to suboptimal culture media that inadequately supported the invitro maturation of oocytes. Additional research is required to enhance the efficacy of this technology prior to its potential clinical implementation, as in vitro maturation of human oocytes remains less effective than the utilisation of in vivo matured oocytes for IVF treatment.

Clinical application of NT to avert mtDNA disorders: The initial human offspring conceived after mitochondrial replacement therapy (MRT) to successfully evade the transfer of a harmful mtDNA mutation was reported in the mainstream media in 2016 and has since been published. The male newborn had low amounts of karyoplast-derived mtDNA from the mother, indicating that MST is an efficacious method for mitigating the risk of severe mtDNA disorders. The clinic is currently intending to conduct an additional 20 MST pregnancies in the first half of 2017 (https://www.newscientist.com/article/2115731-exclusivemexico-clinic-plans-20-three-parent-babies-in-2017/). In the United Kingdom, although regulations were enacted by the government to allow the clinical application of mitochondrial donation in March 2015 (Craven et al., 2017), the HFEA only granted approval for 'cautious clinical use' of the technique in December 2016. Fertility clinics may now seek a licence to utilise mitochondrial donation for the avoidance of severe mtDNA diseases, with applications evaluated individually on a case-by-case basis. A recent assessment in the United States found that clinical trials using NT should be allowed but with restrictions, including the transfer of exclusively male embryos, a stipulation that was rejected in the United Kingdom. Currently, there is an abundance of scientific data endorsing the use of MST and PNT compared to other NT procedures; nevertheless, no evidence suggests a preference for either MST or PNT. In the absence of additional evidence, potentially derived from long-term follow-up studies of children conceived post-NT, the determination will likely rest on the proficiency of the fertility clinic executing the procedure, as well as the perspectives of the prospective parents. The technical elements of various NT approaches are examined below, emphasising MST and PNT due to their recent clinical applications, followed by a concise review of potential safety concerns.

Micromanipulation and technical considerations: The apparatus required for NT, encompassing micromanipulation and optical systems, is commonly available in most IVF laboratories. This renders NT available to numerous clinics; yet, executing these approaches both efficiently and safely necessitates the expertise of a highly skilled embryologist. Crucial factors to consider include oocyte/zygote viability post-NT, blastocyst development pace and the extent of mtDNA carryover. These performance measures may exhibit significant variability across operators, highlighting the necessity for training in NT techniques, and may also vary based on the specific NT approach employed. MST is conducted using unfertilised MII oocytes, which provides certain advantages over PNT by eliminating the need to create zygotes for treatment reasons. The spindle karyoplast is smaller than the PN karyoplast, potentially facilitating less intrusive manipulation and reducing the amount of karyoplast-derived mtDNA. Most investigations on human embryos indicate that both MST and PNT yield comparably low amounts of mtDNA carryover, hence demonstrating equal efficacy in mitigating the risk of mtDNA illness in progeny. In contrast to PNT, the spindle lacks a nuclear membrane, indicating that the chromosomes positioned on the spindle could potentially be lost or damaged during the NT operation. Data from human oocytes indicate that this phenomenon is feasible, with chromosomal scattering observed in oocytes from older women (Battaglia et al., 1996) and in reaction to temperature variations in the surrounding environment. This may result in increased aneuploidy rates following MST, a phenomenon largely unreported in most investigations involving human oocytes, although documented in the majority (80%) of MST blastocysts generated in the current treatment cycle conducted by Zhang et al. (2017). Utilising MST with human oocytes may lead to elevated rates (~50%) of abnormal fertilisation post-ICSI, likely attributable to spindle damage incurred during the procedure, which could markedly diminish the efficacy of MST by potentially yielding fewer normal zygotes suitable for uterine transfer. Moreover, the data about the quality of MST blastocysts is insufficient, complicating the assessment of the probability of obtaining pregnancy post-MST. During the ultimate maturation of oocytes, the GVBD and chromosomes condense. Chromosomes are subsequently organised at the MI stage within the MI spindle–chromosome complex, followed by the shift from anaphase I to telophase I and the ejection of the PBI. Oocytes become halted in the metaphase II stage, containing the metaphase II spindle–chromosome complex, also referred to as the maternal spindle. During the use of MST, meiotic spindles may experience possible harm. The spindle–chromosome complex, devoid of a nuclear membrane, unlike GVs, is susceptible to mechanical manipulation, which can compromise this delicate structure and potentially lead to chromosome loss from the spindle. This is probably the result of manipulation-induced strain causing the disassociation of meiotic spindle chromosomes by decreasing kinase phosphorylation. Moreover, the application of electrofusion to merge the spindle with recipient oocytes may lead to premature oocyte activation induced by electrical currents, which might compromise the integrity of the transferred spindles, as evidenced by several published studies. In a primate model, immunocytochemical analysis revealed that post-MST, the majority of oocytes resulting from electrofusion exhibited indications of premature activation, advancing to the anaphase II stage with sister chromatids migrating towards opposing poles. Conversely, after fusion via HVJ-E, the spindles of reconstructed oocytes demonstrated greater stability at the MII stage, with chromosomes accurately aligned along the equatorial plate. Tachibana et al. (2013) demonstrated that postfertilisation, merely 50% of the reconstructed MST oocytes exhibited normal fertilisation, while 52% of MST zygotes presented abnormal fertilisation (1PN or ≥3PN), potentially signifying spindle damage resulting from extensive micromanipulation. Furthermore, Paull et al. (2013) established that low temperatures during micromanipulation could partially depolymerise isolated spindles, resulting in micro- or multinucleation postfertilisation. Unlike MST, PNT provides the benefit of encapsulating both maternal and paternal genomes within a distinct nuclear membrane, hence diminishing the likelihood of chromosomal anomalies in the reconstituted zygote. Compared to the spindle, the PNs are considerably larger and relocate to the centre of the zygote postfertilisation, potentially rendering their removal more invasive. This is unlikely to be a concern with early PNT, which has been linked to enhanced survival of reconstituted zygotes. The prompt excision of the polar nuclei upon their emergence, when they are diminutive and situated near the zygote's periphery, is expected to reduce the dimensions of the polar nucleus karyoplast; however, it is also feasible to eliminate surplus cytoplasm subsequent to the extraction of the karyoplast from the zygote. Care must be exercised throughout this process, as the zygote must form a proper bipolar spindle for mitosis, necessitating the existence of two centrioles. In humans, centrioles are located adjacent to the PN membrane, indicating that the karyoplast must preserve sufficient cytoplasm to include the centrioles and other spindle assembly components. Current research indicates that this is not an issue, as there is no observed increase in the incidence of aneuploidy in PNT blastocysts compared to both unmanipulated control blastocysts and a reference dataset of age-matched IVF blastocysts. PBT has distinct advantages over both MST and PNT, primarily because polar bodies serve as 'ready-made' karyoplasts containing nuclear DNA and a minimal volume of cytoplasm, all encased within a membrane. This should render their removal significantly less invasive and eliminate the necessity for cytoskeletal inhibitors. Additionally, polar body biopsy is an IVF method currently utilised in certain clinics as an alternative to PGD. Another advantage is that, due to their diminutive size and restricted cytoplasm, polar bodies possess a minimal quantity of mitochondria in comparison to the oocyte, resulting in a low level of karyoplast-derived mtDNA within the reconstituted oocyte post-PBT. Nonetheless, there are drawbacks, including potential time limitations on PBT resulting from the degradation of polar bodies after their ejection from the oocyte. A present drawback is the low blastocyst development rate subsequent to PBT in human oocytes, which must be resolved for this approach to be deemed suitable for clinical application.

Compatibility between mtDNA and nuclear DNA: The functionality of mitochondria is regulated by both nuclear and mitochondrial genomes, raising a significant issue regarding nuclear genome compatibility with the mtDNA from the donor in the reconstituted oocyte or zygote. Prior research has suggested nuclear–mitochondrial incompatibility across various animal models; however, this does not seem to be a uniform observation, as other studies have reported no detrimental effects despite the existence of divergent mitochondrial genomes. It is crucial to acknowledge that this research utilise highly inbred laboratory animals, which significantly differ from predominantly outbred human populations, therefore the applicability to NT remains contentious. Moreover, the health assessments of macaques produced by MST, along with the birth of the first seemingly healthy infant, have indicated no detrimental consequences from NT and imply that nuclear–mitochondrial incompatibility is not a concern. Additional assurance is offered by the typical preimplantation development of NT embryos with various nuclear–mtDNA combinations, as well as the derivation and differentiation of embryonic stem cells (ESCs) from these blastocysts. Current research indicates that mitochondrial function is unchanged after NT in humans. This was shown in both undifferentiated and differentiated ESCs produced from MST blastocysts, which exhibited similar mitochondrial respiratory chain enzyme activity and oxygen consumption rates regardless of the nuclear–mtDNA combination.

Possible augmentation of mtDNA generated from karyoplasts:

During the NT technique, a minor volume of cytoplasm will inevitably be extracted along with the karyoplast and the nuclear genome. The cytoplasm will encompass mitochondria, likely leading to mtDNA carryover in the reconstructed oocyte/zygote. Both human and animal studies have demonstrated low levels of mtDNA heteroplasmy (below 2%) in the majority of reconstituted embryos, irrespective of the NT technique employed. At this level, the transmission of the pathogenic mtDNA mutation would remain well below the threshold for illness manifestation, hence rendering the risk of mtDNA disease in the progeny negligible. Nonetheless, a concern exists over the potential alteration of heteroplasmic mtDNA levels throughout embryonic development, as evidenced in a restricted number of hESC lines obtained from NT blastocysts (Fig. 9.3B) or in certain animal models. The research conducted by Hyslop et al. (2016) noted a gradual escalation in mtDNA carryover from around 4% to 60% in one of five ESC lines obtained from PNT blastocysts. Yamada and et al. (2016) similarly observed an increase in mtDNA carryover from approximately 1% to around 53% in one of eight stem cell lines produced from MST blastocysts, however this level subsequently diminished to 1% following extended culture. Kang et al. (2016) made the same observations following MST with human control oocytes and oocytes containing a pathogenic mtDNA mutation. Despite the mtDNA carryover in preimplantation embryos being less than 1%, three out of 18 ESC lines produced from MST blastocysts exhibited an elevation in mtDNA carryover to homoplasmic levels. The factors contributing to the rise in mtDNA carryover in hESC lines remain unclear. One hypothesis posits that the karyoplast-derived mtDNA possesses a replicative advantage due to its closeness to the nucleus, perhaps enhancing the availability of nuclear-encoded components necessary for mtDNA replication. This is plausible since karyoplast-derived mtDNA is located in the perinuclear mitochondria transferred during NT; however, it remains uncertain whether these mitochondria will persist in the perinuclear region throughout preimplantation embryo development and subsequent ESC culture. Another idea posits that the replicative advantage arises from sequence variants present within the karyoplast-derived mtDNA genotype, which are utilised to ascertain the mtDNA haplotype. Research utilising animal models has demonstrated that the existence of distinct mtDNA haplotypes can cause the dominance of one mtDNA genotype over another (Burgstaller et al., 2014; Burgstaller et al., 2015), prompting the suggestion that prospective mothers and mitochondrial donors be matched according to their mtDNA haplotype. The results from human investigations, however, are incongruous and cast doubt on the necessity for such haplotype matching. The study by Kang et al. (2016) demonstrated that mtDNA carryover in two distinct hESC lines, both possessing identical combinations of mtDNA haplotypes, yielded divergent outcomes; one hESC line attained homoplasmic levels of karyoplast-derived mtDNA, whereas the other maintained less than 1%. Conflicting results were similarly found in hESC lines produced from SCNT blastocysts with identical mtDNA haplotype combinations. The research conducted by Hyslop et al. (2016) demonstrated that a variation in the quantity of karyoplast-derived mtDNA may transpire even when both the nuclear and mitochondrial donors are from the same haplogroup (Table 9.1). Nonetheless, a single haplogroup may encompass numerous distinct mtDNA variants, with a variation of approximately 20 mtDNA variants being relatively frequent. Consequently, the potential exists that a specific variant may account for the proliferative superiority of one mtDNA genotype relative to another. The work by Kang et al. (2016) studied a polymorphism in the noncoding region of the mitochondrial genome (D-loop), identified as a possible variant responsible for the observed effects. While the participation of the D-loop in mtDNA transcription and replication is plausible (Nicholls & Minczuk, 2014), mtDNA variations in this region are not consistently observed. It may be essential to contemplate the possibility that this observation is simply a characteristic of stem cell biology. Considering that pluripotency is transient throughout in vivo embryonic development, the significance of ESCs as a paradigm for postimplantation development remains ambiguous.

Safety considerations for northern territory: In light of the incongruous findings on mtDNA haplotype mismatch, the latest study on the safety and efficacy of mitochondrial donation in the United Kingdom advised that haplotype matching be regarded as a preventive measure when choosing appropriate oocyte donors. Since all oocyte donors will undergo complete mtDNA sequencing to verify the absence of harmful mtDNA mutations, haplotype assignment will be rather uncomplicated. Challenges may emerge in correlating these haplotypes with the women possessing the mtDNA mutation, as achieving an exact match for each mtDNA variant could be difficult, if not unfeasible, potentially restricting the application of the technology unduly. Additional research is necessary to ascertain the significance of the observations in hESCs; nonetheless, this should not impede the therapeutic implementation of NT to mitigate the risk of mitochondrial illness. Simultaneously, NT should be evaluated alongside prenatal screening to detect alterations in mtDNA carryover levels that may elevate the risk of mtDNA illness. This underscores the significance of long-term surveillance of infants born subsequent to NT. Another potential issue to consider is if NT methods could cause alterations in genomic imprinting in the resulting embryos or offspring. To date, the literature contains limited data regarding this subject. Wang et al. (2014) showed in mice that prior to PBT, there are no significant variations in the patterns of epigenetics, DNA methylation and histone modifications between PB1/2 chromosomes and the matching oocyte/zygote. Following the implementation of PBT, the progeny exhibited normal development, prompting the conclusion that 'this approach is unlikely to induce epigenetic and imprinting abnormalities'. Analysis of DNA methylation and imprinted genes in various mice models has consistently shown no substantial epigenetic or imprinting anomalies in the tissues of GVT-derived pups. Paull et al. (2013) demonstrated that embryos subjected to MST exhibited no epigenetic or genomic modifications when compared to unmanipulated controls. Hyslop et al. (2016) utilised genome-wide RNA sequencing to show that global gene expression in early PNT blastocysts was comparable to that of unmanipulated controls. Moreover, analysis of ESCs produced from PB1T and control blastocysts indicated that the patterns of DNA methylation and overall gene expression in PB1T ESCs were analogous to those in the controls. Despite the absence of any deleterious imprinting or epigenetic alterations identified in the aforementioned research, we advocate for prudent clinical application of this technology and subsequent monitoring of the children following the implementation of NT procedures. Uutilisation of NT and other methodologies for infertility treatment. In the last 10 years, the quantity of ART procedures has risen significantly, and traditional technologies, such as IVF and ICSI, have been enhanced through the incorporation of more sophisticated technologies, including time-lapse imaging, '-omic' technologies, assisted oocyte activation, advanced cryopreservation techniques and others. Despite common belief, ART does not ensure success; between 38% and 49% of couples initiating invitro treatment will remain childless, even after undergoing up to six treatment cycles. Nevertheless, these figures appear somewhat bleak, given De Vos et al. (2016) recently documented a cumulative live birth rate above 52% per IVF/ICSI cycle. A recent study indicated that the cumulative live birth rate after three IVF cycles was 42.3%, and after eight IVF cycles, this percentage increased to 82.4%. The likelihood of success is connected with female age, as ART significantly mitigates infertility in younger women but does not entirely counteract the age-related decline in fertility. The function of mitochondria in infertility remains a topic of contention. Research indicates that with ageing, the prevalence of various mitochondrial deletions and point mutations escalates markedly (Cortopassi et al., 1992), with estimates suggesting that as much as 50% of human oocytes include altered or deleted mtDNA. In males, many deletions of mtDNA have been documented in human sperm, particularly in older men or those with subpar sperm quality. Consequently, it is plausible that impaired mitochondria may significantly contribute to certain refractory instances of infertility, which might potentially be addressed with the implementation of NT technology. Nonetheless, this is refuted by the observation that live birth rates in women with mtDNA disease are similar to those of the general population, suggesting normal fertility despite the existence of a harmful mtDNA mutation.

Female infertility and the function of mitochondria:

The quantity of mtDNA molecules in human oocytes exhibits significant heterogeneity, ranging from 20,000 to 600,000 copies, potentially reflecting the diverse quality of human oocytes. The mtDNA copy number in oocytes from older women or those with impaired ovarian reserve is markedly lower than that in younger women or those with normal ovarian reserve. In older women, a reduced mtDNA content is detected in early cleavage-stage embryos; nevertheless, in the blastocyst stage, a higher mtDNA content is paradoxically linked to increased aneuploidy and unsuccessful implantation. Mitochondrial proliferation may serve as a compensating response to mitochondrial malfunction during embryonic stress. It has been proposed that a minimum mtDNA copy number is necessary to facilitate fertilisation and subsequent embryonic development effectively (Schatten et al., 2014). As an egg grows, energy demands escalate, necessitating a transition from glycolysis to oxidative phosphorylation, which peaks at ovulation. The critical activities involved in oocyte maturation, including spindle formation and proper metaphase chromosomal alignment, rely on a sufficient supply of ATP produced from mitochondria (Schatten et al., 2014). The energy required for fertilisation and embryonic development until the blastocyst stage is exclusively supplied by the mitochondria within the mature egg, as mitochondrial replication is not believed to commence until the hatched blastocyst stage. Consequently, any mitochondrial deficiencies may impede essential processes such as maturation, fertilisation, and embryonic development, as evidenced by the disruption of the oxidative phosphorylation system in mouse oocytes, which resulted in meiotic spindle abnormalities and a subsequent decline in embryonic development. Significant evidence indicates that meiotic and embryonic aneuploidies correlate with female age and mitochondrial function (Schatten et al., 2014), suggesting that oocyte competence, encompassing cytoplasmic quality and specifically mitochondrial integrity, may be influenced by ageing. Aneuploidies in oocytes and resultant embryos are regarded as a significant contributor to female infertility, leading to spontaneous abortion and recurrent implantation failure, particularly in advanced maternal age. For additional information regarding infertility and the function of mitochondria, refer to the reviews by Benkhalifa et al. (2014), Schatten et al. (2014), Otten and Smeets (2015), May-Panloup et al. (2016), and Demain et al. (2017).

Ooplasmic transfer

The OT technique entails extracting cytoplasm, which includes mitochondria and various other cellular constituents, from a donor MII oocyte and implanting it into a recipient oocyte. Initial animal experiments established the viability of this approach, which was noted to elevate intracellular ATP levels (Van Blerkom et al., 1998) and/or inhibit apoptosis (Perez et al., 2000) subsequent to cytoplasmic injection into mouse oocytes. In cattle, oocytes purposefully devoid of mtDNA might be salvaged by OT, resulting in normal progeny. In the 1990s, some fertility clinics in the United States supplemented human oocytes with donor cytoplasm from younger oocytes to address female infertility potentially caused by inadequate oocyte cytoplasmic quality. This entailed the injection of a restricted volume (1%–5%) of cytoplasm from healthy MII oocytes into the patient's oocytes. This experimental treatment resulted in nearly 50 pregnancies but was deemed controversial due to the absence of preceding animal studies to validate the benefits and safety of this technology. Additionally, certain chromosomal abnormalities (Turner's syndrome, XO karyotype) and infantile disorders (borderline pervasive developmental disorder) have been documented in infants conceived through ovum transplantation. Due to the restricted sample size, it is impossible to establish that these anomalies resulted from the OT process itself; however, because of these bad consequences, the US FDA banned the application of OT in human oocytes in 2002. Chen et al. (2016) conducted a follow-up study on 13 children aged 13 to 18 who were born following OT, reporting that their general health and cognitive abilities appeared normal, although they exhibited minor skin issues, slight vision impairments, mild asthma and chronic migraine headaches. Oocyte transfer is an inadequate method for preventing the transmission of pathogenic mtDNA mutations, as it would necessitate the transfer of substantial quantities of donor cytoplasm into the patient's oocyte to sufficiently dilute the mutant mtDNA and mitigate the risk of mtDNA disease, a feat that is practically unfeasible in humans.

Mitochondrial substitution: Chapter 14 provides a full account of this technique. The transplantation of mitochondria extracted from somatic or, more recently, stem cells has been proposed as a potential therapeutic approach to address specific types of infertility. Mitochondrial supplementation derived from follicular cells has demonstrated benefits in cattle and in humans. Atresia of oocytes is produced by apoptotic signals from follicular cells, significantly increasing the probability of oocyte degeneration. Furthermore, mitochondria extracted from the identical somatic cells as the recipient oocyte are likely to retain the drawbacks associated with older mitochondria. Consequently, superior autologous mitochondria would be more appropriate for enhancing the energy of oocytes via mitochondrial substitution. The potential utilisation of mitochondria extracted from oocyte precursor cells located in the ovary has been proposed. Nonetheless, debate exists, as several investigations have not detected mitotically active female germline progenitors in postnatal mouse ovaries. Despite their ambiguous function, these precursor cells likely did not undergo prolonged meiotic arrest, suggesting they may possess fewer mtDNA mutations and deletions (Yabuuchi et al., 2012). Currently, there is a dearth of experimental evidence supporting the advantages of mitochondrial substitution from oocyte precursor cells into oocytes, both in animal models and particularly for infertile applications. For women experiencing multiple IVF failures, the technology known as 'Autologous Germline Mitochondrial Energy Transfer (AUGMENTSM)' entails the injection of 1–2 pL of harvested mitochondria into the patient's MII oocytes in conjunction with sperm, and has been utilised in various infertile couples. The limited patient population and the biased study design hinder the validation of this treatment's advantages. Similar to OT, the substantial quantity of healthy mitochondria necessary to sufficiently dilute the mutant mtDNA precludes the application of this AUGMENT technology to mitigate the risk of mitochondrial diseases, which necessitates mitochondrial transfer in a heterologous manner (mitochondria sourced from another individual).

NT for the treatment of infertility: Numerous NT approaches outlined in the section 'The use of nuclear transfer technologies in animal and human models' have been suggested to address some types of female infertility. In summary, these technologies involve transferring the nucleus from an egg obtained from an infertile woman to a donor oocyte from which the nucleus has been excised. Zhang et al. (2016b) recently detailed their therapeutic application of PNT for an infertile couple who had undergone two unsuccessful IVF cycles characterised by recurring embryo arrest at the two-cell stage, as shown in an abstract published in 2003. Five rebuilt zygotes that cleaved to the four-cell stage on Day 2 were transferred, resulting in a triplet pregnancy. Following foetal reduction to a twin gestation, one nonviable foetus was delivered due to early rupture of membranes at 24 weeks of gestation, while the other nonviable foetus was delivered following intrauterine foetal demise caused by cord prolapse at 29 weeks. A standard karyotype was identified in the foetal material, and fingerprinting verified the nuclear origin from the patient and mtDNA from the donor, with no patient mtDNA present. Despite some circumstantial data suggesting that oocyte cytoplasm is a critical determinant for success in ART, it is advised that this NT technology should not be utilised for infertility therapy at present. The primary reason is the absence of proof on which category or indication of infertility this technique may benefit. The availability of appropriate

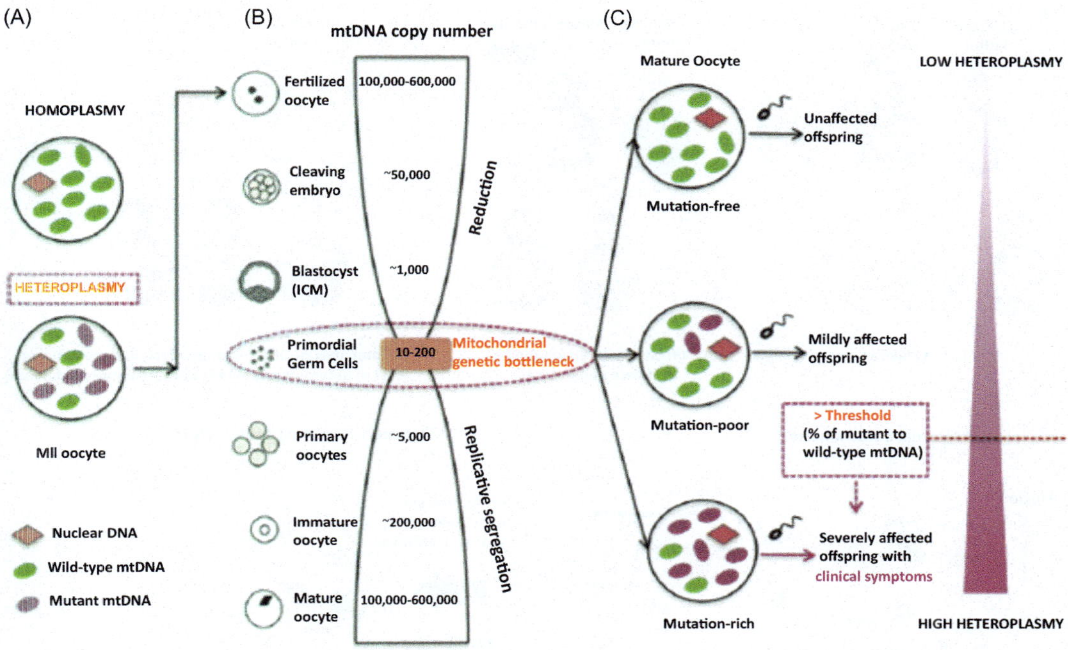

FIGURE 9.6 Presents a schematic representation illustrating mitochondrial DNA (mtDNA) heteroplasmy, the mitochondrial genetic bottleneck and potential variations in mtDNA heteroplasmy. (A) mtDNA genomes may be classified as homoplasmic, indicating the presence of identical mtDNA variants within a cell, or heteroplasmic, denoting the coexistence of wild-type and mutant mtDNA or other variants within the same cell. (B) In humans, the quantity of mtDNA molecules in preimplantation embryos progressively decreases to a minimal level when primordial germ cells are established (10–200 copies). Subsequently, the mtDNA copy number exhibits an increase of around 100-fold to 100,000-fold when the oocytes progress to the metaphase II stage of replicative segregation. (C) During replicative segregation, wild-type and mutant mtDNA may display an uneven distribution due to random segregation, resulting in diverse levels of mtDNA mutations that can be inherited maternally, causing significant variation in mutated mtDNA across affected tissues (oocytes/embryos). The proportion of mutant to wild-type mtDNA will dictate the severity of the problem, and surpassing a specific threshold will result in the clinical manifestation of the mtDNA disease.

infertile animal models to investigate the efficacy and safety of these NT technologies would be beneficial.

Mitochondrial DNA heteroplasmy: the mitochondrial genetic bottleneck and possible variation in mitochondrial DNA heteroplasmy

The figure below depicts how heteroplasmy affects reproductive traits (Fig. 9.6).

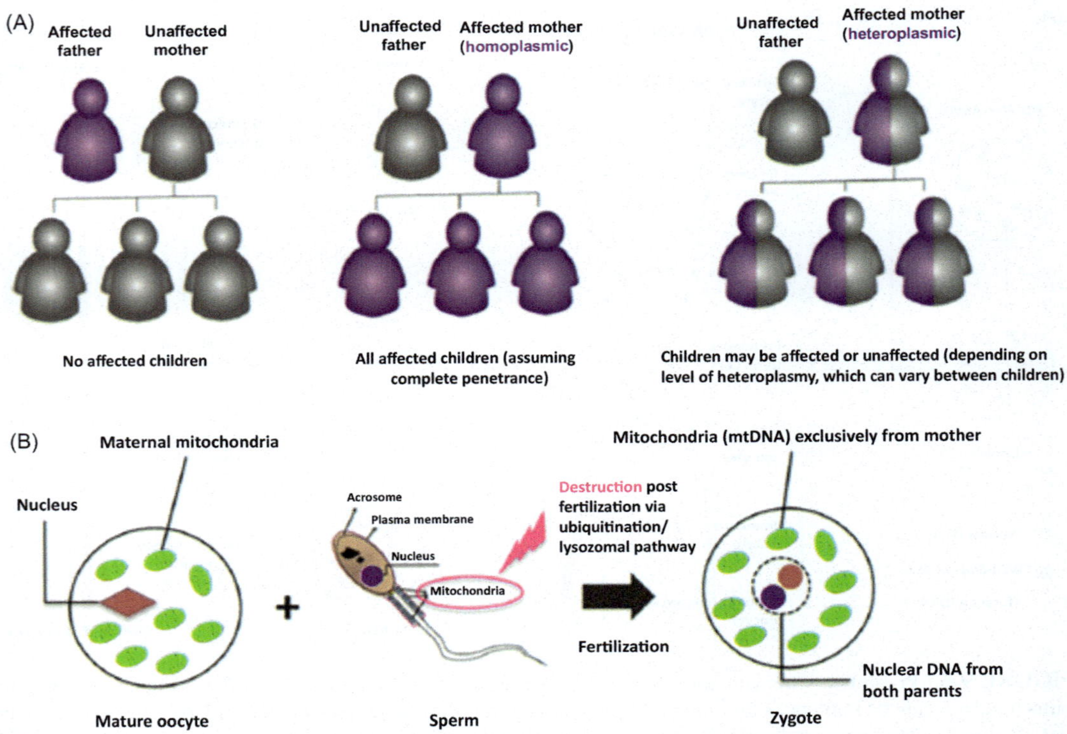

FIGURE 9.7 Schematic representation illustrating the exclusive maternal transmission of mitochondrial DNA (mtDNA). (A) mtDNA demonstrates exclusive maternal inheritance, indicating that mutant mtDNA possessed by a male cannot be passed on to his offspring. If a woman's oocytes possess a homoplasmic mtDNA mutation, all offspring will be carriers and may be impacted, assuming complete penetrance. If a woman's oocytes possess a heteroplasmic mtDNA mutation, all offspring may be carriers, with clinical manifestations contingent upon mutation penetrance, heteroplasmy levels and other aspects that remain to be completely clarified. (B) Schematic diagram illustrating the mechanism of maternal inheritance of mtDNA, resulting from the degradation of paternal mtDNA by either ubiquitination or the lysosomal pathway.

Maternal inheritance of mitochondrial DNA

See Fig. 9.7.

Nuclear transfer techniques and heteroplasmic mtDNA drift/reversion

See Fig. 9.8.

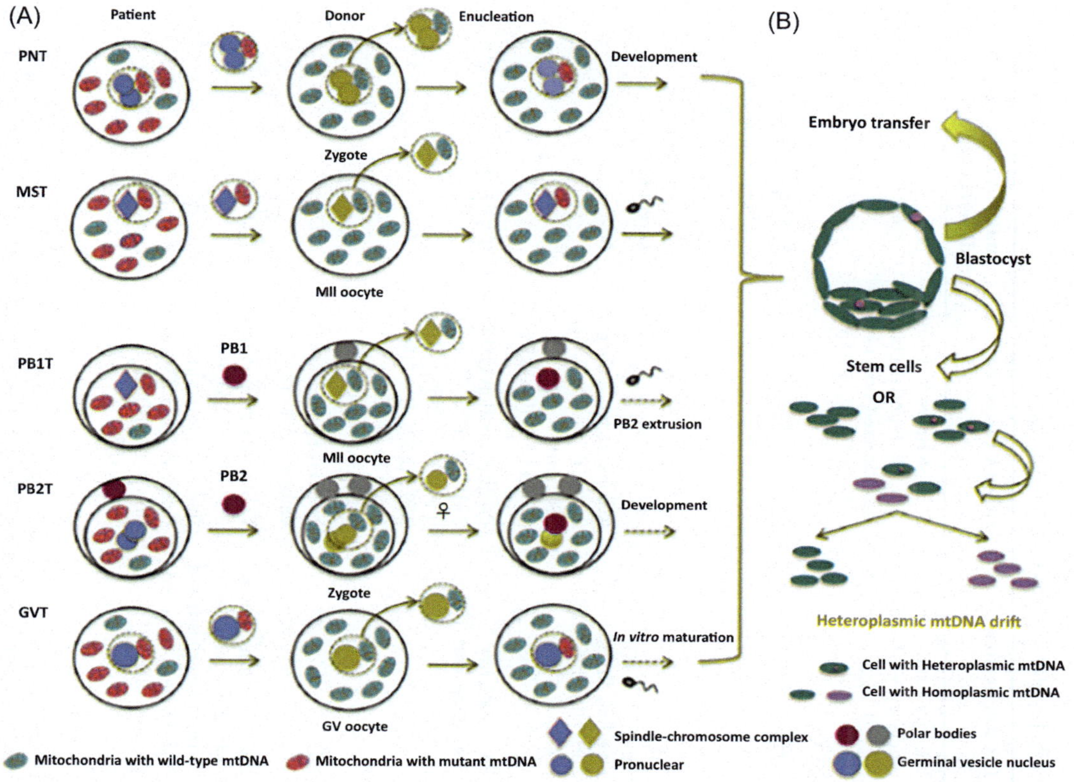

FIGURE 9.8 Schematic representation illustrating nuclear transfer (NT) methodologies and heteroplasmic mtDNA fluctuation/reversion. (A) The methodologies of pronuclear transfer (PNT), maternal spindle transfer (MST), first/second polar body transfer (PB1T/PB2T) and germinal vesicle transfer (GVT) involve the isolation of the karyoplast from the patient oocyte or zygote, which is subsequently transferred into the enucleated recipient oocyte or zygote. (B) In the NT technique, perinuclear mitochondria from the nucleus donor may inadvertently transfer into the recipient cytoplast, potentially resulting in a low level of karyoplast-derived mtDNA heteroplasmy in the reconstructed embryos.

Studies on pronuclear transfer and maternal spindle transfer in animal and mouse model

See Table 9.3.

Important points:

Although NT approaches have demonstrated efficacy in diminishing the transmission of heteroplasmic mtDNA variations in both animal and human studies, there is a residual transmission of mtDNA originating from the karyoplast in most instances. To prevent the potential rise in karyoplast-derived mtDNA levels and ensure the avoidance of mtDNA diseases, one might entirely eradicate all mtDNA within the karyoplast. This may be accomplished by the induction of mitophagy or by employing nucleases directed at specific

TABLE 9.3 Summary of the included studies on maternal spindle transfer in animal and human models.

Species	Origins	Embryonic development	ESCs derivation	mtDNA carryover	mtDNA carryover on differentiation
Animals	MII oocytes from rhesus monkeys MII oocytes from NZB/PlaHsd and B6D2/FI mice	Blastocyst formation: 61% versus 60% (Ctr) two-cell embryos: 91%; blastocyst development 90%~ Ctr	ESC derivation: 25% efficiency ~ Ctr ND	DNA samples of infants: <3% ESCs: undetectable reconstructed oocytes/embryos: 0.29% ± 0.63%	ND
Humans	Donated MII oocytes	ND	ND	ND	ND; but mtDNA copy number: 13,222 ± 5733 (spindle + karyoplasts) revealed no difference with that of pronuclear-karyoplasts
	onated MII oocytes	Blastocyst rate: 62% (13/21) versus 76% (16/21) (Ctr)	Five ESC lines; derivation rate: 15% (3/18)	Reconstructed embryos: 0.5% ± 0.4% (0%–0.9%); ESC lines: 0.6% ± 0.9% (0%–1.7%)	ND
	Donated MII oocytes	Blastocyst rate: 37% (7/18) versus 52% (54/103) (control IVF group) versus 33% (7/21) (control parthenogenetic group)	Four Esc lines; derivation rate: 15% (3/18)	Preimplantation embryos: 0.31% ± 0.27%; three ESC lines: 0% (P2–59); I line 2.79 ± 0.27% (P4–10), 0% (from P14); PSCs: 0.01% ± %0.04%	0%
	Donated MII oocytes	ND	ND	1/10 embryos: 0% 9/10 embryos: 1%–3%	ND
	Donated MII oocytes	Blastocyst 32% (fresh cytoplasm and cryo-nucleus) versus 22% (Ctr with fresh oocyte) versus 9% (Ctr with cryo-oocyte)	Eight lines; derivation rate: 25% (combination of fresh cytoplasm and cryo-nucleus achieved the best rate)	Preimplantation embryos: 0.2% (0%–2.2%); four ESC lines: 0%; three lines: 0.42% (0.2%–1.7%); one line: 1.3% (PI) to 53.2% (P36)	ND

II oocytes from patient with Leigh syndrome (m.8993 T>G)	Blastocyst rate: four out of five reconstructed fertilised oocytes	ND	Blastocyst: 5.10% ± 1.11%; tissue from newborn: 1.60% ± 0.92%	ND
MII oocytes from patients with Leigh syndrome and MELAS mtDNA variants (m.8993 T>G; m.13513 G>A; m.3243 A>G)	Blastocyst rate: ND	Control ST (using healthy oocytes): 15 ESC lines; patient ST (using patient oocytes): three ESC lines; detvation rate: ND	Preimplantation embryos: <1%; control ST. ESCs: 13/15 lines (≤1%); carryover reversed lines 2/15 from P2 (81% ± 9%) to P10 (100%) respectively; patient ST. ESCs: 2/3 lines (≤1%); carryover reversed line: 1/3 from 20% ± 13% (P2) to 100% (P10)	Control reversed ST-ESC line: increased from 81% to 100%; patient reversed ST-ESC line: increased from 4% to 8%

II, metaphase II; Ctr, nonmanipulated controls; ESCs, embryonic stem cells; ND, not defined; mtDNA, mitochondrial DNA; MELAS, mitochondrial encephalomyopathy with lactic acidosis and stroke-like episodes; P, ……….; IVF, in vitro fertilisation. These three studies used parthenogenesis whilst other studies listed in this table used oocyte fertilisation.

mtDNA sequences, such as mitochondrially-targeted zinc finger nucleases (mtZFNs) or mitoTALENs (http://www.hfea.gov.uk/docs/Fourth_scientific_review_mitochondria_2016.PDF). While this necessitates validation, it could mitigate the possible risk of elevated mutant mtDNA in the progeny. This genomic editing technology has been suggested as a potential strategy for addressing mitochondrial disease. A distinct chapter has been incorporated for gene editing of mtDNA. In a proof-of-concept study, Gammage et al. (2014) successfully developed mtZFNs capable of targeting specific mtDNA sequences. The mtZFNs effectively targeted and degraded mtDNA containing harmful point mutations, such as m.8993 T>G and eliminated prevalent deletions, including the 4977-bp deletion. Moreover, Reddy et al. (2015) have evidenced the effectiveness of mitoTALENs in the mammalian germline by reducing the levels of either m.14459 G>A LHON or m.9176 T>C mtDNA mutations. The widely utilised CRISPR/Cas9 genomic editing technique has been employed to target mitochondrial copies in human kidney cell lines. In conclusion, growing evidence endorses the application of NT to avert the spread of human mtDNA disease; however, comprehensive long-term assessment is essential. Furthermore, prenatal screening would continue to be strongly endorsed with NT technology. Currently, there is insufficient scientific data to endorse the use of NT or alternative reproductive procedures for mitochondrial replacement in cases of infertility.

A case analysis: reduced male reproductive efficacy linked to mitochondrial dysfunction. The reproductive success of men with mitochondrial illness has not been documented to yet. We analysed the age- and era-adjusted reproductive success of 94 British male patients with mitochondrial illness in comparison to the general male population of the United Kingdom. The reproductive success of men with mitochondrial illness was 65% of that of the general population (95% confidence interval: 53%; 79%), and the degree of the effect correlated with disease severity. This contrasts with the findings of the two preceding investigations, which indicate that the reproductive success of women with mtDNAmitochondrial DNA disorder is not compromised. The reproductive health of men with mitochondrial illness warrants heightened clinical focus.

Genetic causes of mitochondrial disease (see Table 9.4).

Mitochondrial diseases are largely controlled by numerous genes.

Results: the reproductive success of men with mitochondrial illness was 0.65 (95% CI: 0.53; 0.79) compared to the general UK male population (Table 9.1), and was consistent across mtDNA and nDNA subgroups (Table 9.2; Supplementary Table S1). Men with more severe sickness exhibited decreased relative reproductive success compared to those with milder afflictions (Fig. 9.1; Table 9.1), a phenomenon not found in women with mtDNA variations. The percentage of single males in our cohort was 24.8% (95% CI: 17.1–34.3%), while the percentage of single males in the general population, adjusted for the age and birth year distribution of our cohort, was 16.3% (P=.027). The marriage status-adjusted general fertility rate indicated that, after considering marital status, the reproductive success of our cohort was 0.73 (95% CI: 0.59; 0.88) compared to the general UK male population (Table 9.1). The severity of the condition, as indicated by the NMDAS score, showed no significant difference between males who participated in the trial and those

TABLE 9.4 Genetic causes of mitochondrial disease among the study participants.

Nuclear DNA gene	Function	N	Prevalence	mtDNA variant	Effect	N	Prevalence
All		21	2.9×10^{-5}	All		73	9.6×10^{-5}
POLG	mtDNA polymerase	6	0.3×10^{-5}	m.3243 A>G	tRNA leucine 1	36	3.5×10^{-5}
OPA1	Mitochondrial dynamics	6	0.4×10^{-5}	Single mtDNA deletion		20	1.5×10^{-5}
TWNK	mtDNA helicase	5	0.7×10^{-5}	m.8344 A>G	tRNA lysine	6	0.2×10^{-5}
GFER	Mitochondrial biogenesis	1	n/a	m.7989 T>C	COX subunit	1	n/a
RRM2B	mtDNA maintenance	1	0.2×10^{-5}	m.8839 G>C	ATP synthetase FO subunit 6	1	n/a
SPG7	Mitochondrial protein paraplegin	1	0.8×10^{-5}	m.8993 T>C	ATP synthetase FO subunit 6	1	0.1×10^{-5}
YARS2	Mitochondrial tRNA synthetase	1	n/a	m.8993 T>G	ATP synthetase FO subunit 6	1	0.05×10^{-5}
				m.9185 T>C	ATP synthetase FO subunit 6	1	n/a
				m.12271 T>C	tRNA leucine 2	1	n/a
				m.12283 A>G	tRNA leucine 2	1	0.05×10^{-5}
				m.13051 A>G	Complex I subunit	1	n/a
				m.13513 G>A	Complex I subunit	1	n/a
				m.14674 T>C	tRNA glutamic acid	1	n/a
				m.14709 T>C	tRNA glutamic acid	1	0.1×10^{-5}

Abbreviations: ATP, adenosine triphosphate; COX, cytochrome C oxidase; mtDNA, mitochondrial DNA; prevalence, prevalence among affected adults (Gorman et al); tRNA, transfer RNA.

who did not (P=.269). The absence of age-matching among the low/medium/high NMDAS groups precluded a statistical evaluation of the impact of disease severity on reproductive success by direct comparison. Each group was compared with the age and era-matched general population, revealing an inverse correlation between reproductive success and disease severity (Fig. 9.1; Table 9.1). In this investigation, smoking status (P=1.000), education level (P=.186) and the presence of diabetes mellitus (P=.806) were not correlated with the number of documented births (Fig. 9.9).

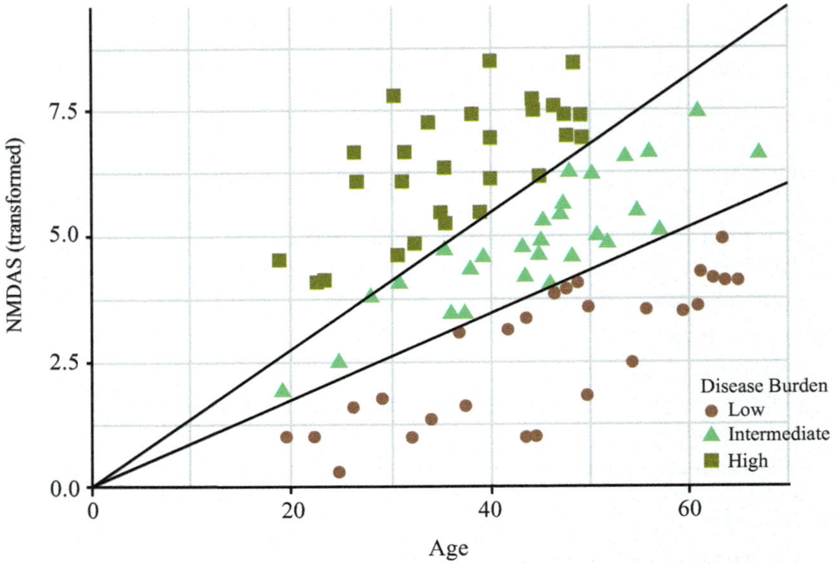

FIGURE 9.9 Reproduction versus disease. Segregation of patients by NMDAS score into three groups according to disease burden. Both age and the square root-transformed NMDAS score are used to define the disease burden groups.

Age specific or overall relative reproductive success of men with mitochondrial disease

See Table 9.5.

Role of mitochondrial genes affecting reproduction and reproductive disorders

In our lab, we are studying the role of mitochondrial genes controlling reproduction and reproductive disorders in animal models—pig and sheep models.

Mutations identified in cytochrome B affects female reproduction in the sheep model: a case study

In our lab, we could identify the role of SNPs in the mitochondrial gene (cytochrome B) controlling litter size in female sheep (ewe), indirectly controlling the ovulation and release of ova. Deleterious mutations cause less number of offspring (i.e. less litter size).

At the same time, we could also explore the role of cytochrome B controlling reproductive disorders, such as stillbirth in lambs. Individuals with deleterious mutations in cytochrome B exhibit stillbirth in lambs.

TABLE 9.5 Age-specific and overall relative reproductive success of men with mitochondrial disease.

Disease burden	No. patients	Age low	Age high	Births	Person-years	GFR pop	Ratio	Ratio L	Ratio U	P value
All patients										
		15	19	4	470	0.0108	0.7860	0.2142	2.0125	0.8500
		20	24	17	468	0.0631	0.5755	0.3353	0.9215	0.0180
		25	29	22	452	0.0996	0.4885	0.3062	0.7396	2.22E-04
		30	34	36	403	0.0950	0.9402	0.6585	1.3017	0.7918
		35	39	13	368	0.0608	0.5806	0.3092	0.9929	0.0464
		40	44	5	325	0.0267	0.5755	0.1869	1.3431	0.2720
		45	49	0	269	0.0091	0.0000	0.0000	1.5011	0.1713
		50	54	0	167	0.0024	0.0000	0.0000	9.2989	1
		55	59	0	110	0.0010	0.0000	0.0000	33.1035	1
		60	64	0	79	0.0005	0.0000	0.0000	97.5063	1
				97	3911	0.0480	0.6495	0.5267	0.7923	6.2×10^{-06}
By disease burden										
Low	27			32	923	0.046	0.7544	0.516	1.065	0.1183
Medium	27			29	978	0.0478	0.6203	0.4155	0.8909	0.0073
High	28			23	712	0.0528	0.6118	0.3878	0.918	0.0146
By genotype										
mtDNA	73			76	2364	0.0489	0.6569	0.5176	0.8222	0.0001
nDNA	21			21	747	0.0450	0.6252	0.3870	0.9557	0.0275

Abbreviations: GRF, general fertility rate; mtDNA, mitochondrial DNA; nDNA, nuclear DNA.

We have detected the mutations in mitochondrial genes initially through PCR amplification followed by Sanger sequencing. In the next step, we have followed next-generation sequencing technique, whole mitochondrial genome sequencing technique for the detection of mutations in the mitochondrial genome in sheep.

Mutations identified in cytochrome B affects female reproduction in the pig model: a case study

We could explore the important role of cytochrome B in female reproduction in the Ghungroo pig. Certain deleterious mutations were detected.

Cytochrome B is an important polypeptide of the mitochondria, helpful in energy metabolism through oxidative phosphorylation. Cytochrome B plays an immense role in the reproduction of animals, and due to its mutation-prone nature, it can affect the basic physiology of animals. Cytochrome B affects the reproductive system in males and equally plays an important role in transferring and providing energy in the development of the embryo, zygote, and oocytes, precisely in females. The present study was conducted on the Ghungroo pig to study their molecular and reproductive traits and the effect of the cytochrome B gene on the female reproduction of the Ghungroo pig. Although studies are available for cytochrome B gene analysis for evolutionary studies through phylogenetic analysis. This is the first report for the study of the cytochrome B gene on reproduction in pigs. The cytochrome B gene was sequenced, and seven SNPs were observed, out of which three were nonsynonymous. An INDEL mutation was detected in variant B, which had led to a frameshift mutation resulting in a stop codon AGA. The effect on the reproductive traits of the sow was studied due to the occurrence of nucleotide substitution. Bioinformatics analysis (I-mutant, PROVEAN and SIFT) had revealed that the mutations were deleterious for the mutant type. Mutation leading to alterations in posttranslational modification sites, such as phosphorylation sites, leucine-rich nuclear export signal, the occurrence of transmembrane helices, arginine and lysine peptide cleavage sites for the mutant variant, have resulted in a reduced physiological response. 3D protein structure (predicted through bioinformatics analysis) for cytochrome B has revealed distinct structural differences in the mutated form with the truncated protein by RMSD analysis through TM-align software. Associated studies of genotype variants with reproductive traits have revealed the significant effect of variants of the cytochrome B gene on reproductive traits, namely litter size at first, second and third farrowing, piglet mortality, age at first farrowing and farrowing interval. Mitochondrial gene as cytochrome B variants might be used as a marker for studying the female reproduction of the Ghungroo sows in the future Fig. 9.10 (Figs. 9.11–9.24).

I-Mutant 2.0:Mut 538

```
******************************************************************************
**                                                      **
**                         I-Mutant v2.0                      **
**       Predictor of Protein Stability Changes upon Mutations          **
**                                                      **
******************************************************************************
    SEQ File: fileseq.txt
    Position  WT NEW    DDG   pH   T
        33   F   L  -2.34  7.0  25

    WT:  Amino acid in Wild-Type Protein
    NEW: New Aminoacid after Mutation
    DDG: DG(NewProtein)-DG(WildType) in Kcal/mol
         DDG<0: Decrease Stability
         DDG>0: Increase Stability
    T:  Temperature in Celsius degrees
    pH:  -log[H+]
```

I-Mutant 2.0: Mut614

```
******************************************************************************
**                                                      **
**                         I-Mutant v2.0                      **
**       Predictor of Protein Stability Changes upon Mutations          **
**                                                      **
******************************************************************************
    SEQ File: fileseq.txt

    Position  WT NEW    DDG   pH   T

        46   L   Q  -1.25  7.0  25
    WT:  Aminoacid in Wild-Type Protein
    NEW: New Aminoacid after Mutation
    DDG: DG(NewProtein)-DG(WildType) in Kcal/mol
         DDG<0: Decrease Stability
         DDG>0: Increase Stability
    T:  Temperature in Celsius degrees
    pH:  -log[H+]
******************************************************************************
```

FIGURE 9.10 Determination of functional stability of the mutated variant (variant B) of the cytochrome B protein of Ghungroo pig.

PROVEAN Result (Download)

PROVEAN Prediction - Job ID: 294157930594534

Query sequence (fasta)

Supporting sequence set used for prediction

Number of sequences: **68360** (fasta, E-values)

Number of clusters: **30**

Score thresholds for prediction

(1) Default threshold is -2.5, that is:

-Variants with a score equal to or below -2.5 are considered "deleterious,"

-Variants with a score above -2.5 are considered "neutral."

(2) How to use a more stringent threshold.

Variant	PROVEAN score	Prediction (cutoff= -2.5)
F33L	-4.118	Deleterious
L46Q	-3.551	Deleterious
T56P	-4.275	Deleterious

FIGURE 9.11 Determination of the deleterious nature of the mutated variant (variant B) of the cytochrome B protein of Ghungroo pig.

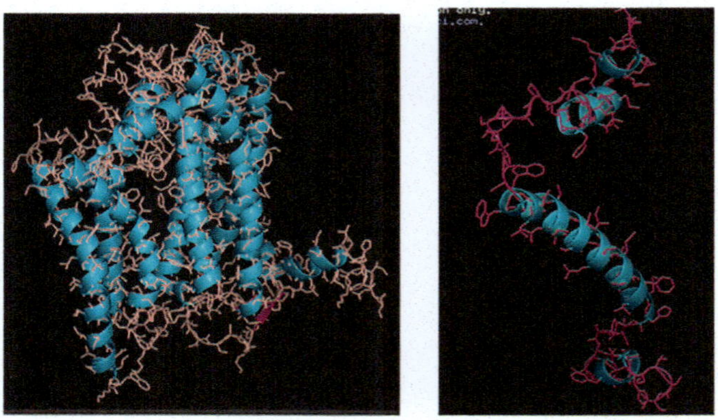

Wild type of cytochrome B Trunkated Cytochrome B

FIGURE 9.12 Deletion mutation produces truncated cytochrome B in contrast to wild type.

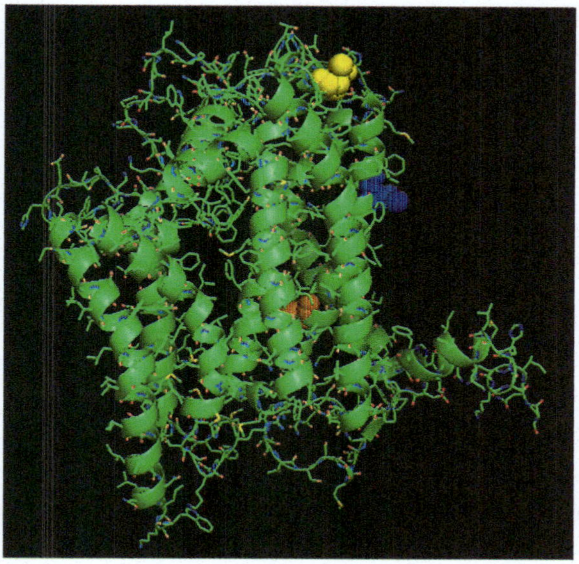

FIGURE 9.13 It depicts the sites for amino acid variations at position 33, 46 and 56 in cytochrome B variant A (wild type of cytochrome B) represented respectively by red, blue and yellow colour, which are the sites prone to mutation leading to reproductive abnormalities.

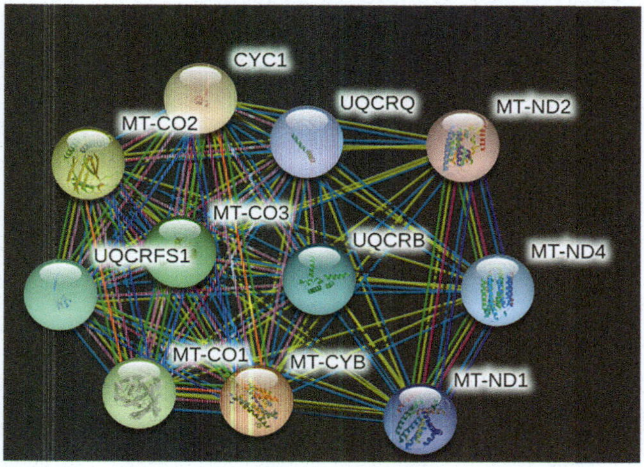

FIGURE 9.14 Cytochrome B interacts with other proteins for the formation of bc1 complex, which is the ultimate functional unit for oxidative phosphorylation. The interaction had been studied by String analysis.

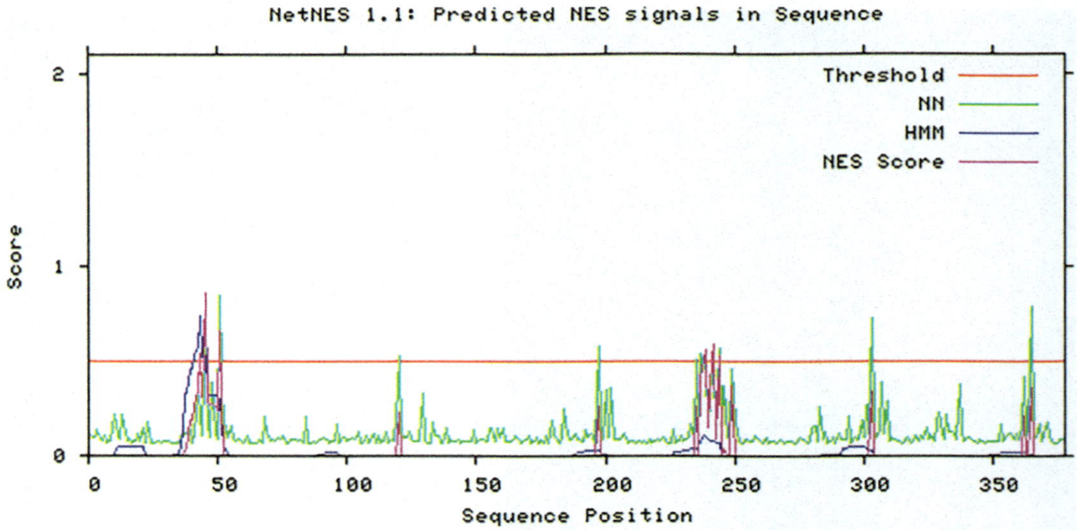

FIGURE 9.15 Prediction of leucin rich nuclear export cytochrome B variant A from NetNes software.

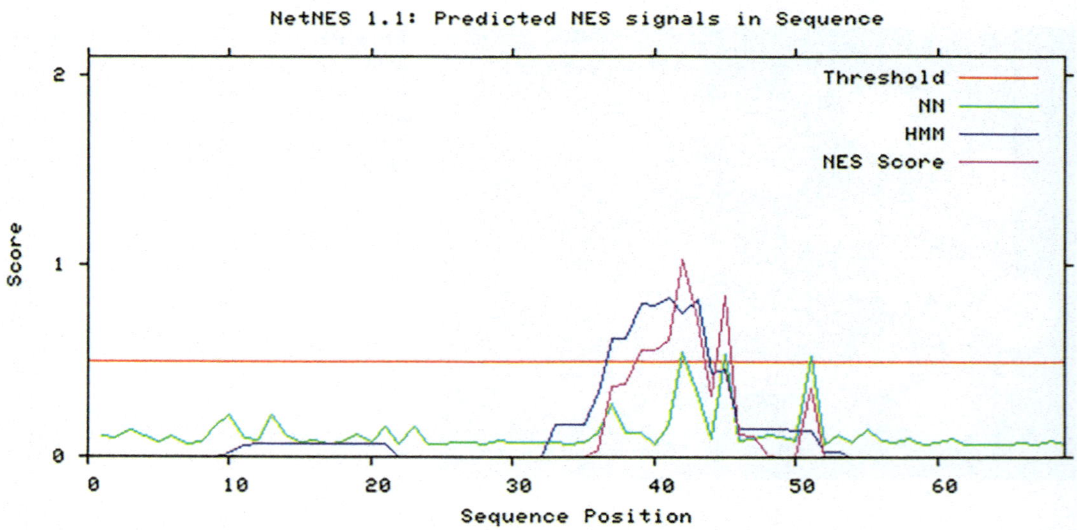

FIGURE 9.16 Prediction of leucin rich nuclear export cytochrome B variant B from NetNes software.

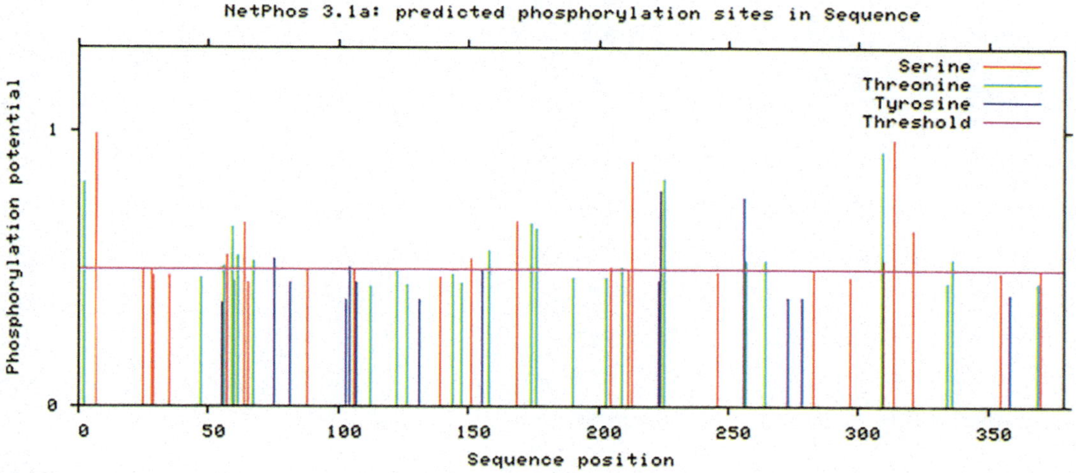

FIGURE 9.17 Prediction of phosphorylation sites of cytochrome B variant A from NetPhos.

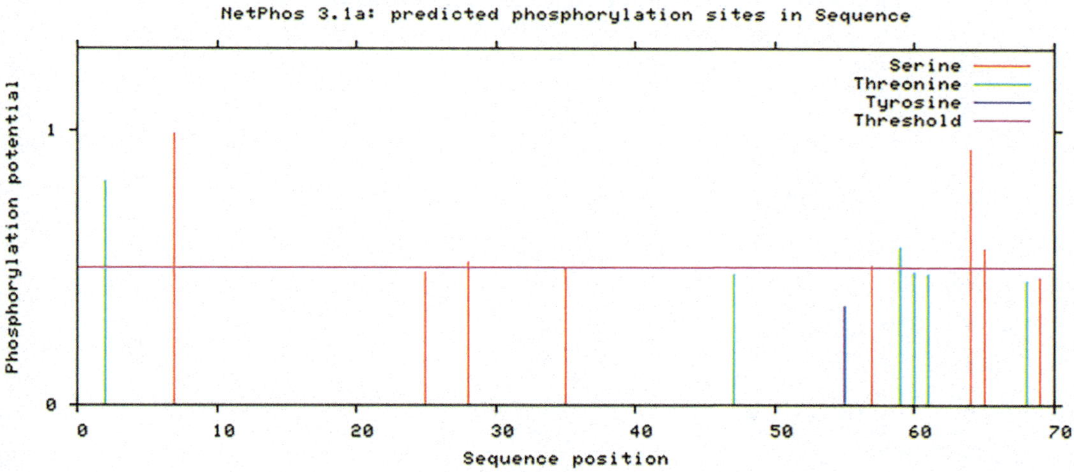

FIGURE 9.18 Prediction of phosphorylation sites of cytochrome B variant B from NetPhos.

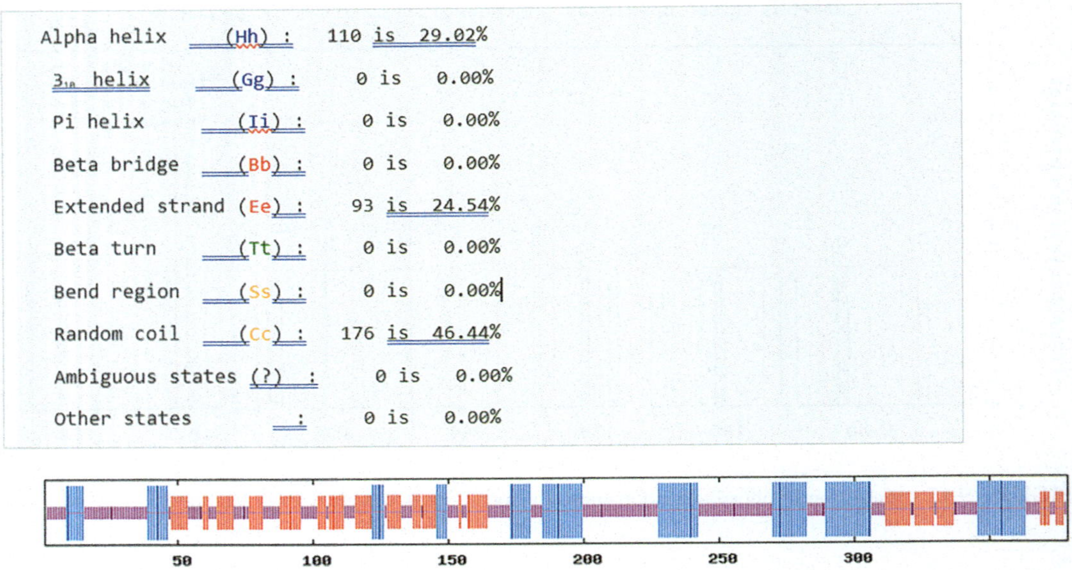

FIGURE 9.19 The secondary structure of cytochrome B variant A using GOR software.

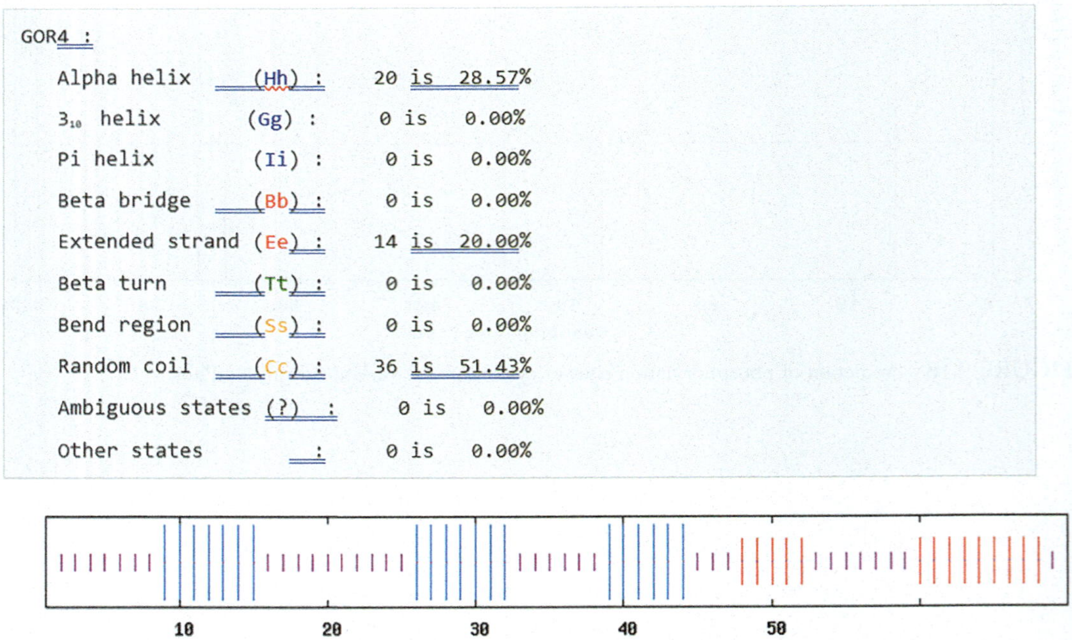

FIGURE 9.20 The secondary structure of cytochrome B variant B using GOR software.

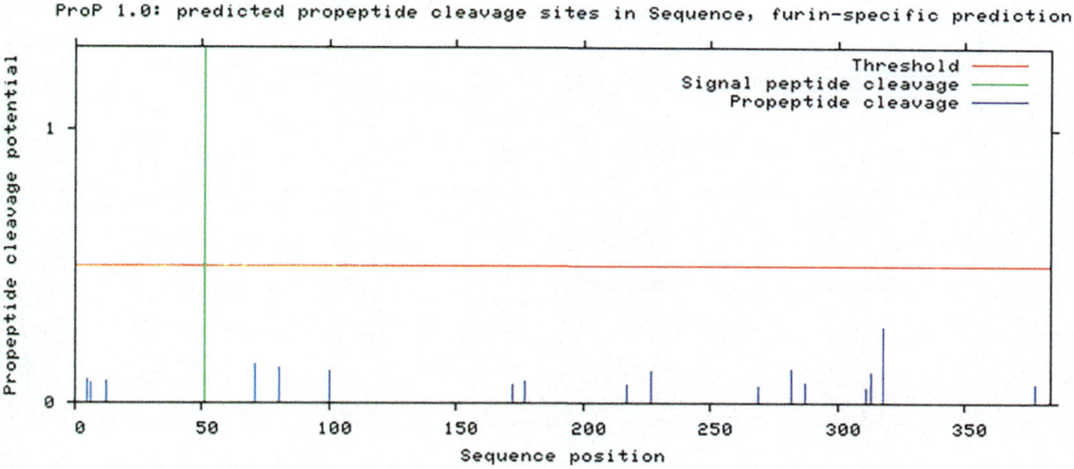

FIGURE 9.21 Prediction of arginine and lysine propeptide cleavage sites in eukaryotic proteins cytochrome B variant A from PropP.

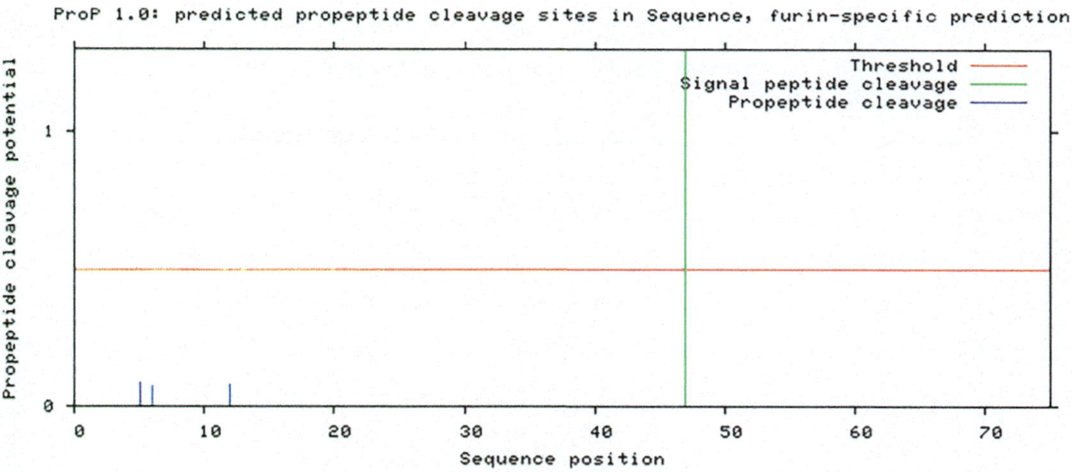

FIGURE 9.22 Prediction of arginine and lysine propeptide cleavage sites in eukaryotic proteins cytochrome B variant B from PropP.

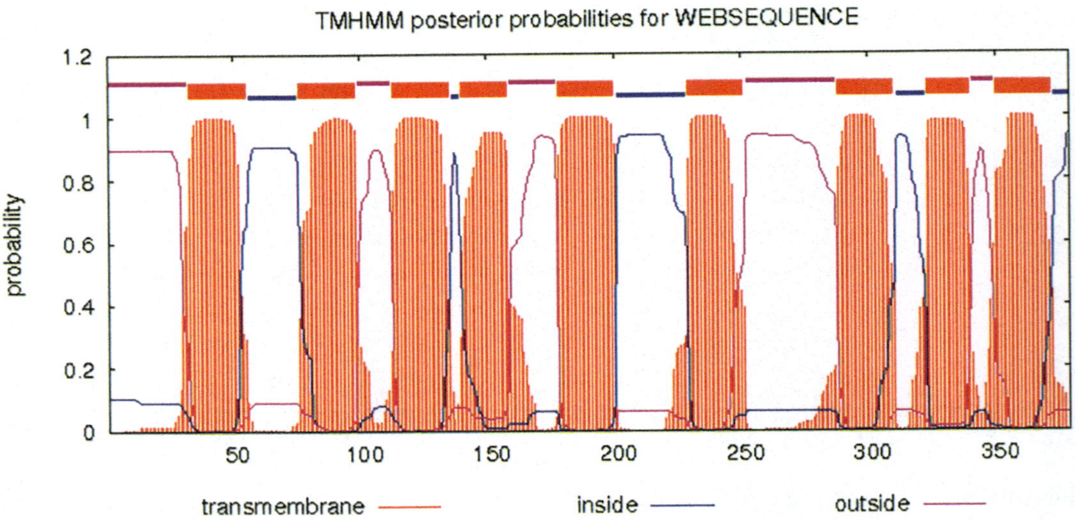

FIGURE 9.23 Determination of transmembrane helices of cytochrome B variant A.

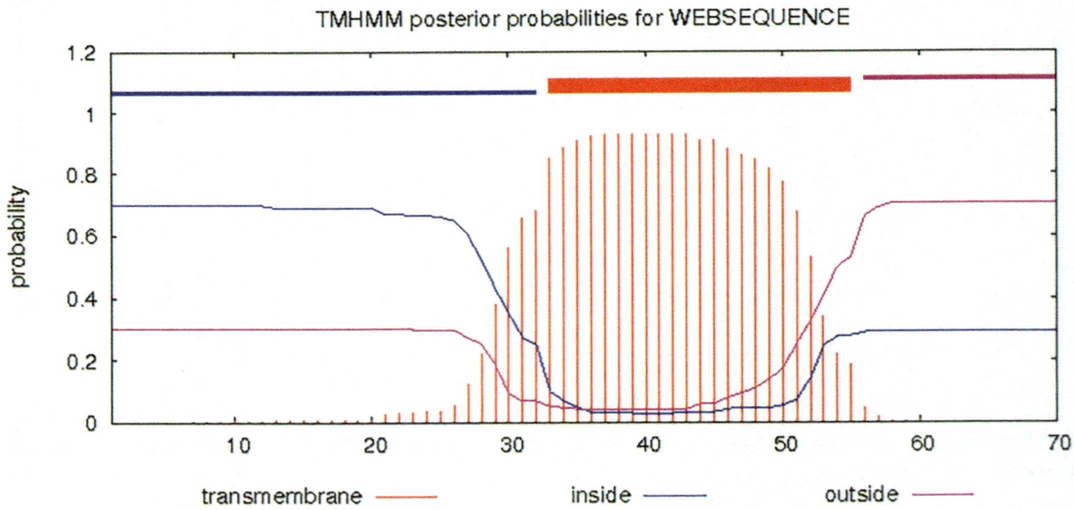

FIGURE 9.24 Determination of transmembrane helices of cytochrome B variant B.

Characterization of identified mitochondrial genes affecting reproduction in sheep

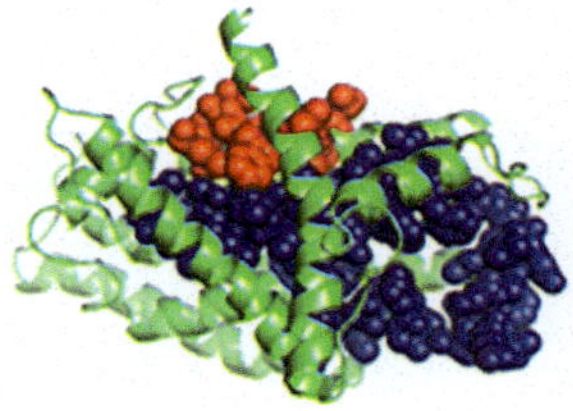

3D structure for NADI, where Red sphere, (87-96aa) represents NADH binding site, while blue sphere (199-247aa) represents FMN binding site.

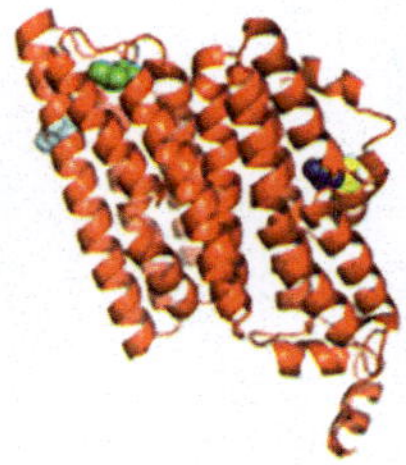

3D structure for ND2 with the sites for mutation at N150D: Green sphere, G259S: Blue sphere, A331S: Yellow sphere, L71P: cyan sphere

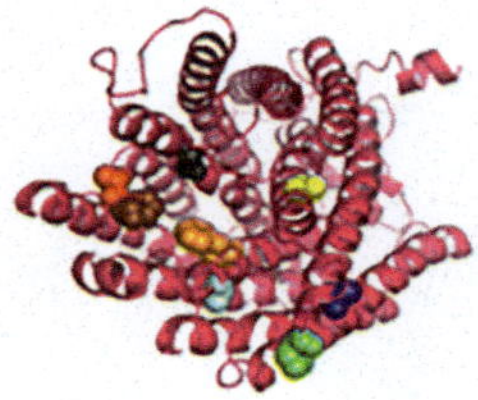

3D structure for CoxI with the binding sites predicted. Calcium binding sites: 40(blue sphere), 45 (green sphere), Iron binding site: 61 (yellow sphere), 376 (cyan sphere), 378 (orange sphere). Copper cation binding site: 240 (deep grey sphere), 290 (red sphere), 291 (chocolate sphere)

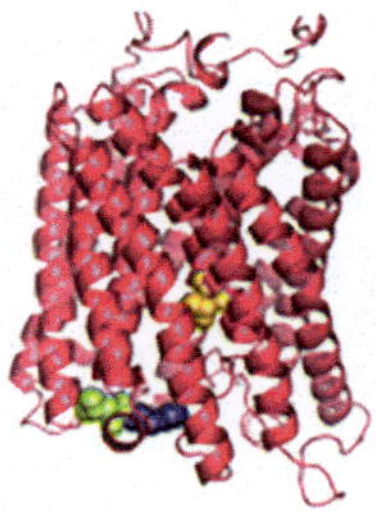

3D structure for CoxI with the binding sites provided. Magnesium binding site: 368 (limon sphere), 369(deep blue sphere)

Oxygen binding site: 244 (bright orange sphere)

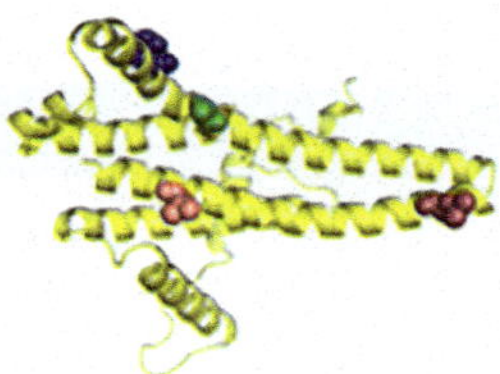

3D structure for ATP6 with the Mutations or SNPs detected in ATP6 resulting into diseases: SNPs detected were L156R (forest green sphere), L156P forest green sphere, I192T (raspberry sphere), S148N (density blue), L217P(deep salmon sphere)

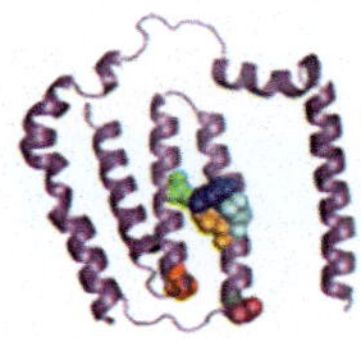

3D structure for ND6 with sites for mutation predicted as I26M (red sphere), G36S green sphere), Y59C blue sphere), L60S yellow sphere), M63V cyan sphere), M64I orange sphere), M64V orange sphere), A72V, grey sphere), A74V warm pink sphere).

FIGURE 9.25 3D structure predicted for sheep mitochondrial genes with important domains identified.

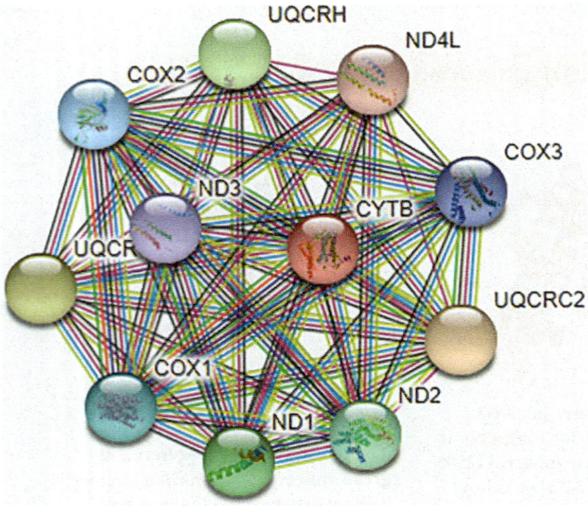

FIGURE 9.26 Molecular interaction among mitochondrial genes (STRING analysis).

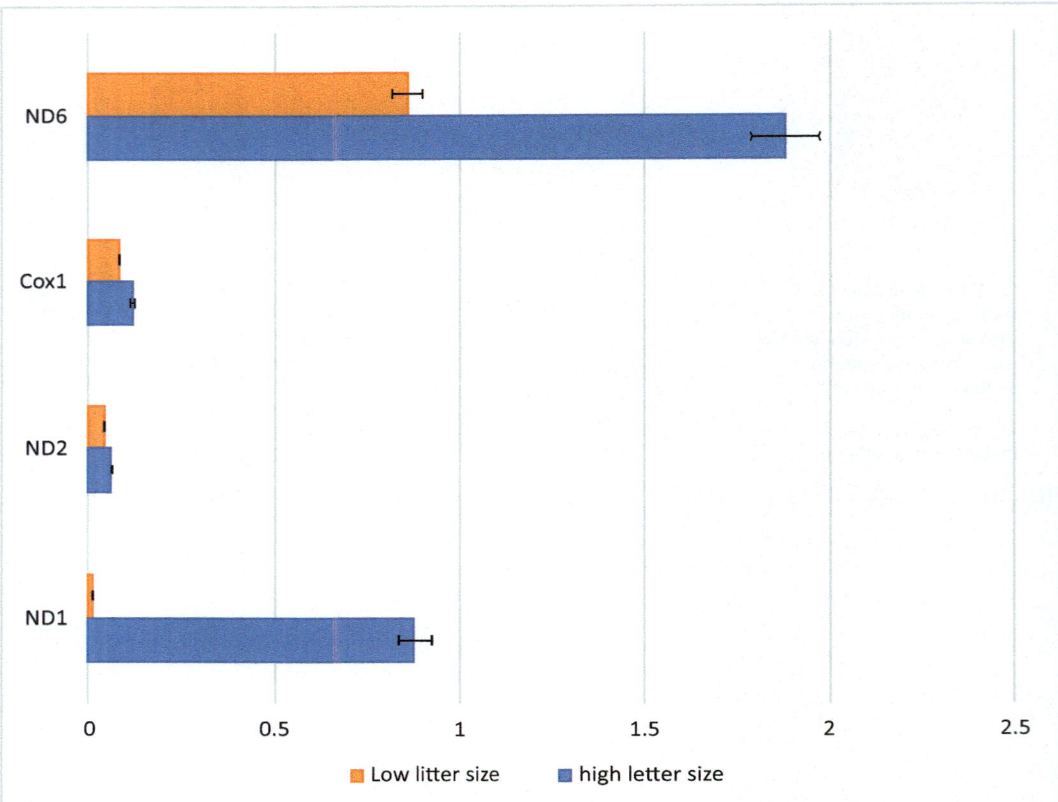

FIGURE 9.27 Differential mRNA expression profile for mitochondrial genes (ND1, ND2, ND6 and Cox1) with respect to average litter size in Garole sheep.

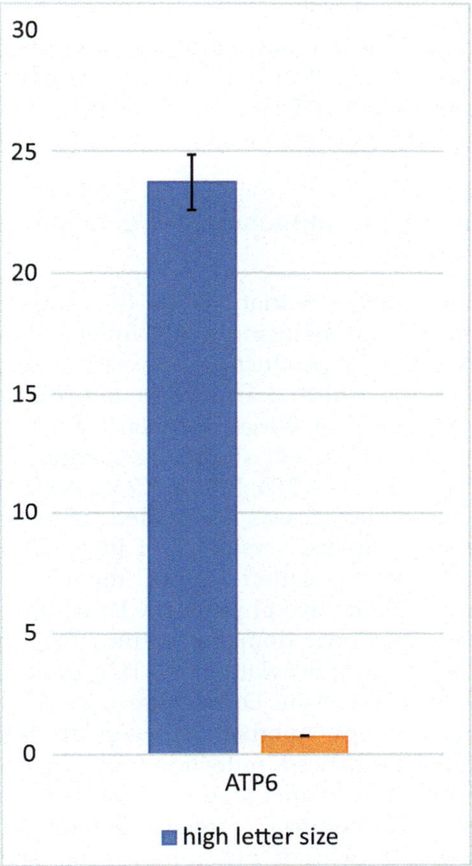

FIGURE 9.28 Differential mRNA expression profile for ATP6 with respect to average litter size in Garole sheep.

PROVEAN result (download)

PROVEAN prediction - Job ID: 294157930594534

Query sequence (fasta); **Supporting sequence set used for prediction.** Number of sequences: **68,360** (fasta, E-values). Number of clusters: **30. Score thresholds for prediction:** (1) default threshold is -2.5, that is: -variants with a score equal to or below -2.5 are considered 'deleterious', -variants with a score above -2.5 are considered 'neutral'. (2) How to use a more stringent threshold.

Variant	PROVEAN score	Prediction (cutoff = -2.5)
F33L	-4.118	Deleterious
L46Q	-3.551	Deleterious
T56P	-4.275	Deleterious

We have detected the mutations in mitochondrial genes initially through PCR amplification followed by Sanger sequencing. In the next step, we have followed next-generation sequencing technique, Whole mitochondrial genome sequencing technique for the detection of mutations in the mitochondrial genome in pigs.

Copy number variations in the mitochondrial genome as detected in the sheep model affecting reproduction

We have detected copy number variations in the mitochondrial genome through differential mRNA expression patterns using qPCR for mitochondrial genes with respect to better-reproducing versus poor reproduction. We have assessed two basic parameters: litter size (number of offspring per birth/parturition) and stillbirth (mortality at birth).

Mitochondria contain 37 genes, of which thirteen encode polypeptides essential for oxidative phosphorylation. In the present study, we identified a significant role for five mitochondrial protein-coding genes—ATP6, NAD1, Cox1, ND2 and ND6—in Garole sheep, where variation within these genes was associated with reproductive performance, measured by higher litter size (triplets) versus lower litter size (singletons). Because litter size is determined by multiple ovulation events, mitochondrial function appears to contribute to the underlying reproductive physiology. Three-dimensional protein modelling enabled us to identify key functional domains within these proteins. All five proteins participate directly in oxidative phosphorylation, the TCA cycle and apoptosis, highlighting their central roles in cellular metabolism. Differential mRNA expression profiling further revealed higher mitochondrial copy number in sheep exhibiting superior growth and reproductive traits, indicating enhanced mitochondrial biogenesis or activity in highly prolific individuals. A key finding of this study is that, beyond their established role in energy metabolism, these mitochondrial proteins exhibit iron- and calcium-binding capabilities, thereby influencing ovarian physiology. Iron and calcium are critical for follicular development, steroidogenesis and ovulation, suggesting that mitochondrial variants affecting mineral metabolism may modulate ovarian function, ultimately contributing to differences in litter size. The candidate genes identified in this study represent promising targets for genomic selection, marker-assisted breeding, or even gene-editing strategies aimed at improving reproductive efficiency and enhancing meat production in sheep. Moreover, Garole sheep, with their natural variation in mitochondrial gene function and reproductive output, may serve as a valuable animal model for studying human reproductive disorders linked to mitochondrial dysfunction (see Fig. 9.25), (Fig. 9.26 (Deb, 2026), Fig. 9.27, Fig. 9.28).

Interpretation: a Significant role of five mitochondrial protein-coding genes—ATP6, NAD1, Cox1, ND2 and ND6 was identified, where variation within these genes was associated with reproductive performance. All five proteins participate directly in oxidative phosphorylation, the TCA cycle and apoptosis, highlighting their central roles in cellular metabolism. Differential mRNA expression profiling revealed higher mitochondrial copy number in sheep exhibiting superior reproductive traits, indicating enhanced mitochondrial biogenesis or activity in highly prolific individuals. These mitochondrial proteins exhibit iron- and calcium-binding capabilities, thereby influencing ovarian physiology since iron and calcium are critical for follicular development, steroidogenesis and ovulation. A higher mitochondrial copy number indicates better reproductive traits in sheep. It is a novel and first report in the sense that these mitochondrial genes are being reported to influence reproduction in the sheep model.

Copy number variations in the mitochondrial genome as detected in the pig model affecting reproduction

We have detected copy number variations in the mitochondrial genome through differential mRNA expression patterns through qPCR for mitochondrial genes with respect to better reproducing versus poor reproduction. We have assessed two basic parameters as litter size (number of offspring per birth/parturition) and stillbirth (mortality at birth) in the Ghungroo pig, maintained in our university farm.

We have studied the mitochondrial genes, such as, ATP6, Cox1, cytochrome B, ND4, for copy number variations to have an important role in controlling litter size (number of offspring during birth) and stillbirth (dead offspring at birth).

Characterisation of other nuclear origin genes involved in mitochondrial function (replication and transcription) in association to reproduction in the sheep model: case study

Certain proteins encoded by nuclear genes serve as a basic function for mitochondrial structural and functional roles, namely cytochrome C, POLG, MPV17, and TFAM, whose basic functions have been described in detail in Chapter 2.

We have observed a significant role for these genes controlling reproduction in Garole sheep. A better mRNA expression profile was observed for sheep with more litter size and no stillbirth was observed.

MPV17

The MPV17 gene encodes an inner mitochondrial membrane protein that is critically involved in regulating mitochondrial deoxynucleotide pool homoeostasis and maintaining mtDNA integrity. This protein is essential for preserving mtDNA stability and ensuring proper mitochondrial genome replication and overall mitochondrial homoeostasis.

TFAM

This gene encodes TFAM, a central regulator of mitochondrial gene expression and a critical contributor to mtDNA replication. TFAM binds to mitochondrial promoter regions to initiate transcription of the mitochondrial genome. Studies in murine models have shown that TFAM is essential for regulating mtDNA copy number and is indispensable for embryonic development. Targeted ablation of TFAM expression in cardiac and skeletal muscle tissues results in a mouse model that recapitulates key features of Kearns–Sayre syndrome.

TFAM is a high-mobility group (HMG) DNA-binding protein containing two HMG-box domains that facilitate DNA binding and bending. This DNA-bending activity is crucial for transcription initiation in mammalian mitochondria, in contrast to yeast mitochondria, where the homologous protein Abf2 does not require such bending. In addition, TFAM is thought to play a role in mitochondrial genome packaging, as its DNA-binding activity is largely nonsequence-specific.

Cytochrome C

Cytochrome C is a small, soluble, haem-containing protein encoded by nuclear DNA that plays a pivotal role in mitochondrial oxidative phosphorylation. It functions as an essential electron carrier within the mitochondrial electron transport chain, facilitating electron transfer between complex III (cytochrome bc_1 complex) and complex IV (cytochrome C oxidase), thereby contributing to efficient ATP production.

It is localised to the mitochondrial intermembrane space, where it functions as a mobile electron carrier, transferring electrons between complex III (cytochrome bc_1 complex) and complex IV (cytochrome C oxidase) of the electron transport chain, thereby supporting ATP generation.

Structurally, cytochrome C is composed of approximately 100–105 amino acids and contains a covalently bound haem prosthetic group with a central iron atom capable of reversible oxidation–reduction. This redox flexibility underlies its efficiency in electron transfer reactions. The protein is peripherally associated with the outer surface of the inner mitochondrial membrane through electrostatic interactions. Cytochrome C is highly conserved among eukaryotes, reflecting its critical and evolutionarily conserved function in cellular energy metabolism.

POLG

DNA polymerase subunit gamma (POLG or POLG1) is a nuclear-encoded enzyme responsible for mtDNA replication and repair in humans. mtDNA polymerase functions as a heterotrimeric complex composed of a catalytic subunit and a homodimer of accessory subunits. The POLG gene encodes the catalytic subunit, which possesses DNA polymerase, 3′–5′ exonuclease and 5′-deoxyribose phosphate lyase activities essential for mtDNA maintenance. Pathogenic variants in POLG are associated with a spectrum of mitochondrial disorders, including progressive external ophthalmoplegia with mtDNA deletions (PEOA1), sensory ataxic neuropathy, dysarthria, and ophthalmoparesis, Alpers–Huttenlocher syndrome and mitochondrial neurogastrointestinal encephalopathy (see Fig. 9.29).

See Fig. 9.30.

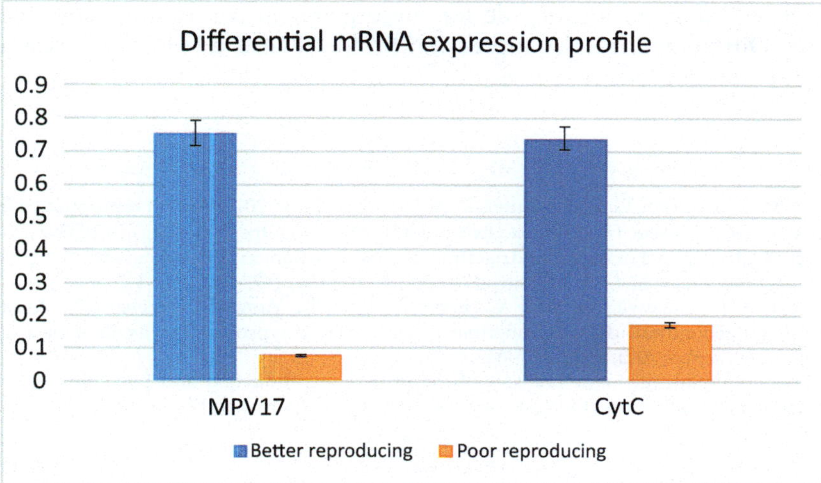

FIGURE 9.29 Differential mRNA expression profiling for MPV17 and cytochrome C gene with respect to high and low reproducing sheep.

Characterisation of other nuclear origin genes involved in mitochondrial function (replication and transcription) in association to reproduction in pig model: case study

Certain proteins encoded by nuclear genes serve basic functions in mitochondrial structural and functional roles, namely cytochrome C, POLG, TWINKLE, MPV17, TFAM, whose basic functions have been described in detail in Chapter 2.

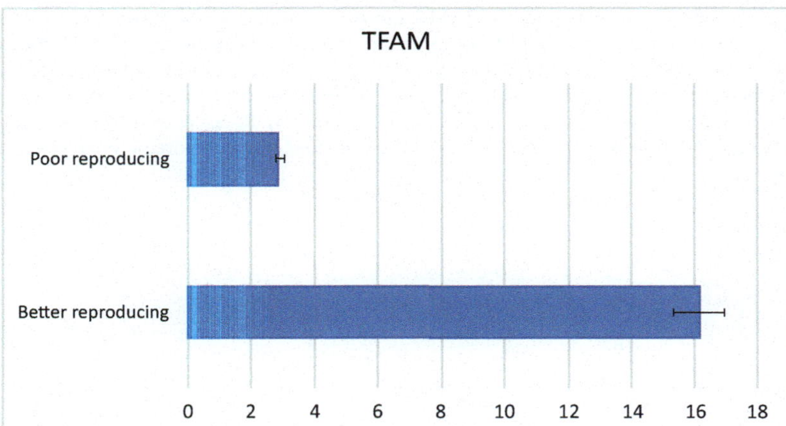

FIGURE 9.30 Differential mRNA expression profiling for the TFAM gene with respect to high and low reproducing sheep.

We have observed a significant role for these genes in controlling reproduction in the Ghungroo pig. Differential mRNA expression profile reveals a better expression profile for these genes in better-reproducing pigs.

References

Craven, L., Tang, M.-X., Gorman, G. S., De Sutter, P., & Heindryckx, B. (2017). Novel reproductive technologies to prevent mitochondrial disease. *Hum Reprod Update, 23*(5), 501–519. https://doi.org/10.1093/humupd/dmx018.

Schatten, et al. (2014). The impact of mitochondrial function/dysfunction on IVF and new treatment possibilities for infertility. *Reprod Biol Endocrinol, 12*, 111. https://doi.org/10.1186/1477-7827-12-111.

Yabuuchi, A., Beyhan, Z., Kagawa, N., Mori, C., Ezoe, K., Kato, K., Aono, F., Takehara, Y., & Kato, O. (2012). Prevention of mitochondrial disease inheritance by assisted reproductive technologies: Prospects and challenges. *Biochim Biophys Acta, 1820*, 637–642.

Deb, D. 2026. Phenotypic and genomic analysis of the growth, reproduction and parasitic resistance in Garole sheep, PhD Thesis to be submitted to West Bengal University of Animal and Fishery Sciences, Kolkata-37. West Bengal, India.

Further Reading

Deb, D., Pal, A., Rahman, S., Samanta, R. 2025. Whole mitochondrial genome sequencing reveal certain mutations responsible for reproductive disorders in sheep model. In Proceedings on National symposium on "Reinforcing BioSecurity: Paving the way for a safer One Health Environment". Jan, 2025. venue: West Bengal University of animal and fishery sciences, 37, K.B. Sarani, Kolkata-37. West Bengal, India.

Martikainen, M. H., Grady, J. P., Ng, Y. S., Alston, C. L., Gorman, G. S., Taylor, R. W., McFarland, R., & Turnbull, D. M. (2017). Decreased male reproductive success in association with mitochondrial dysfunction. *Eur J Hum Genet, 25*, 1162–1164. https://doi.org/10.1038/ejhg.2017.114.

Pradhan, M., Pal, A., Samanta, A. K., Banerjee, S., & Samanta, R. (2018). Mutations in ***cytochrome B gene*** effects female reproduction of Ghungroo pig. *Theriogenology, 119*, 121–130. https://doi.org/10.1016/j.theriogenology.2018.05.015.

Rahaman, S. 2026. Phenotypic and molecular characterization and differential mRNA expression profiling of genes responsible for growth and reproduction in Ghungroo pig. PhD Thesis to be submitted to West Bengal University of Animal and Fishery Sciences, Kolkata-37. West Bengal, India.

Rahman, S., Pal, A., Deb, D., Samanta, R. 2025. Mutations in Mitochondrial genes have role in reproductive disorders in pig model- a study through Whole mitochondrial genome sequencing. In Proceedings on National symposium on "Reinforcing BioSecurity: Paving the way for a safer One Health Environment". Jan, 2025. venue: West Bengal University of animal and fishery sciences, 37, K.B. Sarani, Kolkata-37. West Bengal, India.

Zhong, Z., Hao, Y., Li, R., Spate, L., Wax, D., Sun, Q.-Y., Prather, R. S., & Schatten, H. (2008). Analysis of heterogeneous mitochondria distribution in somatic cell nuclear transfer porcine embryos. *Microsc Microanal. 14*(5), 418–432.

CHAPTER

10

Mitogenomics and neurodegenerative disorder

Introduction

Mitochondrial dysfunction is a central feature in the pathogenesis of various neurodegenerative and neurological disorders, including Huntington's disease (HD), Parkinson's disease (PD), amyotrophic lateral sclerosis (ALS), epilepsy, schizophrenia, multiple sclerosis, neuropathic pain and Alzheimer's disease (AD). It is increasingly recognised as a fundamental contributor to the molecular and cellular mechanisms underlying these conditions. Apart from various other causal factors, mitochondrial disorder is one important factor.

It is true for all the conditions that arise due to mitochondrial dysfunction. There may be mutations in mitochondrial DNA (mtDNA), copy number variations, mitochondrial fusion and fission. Defects may also occur in basic processes for mtDNA replication or transcription, translation process. These may lead to defects in the electron transport chain (ETC)/oxidative phosphorylation (OXPHOS) and defects in mitophagy.

Mitochondrial involvement in neurodegenerative diseases

The sequence of events leading to neurodegeneration is complex. Diverse neurodegenerative diseases represent distinct symptoms and impact certain regions of the brain. Mitochondrial dysfunctions are seen as interconnected characteristics, serving as a convergence point for several disease processes. Mitochondria are cytoplasmic organelles in eukaryotic cells that primarily supply cellular energy. Moreover, they are essential regulators of cellular apoptosis and a significant characteristic of neurodegeneration, and they perform vital functions such as involvement in cellular processes, signalling pathways, calcium homoeostasis, cell cycle regulation, apoptosis, reactive oxygen species (ROS) generation and thermogenesis. Mitochondrial malfunction, elevated ROS generation, and oxidative damage contribute to several neurodegenerative illnesses. Apoptosis and excitotoxicity are the two primary causes of neuronal cell death, with mitochondria playing a critical role in both instances. Elevated ROS generation during neurodegeneration may influence mitochondrial metrics, ATP synthesis, membrane potential, permeability transition pore (MPTP) activation and calcium absorption.

Mitochondrial Genomics
DOI: https://doi.org/10.1016/B978-0-443-26655-3.00008-1

These alterations may cause neuronal injury. The initial evidence of mitochondrial participation in the pathophysiology of neurodegenerative processes was documented upon the identification of complex I deficiency in the substantia nigra and platelet mitochondria of patients with PD. Additional compelling evidence was discovered about ETC deficiencies: complex I and cytochrome c oxidase (complex IV, COX) in AD and complexes II and III in HD. Biochemical examination of postmortem AD brains revealed dysfunction in the citric acid cycle enzymes, specifically pyruvate dehydrogenase, α-ketoglutarate dehydrogenase and isocitrate dehydrogenase. These alterations were associated with the clinical condition, and enzyme activity may be linked to reduced cerebral metabolism.

Compromised mitochondrial dynamics

Mitochondria are highly dynamic organelles capable of adopting a range of morphologies, from elongated tubular networks to discrete spherical structures, through rapid and reversible fission and fusion events. Fusion is facilitated by large GTPase proteins, including ocular atrophy factor 1 (OPA1), which governs inner membrane fusion, and mitofusin 1 (Mfn1) and mitofusin 2 (Mfn2), which are responsible for outer membrane fusion. Fusion facilitates the appropriate distribution of mitochondrial constituents, including lipid membranes, OXPHOS complexes, and mtDNA. Fission is crucial for the correct assembly of mitochondrial ETC complexes, facilitated by dynamin-related protein-1 (Drp1, GTPase), human fission protein-1 (Fis1), mitochondrial fission factor and mitochondrial dynamics proteins (MiD49/MiD51). Changes in the expression of mitochondrial fusion-fission proteins can lead to modified mitochondrial distribution. The failure of mitochondria may come from an imbalance in the dynamics of mitochondrial fission and fusion, which in AD is significantly skewed towards fission, perhaps leading to malfunctioning mitochondria in impaired neurones. Immunoblot research revealed that amyloid precursor protein (APP) expression influenced mitochondrial fusion/fission proteins; Drp1, OPA1, Mfn1 and Mfn2 were diminished, whereas Fis1 was markedly elevated in AD. In a mouse model of AD, mitochondrial dynamics were compromised; there was a reduction in mitochondrial anterograde transport, an increase in mitochondrial fission, a decrease in fusion and reduced mitochondrial activities were noted. In human fibroblasts derived from patients with sporadic AD, mitochondrial distribution is characterised by an accumulation of elongated mitochondria predominantly localised to the perinuclear region. This investigation indicated that heightened oxidative stress and augmented amyloid-β (Aβ) generation may contribute to the decrease of Drp1. Cells with the P301L tau mutation (SY5Y cells overexpressing P301L tau protein) exhibited a deficiency in complex I and reduced ATP levels. Phosphorylated tau (pTau) and Aβ induce heightened nitrosylation of Drp1 protein, resulting in augmented mitochondrial fission and neurodegeneration. Cells lacking mitochondrial fusion exhibited a loss of mitochondrial membrane potential ($\Delta\psi m$) and decreased mitochondrial respiration. Notably, diminished OPA1 was demonstrated to provoke spontaneous cytochrome c (cyt *c*) release and to expedite cyt *c* release in response to apoptotic stimuli. In summary, the reports indicated heightened mitochondrial fission and diminished fusion, augmented interaction between Aβ and pTau with the mitochondrial fission protein Drp1, likely resulting in enhanced mitochondrial fragmentation, compromised axonal transport of mitochondria and synaptic degeneration in neurons impacted by AD.

AD, PD and HD are linked to the deposition of amyloid fibrils. Soluble oligomers of amyloid proteins can permeabilize cellular membranes and lipid bilayers, thereby disrupting membrane activities; nevertheless, the mechanism of this disruption remains poorly understood. They are capable of being inserted. Influence membranes, alter dielectric characteristics, disturb ion gradients and/or inactivate functional proteins. Amyloid oligomers enhanced conductance in a conformation-specific manner; this effect is contingent upon oligomer concentration and can be reversed by antioligomer antibodies. In HD, mutant huntingtin (Htt) interacts with Drp1 and associated GTPases, resulting in increased mitochondrial fragmentation and aberrant mitochondrial distribution. Modified mitochondrial morphogenesis, enhanced mitochondrial fission, diminished fusion and mitochondrial depletion are associated with neuronal dysfunction and cell death. Aberrant mitochondrial dynamics lead to the dysfunction of the ETC complex. In PD, parkin interacts with alpha-synuclein and contributes to the pathogenesis. The hereditary form of PD is associated with the phosphatase and tensin homolog (PTEN)-induced putative kinase 1 (PINK1) and parkin genes, which are crucial for mitochondrial integrity. It has been proposed to enhance mitochondrial fission and to impede fusion. PINK and parkin likely modulate mitochondrial dynamics and facilitate the degradation of impaired mitochondria.

Mitochondrial reactive oxygen species and apoptosis

The disparity between the cellular generation of ROS and the capacity of cells to effectively counteract them is termed 'oxidative stress'. This phenomenon is associated with neurodegenerative disorders and the ageing process; it can lead to cellular damage resulting in necrotic or apoptotic cell death, as ROS oxidise essential cellular constituents, including lipids, proteins and nucleic acids. Impaired OXPHOS function may lead to disruptions in energy metabolism, commonly observed in AD. Impaired energy metabolism leads to a diminished respiratory control ratio and reduced ATP levels. Numerous potential reasons exist for decreased oxidation and ATP production rates that do not involve a deficiency in respiratory chain enzymes. ROS play a significant role in intracellular signalling and the regulation of signal transduction. ROS appear to be pivotal in the processes of brain ageing, and impaired mitochondrial respiration, coupled with elevated ROS generation, substantially influences functional alterations in the brain during ageing. Complex I and complex III are regarded as the principal sources of ROS in the brain under both normal and pathological situations, such as neurodegenerative diseases. Complex I primarily generates and releases superoxide ($O_2^{\cdot -}$) into the mitochondrial matrix, whereas complex III is capable of releasing superoxide to both the matrix and the intermembrane space sides of the inner mitochondrial membrane. Through superoxide dismutase, $O_2^{\cdot -}$ can be transformed into hydrogen peroxide (H_2O_2), which diffuses through membranes and can serve as a source of extremely reactive hydroxyl radicals. The production of physiologically generated H_2O_2 and O_2 from the ETC is contingent upon the magnitude of the proton-motive force (Δp) and the respiratory state of the mitochondria. State 4 is characterised by a high rate of ROS; conversely, states 3 and 5 generate minimal ROS. Both impaired generation and detoxification of ROS contribute to the pathophysiological consequences of mitochondrial dysfunctions. Defective mitochondria emit substantial quantities of ROS; likewise, a reduction in antioxidative enzyme activity, as observed in the elderly, exacerbates ROS production. Adverse effects of ROS can impair the respiratory chain, with complexes I, III and COX being the most impacted, while the function of complex II remains unaffected.

Mitochondria are critical in the intrinsic mechanism of apoptosis

During apoptosis, the mitochondrial network disintegrates, and the outer mitochondrial membrane becomes permeable, resulting in the release of various apoptotic proteins, including cyt c. There are interconnected mitochondrial pathways that promote cell death: (1) the opening of mitochondrial permeability transition pores (MPTPs) can induce mitochondrial swelling and cell death via apoptosis or necrosis; (2) increased permeability of the mitochondrial membrane leads to the leakage of apoptotic factors, such as second mitochondria-derived activator of caspases and cyt c, which activate the caspase cascade, culminating in apoptosis; and (3) the release of the caspase-independent death effector, apoptosis-inducing factor (AIF), initiates chromatin condensation and DNA degradation. Mitochondria experience fragmentation during apoptosis prior to the activation of caspases. In apoptotic cells, the fast decline of the inner mitochondrial membrane potential ($\Delta\psi m$) is correlated with the generation of ROS. Recent focus has been directed on the ROS-induced impairment of ETC complexes, facilitated by the peroxidation and oxidative damage of cardiolipin. Membrane lipids, primarily cardiolipin, are essential for the stability of respiratory supercomplexes and function as a diffusion microdomain for ubiquinone. Cardiolipin actively participates in mitochondrial-mediated apoptosis, can undergo oxidation, and interacts with cyt c and B-cell lymphoma 2 (Bcl-2) proteins. In AD, membrane-associated oxidative stress, elevated free radical generation and disrupted Ca^{2+} homoeostasis have been documented. Increased mitochondrial membrane permeability and cyt c release, triggered by the oligomerization and polymerisation of Aβ and α-synuclein, promote the opening of the MPTP, ultimately leading to the activation of apoptotic pathways. Furthermore, COX activity is diminished, and neurones display mitochondrial impairment and death. The aetiology of mitochondrial changes in AD remains unexplored. Fig. 10.1 illustrates the mechanisms of mitochondrial dysfunction in AD.

Major neurological and mitochondria-related diseases and mitochondria-related neuropathies

Major neurological and mitochondria-related diseases and mitochondria-related neuropathies (see Tables 10.1–10.3).

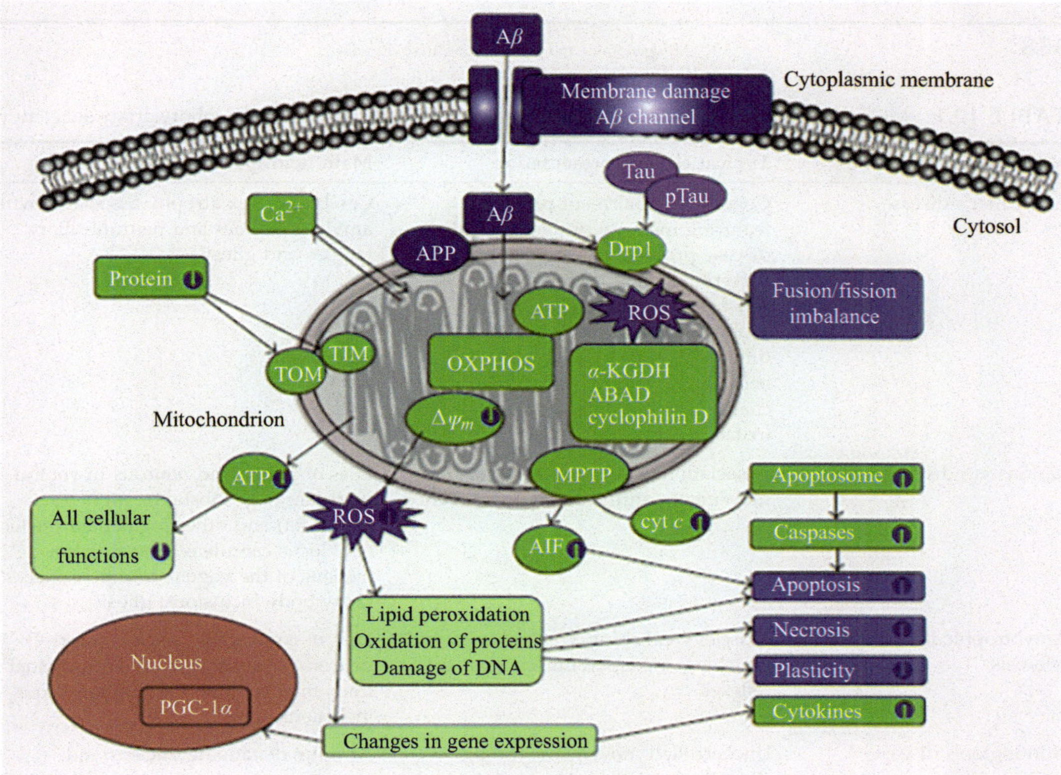

FIGURE 10.1 Mitochondrial dysfunctions in Alzheimer's disease. Amyloid-beta (Aβ) compromises cytoplasmic membrane integrity and induces mitochondrial dysfunction. Aβ inhibits the oxidative phosphorylation (OXPHOS) pathway, leading to reduced ATP generation and elevated reactive oxygen species (ROS) creation. Reduced ATP synthesis results in the dysfunction of ATP-dependent activities, affecting all cellular functions.

Mitochondrial abnormalities lead to neurodegenerative diseases

Mitochondrial dysfunction is prevalent in numerous neurodegenerative disorders, such as PD, AD, HD, Friedreich's ataxia (FRDA) and ALS. AD is a commonly occurring type of dementia related to the ageing process. The characteristic features of hAβ aggregation and tau pathologies are associated with mitochondrial dysfunctions in neurones. Aβ is produced via the cleavage of the Aβ precursor protein by α-, β-, and γ-secretases. Elevated oxidative stress due to mitochondrial malfunction produces the lipid peroxidation byproduct 4-hydroxynonenal, which covalently alters the γ-secretase complex, resulting in enhanced secretase activity. Elevated γ-secretase activity results in enhanced Aβ buildup.

Mitochondrial DNA in Alzheimer's disease

Alterations in mtDNA are primarily accountable for the ageing of phenotypes. Deficiencies in mtDNA have also been observed in non-AD older individuals; numerous

TABLE 10.1 Main clinical syndromes reported to be associated with oxidative phosphorylation deficiency.

Neurological disorder	Typical clinical presentation	Main neuropathological features
Alzheimer's disease	Cognitive impairment primarily featuring memory problems; as the disease progresses, language, perceptual skills, attention, constructive abilities, orientation, problem solving and functional ability difficulties also arise, as well as behavioural and neuropsychiatric changes, including wandering, irritability and labile affect	Cerebral cortex atrophy associated with amyloid plaques and neurofibrillary tangles and gliosis
Parkinson's disease	Association of shaking, slowness of movements, stiffness and poor balance	Loss of pigmented neurons in ventral midbrain (e.g. substantia nigra pars compacta) and other pigmented nuclei (e.g. locus caeruleus, dorsal motor nucleus of the vagus); intraneuronal Lewy body inclusions; gliosis
Amyotrophic lateral sclerosis	Muscles weakening, wasting away and twitching; increased muscle tone, brisk reflexes	Loss of cortical and spinal motor neurons; degeneration of corticospinal tract; multiple forms of proteinaceous inclusions; gliosis
Huntington's disease	Uncontrolled movements (called chorea), abnormal body postures and changes in behaviour, emotion, judgement and cognition	Atrophy of caudate nucleus and putamen, accompanied by mild frontal and temporal atrophy; loss of medium-size spiny neurons in the striatopallidal and striatonigral pathways associated with striatal gliosis
Charcot–Marie–Tooth disease	Progressive muscle wasting and weakness; sensory loss	Loss of large myelinated motor and sensory fibres in peripheral nerves

tissues exhibit diminished respiratory efficiency and reduced COX activity. In AD, mtDNA in the brain exhibits greater oxidative damage than that attributable to ageing, perhaps resulting in heightened mutations and deletions, as well as postgenomic issues with transcriptional regulation. Variations in the expression of mitochondrial and nuclear genes that encode components of COX and complex I enzymes lead to modifications in oxidative metabolism in AD. Downregulation of mitochondrial genes in complex I was seen in both early and definitive AD brain tissues. Research indicated diminished complex I activity in the brains of AD patients, and the gene expression of the ND4 subunit of complex I was observed to be reduced in the temporal cortex of individuals with AD. The differential expression of mitochondrial genes encoding complex I, COX and complex V was assessed in AD brains.

mtDNA may not have a role in the aetiology of AD, but its role starts thereafter. The elevated expression of COX genes may result from heightened oxidative damage and early

TABLE 10.2 Other suspected mitochondria-related diseases.

Neurological disorder	Typical clinical presentation	Main neuropathological features
Leigh's syndrome	Progressive loss of mental and movement abilities (psychomotor regression) with high serum lactic acid levels	Bilateral symmetrical lesions in the brainstem and basal ganglia with gliosis, vacuolation, capillary proliferation and relative neuronal preservation
Mitochondrial encephalomyopathy, lactic acidosis and stroke-like episodes	Episodes of muscle weakness on one side of the body (hemiparesis), altered consciousness, vision abnormalities, seizures and severe headaches; can progress to vision loss, problems with movement and dementia	Multifocal infarct-like lesions in the posterior cortex; recurrent stroke-like episodes; lactic acidosis and ragged-red fibres
Kearns–Sayre syndrome	Progressive external ophthalmoplegia, retinitis pigmentosa; common additional features include deafness, cerebellar ataxia and heart block	Loss of neurons and gliosis of the basal ganglia; spongy change of the white matter in the cerebrum, brainstem and cerebellum; retinal degeneration; muscle ragged-red fibres
Optic atrophy	Bilateral visual loss; central vision affected prior to peripheral vision	Degeneration of the retinal ganglion cell bodies and axonal pathways up to the lateral geniculate nuclei
Friedreich's ataxia	Incoordination of gait and often of hands; other features include gradual loss of strength and sensation in the arms and legs, muscle stiffness (spasticity) and impaired speech, hearing and vision; often the heart is also affected	Thinning of dorsal roots, degeneration of dorsal columns, atrophy of the neurons in Clarke's column and dorsal spinocerebellar fibres, atrophy of gracile and cuneate nuclei and neuropathy of sensory nerves; lesions of the dentate nucleus and the corticospinal tracts are also observed
Spinocerebellar ataxia	Incoordination of gait and often of hands, speech and eye movements	Degeneration of the spinal cord and the cerebellum, as well as many nuclei of the basal ganglia and the brainstem
Primary coenzyme Q10, deficiency syndrome	Syndrome affecting the brain (incoordination and poor balance), muscles (weakness) and kidneys (nephrotic syndrome); other neurological abnormalities include seizures, intellectual disability, hypotonia, dystonia, spasticity, nystagmus, vision loss and deafness	Multisystem neurodegeneration with gliosis
Progressive encephalopathy associated with cytochrome c oxidase (COX) deficiency	Progressive uncoordinated gait, dysarthria, lower limb areflexia, deafness and high serum lactic acid levels, followed by dementia	Cortical atrophy and basal ganglia calcifications as well as severe mitochondrial myopathy with numerous COX-negative ragged-red fibres

(*Continued*)

TABLE 10.2 (Continued)

Neurological disorder	Typical clinical presentation	Main neuropathological features
Hereditary spastic paraplegia	Difficulty in walking and poor balance followed by increased muscle tone, brisk reflexes, muscle weakness, bladder disturbances and paraesthesia	Degeneration primarily in the corticospinal tracts and the fasciculus gracilis, and to a lesser extent in the spinocerebellar tract

TABLE 10.3 Mitochondria-related neuropathies.

Neurological disorder	Typical clinical presentation	Main neuropathological features
Chemotherapy-induced peripheral neuropathy	Burning/shooting pain, tingling, cramping and weakness in hands and feet; impaired balance and walking, falls; loss of heat sensitivity	Peripheral nerve neuronopathy and/or myelinopathy; axonopathy, particularly intraepidermal nerve fibre degeneration; there are differences in the pathobiology depending on the drug used
Diabetic neuropathy	Sensory impairment with loss of sensory function or spontaneous feeling of touch, vibration, pricking and hot and cold pain; bilateral and symmetric damage to nerves of the feet and hands	Retrograde neurodegenerative disease of the peripheral nervous system; damage to the small sensory nerve fibres accompanied by continuous and episodic pain
HIV-induced neuropathy	Debilitating chronic neuropathic pain that is constant and severe; bilateral pain on the soles; the dysesthesias ascend to lower extremities and may involve fingertips	Retrograde axonal degeneration of long axons in distal regions of legs or arms, loss of unmyelinated fibres and macrophage infiltration in peripheral nerves and dorsal root ganglia

modifications in mitochondrial function inside surviving neurones. The expression levels of the mitochondrially encoded COX I and the nuclear-encoded subunit IV (COX IV) were examined in the hippocampal tissue of individuals with AD. The level of mitochondrial-encoded COX IV is associated with the quantity of hyperpTau protein deposited in certain hippocampus regions, but not with the amount of accumulated Aβ. A separate investigation identified a pattern of amyloid plaques that is unique from COX-deficient neurones in the hippocampus. Moreover, COX-defective mice demonstrated a substantial decrease in amyloid plaques, alongside a drop in β-secretase, Aβ-42, and oxidative damage. The expression of mitochondrial- and nuclear-encoded genes responsible for subunits of COX and complex I was examined in discrete brain regions of individuals with AD and age-matched controls. Alterations in the transcript ratios of COX subunits, specifically COX II (mitochondrial-encoded) and COX IV (nuclear-encoded), were observed in the brains of AD patients, indicating dysregulation of COX subunit expression. Alterations in the quantities of these subunits may contribute to the kinetic perturbations observed in COX in AD. A reduction in ND4 and ND15 mRNAs, which encode subunits of complex I, was noted in the hippocampus and inferior parietal lobule of AD patients, but not in the cerebellum. Alterations in genes encoding components of complex I and COX may influence oxidative metabolism in AD. The imbalance between fusion and fission is associated with modified mtDNA; mitochondrial fusion facilitates the exchange of mitochondrial components, including mtDNA. The deletion of Mfn2 inhibited fusion, resulting in a predominance of mitochondria devoid of mtDNA.

Damage to the electron transport chain in Alzheimer's disease

The activity of COX was seen to be diminished in the platelets of AD patients. Likewise, a notable reduction in COX activity was detected in the brain of AD patients. A separate investigation corroborated the diminished activity in the hippocampus of AD patients, indicating anatomical specificity. Mitochondrial deficits were seen in the platelets of AD patients, demonstrating a considerable reduction in complex III and COX function. Research indicates a reduction in acetylcholinesterase (AChE), and it has been proven that AChE may enhance Aβ activity. The ETC activities of human lymphocytes were assessed in AD patients, revealing elevated activities of complexes II and IV, potentially indicating a compensatory mechanism to meet energy demands. Evidence of ETC dysfunctions in AD is summarised in Table 10.1. Mitochondrial abnormalities linked to neurodegenerative disorders result in oxidative stress, energy dysfunction and disrupted cytosolic calcium homoeostasis. The OXPHOS system does not adhere to thermodynamic equilibrium but rather reflects a rate of uncoupling. A diminished $\Delta\psi m$ may cause hydrolysis of cytoplasmic ATP, whereas an elevated $\Delta\psi m$ results in proton leakage and heightened uncoupling. Excessive ROS generation, reduced $\Delta\psi m$, and Ca^{2+}-dependent enhancement of MPT result in apoptosis. Reduced electron transfer rates were identified as a cause of mitochondrial dysfunction associated with ageing, and a decline in complex I and COX was observed with age. Inhibition of complex III and COX is necessary to enhance glutamate release independent of Ca^{2+}. Partial suppression of complex I activity diminished oxygen consumption in nerve terminals and augmented glutamate release from depolarised synaptosomes. Mitochondrial dysfunctions implicated in the pathophysiology of neuropsychiatric disorders encompass disruptions in OXPHOS, elevated mtDNA deletions, mutations or polymorphisms, compromised calcium signalling, and impaired energy metabolism, along with interactions with disease-specific proteins (e.g. Aβ, parkin, PINK1, alpha-synuclein and Htt). Mitochondrial dysfunction may significantly influence the clinical symptoms of neurodegenerative illnesses; hence, treatment strategies aimed at enhancing mitochondrial activities could be highly beneficial. Evidence endorses the utilisation of antioxidants and other mitochondria-targeting compounds with potential efficacy in AD, such as carnitine, vitamin C, vitamin E, alpha-lipoic acid, coenzyme Q10, methylene blue, piracetam, simvastatin, *Ginkgo biloba*, curcumin and omega-3 polyunsaturated fatty acids. Targeting mitochondrial proteins may constitute an innovative therapeutic approach for AD; for instance, many mitochondria-targeted antioxidants have been formulated.

Disruptions in mitochondrial dynamics, particularly the enhanced mitochondrial fission observed in AD, detrimentally impact multiple aspects of mitochondrial function and are proposed to play a critical role in the pathogenesis of the disorder. Consequently, strategies to alter aberrant mitochondrial dynamics may represent a compelling therapeutic intervention target for AD. Therapeutics aimed at diminishing the expression of the mitochondrial fission protein Drp1, Aβ and pTau may safeguard neurones from the detrimental effects of these factors and their interactions (see Fig. 10.1).

A decline in mitochondrial membrane potential (Δψm) typically precedes the opening of the MPTP. This event facilitates the release of cyt c and other pro-apoptotic factors from the intermembrane space into the cytosol, initiating apoptosome assembly and subsequent activation of caspases, ultimately leading to apoptosis. Additionally, AIF, a mitochondrial pro-apoptotic protein, is translocated to the nucleus upon its release, where it promotes caspase-independent apoptotic pathways.

pTau protein and Aβ induce heightened nitrosylation of Drp1, resulting in compromised mitochondrial dynamics, augmented mitochondrial fission and neurodegeneration. Additionally, Aβ obstructs the importation of proteins into mitochondria and diminishes the function of mitochondrial amyloid-beta binding alcohol dehydrogenase, α-ketoglutarate dehydrogenase complex, and cyclophilin D. The capacity of mitochondria to manage Ca^{2+} is compromised by Aβ and APP; thus an excess of mitochondrial calcium results in a reduction in $\Delta\psi m$, the opening of MPTPs, the release of pro-apoptotic factors, elevated ROS production, and diminished ATP synthesis. Peroxisome proliferator-activated receptor-gamma coactivator-1-alpha (PGC-1α); translocase of the inner membrane; translocase of the outer membrane.

Parkinson's disease

PD is a prevalent neurological disorder. The principal characteristic of PD is the degeneration of dopaminergic neurones in the substantia nigra. Numerous significant genes, such as PARK7 (which encodes DJ-1), α-synuclein, parkin, PINK1 and leucine-rich repeat kinase 2 (LRRK2), possess pathogenic mutations in PD that result in impairments in mitochondrial dynamics and functionality. The mutation of α-synuclein results in its aggregation, which subsequently hinders the fusion of phagosomes with lysosomes during the mitophagic process. Concurrently, the absence of PINK1 results in heightened oxidative stress in mitochondria. Moreover, environmental contaminants induce mitochondrial dysfunction and are considered risk factors for PD, as evidenced by studies in animal and cellular models of the condition. Complex I inhibitors such as MPTP, rotenone, pyridaben and fenpyroximate can replicate the clinical characteristics of PD, at low dosages, resulting in neurodegeneration in models including flies, rodents and mammalian cell cultures. Recently, the MitoPark mouse, a model that particularly lacks the mitochondrial transcription factor A (TFAM) gene, has emerged as the novel genetic model for PD. Mitochondrial failure in dopaminergic neurones of MitoPark mice replicates numerous specific features of PD, such as the gradual and selective degeneration of dopamine neurones, motor impairments and the buildup of inclusion bodies.

Hyperphosphorylation

The presence of pTau protein is another marker of AD alongside amyloid-beta (Aβ). Increased Ca^{2+} and ROS levels during mitochondrial failure both facilitate the generation of pTau aggregates. Melov et al. demonstrated that the loss of mitochondrial superoxide dismutase 2 (SOD2) in mice leads to the accumulation of hyperpTau aggregates, a pathological condition that can be attenuated through the administration of antioxidant compounds. Sorrentino et al. have revealed that amyloid accumulation in cells can be diminished by pharmacologically and genetically manipulating the mitophagy mechanism.

Huntington's disease

HD is a neurological disorder resulting from CAG repeat expansion in the mutant HTT (or IT15) gene, leading to an elongation of the polyglutamine (polyQ) tract in the N-terminal region of the Htt protein. The mutant Htt protein with an enlarged polyQ sequence generates aggregates. This aggregation draws more proteins and mitochondria, negatively impacting the mitochondrial fission-fusion process and impairing mitochondrial transport throughout axons and dendrites. Mutant Htt may directly interact with the outer mitochondrial membrane, destabilising it and enhancing sensitivity to Ca^{2+} or other apoptotic inducers. Concurrently, ROS generation escalates in HD patients in the murine models, resulting in mitochondrial dysfunction. Mitochondrial fragmentation increases in HD patients and mouse models, whereas motility and respiration decline. 3-nitropropionic acid, a mitochondrial citric acid cycle inhibitor, induces selective striatal degeneration and replicates the progressive locomotor decline characteristic of HD. Glyceraldehyde 3-phosphate dehydrogenase (GAPDH) is an essential enzyme in the glycolytic process. In normal cells, oxidised inactive GAPDH facilitates the engulfment of damaged mitochondria into the lysosome for destruction. Hwang et al. demonstrated that enlarged polyQ repeats in HD cell models exhibited aberrant interactions with GAPDH, which inhibited the GAPDH-mediated mitophagy.

Friedreich's ataxia

FRDA, akin to HD, is a hereditary neurological ailment resulting from a homozygous DNA repeat expansion mutation. The guanin, adenine, adenine (GAA) expansion mutation in FRDA results in a deficit of the frataxin protein, leading to mitochondrial dysfunction by increasing ROS generation. Recent research by Abeti and co workers demonstrated that FRDA animal models exhibit a reduction in mitochondrial membrane potential due to an activity imbalance between complexes I and II in the ETC. This imbalance induces ROS production in the mitochondrial intermembrane space and matrix, leading to lipid peroxidation and subsequent neuronal degeneration. PGC-1α, the principal regulator of mitochondrial biogenesis, also contributes to the pathogenesis of FRDA. Lin et al. found that in diseased FRDA animal models, both PGC-1α and its downstream effectors are markedly diminished relative to healthy controls. This dysfunction arose early in the mitochondrial biogenesis pathway and is seen as a possible therapeutic target for the treatment of FRDA.

Amyotrophic lateral sclerosis

ALS is a degenerative condition that impacts motor neurons in the spinal cord and brain. More than 90% of ALS cases are sporadic. Patients with ALS exhibit motor dysfunction alongside a gradual decline in cognitive abilities. Mitophagy, the mechanism for removing damaged mitochondria, is crucial in the pathophysiology of ALS. The neuromuscular junction (NMJ) of SOD1G39A mice, a model for ALS, exhibits a markedly reduced number of phagosomes compared to wild-type mice, suggesting a disturbance in mitophagy. The mitophagy-related proteins PINK1 and Parkin are likewise downregulated. In PINK1-Parkin double-knockout models, mice exhibit intensified NMJ degradation and axonal swelling, correlating with ALS symptoms. Moreover, double-knockout animals exhibit an elevated quantity of ATP synthase beta subunit. This indicates that the elevated number of mitochondria at the junction stemmed from impaired mitophagy.

The eradication of aberrant mitochondria: mitophagy preserves mitochondrial integrity inside cells

Mitophagy constitutes a component of the mitochondrial quality control system and is modulated by proteins that promote mitochondrial fission and fusion. During these processes, dysfunctional mitochondria are sequestered into an autophagosome, which subsequently merges with a lysosome for destruction by lysosomal enzymes. Mitochondrial dynamics is essential for preserving mitochondrial shape, controlling mitochondrial function, responding to apoptotic stimuli and assessing mitochondrial quality. The connection between PINK1 and Parkin is essential for the regulation of mitophagy in mammalian cells. PINK1 is a serine/threonine protein kinase located in the cytosol and also associated with the outer mitochondrial membrane. Mitochondria possessing a positive mitochondrial membrane potential import and destroy PINK1, hence inhibiting its accumulation on the outer mitochondrial membrane. PINK1 preferentially accumulates on the outer mitochondrial membrane of dysfunctional mitochondria exhibiting a reduced mitochondrial membrane potential ($\Delta\psi m$). Parkin, a constituent of a multiprotein E3 ubiquitin ligase complex, associates with PINK1 accumulated on dysfunctional mitochondria and marks the damaged mitochondria with ubiquitin for disposal by mitophagy.

Furthermore, mounting data indicate that mitophagy can potentially transpire in a Parkin-independent manner. In the absence of Parkin, some proteins, including NIX, FUN14 domain containing 1 (FUNDC1), and BCL2/adenovirus E1B 19 kDa protein-interacting protein 3 (BNIP3), together with cardiolipin, directly associate with LC3 protein to sequester defective mitochondria into autophagosomes. Additionally, several E3 ubiquitin ligases, including SMURF1 and mitochondrial E3 ubiquitin protein ligase 1 (MUL1), are capable of ubiquitinating damaged mitochondria and facilitating mitophagy. Endosomes also contribute to mitochondrial degradation. Dysfunctional mitochondria are tagged with ubiquitin by parkin, confined into ras-related protein rab-5 (Rab5)-positive early endosomes, and finally transported to lysosomes for destruction. In specific cell types, impairment of the endosomal pathway via loss of Rab5 activity elevates the probability of cell death resulting from mitochondrial stress.

Aberrant mitochondrial function influences lysosomal activity

The maintenance of mitochondrial homoeostasis requires the coordinated regulation of mitochondrial biogenesis, fusion, fission and mitophagy. The lysosome serves as the primary organelle for degradation and recycling. Mitophagy marks and transports defective mitochondria to the lysosome for destruction. Concurrently, mitochondrial dysfunction impacts the shape and functionality of lysosomes.

Mitofusin 2 facilitates lysosomal autophagocytosis

Depletion of Mfn2 in cardiomyocytes impairs the fusion of autophagosomes with lysosomes, thereby disrupting the autophagic flux. Phagocytosis may be compromised by the depletion of proteins AIF, OPA1, or PINK1, as well as by chemical disruption of the ETC, resulting in larger lysosomal vacuoles. Carbonyl cyanide *m*-chlorophenyl hydrazone (CCCP) elevates mitochondrial ROS levels by membrane depolarisation and selectively activates lysosomal transient receptor potential cation channel, mucolipin subfamily transcription factor EB (TFEB) (TRPML1) channels, resulting in lysosomal Ca^{2+} release. This activation induces the nuclear translocation of transcription factor EB. This calcineurin-dependent transcription factor can initiate the autophagy/lysosome pathway by modulating the synthesis of autophagy/lysosome organelles.

Strategies to improve mitochondrial performance: pharmacological approaches

Numerous pharmaceutical approaches are designed to stimulate mitochondrial biogenesis for the treatment of neurodegenerative disorders. Rosiglitazone and bezafibrate have been shown to activate the peroxisome proliferator-activated receptor–PGC-1α signalling axis, whereas sirtuin-1 (SIRT1) agonists such as quercetin and resveratrol can stimulate SIRT activity and activate adenosine mono phosphate (AMP)-activated protein kinase (AMPK). Bezafibrate enhances mitochondrial proteins and ATP production capacity, resulting in a neuroprotective effect in this mouse model of mitochondrial encephalopathy. Rosiglitazone enhances mitochondrial mass and attenuates mitochondrial dysfunction in cells expressing mutant Htt.

Various antioxidants, including creatine, coenzyme Q10 and mitochondrially-targeted antioxidants/peptides, appear to enhance ankle mobility in patients during clinical trials for PD. Nevertheless, the majority of antioxidants, including dimebon, appear to exert advantageous effects in the AD preclinical studies, but are unsuccessful in AD clinical trials. Given that mitochondrial dysfunction manifests in the early stages of illnesses, pharmaceutical intervention strategies should be contemplated.

In addition, to promoting mitochondrial biogenesis and the elimination of ROS, stimulating mitophagy represents another approach to sustain mitochondrial homoeostasis. Urolithin A, a metabolite of ellagitannins generated by the human gut microbiota, stimulates mitophagy, inhibits the buildup of defective mitochondria with ageing, extends lifespan in *Caenorhabditis* worms, and enhances muscle performance in rodents. Urolithin A is capable of crossing the blood–brain barrier and may confer neuroprotective effects in the context of neurodegenerative diseases. Spermidine is biosynthesised from putrescine and can function as a metabolic precursor for the synthesis of spermine. Spermidine functions as an acetylase inhibitor and promotes autophagy independently of SIRT1. Spermidine enhances basal autophagic flux and increases mitophagy, hence restoring mitochondrial function in aged cardiomyocytes. Administration of spermidine mitigates age-related memory deficits through an autophagy-dependent mechanism. Spermidine promotes the production of mitophagosomes and reduces the aggregation of defective mitochondria via the PINK1/Parkin pathway.

Energy restriction

Caloric restriction (CR) involves reducing caloric consumption while preserving necessary nutrients and avoiding malnutrition. CR diminishes the generation of ROS and mitigates oxidative DNA damage, while decelerating the transcriptional alterations linked to ageing. SIRTs are regarded as pivotal in facilitating the advantageous effects of CR, on lifespan extension. Analogous to CR, SIRT1 overexpression is beneficial for prolonging lifespan and mitigating the symptoms of neurodegenerative disorders. CR stimulated the expression of SIRTs, including SIRT1, SIRT3, SIRT5, and SIRT7. SIRT1 deletion mice do not exhibit increased longevity, even when subjected to a CR diet. CR delayed the onset of prion disease in wild-type mice; however, this protective effect was absent in SIRT1 knockout mice, indicating a SIRT1-dependent mechanism.

Additionally, the ablation of mitochondrial SIRT3 negated the protective effect of CR against hearing loss. CR increases the quantity of mitochondrial cristae and the number of mitochondria per cell. CR also mitigates excitotoxic circumstances by indirectly reducing mitochondrial permeability and calcium retention. These are facilitated via CR-activated SIRT3 deacetylation and the suppression of cyclophilin D, a peptidylprolyl isomerase. Furthermore, CR suppresses the phosphoinositide 3-kinase (PI3K)/AKR thymoma (AKT) pathway and activates autophagy, perhaps enhancing mitophagy and preserving mitochondrial homoeostasis.

Physical activity

Research indicates that exercise may aid in slowing the progression of neurodegenerative illnesses, potentially linked to the restoration of mitochondrial function by physical activity. Physical activity enhances mitochondrial function in the brain. Exercise enhances tolerance to rotenone, an inhibitor of complex I activity, and elevates mRNA expression of TFAM and NADH dehydrogenase [ubiquinone] 1 alpha subcomplex subunit 6 (Ndufa6), subunits of mitochondrial complex I. Concurrently, physical exercise enhances mtDNA repair capacity in the hippocampus of mice and upregulates the expression of mitochondrial uncoupling proteins (UCPs), which are involved in the regulation of mitochondrial biogenesis and the modulation of mitochondrial ROS production. Exercise enhances UCP2 levels in the hippocampus, reduces cellular oxidative stress and can stimulate autophagy, which aids in preserving muscle mass. Our findings indicate that exercise may mitigate the adverse impact of chloroquine on skeletal muscles by restoring autophagic flux and activating the autophagy/lysosomal pathway via the AMPK pathway in the cerebral cortex and striatum. Enhanced mitophagy reduces the quantity of defective mitochondria, hence preserving mitochondrial quality inside cells. Mitochondrial malfunction, subsequent oxidative stress and compromised autophagy/lysosomal activity are the primary causes contributing to neurodegeneration. Consequently, pharmaceuticals that promote mitochondrial activity, eliminate excessive ROS, or augment autophagic flux may possess the potential to cure neurodegenerative disorders. Nonetheless, pharmacological agents that enhance mitochondrial integrity for the treatment of neurodegenerative disorders have yet to be developed. Currently, both CR and exercise, which can promote mitochondrial biogenesis and the autophagy/lysosome pathway (including mitophagy), may be beneficial in delaying the onset and progression of neurodegenerative disorders.

Further Reading

Area-Gomez, E., Schon, E. A., & Przedborski, S. (2019). Mitochondria, OxPhos, and neurodegeneration: cells are not just running out of gas. 2019 *J Clin Invest, 129*(1), 34–45. 2019 https://doi.org/10.1172/JCI120848 2019 .

Hroudová, J., Singh, N., & Fišar, Z. (2014). Mitochondrial dysfunctions in neurodegenerative diseases: relevance to Alzheimer's disease. *Biomed Res Int, 2014*, 175062. https://doi.org/10.1155/2014/175062.

Marde, V. S., Tiwari, P. L., Wankhede, N. L., et al. (2021). Neurodegenerative disorders associated with genes of mitochondria. *Futur J Pharm Sci, 7*, 66. https://doi.org/10.1186/s43094-021-00215-5.

Tapias, V. (2019). Editorial: Mitochondrial Dysfunction and Neurodegeneration. *Front Neurosci, 13*, 1372. https://doi.org/10.3389/fnins.2019.01372.

Wang, Y., Xu, E., Musich, P. R., & Lin, F. (2019). Mitochondrial dysfunction in neurodegenerative diseases and the potential countermeasure. *CNS Neurosci Ther, 25*(7), 816–824. https://doi.org/10.1111/cns.13116.

Xu, S., Zhang, X., Liu, C., Liu, Q., Chai, H., Luo, Y., & Li, S. (2021). Role of Mitochondria in Neurodegenerative Diseases: From an Epigenetic Perspective. *Front Cell Dev Biol, 9*, 688789. https://doi.org/10.3389/fcell.2021.688789.

CHAPTER

11

Mitogenomics and cancer

Introduction

The prime function of mitochondria in eukaryotic cells was documented over a century ago by Otto Warburg. He performed the initial investigations on mitochondrial respiration. In healthy replicative eukaryotic cells, mitochondria govern critical cellular functions, including proliferation, apoptosis, metabolic adaptability and Ca^{2+} homoeostasis.

The rediscovery and subsequent reinterpretation of the Warburg effect in the early 2000s led to a period during which the critical roles of mitochondria in cancer cell biology were largely overlooked. For nearly a decade, the scientific focus centred predominantly on constitutive glycolysis as a defining metabolic hallmark of cancer cells—an oversimplification that neglected the significant contribution of mitochondrial function to the malignancy of both oxidative and glycolytic cancer phenotypes, whether Warburgian or hypoxia-adapted.

Mitochondria serve as the locus for critical events, encompassing fatty acid oxidation (FAO), the tricarboxylic acid (TCA) cycle, oxidative phosphorylation (OXPHOS), the initial phase of gluconeogenesis, ketogenesis, haem production and iron-sulphur cluster formation. They possess DNA that can undergo evolutionary variation, mutation, or partial deletion. Considering the myriad activities of mitochondria, it is unsurprising that mitochondrial dysfunctions contribute to several diseases, including cancer. In this chapter, we will discuss the mitochondrial functions and their role in carcinogenesis and cancer progression. Mitochondria are critical in cancer via macromolecular synthesis and energy generation. Malignant tumours preferentially preserve mitochondrial genomes and electron transport chain (ETC) functionality, while tumours harbouring pathogenic mitochondrial DNA (mtDNA) mutations are benign, underscoring the significance of respiration in cancer progression.

Can mitochondria combat cancer?

Mitochondria not only provide essential bioenergetic functions but also supply precursors for tumour anabolism, regulate redox and calcium homoeostasis, engage in transcriptional regulation, and orchestrate cell death.

Mitochondria represent attractive targets for the creation of innovative anticancer drugs. The common signs and symptoms of cancer are as follows:

Mitochondrial Genomics
DOI: https://doi.org/10.1016/B978-0-443-26655-3.00016-0

- Exhaustion or profound lethargy that does not improve with rest.
- Weight fluctuation of 10 pounds or more, without any known cause.
- Feeding difficulties, including lack of appetite, dysphagia, abdominal pain, or nausea and emesis.
- Localised oedema or masses in the body.

Warburg effect

In oncology, the Warburg effect refers to the phenomenon, when the majority of cancer cells prefer to generate energy through aerobic glycolysis coupled with lactate fermentation, although adequate oxygen is present. This is in contrast to utilising mitochondrial OXPHOS, as is in most nonmalignant cells. This metabolic reprogramming was first reported by Otto Heinrich Warburg, who was awarded the 1931 Nobel Prise in Physiology or Medicine for his elucidation of the nature and functional mechanism of the respiratory enzyme.

Mitochondria participate in cancer development

Mitochondria are thought to play a pivotal role in the acquisition of the malignant phenotype through at least five distinct processes linked to the metabolic reprogramming characteristic of cancer cells. Numerous diseases are significantly linked to DNA mutations impacting mitochondria, primarily resulting from modifications in the subunits of the ETC. Subsets of hepatocellular carcinomas and prostate cancers are linked to mutations in the D-loop area of Complex I, while certain neurological malignancies have mutations in succinate dehydrogenase (SDH; Complex II). Secondly, oxidative stress caused by reactive oxygen species (ROS) is the primary catalyst for cancer development and progression towards malignancy. ROS are predominantly generated by mitochondria, which emit superoxide as a consequence of oxidative respiration. Mitochondrial reactive oxygen species (mtROS) can be produced in either the TCA cycle or the ETC. Due to their elevated reactivity, ROS function as harmful entities for cellular macromolecules and, at low concentrations, as intracellular signalling agents that regulate metabolic pathways. Elevated amounts of ROS are frequently observed in cancer cells as a result of heightened metabolic activities and modified antioxidant capabilities. Mitochondria are intimately involved in regulating cell death, encompassing apoptosis and necrosis. B-cell lymphoma-2 (Bcl-2) family proteins promote apoptosis by interacting with mitochondria and binding to the voltage-dependent anion channel (VDAC), facilitating its opening and the subsequent release of cytochrome c (cyt c). Consequently, these proteins function as oncogenic or oncosuppressive agents, contributing to cancer progression and treatment resistance. Myeloid leukaemia cell differentiation protein-1 (MCL-1), an antiapoptotic member of the Bcl-2 family, is often overexpressed in human cancers and correlates with tumour aggressiveness. MCL-1 and Bcl-xL have been identified in various mitochondrial subcompartments. They perform their antiapoptotic functions by opposing the pro-apoptotic members of the Bcl-2 family at the outer mitochondrial membrane (OMM), and, within the mitochondrial matrix, by maintaining mitochondrial homoeostasis and bioenergetics through the preservation of the inner mitochondrial membrane (IMM) integrity and facilitating the assembly of ATP-synthase oligomers at the ETC. Mitochondria regulate necroptosis, a controlled type of necrosis that requires mtROS production and is contingent upon mitochondrial permeability transition. Fourth, metabolic reprogramming involves various mutations in genes that encode TCA cycle enzymes, facilitating malignant transformation. Certain intermediates of the TCA cycle, including fumarate, succinate, aspartate and D-2-hydroxyglutarate (2HG, a de novo metabolite arising from mutations in isocitrate dehydrogenases (IDHs)), exhibit significant pro-carcinogenic effects when they accumulate within cells as a result of genetic mutations and/or cancer-associated dysregulation of protein expression. The fifth characteristic of all tumours is persistent cellular growth due to various molecular changes. One alteration is the avoidance of telomere degradation through constitutive telomerase production, which maintains telomere length. Telomerase reverse transcriptase (TERT) translocates from the nucleus to mitochondria in response to oxidative stress, thereby maintaining mitochondrial functions and reducing oxidative stress, which safeguards mtDNA and nuclear DNA from oxidative damage in order to prevent apoptosis. TERT was observed to accumulate in the mitochondria of murine brain cells following food restriction and rapamycin administration.

Mitochondria serve functions beyond only being the cell's powerhouses

Although mitochondria are acknowledged for their active involvement in cancer progression, their specific functions in the clinical outcomes of cancer patients remain unclear. Over the past 50 years, scientists' interest in mitochondria has surged due to findings regarding their influence on various essential activities in eukaryotic cells. Mitochondria are tubular organelles of around 0.5 to 3 μm in length that continuously reconfigure their network through fusion and fission events. Textbooks first characterise mitochondria as the primary locus of cellular energy production, and certainly, mitochondria serve as a significant site for the synthesis of ATP and macromolecules. The reactions of the TCA cycle take place in the mitochondrial matrix. In conjunction with CO_2 and protons, they produce reducing equivalents (nicotinamide adenine dinucleotide (NADH) and flavin adenine dinucleotide (FADH2)) and precursors for the biosynthesis of lipids, carbohydrates, proteins and nucleotides. Electrons that reduce equivalents power the ETC to create an electrochemical gradient necessary for ATP synthesis and the active movement of certain metabolites, including pyruvate and ATP, across the IMM. Besides this crucial function, mitochondria are involved in various other roles pertaining to mitochondrial dynamics and architecture, which affect several vital cellular activities. The mitochondrial architecture (Fig. 11.1A) is fundamentally linked to mitochondrial functions, including ATP synthesis, regulation of the cell cycle, control of programmed cell death, proliferation and cellular signalling. Mitochondria consist of two membranes, the OMM and the IMM, which enclose an intermembrane space (IMS) and the mitochondrial matrix within the organelle. The OMM serves as a platform for exchange and signalling, functioning as the location where proteins phosphorylate substrates and modulate the immune response following viral infection through the activation of mitochondrial antiviral signalling (MAVS) proteins. The IMM exhibits reduced permeability and serves as the location for ETC complexes involved in ATP synthesis and superoxide production. The mitochondrial matrix serves as the site for mtDNA replication, transcription and the biosynthesis of macromolecules, and it also hosts the amphibolic reactions of the TCA cycle. The investigation of mitochondrial systems

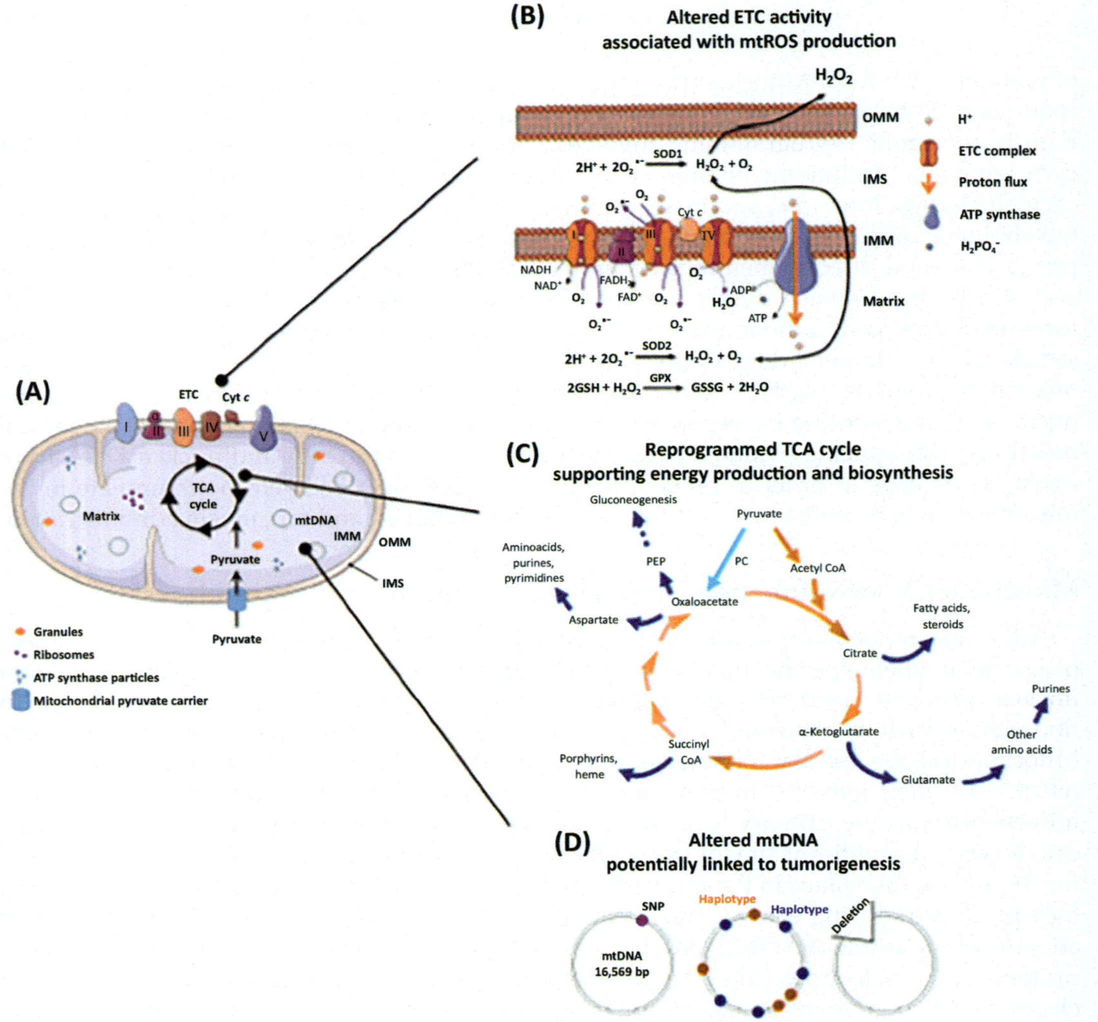

FIGURE 11.1 Cancer is linked to modifications in mitochondrial functioning. (A) Mitochondria in both normal and cancerous cells consist of three compartments. They are delineated from the cell cytoplasm by an outer membrane (OMM), an intermembrane space (IMS) and an inner membrane (IMM) that creates invaginations known as 'crests'. The IMM defines the mitochondrial matrix, a viscous substance comprising mitochondrial DNA (mtDNA), granules, ribosomes and ATP-synthase complexes. The mitochondrial matrix contains the tricarboxylic acid (TCA) cycle, whereas the IMM houses the electron transport chain (ETC). (B) In cancer cells that are extremely metabolically active or hypoxic, electrons escape during mitochondrial electron transport at Complexes I and III, resulting in the generation of superoxide (O_2^-) from oxygen (O_2) in both the IMS and the matrix, significantly more than in normal cells under standard conditions. O_2^- is promptly dismutated to H_2O_2 either spontaneously or catalysed by superoxide dismutases 1 (SOD1) (in the IMS) or SOD2 (in the matrix). Within the matrix, hydrogen peroxide (H_2O_2) can be neutralised by glutathione (GSH). It can also send signals to the cytosol. (C) In cancer cells, the TCA cycle functions to make reducing equivalents for the ETC (*green arrows*) and to generate biosynthetic intermediates essential for cellular proliferation (*pink arrows*). The principal anaplerotic process generates oxaloacetate directly from pyruvate, catalysed by pyruvate carboxylase (PC) (*blue arrow*). Oxaloacetate can be converted to phosphoenolpyruvate (PEP) via PEP carboxykinase), thus contributing to gluconeogenesis. mtDNA variations, encompassing single-nucleotide polymorphisms (SNPs), maternally inherited haplotypes, and deletions, have been examined for their correlation with cancer. Only certain mtDNA deletions were observed to be linked to malignancies. Cyt c—cytochrome c; Gpx—glutathione peroxidase; Q—coenzyme Q10.

governing metabolic changes and mitochondrial morphology has yielded evidence that, in diseases such as cancer, they may serve as promising therapeutic targets.

Cancer is associated with alterations of mitochondrial functions Fig. 11.1

The mitochondrion: a compartmentalised organelle, the outer mitochondrial membrane— a platform for signalling

The primary role of mitochondria is to prevent the cell from engaging in processes for which they are ill-suited, so avoiding a mismatch between cellular metabolic demands and mitochondrial capabilities. Mitochondria can communicate with the cell through many mechanisms. One method is via the OMM, which can be likened to a signalling platform. The OMM links mitochondria to many organelles in the cell, including the endoplasmic reticulum (ER), lysosomes and the plasma membrane. VDAC, which creates pores made of various subunits, renders the OMM permeable to tiny ions and proteins. VDAC facilitates the transport of nucleotides, ions and metabolites between the cytosol and the IMS, serving as an intracellular signalling platform for metabolic modulation and the regulation of apoptosis. In terms of metabolic regulation, VDAC serves as a platform for the attachment of hexokinase 2 (HK2), the embryonic form of the initial enzyme in glycolysis, to the OMM to enhance ATP utilisation by HK2 in cancer cells. The phosphorylation of glucose by HK2 sequesters glucose-6-phosphate within cells, indicating that VDAC has a role in regulating both glycolysis and mitochondrial respiration. The interaction of Bcl-2 family members (Bax, Bak, Bok, Bad, Bid, or Bim) with VDAC induces pore formation, facilitating the release of cyt c, a tiny hemoprotein that can disseminate throughout several mitochondrial compartments. Cyt c, in addition, to its role in the ETC essential for ATP synthesis, also triggers caspase-dependent apoptosis in response to pro-apoptotic signals.

Mitochondria also regulate the innate immune system. The OMM is the locus of phosphorylation for many proteins. Protein kinase A (PKA), a tetramer consisting of two cAMP-binding subunits and two catalytic subunits, is localised in the mitochondria. Mechanistically, a-kinase-anchoring proteins are located in the OMM, where they enable the binding of PKA catalytic subunits to the organelle membrane, hence enhancing PKA localisation for protein phosphorylation. Upon cAMP binding to PKA, the two catalytic subunits separate, activating them to phosphorylate several target proteins exhibiting the arginine-arginine-X-serine motif, including splicing factors SRSF1, SRSF2, and SRSF9. The OMM contains antiviral signalling regulators that trigger the immune response via MAVS, as recently detailed by Mohanti A et al.

OMM signalling represents but one method by which mitochondria interact with the remainder of the cell. Mitochondria are necessary for various significant signalling pathways.

1. Initially, they facilitate the synthesis of acetyl-coenzyme A (CoA) and S-adenosyl-methionine, which both modulate epigenetics by governing the signal transduction of DNA and histones via acetylation and methylation, respectively.
2. Secondly, mtROS generation and release occur as a reaction to cellular stress and serve as a signalling mechanism to significantly activate transcription factors such as hypoxia-inducible factors (HIFs) and nuclear factor erythroid-derived 2-like 2 (Nrf2), along with downstream gene expression.

3. Mitochondria regulate Ca^{2+} homoeostasis. Calcium (Ca^{2+}) acts as a signalling mediator between mitochondria and the ER through specialised contact sites referred to as mitochondria-associated membranes.

This is a crucial signal utilised by these organelles for communication. Ca^{2+} functions as a bidirectional signalling molecule, with mitochondrial Ca^{2+} intake influencing mitochondrial metabolism and mitochondrial Ca^{2+} release affecting death. Increased Ca^{2+} concentration within mitochondria activates several TCA cycle enzymes and boosts cAMP generation, hence enhancing ATP synthesis and facilitating metabolic adaptability. Conversely, Ca^{2+} serves as a signal for programmed cell death, with elevated levels of Ca^{2+} prompting the opening of mitochondrial permeability transition pores, leading to the release of cyt c and the initiation of apoptosis. Calcium ions (Ca^{2+}) released from ER-mitochondria contact sites can also induce apoptosis via Bcl-2 family proteins, contingent upon precise regulation of calcium homoeostasis. Moreover, alterations in ATP production by mitochondria serve as a signal relayed to the cytosol as AMP; AMP activates the energy sensor AMP kinase (AMPK), hence reducing anabolic cellular processes to prioritise catabolic reactions for ATP synthesis.

Mitochondrial signalling is a dynamic and intricate process that influences numerous cellular processes. By comprehending these processes, it may be feasible to treat diseases such as cancer more efficiently.

The inner mitochondrial membrane: the ATP manufacturing facility

In contrast to the OMM, the IMM lacks porins and serves as a highly impermeable barrier to ions and molecules, necessitating specialised membrane transport proteins for bidirectional exchanges. The IMM possesses an electrochemical membrane potential of around 180 mV that governs its transport. It is also more widespread than the OMM, as it is structured in invaginations known as 'mitochondrial cristae' to facilitate the assembly of ETC complexes. The extent of cristae invaginations is dependent on the energetic demands of the tissue. Mitochondria within muscle cells exhibit a high density of cristae to support elevated energy requirements. The IMM is distinguished from the OMM by its elevated protein content, exhibiting a protein-to-lipid ratio of 80:20 in the IMM compared to 50:50 in the OMM. The elevated protein concentration is indicated by all complexes involved in OXPHOS, along with transport proteins and proteins that govern fusion and fission processes. The respiratory chain comprises a number of multienzymatic complexes, including both hydrophilic and lipophilic substances that sequentially transport electrons through the complexes to the final electron acceptor, O_2. The passage of electrons generates a motor force that transports protons from the mitochondrial matrix, their production site, to the IMS against their concentration gradient, thus establishing an electrochemical gradient across the IMM. The complexes are organised in ascending order of reduction potentials, allowing electrons to transition from a higher energy state to a lower energy one, resulting in energy release that contributes to ATP production. A portion of the generated energy is utilised for thermoregulation. The respiratory chain consists of five complexes, three of which function as proton pumps, together with two carriers that serve as cosubstrates and electron transporters.

Complex I—NADH dehydrogenase, commonly known as coenzyme Q (CoQ) reductase.

Complex I consists of 45 subunits. Among the 14 catalytically active subunits, seven are hydrophobic—comprising ND1, ND2, ND3, ND4, ND4L, ND5 and ND6—and are encoded by the mitochondrial genome, whereas the remaining seven subunits are hydrophilic and nuclear-encoded. The residual subunits are referred to as 'non-essential subunits'. They are critical for the assembly and stability of Complex I and these are coded by the nuclear genome. Complex I accepts two hydrogen atoms from NADH, which undergoes oxidation, resulting in the release of two electrons and the reduction of flavin mononucleotide (FMN) to $FMNH_2$. The former transmits electrons to the second carrier in the ETC, CoQ, via its Fe/S centres. Complex I utilise the energy derived from electron transfer to translocate four protons from the mitochondrial matrix to the IMS, corresponding to a movement of two protons per NADH oxidised. Dysfunctions of Complex I have been linked to malignancy, since diminished activity of the complex has been noted in renal oncocytomas and thyroid adenomas.

CoQ: CoQ is a lipophilic ubiquinone transporter integrated within the IMM lipid bilayer. It is capable of separating the protons released into the mitochondrial matrix from the electrons donated by $FMNH_2$.

Complex II—SDH: Complex II consists of four nuclear-encoded subunits: two hydrophilic catalytic subunits, SDHA/SDH1 and SDHB/SDH2, and two hydrophobic subunits, SDHC/SDH3 and SDHD/SDH4. It consists of a haem b group and two CoQ-binding sites. It is also a component of the TCA cycle. Complex II facilitates electron transmission; however, it does not engage in proton translocation to the IMS. The two electrons generated from the oxidation of succinate to fumarate are directly sent to CoQ. The diminished activity of Complex II has been linked to human cancers, including renal carcinoma and breast cancer, where its activity is reduced in comparison to the adjacent normal tissues.

Complex III—Cytochrome bc1 complex, alternatively known as Cyt c reductase. Complex III is a symmetrical dimer, with each subunit consisting of three catalytic cores (MT-CYB, CYC1, and UQCRFS1) and seven supernumerary subunits. This complex accepts electrons from CoQ and transfers them to cyt c; subsequently, it translocates four protons into the IMS. Elevated physiological activity of Complex III has been observed in breast cancer.

Cyt c: Cyt c is a hydrophilic haem protein situated on the outer surface of the IMM. It facilitates electron transport between Complexes III and IV.

Complex IV—Cyt c oxidase: Complex IV consists of 13 or 14 subunits and is unique among the OXPHOS complexes in that it contains isoforms that are regulated in a tissue-specific and developmentally dependent manner. This complex transfers four electrons, supplied by four molecules of cyt c, directly to O_2, sourced from the blood, reducing it to two molecules of H_2O, which utilises four protons extracted from the mitochondrial matrix. A diminished activity of Complex IV has been noted in malignancy. In renal carcinomas, Complex IV expression was observed to be five times lower than in healthy kidney tissues.

Complex V—ATP synthase: Complex V consists of two separate domains. The F1 domain is external to the IMM and resides in the matrix, whereas the Fo domain is integral to the IMM. F1 comprises nine subunits, whereas Fo consists of two subunits. ATP synthase is the complex responsible for synthesising ATP from the substrates ADP, $H_2PO_4^-$, and H^+ during the last reaction of the OXPHOS process. Electrons traversing the ETC are ultimately converted to O_2 at Complex IV, while the majority of protons moved from the mitochondrial matrix to the IMS return through the ATP-synthase complex, as a result, neutralise the electrochemical gradient. The energy released by protons returning according to their electrochemical potential is utilised as mechanical energy to facilitate the operation of ATP synthase.

Recent results reveal that ETC complexes are organised into supramolecular structures known as 'supercomplexes'. These structures seem to be arranged in various configurations and to exhibit stoichiometry. For instance, Complex I is often associated with a dimer of Complex III and Complex IV. This super complex is designated as 'respirasome' as it comprises of all the complexes involved in the transfer of electrons from NADH to O_2. Complex I is consistently associated with supercomplexes, whereas Complexes III and IV can alternatively exist as independent oligomeric enzymes. This indicates that the assembling of the complexes into a singular structure is essential for the stabilisation and proper functioning of Complex I in the respiratory chain. A lot of experiments have shown that the depletion of Complexes III and IV lead to the concomitant loss of Complex I. ETC activity is linked to cancer through many mechanisms. Increasing experimental evidence suggests that OXPHOS influences ATP production more profoundly in cancer cells compared to normal cells. Specifically, OXPHOS facilitates substantial ATP synthesis in invasive and metastatic cells, as well as in circulating cancer cells. Additional data indicate that cancer stem cells (CSCs) preferentially employ OXPHOS for energy metabolism, exhibiting elevated rates of oxygen consumption, ROS formation and enhanced mitochondrial activities relative to nonstem cancer cells. Furthermore, mtROS facilitates KRAS-induced anchorage-independent growth.

Scientists found that an ETC overload might enhance the migration, invasion, clonogenicity and metastatic potential of cancer cells via a mitochondrial superoxide-dependent mechanism that activates the transforming growth factor β (TGFβ) pathway at the src kinase level. Although mitochondrial respiration is regarded as a compelling target for cancer therapy, it has been extensively shown that tumours exhibit metabolic heterogeneity and are affected by several substrates and environmental variables. Metabolism varies not only among different cancer types but also among subpopulations of cancer cells within the same tumour. Distinct cellular subpopulations exhibiting varying metabolic characteristics (enhanced OXPHOS or impaired glycolysis) have been discovered in melanoma, lymphoma, pancreatic and breast tumours.

The mitochondrial matrix: the epicentre of metabolism

The mitochondrial matrix is the innermost compartment of the mitochondrion, enclosed by the IMM. It mostly contains mtDNA, ribosomes, enzymes, tiny chemical compounds, nucleotide cofactors and inorganic ions. The mitochondrial matrix is a highly viscous compartment distinguished by a substantially higher protein concentration relative to the IMS and a more alkaline pH (approximately 7.8 in the matrix vs 7.0–7.4 in the IMS). It serves as the site for numerous enzymatic processes, including the TCA cycle, anaplerotic and cataplerotic reactions, segments of the urea cycle, transamination reactions, and also houses components of the ETC, notably Complex II.

The tricarboxylic acid cycle

The TCA cycle, commonly referred to as the Krebs cycle or citric acid cycle, is the principal process occurring in the mitochondrial matrix. The process consists of a sequence of chemical processes that convert two-carbon units from carbohydrates, amino acids and fatty acids into acetyl-CoA, producing GTP and reducing equivalents (NADH and FADH2) that power the mitochondrial ETC to generate ATP. Acetyl-CoA undergoes oxidation in a cyclic metabolic route to yield CO_2, resulting in the net synthesis of one CoA-SH, two CO_2, three NADH, one FADH2, one GTP/ATP and three H^+ for each molecule of acetyl-CoA used.

The initial step of the cycle is the condensation of acetyl-CoA with oxaloacetate, catalysed by citrate synthase (CS) to produce citrate. Citrate is transformed to its isomer, isocitrate, by mitochondrial aconitase (ACO2) and then decarboxylated to α-ketoglutarate (α-KG) via mitochondrial IDH. In this reaction, one molecule of carbon dioxide (CO_2) is released, and one molecule of NAD^+ is reduced to NADH with the concomitant release of a proton (H^+). A subsequent decarboxylation transpires as α-KG is transformed into succinyl-CoA by α-KG dehydrogenase, concurrently generating NADH + H^+.

The subsequent reactions facilitate the regeneration of oxaloacetate through,

1. the oxidation of succinyl-CoA to succinate by succinate-CoA synthetase, yielding GTP;
2. the oxidation of succinate to fumarate by SDH, resulting in the production of FADH2;
3. the hydration of fumarate to malate by fumarate hydratase (FH); and, ultimately,
4. the oxidation of malate to regenerate oxaloacetate is catalysed by malate dehydrogenase, resulting in the reduction of NAD^+ to NADH and the release of a proton (H^+). The TCA cycle is integral to cellular metabolism, serving a central role in both energy production and the generation of biosynthetic precursors. Consequently, to preserve cellular metabolic homoeostasis, a balance between the production and consumption of intermediates must be sustained. Anaplerotic reactions are enzymatic processes that generate metabolic intermediates to refill the TCA cycle, while cataplerotic reactions are biosynthetic processes that use TCA cycle intermediates as substrates for macromolecule synthesis. The TCA cycle generates NADH and FADH2, which are essential for ATP production in the ETC, while also supplying chemicals necessary for many metabolic processes, including gluconeogenesis and the synthesis of fatty acids and nucleotides.

The primary anaplerotic reaction generates oxaloacetate directly in the mitochondria from pyruvate, a process catalysed by pyruvate carboxylase. This process is subject to allosteric regulation by acetyl-CoA and aspartate, which serve as indicators of oxaloacetate deficiency. Various anaplerotic processes yield distinct intermediates of the TCA cycle, such as α-KG generated from glutamate by glutamate dehydrogenase, succinyl-CoA synthesised through FAO and oxaloacetate formed from aspartate by aspartate transaminase.

When the synthesis of TCA cycle intermediates is adequate for its proper operation, cataplerotic events occur to regulate metabolite concentrations. Consequently, citrate can be exported to the cytosol and converted into acetyl-CoA by the enzyme ATP citrate lyase, thereby initiating the biosynthesis of fatty acids.

In a reversible mechanism, α-KG and aspartate yield glutamate and oxaloacetate by the action of aspartate transaminase, contributing to purine production. Oxaloacetate can be transformed into phosphoenolpyruvate (PEP) by PEP carboxykinase (PEPCK), hence facilitating gluconeogenesis.

Modifications in the tricarboxylic acid cycle in neoplasia

Genetic mutations in TCA cycle enzyme-producing genes and the excessive buildup of TCA cycle intermediates can facilitate carcinogenesis. The primary enzymes identified as changed in cancer include SDH, FH, IDH, CS, and ACO2, indicating a significant field for further research. Mitochondrial defects prompt metabolic reprogramming, leading cells to increase their independence on glycolysis, hence facilitating the tumorigenic process.

Initial research on the role of SDH mutations in cancer indicated that patients with hereditary paragangliomas and pheochromocytomas, two uncommon neuroendocrine tumours, exhibited inactivating mutations in the subunits of SDH enzymes. Consequently, succinate accumulates in these tumours.

Inactivating mutations of FH predispose individuals to hereditary leiomyomatosis, renal carcinoma and numerous cutaneous and uterine leiomyomas. Fumarate, when accumulated, functions as an oncometabolite that inhibits α-KG-dependent dioxygenases responsible for DNA and histone demethylation. It additionally facilitates the epithelial to mesenchymal transition (EMT) by obstructing ten–eleven translocation-dependent DNA demethylation of a regulatory region of the antimetastatic miR-200, resulting in diminished expression of miR-200 and E-cadherin, alongside heightened expression of Twist1 and vimentin, among other target genes. Deficiencies in FH enhance the nuclear translocation and activity of Nrf2, leading to the transcription of antioxidant genes via antioxidant response elements. These genes include mitochondrial components such as thioredoxin 2 (Trx2), thioredoxin reductase 2 (ThxR2), peroxiredoxins (Prxs) and superoxide dismutase 2 (SOD2), along with transporters and enzymes that participate in glutathione (GSH) manufacture and redox recycling. Nrf2 activity is inhibited by Kelch-like ECH-associated protein 1, which binds to Nrf2 in the cytosol, leading to its polyubiquitylation and subsequent proteasomal destruction, as well as by transcription modulators that heterodimerise with Nrf2 in the nucleus. In contrast to its stimulation of antioxidant mechanisms, fumarate can directly interact with GSH to generate succinated GSH, which resembles the substrate of GSH reductase. This results in NADPH consumption without antioxidant benefits, hence elevating oxidative stress.

IDH exhibits three isoforms: NADP+-dependent IDH1 located in the cytoplasm and peroxisomes, and both NADP+-dependent IDH2 and NAD+-dependent IDH3 situated in the mitochondrial matrix. IDH1 and IDH2 mutations have been documented in 70% of grade II and III gliomas and glioblastomas, as well as in angio-immunoblastic T-cell lymphomas, acute myeloid leukaemia (AML), and other prevalent malignancies, including thyroid, colorectal and prostate cancers. Mutations in IDH1 and IDH2 reduce NADPH and GSH levels, resulting in increased activity of the PI3K-AKT-mammalian target of rapamycin (mTOR) signalling pathway and improved motility of cancer cells. Numerous studies indicate that IDH mutations influence carcinogenesis and cancer progression, with reports suggesting these mutations elevate ROS levels in cancer cells. Elevated ROS levels and activation of the PI3-AKT pathway enhance signalling activity conducive to cancer development and progression. Mutated IDHs frequently acquire the ability to generate 2HG, the R enantiomer of α-KG. 2HG is an oncometabolite that, when present in elevated concentrations, inhibits α-KG-dependent dioxygenases, including histone lysine demethylases, resulting in a hypermethylated condition of DNA and histones. It activates HIF-1 by blocking the degradation of HIF-1α via prolylhydroxylases (PHDs).

CS is a crucial enzyme frequently regarded as the rate-limiting enzyme of the TCA cycle. The loss leads to a metabolic transition from oxidative to glycolytic metabolism and triggers EMT, hence enhancing tumour malignancy. Nevertheless, elevated levels of CS expression have been noted in pancreatic, renal and ovarian malignancies.

A reduced expression of ACO2 has been associated with gastric and prostate malignancies. FH-deficient cells have reduced ACO2 activity due to fumarate accumulation, which inhibits ACO2 function in a dose-dependent manner through the succination of essential cysteine residues.

Inactivating mutations in TCA cycle enzymes can lead to the accumulation of metabolic intermediates, distinct from the direct substrates of the affected enzymes, which may function as oncometabolites. Citrate is an oncometabolite that, upon accumulation, disturbs the balance of the TCA cycle and facilitates cancer progression. High citrate concentrations inhibit pyruvate dehydrogenase (PDH) activity, leading to pyruvate buildup. In this context, cells convert pyruvate to lactate while regenerating NAD+, resulting in a metabolic shift towards glycolysis, a recognised occurrence in cancer progression. Additionally, citrate serves as a substrate for acetyl-CoA carboxylase, facilitating the production of acetyl-CoA and malonyl-CoA, which are precursors for lipid and steroid biosynthesis. Lipids serve as the primary structural constituents of biological membranes and are crucial for cell signalling, facilitating cell proliferation and transformation. Conversely, acetyl-CoA sourced from citrate is an essential substrate for histone acetylation, influencing chromatin architecture and, consequently, gene transcription.

Histone acetylation is a dynamic and reversible process of acetyl group transfer regulated by histone acetyltransferases and histone deacetylases (HDACs). The acetylation of particular histones promotes chromatin relaxation, thereby enhancing the transcription of crucial genes that govern cell proliferation, cell-cycle transition, differentiation and apoptosis, all of which are associated with tumorigenesis and affect the growth programme of cancer cells. HDAC inhibitors have demonstrated efficacy in restoring the altered cellular phenotype, and multiple HDAC inhibitors have received FDA approval for cancer treatment. Additionally, histones, oncoproteins, tumour suppressors and enzymes are subject to acetylation, which regulates their functions. This pertains to HIFs, Myc, KRAS, p53, retinoblastoma protein pRb, PTEN and enzymes associated with glucose, fatty acid and glutamine metabolism.

Modifications of mitochondrial DNA in neoplasia

Mutations and deletions in mitochondrial DNA have been linked to cancer

mtDNA single-nucleotide polymorphisms (mtSNPs) were assessed for their potential correlation with increased cancer susceptibility and/or enhanced disease aggressiveness. A study indicated an association between a reduction in mtDNA copy number and colorectal cancer malignancy; however, another study found no correlation between mtDNA copy number and mtSNPs regarding cancer advancement in colorectal cancer patients. This study analysed six mtSNPs (MitoT479C, MitoT491C, MitoT10035C, MitoA13781G, 10398A/G, and 16189T/C) in 536 patients, along with mtDNA copy number in 274 individuals, comparing tumour and healthy tissues. A possible association between mtSNPs and prostate cancer risk has been investigated. After comparing 350 mtSNPs among 4086 cancer patients and 3698 healthy individuals in a multiethnic population, the authors found no correlation between mtSNPs and prostate cancer risk.

Ethnicities are distinguished by several haplogroups, defined as groupings of mtDNA sequence differences that are statistically inherited from the ancestral maternal mitochondrial genome. For instance, Europeans exhibit nine main haplogroups (H, U, J, T, K, W, I, V and X). Variations in the mtDNA sequence can lead to modifications in mitochondrial proteins, primarily components of the ETC, which affect their functionality, ETC efficiency and ROS generation. In oncology, haplogroups have been thoroughly investigated, particularly concerning their potential association with the risk of getting certain cancer types. Analyses indicated that certain haplogroups are associated with an elevated risk for particular cancer types, while concurrently diminishing the risk of other cancer types. Haplogroup K is statistically associated with a reduced incidence of thyroid cancer among Southeastern Europeans, but it correlates with an elevated risk of breast cancer in European Americans. A study involving approximately 7700 European adults revealed no association between haplogroups and breast cancer risk factors in either mothers or their offspring. A significant challenge in correlating haplogroups with cancer risks arises from the methodology employed in data analysis. One study indicated that haplogroup N is associated with an elevated risk of breast cancer, while another study found that haplogroup U is linked to a reduced risk and haplogroup K to an increased risk of breast cancer, respectively. According to phylotree.org, haplogroup N encompasses haplogroup U, which subsequently includes haplogroup K, complicating the assessment of the impact of these variations on cancer risk and rendering the comparison of various studies challenging. From a molecular perspective, it remains ambiguous how an identical collection of mtDNA mutations might concurrently exhibit both pro- and anticarcinogenic properties.

Extensive mtDNA deletions typically lead to significant disruptions in the ETC and, as a result, to OXPHOS deficiency. Three primary studies have examined the rates of these occurrences in malignancies. The initial investigation examined a 4977 bp mtDNA deletion in over 1600 samples of tumour and adjacent healthy tissues. The findings indicated that this specific deletion may serve as a cancer biomarker, particularly in breast cancer. The second and third investigations examined the prevalence of a 3400 bp mtDNA deletion in prostate cancer. This deletion is known to be prevalent in prostate cancer and in the neighbouring healthy tissue, but it is missing in normal prostate epithelium. The 3400 bp mtDNA deletion was proposed as a biomarker for prostate cancer.

Overall, with the exception of significant mtDNA deletions, the plethora of conflicting data concerning mtDNA variants and their associations with malignancies renders it impossible to draw any therapeutically relevant conclusions at present. Besides conflicts, the biological ramifications of these modifications are inadequately defined at both molecular and functional levels. As a result, targets are not readily identifiable, and pharmaceutical strategies cannot be formulated.

Mitochondrial reactive oxygen species

Mitochondria as a producer of reactive oxygen species

The word 'ROS' denotes a category of extremely reactive oxygen-containing compounds characterised by a brief average half-life. ROS encompass superoxide ($O_2^{\circ-}$), hydroxyl radical (OH°) and hydrogen peroxide (H_2O_2). Mitochondria are regarded as the primary generators of superoxide within cells.

Mitochondrial superoxide, the primary ROS, is predominantly generated at the ETC due to electron leakage at the ubiquinone-binding sites of Complex I (IQ site) and Complex III (IIIQo site) during electron transport. Superoxide generation predominantly transpires by the transfer of a solitary electron from fully or partially reduced electron carriers, such as semiquinone, to O_2. The kinetics of $O_2^{\circ-}$ formation are influenced by the average duration of the carrier enzyme in its reduced state and its concentration, emphasising that the concentration of enzymes involved in $O_2^{\circ-}$ generation modulates the roles of ROS, within cells. A critical variable is the O_2 content within cells. Accordingly, experimental modulation of the local oxygen concentration can either augment or attenuate the rate of mitochondrial oxygen consumption, thereby influencing the generation of superoxide ($O_2^{\circ-}$). Due to its high reactivity, $O_2^{\circ-}$ is promptly dismutated to H_2O_2, either spontaneously or catalysed by mitochondrial SODs. SOD1 is located in the IMS and cytosol, where it neutralises superoxide generated at ETC Complex III, whereas SOD2 is situated in the mitochondrial matrix, where it neutralises superoxide produced at Complexes I and III.

Besides the ubiquinone-binding sites in ETC Complexes I and III, additional sites that produce ROS have been identified in mitochondria: flavin in Complex I (IF site), the electron transferring flavoprotein Q oxidoreductase in FAO, glycerol 3-phosphate dehydrogenase, PDH and 2-oxoglutarate dehydrogenase. While the processes regulating the operation of each site remain incompletely elucidated, research indicates that all sites generate ROS in the mitochondrial matrix; however, only the IIIQo site and glycerol 3-phosphate dehydrogenase are capable of releasing superoxide ($O_2^{\circ-}$) in the IMS. In some instances, upon synthesis in the IMS, H_2O_2 can traverse the OMM and perform signalling functions in the cytosol. Compelling evidence supports the significance of mtROS as a signalling mechanism facilitating communication between mitochondria and cellular processes that regulate homoeostasis and respond to stress circumstances. mtROS provide distinct signalling functions in cells by facilitating cell survival or promoting apoptosis, as well as contributing to mitochondrial recycling.

Mitochondrial H_2O_2 modulates posttranscriptional protein modification by oxidising the thiol groups of cysteines, thereby altering the activity of target proteins and consequently influencing the final response. The interaction of H_2O_2 with phosphatases, including mitogen-activated protein kinase phosphatase, impedes their dephosphorylative function, resulting in the activation of c-Jun N-terminal kinase. Secondly, an augmented production of mtROS constitutes a cellular adaptive response to hypoxia that activates HIFs. ROS oxidise the iron component of hypoxia-inducible factor PHDs, thereby reducing the enzymatic activity that typically initiates the degradation of HIF-1α by the hydroxylation of this subunit. ROS further oxidise vitamin C, which is an essential cofactor for the recycling of PHD domain by the reduction of Fe^{3+}. HIF-1 is recognised for upregulating the expression of glycolytic enzymes and transporters to sustain adequate ATP levels in cells and for augmenting the production of vascular endothelial growth factor to stimulate angiogenesis. Third, mtROS serve as a signal to initiate autophagy, which can have either pro-survival or proapoptotic effects. In starvation, mtROS oxidise and deactivate cysteine protease Atg4, initiating the lipidation of Atg8, a critical step in autophagy, hence enhancing autophagy and aiding in the recycling of intracellular molecules. Notably, autophagy might additionally facilitate the buildup of ROS by the selective destruction of catalase. Fourth, mtROS induce signalling responses at the level of nuclear gene expression by creating a perinuclear clustering of mitochondria, resulting in an accumulation of ROS in the nucleus and subsequent modifications of gene transcription.

Mitochondrial antioxidant defences

To sustain equilibrium between ROS generation and their detrimental effects, cells possess particular defence systems. They safeguard cells from the harmful effects of ROS, such as damage to lipids, proteins and nucleic acids, as well as the production of double-strand DNA breaks. Superoxide generated in mitochondria is converted to H_2O_2 by SOD1 (copper-zinc superoxide dismutase (CuZnSOD)) in the IMS and cytosol, and by SOD2 (manganese superoxide dismutase) in the mitochondrial matrix (Fig. 11.1B). SOD3 is a member of the CuZnSOD family and is located extracellularly. Following the conversion of superoxide to H_2O_2 by SODs, peroxidases catalyse the further transformation of two H_2O_2 molecules into O_2 and two H_2O. These peroxidases include catalase, found in peroxisomes, the Trx system and glutathione peroxidase. The Trx system, in conjunction with Prxs, facilitates the reduction of H_2O_2 to H_2O utilising NADPH. It comprises Trx peroxidases, ThxR and the substrates Trx and Prx (Fig. 11.2). Mitochondria possess a distinct ThxR2 for redox control and Prx for peroxide reduction. Ultimately, in the cytosol, GSH peroxidase facilitates the transformation of H_2O_2 into H_2O utilising GSH as a substrate, which is then reduced to GSSG and finally restored with NADPH by GSH reductase. Consequently, the synthesis of GSH (via glutaminolysis and the serine route) and the creation of NADPH effectively facilitate the inactivation of H_2O_2.

Regulation of cell death by metabolic and mitochondrial mechanisms

A distinct relationship exists between metabolism and cellular apoptosis, facilitated by diverse signal transduction mechanisms. In cancer cells, p53 orchestrates a prevalent core route. p53 is the paramount pro-apoptotic protein and is mutated or inactivated in around 50% of tumours. Nonetheless, whereas p53 plays a pivotal role in tumour suppression, it also participates in the regulation of cancer metabolism. Two prominent p53 targets that modulate cellular metabolism are TP53-induced glycolysis and apoptosis regulator and the synthesis of cytochrome oxidase 2 (SCO2), a protein involved in the assembly of cytochrome oxidase 2. TI-GAR reduces glycolytic flow by dephosphorylating fructose-2,6-bisphosphate,

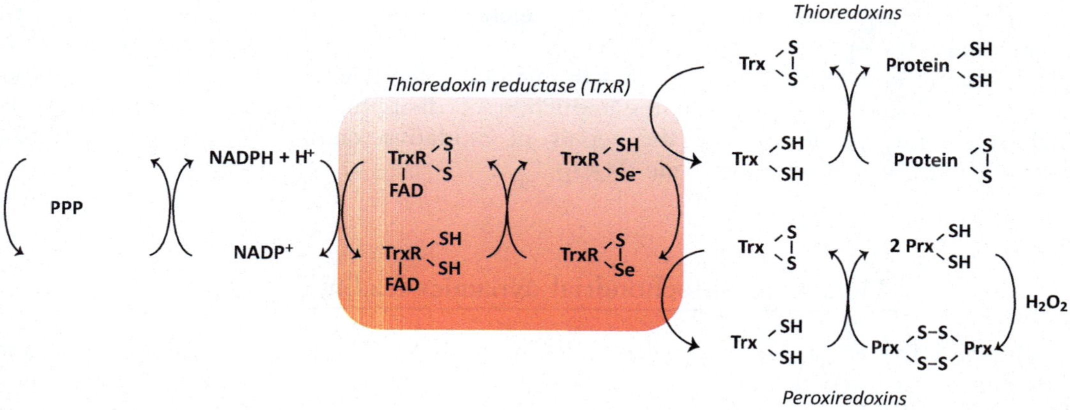

FIGURE 11.2 Mitochondrial antioxidant defence mechanisms encompass the thioredoxin (Trx) and peroxiredoxin (Prx) systems. The graphic illustrates redox reactions facilitated by the Trx and Prx systems, which include thioredoxin reductases (TrxR, specifically ThxR2 expressed in mitochondria), Trxs, Prxs, and NADPH. NADPH serves as the electron source, primarily derived from the pentose phosphate pathway (PPP). Oxidised thioredoxins (Trx-S2) are reduced by NADPH and selenoenzymes known as TrxRs. Electrons are transferred in a sequential manner from NADPH to FAD, then to the N-terminal redox-active disulphide of one TrxR subunit, and finally to the C-terminal active site of the opposing subunit. Reduced thioredoxins (Trx-SH2) facilitate the reduction of disulphide bonds in many proteins, including Prxs, so ensuring the repair of oxidative damage in proteins and the detoxification of hydrogen peroxide (H_2O_2).

whereas SCO2 facilitates ETC assembly and OXPHOS. In the next step, p53 can inhibit glycolysis and consequently encouraging OXPHOS and FAO.

The availability of nutrients is frequently modified throughout tumour development. Irregular tumour perfusion and, as a result, nutritional deprivation, influence cellular apoptosis. A prevalent microenvironmental change in tumours is hypoxia, which can trigger p53 activation and induce cell death. The metabolic condition of the cell serves as a signal for p53 activation: a reduction in cellular ATP levels leads to a decreased ATP/AMP ratio, which activates AMPK. Subsequently, AMPK phosphorylates and activates p53 at serine 15, thereby triggering an AMPK-dependent cell-cycle arrest.

AMPK predominantly functions as a tumour suppressor, primarily by causing cell-cycle arrest and decreasing the synthesis of several cellular macromolecules. AMPK activation suppresses the mammalian target of rapamycin Complex 1 by phosphorylating its upstream regulator, tuberous sclerosis Complex 2, hence suppressing cell proliferation. ATP levels and AMPK provide a crucial link between p53-mediated regulation of energy metabolism and programmed cell death. Adenosine signalling can promote apoptosis by activating adenosine receptor A2B, which triggers a caspase- and p53-upregulated modulator of apoptosis-dependent apoptotic response that involves the down-regulation of Bcl-2 expression.

The repertoire of mitochondrial antioxidant defences includes the Trx and Prx pathways. The graphic illustrates redox reactions facilitated by the Trx and Prx systems, which include TrxR, specifically ThxR2 expressed in mitochondria, Trxs, Prxs and NADPH. NADPH serves as the electron source, primarily derived from the pentose phosphate pathway.

Oxidised Trxs are reduced by NADPH and selenoenzymes known as TrxRs. Electrons are sequentially transported from NADPH to FAD, then to the N-terminal redox-active disulphide in one subunit of TrxR, and ultimately to the C-terminal active site of another subunit. Reduced Trxs facilitate the reduction of disulphide bonds in many proteins, including Prxs, so promoting the repair of oxidative damage in proteins and the detoxification of hydrogen peroxide (H_2O_2).

Defects in mitochondrial dynamics leading to cancer

Mitochondria are active organelles that traverse the cell, undergo division, fusion and experience regulated turnover via mitophagy. They may also be transferred across cells. Mitochondrial dynamics indicate metabolic changes.

Mitochondrial division, fusion and fission w.r.t. cancer

Mitochondrial fission is the process by which mitochondria divide into two or more separate entities, facilitating the formation of new mitochondria. Mitochondrial fusion involves the merging of adjacent mitochondria, enabling the exchange and redistribution of mitochondrial contents, thereby promoting functional complementation between partially damaged organelles. Both processes necessitate meticulously regulated mechanisms to facilitate the development of new, functional organelles and to prevent the loss of intramitochondrial material.

Mitochondria undergo fission, resulting in their division into smaller fragments, a crucial mechanism for augmenting their quantity in proliferating cells. Fission also enhances mitochondrial elimination via mitophagy, aids mitochondrial transport across the cytoskeletal network, and modulates apoptosis and Ca^{2+} homoeostasis. The process commences with actin and the ER, which initially designates the division location on the OMM. Upon the establishment of contact between the ER and mitochondria via the VDAC on the mitochondrial surface and the inositol 1,4,5-trisphosphate receptor on the ER surface, the ER releases Ca^{2+} into the mitochondrion, initiating actin polymerisation at the constriction point. This facilitates the recruitment and assembly of ***dynamin-related protein 1*** **(*DRP1*)**, a cytosolic GTPase that translocates to the OMM during activation. DRP1 subsequently recruits the membrane-anchored receptor protein Fis1, the tail-anchored *mitochondrial fission factor (Mff)*, and the ***anchored mitochondrial dynamic proteins***. Collectively, these proteins helically wrap around the mitochondrion, constricting and cleaving it into two segments.

Elevated DRP1 expression has been linked to various cancer types, including glioblastomas, thyroid, lung and breast tumours, and is correlated with enhanced metastatic capability of cancer cells.

Notably, the silencing of DRP1 has been demonstrated to diminish the metastatic potential of breast cancer cells by inhibiting lamellipodia production, a crucial process that facilitates cell movement. DRP1 was shown to be overexpressed in squamous cell carcinomas, and its absence in this tumour type results in mitochondrial elongation, leading to inhibited cell proliferation and G2 arrest. DRP1 expression was mechanistically discovered to be positively linked with the expression of cell-cycle genes that govern mitosis in epithelial ovarian cancer (EOC). Increased DRP1 expression enhances mitosis, hence facilitating cell proliferation in the progression of main and recurrent EOC. DRP1 is associated with cancer cell metabolism, since the depletion of DRP1 in HeLa cells has been demonstrated to reduce the activity of ETC complexes, mitochondrial respiration, mitochondrial membrane conductance and ATP production. A compelling association between mitochondrial fission and cancer arises from the functional relationship between DRP1 and cellular stress, with DRP1 suggested as a transcriptional target of p53 and oncogenic RAS/MAPK signalling, enhancing DRP1 mRNA levels. The various mechanisms through which DRP1 participates in cancer indicate that this protein fulfils significant functions that may be distinct from its primary role of mitochondrial fission.

A multitude of studies have examined the metabolism of CSCs. Mitochondrial fission is identified as crucial for the maintenance of stemness, despite variations in metabolic activity among different tumour types. Consequently, the suppression of DRP1 or its pharmacological inhibition in brain tumour-initiating cells diminished their tumorigenicity and induced apoptosis. The pharmacological suppression of DRP1 by mitochondrial division inhibitor-1 (mdivi-1) resulted in impaired tumour sphere formation by breast CSCs, as well as reduced migration and survival. Civenni et al. recently observed that the depletion of Mff in prostate CSCs swiftly diminished their tumorigenic capacity, correlating with the onset of CSC senescence. The elevated rate of mitochondrial fission in CSCs is posited to enable mother stem cells to maintain a complete mitochondrial content, while daughter cells destined for differentiation would get a mixture of intact and deficient mitochondria. This asymmetric segregation contributes to the maintenance of cancer stemness despite repeated cell divisions.

During fusion, subsequent to mitochondrial replication (termed mitochondrial biogenesis, Fig. 11.3B), the OMM and IMM converge, resulting in the mixing of the IMS and mitochondrial matrix. At the molecular level, fusion is regulated by three dynamin family GTPases: mitofusin (Mfn) 1 and 2 in the OMM and optic atrophy protein 1 (Opa1) in the IMM. Opa1 exists in two long isoforms and three short isoforms. Fusion of the OMMs is closely coordinated with the fusion of the IMMs. The activity of mitochondrial fusion proteins is regulated through ubiquitination of proteins localised to the OMM and proteolytic processing of proteins within the IMM. When two mitochondria are adjacent, Mfn1 and Mfn2 initiate dimerisation through homo- and heterotypic interactions, forming antiparallel connections that facilitate the fusing of the OMM by merging the lipid bilayers. The subsequent interaction of the two fusing mitochondria with the ER is thought to enhance the process, as Mfn2 has also been identified in the ER, where it facilitates the interface between mitochondria and the ER, enabling the exchange of Ca^{2+} for signalling purposes. The fusion of IMMs is facilitated by Opa1 in conjunction with Mnf1.

Mitochondrial fusion in cancer cells seems to counteract the effects of fission regarding tumour growth, metastatic potential and metabolic functions. Reduced expression of Mfn1/2 has been observed in mouse medulloblastoma cells compared to nontransformed counterparts. Pharmacological inhibition of DRP1 using Mdivi-1 enhanced mitochondrial fusion and triggered mtDNA replication. Furthermore, Mfn2 overexpression diminished

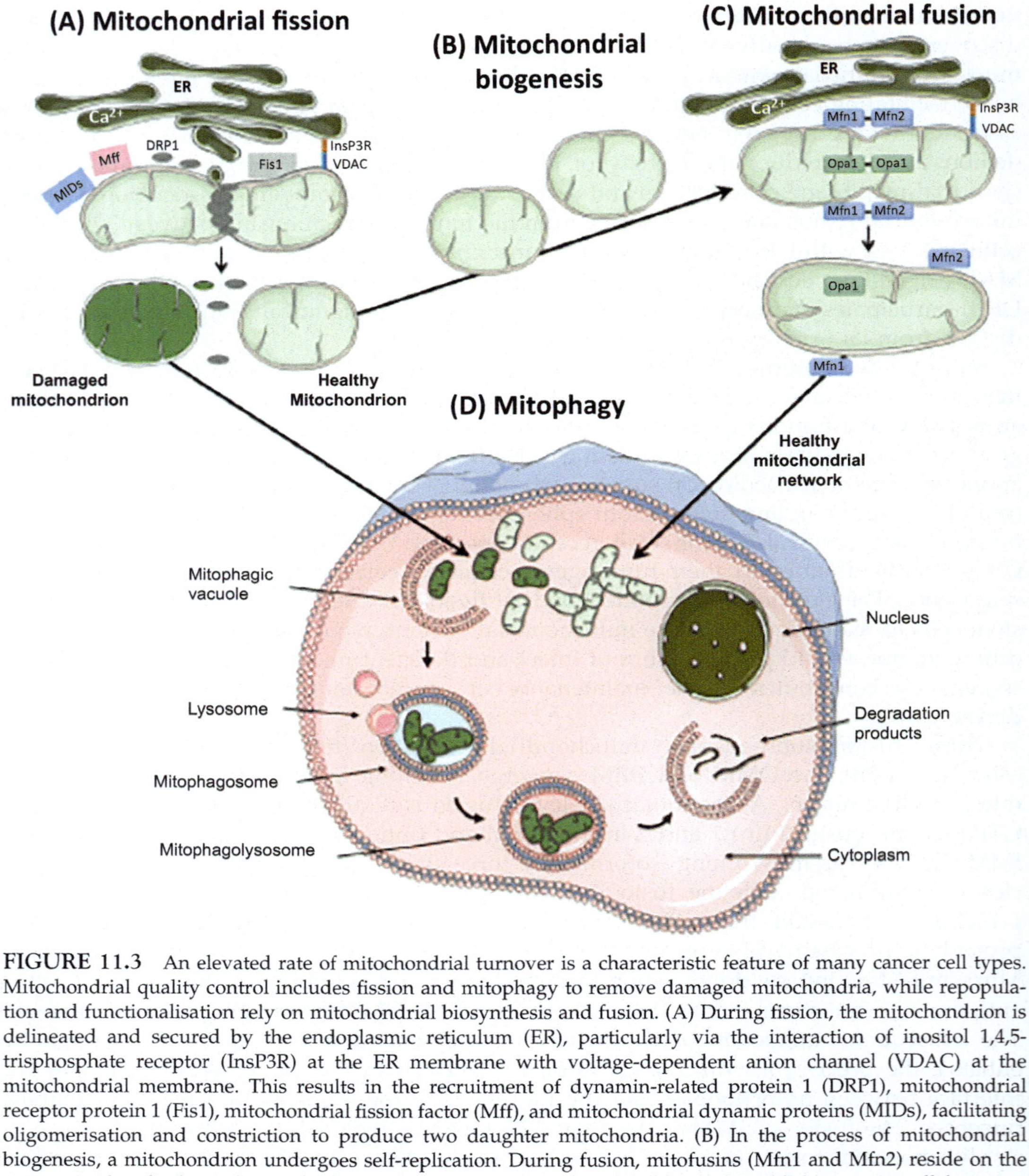

FIGURE 11.3 An elevated rate of mitochondrial turnover is a characteristic feature of many cancer cell types. Mitochondrial quality control includes fission and mitophagy to remove damaged mitochondria, while repopulation and functionalisation rely on mitochondrial biosynthesis and fusion. (A) During fission, the mitochondrion is delineated and secured by the endoplasmic reticulum (ER), particularly via the interaction of inositol 1,4,5-trisphosphate receptor (InsP3R) at the ER membrane with voltage-dependent anion channel (VDAC) at the mitochondrial membrane. This results in the recruitment of dynamin-related protein 1 (DRP1), mitochondrial receptor protein 1 (Fis1), mitochondrial fission factor (Mff), and mitochondrial dynamic proteins (MIDs), facilitating oligomerisation and constriction to produce two daughter mitochondria. (B) In the process of mitochondrial biogenesis, a mitochondrion undergoes self-replication. During fusion, mitofusins (Mfn1 and Mfn2) reside on the outer mitochondrial membrane, facilitating calcium exchange for signalling and establishing antiparallel connections between two fusing mitochondria. Optic atrophy 1 (Opa1) and Mnf1 collaborate in the fusing of the inner mitochondrial membrane. Fusion facilitates the creation of mitochondrial networks.

lung cancer proliferation and inhibited the migration and invasion of breast cancer cell lines. Hyperfused mitochondria are present during the G1/S phase of the cell cycle, correlating with enhanced oxidative capability and increased ATP production. Mfn2 gene therapy, which entails the introduction of genetic material (DNA) into cells to restore Mfn2 gene expression, has been documented to diminish the growth of A549 human lung cancer cells by facilitating apoptosis. Fusion and fission contribute to subcellular specialisation of mitochondria. Mitochondria distribute subcellularly based on the specific metabolic demands for ATP, Ca^{2+} buffering, and other activities. At the axonal level in neurones, mitochondria exhibit fragmentation due to their movement along the cytoskeleton, interacting with desmin, vimentin and tubulin. Conversely, at the dendritic level near synapses, they appear elongated, as this region has a higher demand for ATP to facilitate the re-pumping of ions released from synaptic vesicles. Fission enables the distribution of mitochondria inside the cell and apoptosis via the release of cyt c. Conversely, fusion advantages the cell through mitochondrial complementation, enabling injured mitochondria or those with mutated mtDNA to merge with healthy mitochondria to mitigate the detrimental effects of defective organelles.

Mitophagy and mitochondrial transference

The quality control of mitochondrial dynamics encompasses mitophagy, a degradation process that eliminates malfunctioning or damaged mitochondria (Fig. 11.3D). This mechanism is essential to ensure the physiological activities of cells and tissues and to prevent the emergence of illnesses such as cancer. It also modulates the quantity of mitochondria in accordance with the metabolic demands of cells and during certain phases of cellular development, such as erythrocyte differentiation.

Mitochondrial elimination predominantly employs the molecular apparatus of macroautophagy, a distinct form of autophagy marked by the creation of mitophagosomes, which are double-membrane structures that encapsulate mitochondria. It may occur selectively or nonselectively, wherein autophagosomes sequester mitochondria together with cytosolic components and other organelles.

Selective mitophagy commences with the assessment of intact and impaired mitochondria by PTEN-induced kinase 1 (PINK1). PINK1 is directed to mitochondria via its N-terminal mitochondrial targeting sequence, which facilitates its import under normal mitochondrial membrane potential conditions. When the mitochondrion is in a healthy state (polarised), PINK1 is translocated from the cytosol to the mitochondrial matrix via the translocases of the outer mitochondrial membrane (TOM) and the inner mitochondrial membrane. Upon traversing the IMM membrane, it is cleaved and then released into the cytosol by the mitochondrial protease presenilin-associated rhomboid-like protein and matrix processing peptidase. If the mitochondrion is defective, the IMM gets depolarised, PINK1 remains uncleaved, and cannot be transferred to the IMS. It aggregates in the OMM. It can recruit and activate Parkin by phosphorylating OMM-resident ubiquitins and Parkin itself. Parkin is a cytosolic E3 ubiquitin ligase that, upon activation, ubiquitinates proteins in the OMM, including Mfn1, Mfn2, VDAC1, TOM20 and mitochondrial Rho GTPase 1, thereby initiating mitophagy. In the cascade, autophagy receptors (e.g. optineurin) and autophagy initiators (e.g. unc-51-like autophagy activating kinase 1 (ULK1)) are recruited and activated, as recently elucidated in reference. Protein ubiquitination subsequently activates microtubule-associated protein light chain 3 (LC3), which typically exists in the cytosol as LC3-I. Following activation, LC3 undergoes lipidation through conjugation to phosphatidylethanolamine, generating the membrane-bound form LC3-II, which is recruited to autophagosomal membranes during the formation of the characteristic double-membrane structure. In addition, to this classic pathway, other paths have been identified that involve a direct contact between the activated form of the OMM protein FUN14 domain containing 1 (FUNDC1) and LC3, or BH3-only domain proteins (BNIP3 and NIX), beclin and ULK1. A study by Soubannier et al. demonstrated that mitochondria-derived vesicles can develop during the initial oxidative stress response, and these vesicles are transported to lysosomes independently of LC3 as a means of mitophagy complementation.

Dysfunctional mitophagy is linked to tumour start and progression in various cancer types. In reaction to challenges like hypoxia and nutritional deprivation, mitophagy is initiated to diminish the total mitochondrial mass, hence conserving essential nutrients and averting excessive mtROS production. Although the precise molecular process remains inadequately elucidated, a portion of the reaction may be facilitated by receptors possessing C-terminal transmembrane domains situated at the OMM, such as BNIP3, NIX and FUNDC1.

Parkin is hypothesised to function as a tumour suppressor: its expression enhances oxidative metabolism, mitigates the Warburg effect and modulates the levels of cyclin D1, cyclin E and cyclin-dependent kinase 4 in cancerous conditions. Mitophagy exhibits a dual role in cancer treatment resistance; its suppression in colorectal CSCs and its activation in preclinical AML models both enhance susceptibility to chemotherapeutic agents, including doxorubicin. Mitophagy may play a contributory role in mediating the radioresistance of head and neck cancer cells to γ-radiation.

Mitophagy associated with mitochondrial biogenesis is a primary mechanism for sustaining a reservoir of functioning mitochondria in cancer cells; however, cancer cells can also obtain intact mitochondria from nonmalignant cells within the tumour microenvironment. The mitochondrial transfer entails the creation of intercellular tunnelling nanotubes (TNTs) and bigger microtubes, which have been shown in several cancer cell types, stemming from the first discovery that mitochondria-deficient $\rho 0$ cells can take in mtDNA from host cells. Subsequent investigations revealed that mtDNA transfer entails the complete transfer of mitochondria from nonmalignant donor cells to malignant recipient cells. In AML, ROS have been recognised as catalysts for the development of TNTs that facilitate complete mitochondrial exchange between bone marrow stromal cells and AML blasts, thus enhancing the metabolic capability of cancer cells. Although the molecular mechanisms of TNT generation have been thoroughly examined in other sources, we believe it is crucial to highlight that mitochondrial transfer can also convey malignant properties between cancer cells. The experimental transfer of mitochondria from an invasive bladder cancer cell line (T24 cells) to a less invasive cell line (RT4 cells) resulted in increased invasiveness of the RT4 cells. Another example demonstrated that mtDNA transfer between poorly and highly metastatic lung cancer cells concurrently conveyed the high metastatic characteristics. Pasquier et al. demonstrated that MCF7 breast cancer cells, upon receiving mitochondria from endothelial cells, exhibited increased resistance to doxorubicin treatment. Collectively, although the in vivo documentation of TNT production is notably challenging, these data underscore the possible therapeutic relevance of intercellular mitochondrial transport.

Mitochondrial dynamics indicate metabolic changes

Owing to recurrent fission and fusion events, many types of mitochondria may be present inside a cell. They may morphologically resemble small vesicles, short bars, or reticular networks, which provide a snapshot of forms arising from a continuously shifting equilibrium. Their organisation is contingent upon cellular types and particular stress situations. Mitochondrial dynamics are intricately associated with their functions, and modifications in mitochondrial morphology have been connected to changes observed in cancer. Numerous studies have established a correlation among energy sources, oxygen availability and mitochondrial structure. Cellular metabolic dysfunctions are linked to heightened mitochondrial fragmentation, while hyperfused mitochondria exhibit greater resilience to metabolic stress and, due to their fusion, can safeguard cellular integrity. Molina et al. demonstrated that in the β cells of the pancreas, mitochondria in a nutrient-rich environment undergo fission and are separated, while those in a nutrient-deficient environment remain elongated and form networks for extended periods. Furthermore, hunger induces the formation of fatty acid droplets within mitochondria, redirecting cellular metabolism towards FAO for ATP synthesis. Nutrient deprivation mechanistically promotes mitochondrial elongation via cAMP-activated PKA, which prevents mitochondrial fission and safeguards the organelles from autophagosomal degradation. In cancer, current data indicate that a hyperfused state of mitochondria aids in the survival of cancer cells by preserving ATP production, compensating for impaired mitochondria, facilitating intramitochondrial fatty acid exchanges and preventing metabolic reprogramming towards autophagy (Fig. 11.2 and Fig. 11.3).

The mitophagic process involves the engulfment of damaged mitochondria within a vacuole known as a 'mitophagic vacuole.' The subsequent amalgamation of the mitophagosome with lysosomes, resulting in the formation of a mitophagolysosome, facilitates the breakdown of mitochondria into macromolecules that are transported to the cytoplasm. Mitophagy may be either nonselective or selective, employing canonical and noncanonical mechanisms. It inhibits the buildup of impaired mitochondria that could jeopardise or terminate the cell if apoptosis and/or the generation of ROS were to deviate.

Therapeutic approaches focussing on mitochondria in cancer cells

Considering the essential roles of mitochondria in cancer cells, various techniques have been conceptualised and evaluated that see mitochondria as targets for anticancer therapies.

This document provides a concise summary of key strategies designed to regulate mitochondrial anaplerosis, mitochondrial turnover, the TCA cycle, the ETC, mtROS and mitochondria-mediated apoptosis.

Focussing on mitochondrial anaplerosis

Targeting processes that facilitate mitochondrial anaplerosis, such as glycolysis, glutaminolysis and FAO, is regarded as a potential anticancer strategy. Although these anaplerotic pathways may partially facilitate cancer cell survival and proliferation independent of mitochondrial metabolism, their blockage leads to a deficiency in TCA cycle substrates. 2-Deoxy-D-glucose (2DG) competes with glucose for glucose transporters (GLUTs) and HKs, which are frequently overexpressed in cancer cells, providing a degree of anticancer selectivity for treatment. By obstructing glucose uptake and phosphorylation, 2DG diminishes the supply of pyruvate for mitochondria, hence hindering mitochondrial anaplerosis and OXPHOS. Alongside its antimetabolic properties, preclinical studies indicate that 2DG demonstrates further anticancer benefits, including antiangiogenic action, suppression of cancer metastasis, and inhibition of viral replication of Kaposi's sarcoma-associated herpes virus. It may also enhance the effects of autophagy inhibition. Nonetheless, clinical trials were predominantly equivocal, since 2DG, when administered as a solitary treatment, failed to demonstrate substantial anticancer efficacy at tolerable levels for patients. Further in vitro and in vivo preclinical investigations indicated its potential use in conjunction with standard chemotherapy agents, such as cisplatin and doxorubicin. Encouraging outcomes have also been achieved in conjunction with radiotherapy for glioblastoma treatment, and a Phase III clinical trial is presently underway. 3-Bromopyruvate (3BP) is an alkylating chemical that has been documented to impede glycolysis, among other effects. 3BP effectively inhibits HK2 and GAPDH among a broader array of targets. Despite the publication of promising case studies indicating that 3BP demonstrates anticancer effects in humans, to our knowledge, no clinical investigation has been finalised so far. Despite being a promising molecule, 3BP encounters multiple constraints regarding its therapeutic applicability, such as fast deactivation by GSH and a burning sensation upon intravenous administration. To minimise side effects and enhance its anticancer efficacy, 3BP can be encapsulated in liposomes, polyethylene glycol (PEG)-liposomes, or various targeted or nontargeted nanocarriers. Glutaminolysis has been demonstrated to facilitate the proliferation of cancer cells, particularly in human pancreatic ductal adenocarcinoma and triple-negative breast cancer. Glutaminase 1 (GLS1) is proposed as a target for anticancer therapy; however, no pharmacological therapies are currently available. Targeting FAO represents an additional possible anticancer approach. Particular emphasis has been placed on carnitine palmitoyltransferase I (CPT1), an enzyme situated at the OMM that catalyses the conversion of acyl-CoA, the final result of FAO, to acyl-carnitine, that traverses mitochondrial membranes to provide the TCA cycle in the mitochondrial matrix. The irreversible CPT1 inhibitor etomoxir has demonstrated the ability to enhance the efficacy of radiotherapy, and it may be utilised in the treatment of cachexia, a symptom linked to increased FAO in cancer patients. Etomoxir has undergone clinical trials for type 2 diabetes and congestive heart failure, demonstrating a safe usage profile for the medicine. To our knowledge, no clinical trial involving cancer patients has been initiated to date.

Focussing on the tricarboxylic acid cycle

The selective targeting of the TCA cycle in cancer cells has been thoroughly investigated in recent years. Significantly, numerous Phase I and II clinical trials have been executed with pharmaceuticals that can inhibit dysregulated pathways associated with the TCA cycle. The compounds include CB-839, a selective inhibitor of GLS; CPI-613, a lipoate analogue that inhibits PDH and α-KG dehydrogenase, recently evaluated in Phase I and II clinical trials as a monotherapy or in conjunction with standard chemotherapy for various cancers (NCT02168140, NCT02232152, and NCT01766219); and enasidinib/AG-221, an orally bioavailable inhibitor of mutant IDH2-R140 and IDH2-R172, currently undergoing Phase I/II clinical trials as a monotherapy for AML, angio-immunoblastic T-cell lymphoma, and glioma (NCT01915498 and NCT02273739). Moreover, AG-881 is a promising orally administered dual inhibitor of mutant IDH1 and IDH2, which was undergoing Phase I/II clinical trials this year, enroling AML patients with mutant IDH1/2 and glioma patients (NCT02492737 and NCT02481154).

Focussing on the electron transport chain

As discussed in the section 'Antioxidant Defences in Mitochondria', the mitochondria of cancer cells may exhibit modifications in the functions of ETC complexes, which represent promising anticancer targets. Complex I: A variety of drugs have been evaluated in vitro and in vivo for their ability to inhibit Complex I, including carboxy-amidotriazole and fenofibrate. The most advanced medications targeting Complex I are the biguanides metformin and phenformin. These two substances are FDA-approved pharmaceuticals commonly utilised for diabetic management. Notably, they can also hinder the proliferation of many cancer cell lines, and in vivo investigations have further shown that metformin and phenformin impede tumour growth and metastasis in multiple animal models. Although these two medications were thought to largely function via AMPK activation, two studies published in 2000 revealed that metformin predominantly inhibits ETC Complex I in cancer cells, hence obstructing OXPHOS and activating AMPK. The mechanism of action of these biguanides involves the inhibition of ubiquinone reduction. This discovery unveiled new possibilities for the application of biguanides in oncology. Metformin is presently undergoing evaluation in multiple clinical trials for cancer patients. To date, the most encouraging outcomes have been achieved in Phase III clinical study, which evaluated the efficacy of metformin in treating breast cancer, other nonmetastatic malignancies and cancers with a lower degree of malignancy. Patients who received metformin bi-daily for five years postdiagnosis demonstrated a notable enhancement in progression-free survival. Papaverine, a nonnarcotic opioid commonly used to treat vasospasms and erectile dysfunction via inhibition of phosphodiesterase 10 A, has also been identified as an inhibitor of mitochondrial Complex I in cancer cells. Benej et al. demonstrated that papaverin can radiosensitize lung and breast cancer cells and tumours in vivo by means of oxygen sparing. A Phase I clinical trial is currently underway to evaluate papaverine hydrochloride as a radiosensitizer in patients with nonsmall cell lung cancer (NCT03824327), following its documented anticancer effects on prostate, breast and hepatocarcinoma cancer cells, as well as glioma xenografts in mice.

Complex II: In comparison to Complex I, there are yet no well-characterised inhibitors of Complex II. Lonidamine has recently demonstrated the ability to inhibit Complex II in isolated mitochondria and in DB-1 melanoma cells. Nevertheless, the drug showed no advantages in two randomised Phase III clinical trials when combined with chemotherapy, leading to its withdrawal from clinical investigations.

Complex III: Atovaquone is an FDA-approved medication utilised for the treatment of pneumocystis pneumonia and malaria. This medication is a ubiquinone analogue that functions as a Complex III inhibitor in parasites, cancer cell lines and breast CSCs, decreasing the oxygen consumption rate and alleviating tumour hypoxia at pharmacologically tolerable concentrations. Atovaquone is presently undergoing evaluation in a Phase I clinical trial for its impact on hypoxia in nonsmall cell lung cancer within a preoperative window of opportunity research. Although recruiting has concluded, no results have yet been disseminated concerning this trial.

Cyt c oxidase: Arsenic trioxide is characterised as a Complex IV inhibitor and is licensed by the FDA for the treatment of acute promyelocytic leukaemia. In recent years, it has been investigated in different forms of cancer. Preclinically, arsenic trioxide demonstrated a reduction in hypoxia in Lewis lung carcinoma and transplantable mouse liver tumours by inhibiting cellular respiration, resulting in enhanced radiation response in mice. Nitric oxide and hydrocortisone are additional substances that, among other effects, can inhibit Complex IV and enhance the efficacy of radiation in preclinical studies.

Complex V: To our knowledge, no investigations involving in vivo or clinical trials targeting ATP synthase in cancer models have been described.

Focussing on mitochondrial reactive oxygen species generation

The application of mitochondria-targeted antioxidants such as mitoTEMPO and mitoQ has demonstrated efficacy in inhibiting the migratory, invasive and metastatic characteristics of cancer cells. Both of them inhibit mtROS-induced activation of the TGFβ pathway. MitoTEMPO is a superoxide scavenger that can emulate mitochondrial SODs. It readily traverses lipid bilayers and specifically accumulates in negatively charged mitochondria due to a positively charged triphenylphosphonium group. MitoQ is a molecule analogous to ubiquinone, which is covalently attached to triphenylphosphonium. Like mitoTEMPO, it may swiftly traverse biological membranes and accumulate up to 100-fold in mitochondria. It can penetrate the membrane core of mitochondria, functioning as a chain-breaking antioxidant, which facilitates the reduction of MitoQ to its ubiquinol form through ETC Complex II. In addition, to its antimetastatic properties, MitoQ, a mitochondria-targeted antioxidant, effectively reduces KRAS-induced pancreatic cancer in vivo.

Reinstating mitochondria-mediated apoptosis

Methods that promote cyt c release can be employed to trigger apoptotic death in cancer cells. Numerous studies have postulated that photodynamic therapy (PDT) can inflict harm on cancer cells by disrupting the mitochondrial membrane potential, hence initiating the release of cyt c and promoting caspase-dependent apoptosis. Numerous photosensitisers utilised in PDT have received clinical approval or are currently undergoing clinical trials.

Furthermore, the application of natural substances to induce apoptosis in cancer cells has yielded intriguing outcomes. For instance, resveratrol, curcumin, aloe-emodin and betulin have been demonstrated to induce apoptosis in several cancer cell lines by enhancing cyt c release from mitochondria. Resveratrol is undergoing Phase I clinical trials involving colon cancer patients (NCT00256334, NCT00433576). Promising preclinical outcomes achieved with FDA-approved curcumin in oncology prompted the initiation of multiple clinical trials involving cancer patients (NCT02064673, NCT02944578, NCT02782949). The nontoxicity of these natural chemicals towards noncancer cells, together with the minimal toxicity induced in normal cells by PDT, indicates that targeting cancer cells through the restoration of cyt c-driven apoptosis warrants further exploration.

Focussing on mitochondrial turnover

Mitochondrial fission, fusion and mitophagy have been examined as prospective anticancer targets. Nevertheless, limited research has explicitly examined mitophagy in contrast to broader autophagy. Mdivi-1 can obstruct fission by decreasing the GTPase activity of DRP1, hence averting mitophagy. For treatments exceeding 24 hours, mdivi-1 exhibits a cytostatic impact, which may provide an extra benefit over mitochondrial impairment in cancer therapy. Dynasore is another GTPase inhibitor of DRP1. The inhibition of mitochondrial fission by dynasore has been demonstrated to reduce the proliferation of cancer cells and induce apoptosis in A549 lung cancer cells. It inhibits migratory and/or invasive capacities in multiple cancer cell lines, including the bladder cancer cell line T24, the lung cancer cell line H1080 and the osteosarcoma cell lines MNNG/HOS, MG-63, and U2OS. In vivo studies in mice demonstrated that dynasore exhibited synergistic effects when administered in conjunction with cisplatin. Induction of mitochondrial fusion may exert anticancer effects, as exemplified by S3, a small natural compound that promotes mitochondrial fusion and consequently suppresses mitophagy. The mechanisms of action have yet to be elucidated, requiring in vivo investigations. Significantly, the inhibition of mitophagy may be one of the mechanisms of temsirolimus, a mTOR inhibitor now employed for clinical purposes beyond oncology.

Alternative mitochondrial modulators

Besides mitochondrial inhibitors, various chemicals can indirectly influence mitochondrial processes. Included is α-tocopheryl succinate (α-TOS), a derivative of α-tocopherol (vitamin E). α-tocopherol is a powerful inhibitor of cytosolic GSH S-transferases GSTP1-1 and GSTO1-1, which are enzymes that typically detoxify endogenous and exogenous chemicals by catalysing the conjugation of electrophilic centres to GSH. The analogue α-TOS preferentially induces apoptosis in cancer cells by destabilising mitochondrial integrity. This activity appears to be independent of glutathione S-transferases (GST) inhibition and predominantly involves the inhibition of ETC Complex I or Complex II. The catalytic activity of GSTs can be utilised to bioactivate prodrugs, facilitating their selective accumulation in cancer cells that exhibit elevated expression of specific GST isoenzymes. Canfosfamide/TLK286, a modified GSH analogue and nitrogen mustard prodrug, is bioactivated by GSTP1-1 into an alkylating metabolite that can covalently bind DNA. This drug advanced to Phase III clinical studies with favourable tolerability. Brostallicin exemplifies a GST-activated prodrug. It is presently undergoing Phase II clinical studies. Ketogenic diets (K.Ds), characterised by high fat and low carbohydrate intake, aim to restrict glucose availability for tumours. They were originally created to treat uncommon metabolic disorders, including GLUT1 and PDH complex deficits. K.Ds have been documented to lower blood glucose levels and enhance the utilisation of ketone bodies, resulting in a transition from glycolysis to OXPHOS. Their anticancer efficacy is predicated on the premise that cancer cells with modified mitochondria are incapable of utilising ketone bodies. Numerous research linking mitochondrial modulators to K.Ds for cancer treatment have recently been reviewed. The most notable findings relate to case reports of patients diagnosed with glioblastoma. Although anticancer effects have been documented, numerous clinical trials remain in progress. The findings of these investigations are anticipated to validate whether K.Ds, and which specific type, may serve as nutritional support to enhance the prognosis of certain cancer types. Introductory statements in scientific literature frequently characterise cancer cells as being inherently glycolytic. This reductionist perspective is founded on the reexamination of the Warburg effect, which has significantly enhanced our understanding of cancer biochemistry since 2000, but at the expense of investigating other metabolic pathways, which have been neglected for over a decade. If the Warburg phenotype confers a metabolic advantage for cancer cell proliferation, it is clear that not all cancer cells proliferate concurrently in experimental and clinical tumours. Consequently, tumour metabolism typically exhibits the same heterogeneity as the phenotypic diversity of the cancer cells that constitute the tumour. Notable instances include the metabolic cooperativity between oxidative and glycolytic cancer cells across various tumour types, as well as the finding that metastatic progenitor cells and CSCs exhibit distinct metabolic behaviours compared to the predominant population of cancer cells within a tumour. Mutations in genes encoding TCA cycle enzymes or ETC subunits significantly constrain metabolic plasticity; nonetheless, these occurrences are infrequent in uncommon tumour types. We want to reposition mitochondria into the comprehensive metabolic framework of malignancies. We aimed to deliver a thorough examination of the various roles these organelles play in cancer cells, functioning not only as energy producers but also as dynamic signalling entities that regulate cell survival and apoptosis, motility, stemness and treatment resistance. Nevertheless, critical enquiries remain unresolved by the scientific community. We believe a critical problem is to ascertain if metabolic abnormalities are a cause or a consequence of the malignant process. Literary elements endorse both possibilities. For instance, anaerobic glycolysis associated with lactic fermentation and enhanced mitophagy are adaptive survival mechanisms in response to hypoxia, thereby resulting from hypoxic conditions. The circumstances surrounding aerobic glycolysis are ambiguous, as allosteric regulation of pyruvate kinase M2 by fructose-1,6-bisphosphate (which activates the enzyme) and alanine (which inhibits the enzyme) may influence cell-cycle progression, or conversely, the cell cycle may impose metabolic cycles on cancer cells. These cycles would be characterised by alternations of energy production and biosynthesis. Experiments on mitochondrial transfer have unequivocally shown that mitochondria may convey and transmit malignant information, including the ability to metastasise. Determining whether particular metabolic changes initiate or simply accompany the phenotypic evolution of cancer cells is complex; in the former scenario, targeting these alterations may impede phenotypic progression, while in the latter, metabolic plasticity could swiftly counteract therapeutic measures. Outline of the most sophisticated therapeutic agents aimed against mitochondria in cancer treatment. It is evident to specialists that the list is limited and that the majority of the medications lack distinct objectives. This likely indicates that the area remains nascent, as numerous critical foundational findings are yet to be uncovered, which would delineate specific mitochondrial changes in cancer, thereby facilitating targeted anticancer therapies. The molecular characterisation of 'oncogenic mitochondria', defined as mitochondria that harbour and potentially disseminate malignant information, should be prioritised in fundamental research. Due to their possession of independent genetic material and their susceptibility to cellular environmental variations (including pH, pO_2, metabolite availability and treatment exposure), as well as metabolic dynamics. Scientists propose that mitochondria serve as sensors of the extracellular milieu in cancer, suggesting that they may experience Darwinian selection throughout tumour progression. Several studies have discovered mutations in mtDNA with functional implications, supporting this idea, particularly in renal oncocytomas and pancreatic malignancies. The advancement of mtDNA editing technologies is anticipated to offer experimental methodologies for tracking, characterising and repairing oncogenic mitochondria, necessitating a comprehensive understanding of mitochondrial epigenetics in cancer.

Further Reading

Beadnell, T. C., Scheid, A. D., Vivian, C. J., & Welch, D. R. (2018). Roles of the mitochondrial genetics in cancer metastasis: not to be ignored any longer. *Cancer Metastasis Rev. 37*(4), 615–632. https://doi.org/10.1007/s10555-018-9772-7.

Carrasco, A. G. M., Chammas, R., & Furuya, T. K. (2025). Mitochondrial DNA alterations in precision oncology: Emerging roles in diagnostics and therapeutics. *Clinics, 80*, 100570. https://doi.org/10.1016/j.clinsp.2024.100570.

Hossain, M. S., Islam, M. T., Hossain, M. A., Rifat, K., Maeesa, S. K., Rahman, M., Islam, M. R., Edros, R., Raihan, S. Z., Ardianto, C., Ming, L. C., Goh, B. H., Yunus, M. Y. B. M., & Shariffuddin, J. H. (2025). Role of abnormal mitochondrial DNA in Cancer: A review of molecular changes and therapeutic opportunities. *Adv. Redox Res. 16*, 100134. https://doi.org/10.1016/j.arres.2025.100134.

Kopinski, P. K., Singh, L. N., Zhang, S., et al. (2021). Mitochondrial DNA variation and cancer. *Nat. Rev. Cancer, 21*, 431–445. https://doi.org/10.1038/s41568-021-00358-w.

Kowal, K., Ziółkowska-Twarowska, K., Tkaczyk-Wlizło, A., Grzybowska-Szatkowska, L., & Ślaska, B. (2024). Defects in the Mitochondrial Genome of Dogs with Recurrent Tumours. *Int. J. Mol. Sci. 25*(24), 13414. https://doi.org/10.3390/ijms252413414.

Lin, Y. H., Lim, S. N., Chen, C. Y., Chi, H. C., Yeh, C. T., & Lin, W. R. (2022). Functional Role of Mitochondrial DNA in Cancer Progression. *Int. J. Mol. Sci. 23*(3), 1659. https://doi.org/10.3390/ijms23031659.

Mauro-Lizcano, M., Di Pisa, F., Larrea Murillo, L., et al. (2024). High mitochondrial DNA content is a key determinant of stemness, proliferation, cell migration, and cancer metastasis in vivo. *Cell Death Dis. 15*, 745. https://doi.org/10.1038/s41419-024-07103-9.

Reichert, A. S., & Neupert, W. (2004). Mitochondriomics or what makes us breathe. *Trends Genet. 20*(11), 555–562. https://doi.org/10.1016/j.tig.2004.08.012.

Vodicka, P., Vodenkova, S., Danesova, N., Vodickova, L., Zobalova, R., Tomasova, K., Boukalova, S., Berridge, M. V., & Neuzil, J. (2025). Mitochondrial DNA damage, repair, and replacement in cancer. *Trends Cancer, 11*(1), 62–73. https://doi.org/10.1016/j.trecan.2024.09.010.

Wallace, D. C. (1999). Mitochondrial diseases in man and mouse. *Science, 283*(5407), 1482–1488. https://doi.org/10.1126/science.283.5407.1482.

Warburg, O., Wind, F., & Negelein, E. (1927). The metabolism of tumors in the body. *J. Gen. Physiol. 8*(6), 519–530. https://doi.org/10.1085/jgp.8.6.519.

CHAPTER

12

Numtogenesis and its association to disease

Introduction

Nuclear mitochondrial DNA segments (NUMTs) have been identified in many eukaryotic nuclear genomes. It is the introduction of cytoplasmic DNA into the nuclear genome of the eukaryotic cell. The structural, morphological and length variations of mitochondria across various eukaryotic species correspondingly influence the variability of NUMT sequences among different organisms. Following mitochondrial modifications and morphological changes subsequent to the release of mtDNA into the cytoplasm, it is translocated to the nucleus. NUMT is present in several eukaryotic organisms. Recent research, reports and many experiments have validated the significance of NUMT in human and animal health, environmental issues, diseases and genetic disorders. Although the mechanism of action and its characteristics remain unclear, numerous reports assert its critical involvement in both biological and environmental systems. This chapter on NUMT elucidates its characteristics, provides a concise history, outlines its function, specifically the NUMT pathway and examines its association with illnesses. We have also emphasised its significance in diseases (cancer and ageing) and disorders (genetic disorders and metabolic diseases). The process of numtogenesis leads to innovative concepts explaining several unidentified diseases and abnormalities in the health of individuals, leading to mitochondrial dysfunction.

What is nuclear mitochondrial DNA segment?

NUMT have been identified in many eukaryotic nuclear genomes. It is the introduction of cytoplasmic DNA into the nuclear genome of the eukaryotic cell.

Numtogenesis refers to the process of transferring mitochondrial DNA (mtDNA) into the nuclear genome, or more broadly, the transfer of mitochondria or mitochondrial components into the nucleus.

NUMT sequences exhibit variability analogous to the variances in structure, morphology and length of mitochondria among various eukaryotic organisms. Following mitochondrial modifications and morphological changes resulting from the release of mtDNA into the

Mitochondrial Genomics
DOI: https://doi.org/10.1016/B978-0-443-26655-3.00001-9

cytoplasm, it is subsequently translocated to the nucleus. Eukaryotic organelles have evolved from prokaryotes. Consequently, in eukaryotic cells, these organelles function as supplementary DNA-containing compartments. Organelle DNA is continuously transferred to the nucleus via evolutionary processes. The natural integration of mtDNA into the nuclear genomes of eukaryotic cells is a continuous process; these nuclear copies of mtDNA are referred to as NUMTs (nuclear mtDNA) Lopez (1996). Singh et al. (2017) noted that intact mitochondria, which encompass mtDNA, mitochondrial RNA and mitochondrial proteins, also localise within the nucleus.

Srinivasainagendra et al. (2017) termed the process numtogenesis, defined as the transfer of mtDNA into the nuclear genome, or the translocation of mitochondria or mitochondrial components into the nucleus.

Factors affecting numtogenesis:

1. Srinivasainagendra et al. (2017) also identified mitostress (including ionising radiation, endotoxins, reactive oxygen species (ROS), etc.) as a requirement for numtogenesis.
2. mtDNA fragments may escape from the compromised mitochondrion during flawed mitophagy, fission, fusion and related processes. In the cytoplasm, these mtDNA sequences are safeguarded against nucleases by their association with DNA-binding histone-like proteins. These mtDNA fragments enter the nucleus through membrane fusion (Puertas and González-Sánchez, 2020) and (Hazkani-Covo et al., 2010).
3. Charlotte A. et al. (2021) reported that DNA strands replicate during the cell cycle and division. Throughout the majority of the cell cycle, the occurrence of DNA double-strand breaks (DSBs) is possible. In preserving the integrity of newly synthesised strands, non-homologous end-joining (NHEJ), a repair mechanism for DSBs, is operational during DNA replication, addressing replication-induced DSBs and reinitiating replication forks.
4. Throughout evolutionary history, segments of the mitochondrial genome, referred to as NUMTs, have been incorporated into the nuclear genome. These are passively *i*ncorporated into random chromosomal breaks, leaving sequence remnants in genomes (Hazkani and Covo, 2008).
5. Goldin et al. (2004) and Turner and Killoran (2003) indicated that NUMTs are present on all chromosomes, and the integration of NUMTs into genes has been linked to illnesses.

Blanchard and Schmidt (1996); Ricchetti M. et al. (1999); Ricchetti M. et al. (2004) and Yu and Gabriel (1999) indicated that:

1. NUMTs possess distinct characteristics that render them particularly suitable for investigating the evolutionary signatures of double-strand break repair (DSBR).
2. They do not integrate actively into the genome but are incorporated into the nuclear genome through a NHEJ mechanism.
3. They lack any inherent capability postinsertion via NHEJ, hence serving as a stable indicator of a repaired DSB.
4. NUMTs may be readily detected and differentiated from nuclear DNA through sequence analysis, as they originate from the mitochondrial genome.

NUMT insertion: The physical transport of mtDNA is evaluated to have transpired in the nucleus, succeeded by NHEJ during the DSB repair procedure. The precise mechanism remains unclear and not fully comprehended. The transfer of mtDNA to the nucleus involves its initial release into the cytoplasm. Scientists Thorness and Fox demonstrated the rate of mtDNA translocation into the nucleus utilising the URA3 plasmid, which is critical for uracil production in mitochondria. In the development of a yeast strain with a URA3 mutation, the plasmid DNA migrating from the mitochondria to the nucleus enhances uracil biosynthesis. The DNA transfer rate is significantly greater in this instance compared to the cox2 mutant example.

Mechanism of nuclear mitochondrial DNA segment insertion

Mitochondrial DNA can enter the nucleus through several distinct mechanisms

The predominant process identified thus far pertains to the breakdown of dysfunctional mitochondria. Numerous yeast mitochondrial escape (yme) strains have a significant degree of DNA translocation to the nucleus. The yme1 mutation results in the inactivation of the yme1p protein, a mitochondrial ATP-dependent metalloprotease, which consequently increases the escape rate of mtDNA to the nucleus. The mitochondria of the yme1 strain are subjected to destruction by the vacuole more frequently than those of the wild-type strain.

Alternative mechanisms for transferring mtDNA into the nucleus have been proposed, including:

1. lysis of the mitochondrial compartment,
2. encapsulation of mtDNA within the nucleus,
3. direct physical interaction between mitochondria and the nucleus and membrane fusion events.

mtDNA that penetrates the nucleus can assimilate into nuclear chromosomes. The integration of mtDNA into the chromosome during DSB repair is referred to as NHEJ. The insertion entails two DSB repair events. The involvement of short microhomology in each repair is uncertain. In NHEJ mediated by microhomology, base-pair complements are available between the NUMT and the chromosomal ends, analogous to the sticky ends produced by restriction enzymes. Singh et al. (2017) noted that whereas the majority of NUMT insertions transpire in the noncoding areas of nuclear genes, several findings indicate their occurrence in coding sequences as well. This may lead to the formation of protein-coding exons within existing genes. Ricchetti et al. (2004) reported the insertion of NUMTs in the human genome within coding areas, subsequently altering the layouts of exons and introns in specific genes.

The invasion of NUMTs results in three significant consequences:

1. harmful mutations that may modify or deactivate gene function,
2. incorporation of exons that can reshape existing genes and potentially introduce new functions to the encoded proteins, and
3. no modifications to gene sequence and no changes in gene regulation.

The mechanisms of mtDNA insertion and the repair of DSBs comprise the following steps:

1. DNA segment alignment,
2. DNA end-processing,
3. DNA synthesis and
4. DNA ligation.

Mechanism depicting nuclear mitochondrial DNA segment insertion pathway

Fig. 12.1 (1) illustrates the various *yme* strains migrating to the nucleus from the DNA, with this mutant strain responsible for the inactivation of the Yme1 protein, which is crucial for the mtDNA escape rate to the nucleus, (2) illustrates the lysis of the mitochondrial compartment, (3) illustrates the encapsulation of mtDNA within the nucleus, and (4) illustrates the physical integration of mitochondria with the nucleus and membrane fusions. Finally, the mtDNA that enters the nucleus can integrate into the nuclear chromosomes after the repair of DSBs by NHEJ. The procedure entails two DSBs healing processes. The NHEJ base pairs are complementary between the NUMT and the chromosomal ends, similar to the sticky ends generated by restriction enzymes. The insertion of NUMTs in the genome is seemingly linked to several disorders. The incorporation of the nuclear genome can negatively impact human health in certain instances. A significant aspect of numtogenesis is the extent of variety within the NUMT. Genetic allele polymorphisms of NUMTs may serve as markers for human phylogenetic study and influence susceptibility to cancer and other genetic disorders. Richetti et al. studied the extent of polymorphism in NUMTs and detected 41 NUMTs, of which six were

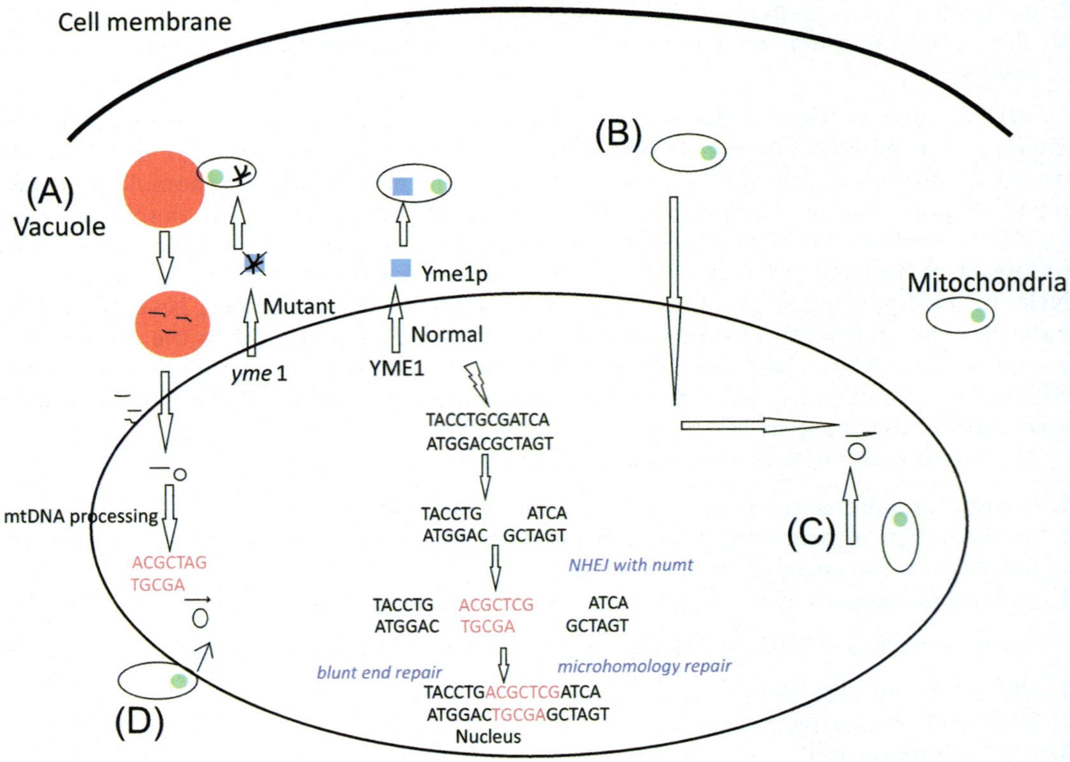

FIGURE 12.1 Mechanism depicting nuclear mitochondrial DNA segment (NUMT) insertion pathway.

associated with insertion polymorphisms. Other analogous investigations have demonstrated the genetic diversity and variation of NUMTs. Numtogenesis resulting in multiple genetic illnesses has been investigated by numerous distinguished scientists. Disorders and diseases identified by scientists due to NUMT insertion include Plasma factor VII deficiency, with the insertion site at FVII splice site measuring 251 base pairs; Pallister—Hall syndrome, where the insertion site is GLI family zinc finger 3 (GLI3) exon 14 with a length of 72 base pairs; mucolipidosis IV, characterised by a NUMT insertion at MCOLN1 exon 2, measuring 93 base pairs; and Usher syndrome type IC, which posess a NUMT insertion at USH1C exon 9, with a length of 36 base pairs. Somatic abnormalities in mtDNA copy quantity, D loop and mitochondrial gene modifications may play a significant role in cancer. mtDNA is transferred from the mitochondria to the nuclear genome, resulting in various alterations, which include: (1) inactivation of a tumour suppressor gene, (2) activation of a proto-oncogene, (3) epigenetic dysregulation of either type of gene, and (4) introduction of an active oncogene. A progressive increase of mtDNA in the nuclear genome can induce active alterations in the total cellular proteome and contribute to cancer development. The work on HeLa cells was the primary indication of mitochondrial insertions in carcinogenesis. This type of pseudogenes was identified in human cell lines and in the human placenta. This was the inaugural report linking mtDNA to malignant transformation. Nielsen et al. subsequently investigated that the NUMT insertion exhibited autonomous replication capacity upon transfection into HeLa cells. Ju et al. indicate that the earliest somatic NUMTs coexisted with other somatic rearrangements of the nuclear genome. The interactions were particularly pronounced in breast cancer tissues at the insertion site within the fifth intron of the potassium channel potassium calcium-activated channel subfamily M alpha 1 (KCNMA1) gene, which is predominantly amplified in breast and prostate malignancies. mtDNA was identified in malignant gastric mucosal cells; however, he could not ascertain a robust mechanism associated with this observation. A further investigation identified a fragment of the mitochondrial cytochrome-c oxidase subunit III incorporated into exons 2 and 3 of the c-myc gene. The prevalence of c-myc changes is contingent upon the concentration of NUMTs. The group posited that the incorporation of mtDNA fragments into the nucleus of cervical epithelial cells affects carcinogenesis through the activation of the c-myc gene, while lacking evidence of direct interaction between NUMTs and c-myc. Ju et al. reveal that NUMTs are located in the nuclear genome with other somatic configurations. The largest effects were observed in breast cancer tissues, where the NUMT insertion in the fifth intron of the potassium channel KCNMA1 gene typically enhances breast and prostate malignancies.

The nonfunctional NUMT genes in somatic cells may contribute to carcinogenesis and ageing. Cheng and Ivessa demonstrated that the mutant strains of *S. cerevisiae* exhibit an elevated frequency of mtDNA migration. Thorsness and Fox emphasised that yeast strains exhibiting a faster rate of mtDNA migration to the nucleus increase subsequent ageing. Caro and his colleagues conducted an experiment on eukaryotic organisms using rats as a model, employing RT-PCR quantification and in situ hybridisation of mtDNA to compare young and old rats. They discovered a significant increase in the quantity of mitochondrial sequences in nDNA as the rats aged. Consequently, these evidences suggest that mitochondria may pose a hazard and contribute to the ageing process.

Association of nuclear mitochondrial DNA segment with diseases

Recent research has clarified the relationship between NUMTs-induced mutations and illnesses. Turner and Killoran (2003) reported a case study of Pallister–Hall syndrome, identifying a 72-bp NUMT insertion in exon 14 of the GLI3 gene as the causal mutation. Hassanin et al. (2010) were the first to report NUMTs utilising a low-quality genome from *Capra hircus*, the Vietnamese domestic goat. Ning et al. (2017) discovered 118 extremely homogeneous NUMT segments across the mitochondrial and nuclear genomes in the Yunnan black goat. Srinivasainagendra et al. (2017) observed enhanced somatic transfer of mtDNA in colorectal tumours in their study.

They disclosed that there are sex-based disparities in the frequency of NUMT occurrences and indicated that the existence of NUMT in the blood correlates with its presence in tumours, implying that NUMT may serve as a biomarker for carcinogenesis. They also disclosed that numtogenesis plays a significant role in cancer development. Chen S. et al. (2022) conducted a study on the correlation between mtDNA and thyroid cancer, reporting for the first time that mtDNA mutations contribute to tumour growth.

History of nuclear mitochondrial DNA segment

The process of recombining, amalgamating and properly regenerating cytoplasmic mtDNA into nuclear DNA is known as NUMT. The term NUMT was initially utilised in reference to a domestic cat, whereby the mitochondrial gene quantity and composition were amplified 38–76 times within the animal's nuclear genome. In this instance, 7.9 kb of a 17 kb mitochondrial genome was translocated to a designated nuclear locus. Whole genome sequencing has been conducted in numerous eukaryotes, encompassing both vertebrates and invertebrates, revealing the presence of NUMTs in diverse organisms such as yeast, sea urchins, locusts, honey bees, rats, maize, rice and primates. In several taxa, such as Plasmodium, Anopheles gambiae and Aedes aegypti mosquitoes, NUMT is scarcely detectable and has not been thoroughly examined. In contrast, the conserved portions of NUMT have been detected infrequently in the genomic data of Ciona intestinalis, Neurospora crassa, Schizosaccharomyces pombe, Caenorhabditis elegans, Drosophila melanogaster and Rattus norvegicus. NUMT was initially identified in the fish genome in 2005 by the application of several bioinformatics technologies, including BLASTN, MAFFT, genome mapping and phylogenetic analysis.

Multiple parameters are significant for the rate of mitochondrial DNA transfer

One significant aspect is the ROS produced in the mitochondria. Certain mutagens and cellular stressors can disrupt the mitochondrial membrane, whereas external factors such as ionising radiation and chemical genotoxic ants elevate the frequency of mtDNA translocation to the nucleus. Researchers have conducted additional investigations on 21 nuclear mutants exhibiting various combinations of mutations across 12 nuclear loci identified as yme, which can disrupt normal mitochondrial activities due to alterations in genetic products. The disruption of pep4, a significant proteinase, facilitates the translocation of mitochondria to the nucleus. The disruption of the protein protein regulator of cytokinesis 1 (PRC1), which functions as a carboxypeptidase, leads to a reduction in the rate of mtDNA escape into the nucleus. Other processes that may be involved in this transfer include mitophagy, lysis of the mitochondrial membrane and fusion with the nucleus, among others.

Preinsertion arrangements

Following the arrival of mtDNA in the nucleus, it must be integrated into the nuclear genome. The incorporation of mtDNA is contingent upon the occurrence of DSBs in the DNA and the rate of mtDNA release. Throughout this entire process, mtDNA must be accurately edited and sequenced, contingent upon the timing of its arrival in the nucleus. The incorporation of unedited mtDNA into the nucleus is quite achievable in plants, such as Arabidopsis, and mammals, utilising many ways, including bioinformatic analyses like BLAST-based assessments. The editing of mtDNA transpires subsequent to its entry into the nucleus. When this occurs, a gene transferred to the nucleus in one lineage prior to mitochondrial editing will be less distinguishable from the revised transcript than from the remaining mitochondrial copies at the edited segments, while it remains in the organelle in other lineages where editing takes place. In an alternative cDNA-mediated hypothesis, which is less compelling, the intron-containing mtDNA translocates to the nucleus, where it undergoes reverse transcription of the spliced mitochondrial transcript, subsequently integrating into the nuclear DNA. The intronless mtDNA insertion transpires in the nucleus, where editing and introns in the mitochondria fluctuate over evolution. The commencement and removal of the intron, together with reverse transcription, occur within the mitochondria, resulting in an unedited and intronless mtDNA that is subsequently integrated into the nDNA after translocation to the nucleus.

Conclusive integration into the nuclear genome

According to the data and prior discussion, mtDNA is integrated into the DSB by the mechanism of NHEJ. This hypothetical evidence is widely acknowledged. This methodology was subsequently validated in the context of NUMT insertion in humans. This occurs in both somatic and germline cells. In both animals and humans, similar processes occur in somatic and germline cells. The capacity for DSB repair in germline cells of animals and humans is contingent upon the oogenetic and spermatogenetic stages, owing to their limited restorative power, and mature sperm lack the capability to repair DSBs. DSBs can be repaired through homologous recombination, which is notably precise and exhibits a reduced likelihood of error during the process (see Table 12.1).

TABLE 12.1 Diseases associated with nuclear mitochondrial DNA segment (NUMT).

S. no	Disease	NUMT sequence insertion site	References
1.	Plasma factor VII deficiency	NUMT insertion site is FVII, splice site and is of 251 base pairs	Borensztajn et al. (2002)
2.	Pallister–Hall Syndrome	NUMT insertion site is GLI3, exon 14, with a 72 base-pair length	Turner and Killoran (2003)
3.	Mucolipidosis IV	MCOLN1, exon 2 and the NUMT size is 93 base pairs	Goldin et al. (2004)
4.	Usher syndrome type IC	USH1C, exon 9 and the length of the NUMT is 36 base pairs	Ahmed and Smith (2002)

Extrinsic influences influencing numtogenesis

Environmental variables also facilitate the transport of mtDNA to the nucleus. These stressors are termed 'numtogens'. The initial phase of numtogenesis involves mitochondrial degradation. Mitostress or mitotoxicity is induced by numtogens. Elevated radiation levels are regarded as a mutagen, causing mtDNA fragments to escape into the cytoplasm at an increased rate in cells subjected to radiation. The incorporation of mtDNA into the nuclear genome entails the mechanism of DSB in DNA caused by ionising radiation. Subsequently, they investigated that the prevalence of numtogenesis was greater in the Gallus egg nuclear genome subjected to X-rays. Moreover, it was observed that ionising radiation elevated the prevalence of NHEJ in irradiated yeast cells, indicating that the occurrence of NUMTs is twice as high compared to nonirradiated cells. Cheng and Ivessa indicated that the rate of mtDNA transfer to the nucleus escalated with the age of the yeast cells. The research indicated that the numtogens implicated were free radicals, which grow with age and might result in mitochondrial defects that may contribute to cancer. Heat stress is regarded as a mutagen for tobacco, as temperatures exceeding 55°C induce DSBs in cytoplasmic DNA, ultimately leading to its transfer and integration into the nuclear genome. Only a single instance of numtogenesis in Pallister–Hall disease has been documented. NUMT—paternally inherited mtDNA: Initially, we believed that mtDNA is only inherited maternally. Subsequently, specific evidence was obtained about paternal inheritance of mtDNA. Paternal mitochondria are localised in the midpiece of spermatozoa. It has also been proposed that the elevated motility of sperm required for fertilisation necessitates substantial energy provided by the sperm's mitochondria (see Fig. 12.2).

Recent results indicate biparental transfer of mtDNA. Analysis of entire mitochondrial genome sequencing and whole nuclear genome sequencing revealed that rare cryptic mega NUMTs can mimic paternal DNA heteroplasmy, suggesting paternal inheritance of nuclear mtDNA. NUMT and its detection methods: Figs. 12.3 and 12.4 illustrate the diagrammatic representation of NUMT and its detection methodology (see Figs. 12.3 and 12.4).

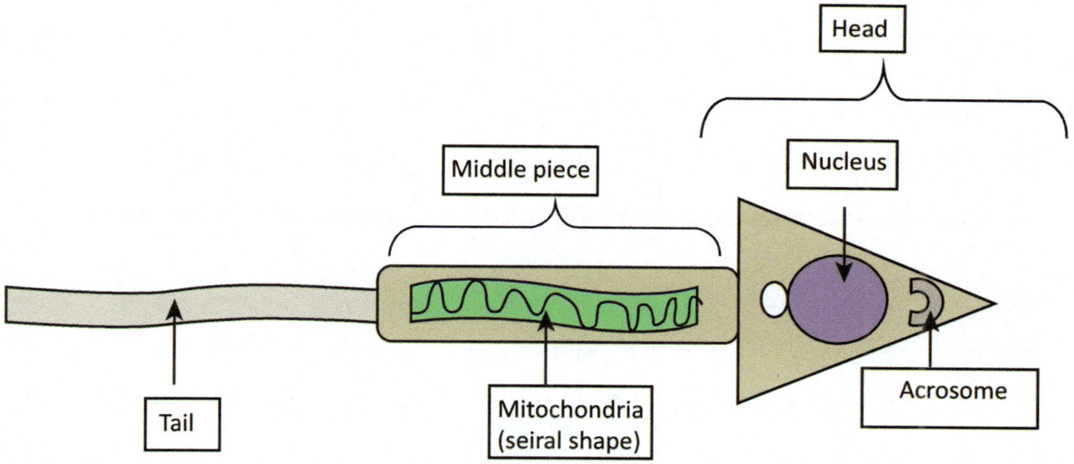

FIGURE 12.2 Schematic diagram for spermatozoa reflecting its different parts and the spiral-shaped mitochondria (green) in the mid-piece.

Conclusion and prospective developments

We have endeavoured to consolidate many facts and details concerning NUMT and the process of numtogenesis. Mitochondria and mtDNA have a significant role and can regulate various facets of the immune system. The intricate process of mtDNA insertion into the nucleus and subsequently into the nuclear genome remains somewhat opaque; however, it has been hypothesised, experimented upon and observed by scientists, who elucidate the underlying theories to raise awareness about this phenomenon. We have demonstrated the endeavours and research conducted by several scientists who have documented significant information concerning its involvement in illness occurrence and immunity. The significance of numtogenesis in cancer development and tumorigenesis has been documented, potentially advancing cancer biology through the discovery of biomarkers, accurate detection and therapy of cancer. Comprehending the process of mtDNA insertion into the nucleus may facilitate the understanding of diseases, defects and numerous health difficulties. NUMTs provide insights into species history and enhance our understanding of population genetics. It presents an uncharted possibility to integrate nuclear markers with mitochondrial markers, hence facilitating the evolution of mitochondrial and nuclear genomes and their associated immunology. The analysis of mtDNA across an expanding array of species reveals increasing complexity, with numerous surprises still awaiting discovery. The relationship between mitochondrial-nuclear signalling and its association with pharmacological reactions, physical fitness and disease development. Mitochondrial retrograde signalling refers to the communication and regulation from mitochondria to the nucleus. The phenomenon was detected in the model yeast *S. cerevisiae*, where two significant mitochondrial-nuclear signalling genes, RTG1 and RTG2, were recognised (although these genes are not uniformly conserved in fungi, as evaluated by BLAST against

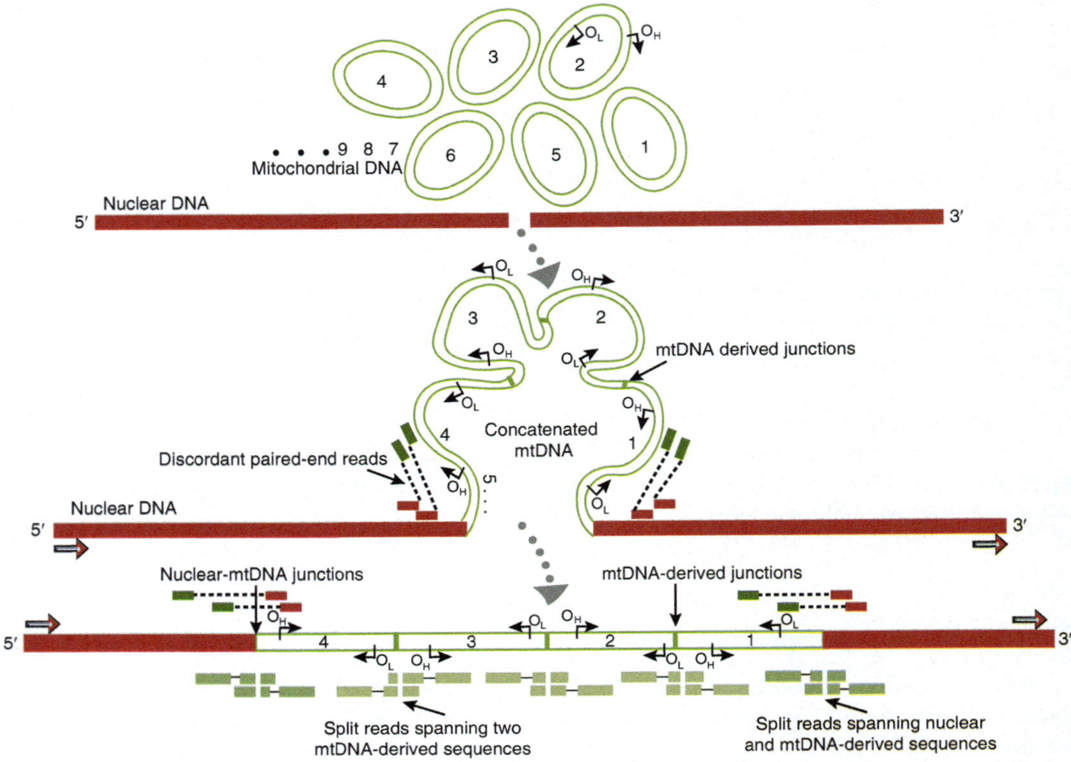

FIGURE 12.3 Nuclear mitochondrial DNA segment (NUMT) and its detection methodology.

NCBI and fungal-specific genome databases). Retrograde signalling in *S. cerevisiae* has predominantly been examined concerning modified food metabolism in reaction to variations in mitochondrial activity. Nonetheless, numerous other significant facets are linked to retrograde signalling. Fig. 12.5 illustrates a genetic screen for the negative regulators of pleiotropic drug resistance genes with a microarray-based genome-wide transcriptional analysis of mitochondria.

1a. Intraspecific (in-group) crosses: Two mating scenarios are observed within the same species: (ai) Same-sex mating (a–a): This results in progeny with recombinant mitochondrial genomes while retaining isogenic nuclear genomes, due to lack of nuclear exchange. (aii) Opposite-sex mating (a–a): Both mitochondrial and nuclear genomes undergo recombination, leading to progeny with recombinant mitochondrial and nuclear backgrounds. Mitochondrial inheritance in these intraspecific crosses is predominantly biparental, involving mitochondrial contribution from one parent at a time. However, in crosses between isogenic strains, mitochondrial transmission is strictly uniparental.

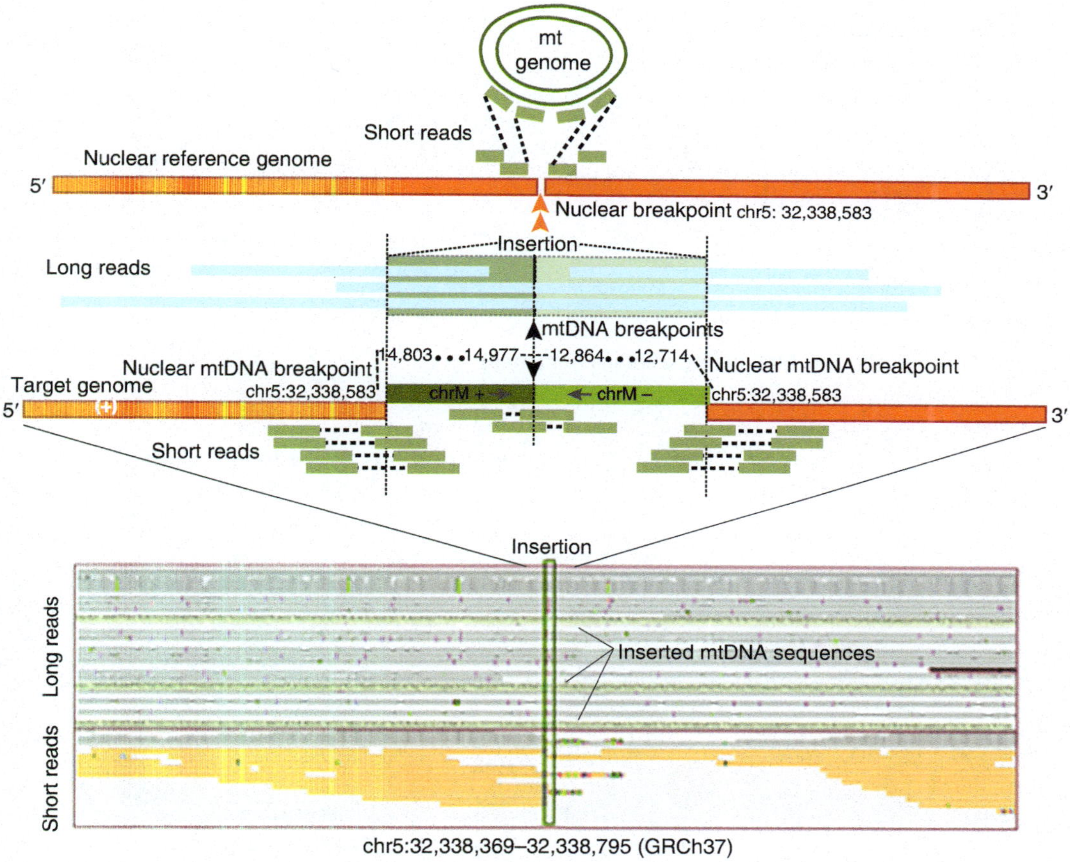

FIGURE 12.4 Validation of nuclear mitochondrial DNA segments (NUMTs) using long read (Oxford nanopore promethylON sequencing).

1b. Interspecific (outgroup) crosses: In crosses between different species, incompatibilities between divergent nuclear genomes can result in postzygotic meiotic arrest or failure. Nonetheless, mitochondrial genomes may still undergo recombination. As a result, viable progeny may contain recombinant mitochondrial genomes within a nuclear background that is isogenic to one of the parental strains.

1c. Potential mitochondrial and nuclear genomic combinations in progeny:

(i) Recombinant mitochondrial genome within an isogenic nuclear background

(ii) Recombinant mitochondrial genome within a recombinant nuclear background

(iii) Nonrecombinant mitochondrial genome from one parent paired with a nonrecombinant nuclear genome from the other parent

(iv) The reciprocal of (iii), that is nonrecombinant mitochondria from the second parent with nonrecombinant nuclei from the first

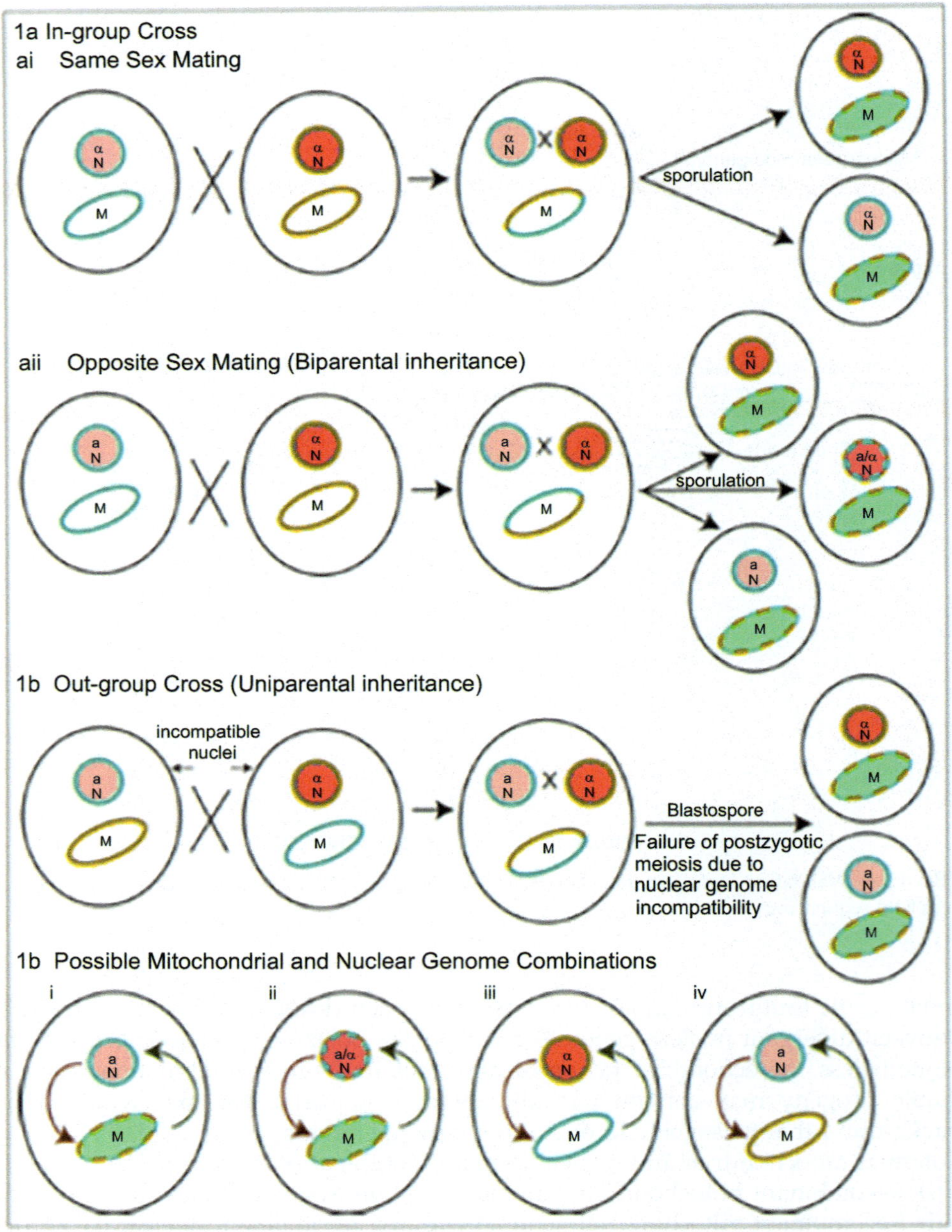

FIGURE 12.5 Possible nuclear and mitochondrial genetic genotypes in *Cryptococcus* species in progeny from crosses contributing to their virulence.

Bidirectional regulatory signalling between mitochondria and nuclei is illustrated as follows:

- Light blue nuclei with green outlines and dark blue nuclei with red outlines represent distinct parental nuclear genotypes. Nuclei with mixed (dotted) outlines denote recombinant nuclear genotypes.
- Mitochondria from different parental backgrounds are depicted as ellipses with either green or red outlines, while recombinant mitochondria are shown with red-green dotted outlines and yellow fill.
- Purple arrows indicate nucleus-to-mitochondrion signalling, whereas brown arrows represent mitochondrion-to-nucleus signalling, denoting interorganelle regulatory interactions.

References

Ahmed, Z. M., & Smith, T. N. (2002). Nonsyndromic recessive deafness DFNB18 and Usher syndrome type IC are allelic mutations of USHIC. *Hum Genet, 110*, 527–531.

Blanchard, J. L., & Schmidt, G. W. (1996). Mitochondrial DNA migration events in yeast and humans: integration by a common end-joining mechanism and alternative perspectives on nucleotide substitution patterns. *Molecular Biology and Evolution, 13*(6), 893. https://doi.org/10.1093/oxfordjournals.molbev.a025650.

Borensztajn, K., et al. (2002). Plasma factor VII deficiency and bleeding diathesis. *Br J Haematol, 117*, 168–171.

Goldin, E., Stahl, S., et al. (2004). Transfer of a mitochondrial DNA fragment to MCOLN1 causes an inherited case of mucolipidosis IV. *Hum Mutat, 24*, 460–465.

Hassanin, A., Bonillo, C., Nguyen, B. X., & Cruaud, C. (2010). Comparisons between mitochondrial genomes of domestic goat (*Capra hircus*) reveal the presence of numts and multiple sequencing errors. *Mitochondrial DNA, 21*(3–4), 68–76. https://doi.org/10.3109/19401736.2010.490583.

Hazkani-Covo, E., & Covo, S. (2008). Numt-mediated double-strand break repair mitigates deletions during primate genome evolution. *PLoS Genetics, 4*(10), e1000237. https://doi.org/10.1371/journal.pgen.1000237.

Hazkani-Covo, E., Zeller, R. M., & Martin, W. (2010). Molecular Poltergeists: Mitochondrial DNA Copies (numts) in Sequenced Nuclear Genomes. *PLoS Genet, 6*(2), e1000834. https://doi.org/10.1371/journal.pgen.1000834.

Lopez, J. V., et al. (1996). Complete Nucleotide Sequences of the Domestic Cat (Felis Catus) Mitochondrial Genome and a Transposed MtDNA Tandem Repeat (Numt) in the Nuclear Genome. *Genomics, 33*(2), 229–246.

Puertas, M. J., & González-Sánchez, M. (2020). Insertions of mitochondrial DNA into the nucleus-effects and role in cell evolution. *Genome, 63*(8), 365–374. https://doi.org/10.1139/gen-2019-0151.

Ricchetti, M., Fairhead, C., & Dujon, B. (1999). Mitochondrial DNA repairs double-strand breaks in yeast chromosomes. *Nature, 402*, 96–100.

Ricchetti, M., Tekaia, F., & Dujon, B. (2004). Continued colonization of the human genome by mitochondrial DNA. *PLoS Biology, 2*(9), E273. https://doi.org/10.1371/journal.pbio.0020273.

Singh, K. K., Choudhury, A. R., & Tiwari, H. K. (2017). Numtogenesis as a mechanism for development of cancer. *Seminars in Cancer Biology, 47*, 101–109. https://doi.org/10.1016/j.semcancer.2017.05.003.

Srinivasainagendra, V., Sandel, M. W., Singh, B., et al. (2017). Migration of mitochondrial DNA in the nuclear genome of colorectal adenocarcinoma. *Genome Med, 9*, 31. https://doi.org/10.1186/s13073-017-0420-6.

Turner, C., & Killoran, C. (2003). Human genetic disease caused by de novo mitochondrial-nuclear DNA transfer. *Hum Genet, 112*, 303–309.

Yu, X., & Gabriel, A. (1999). Patching broken chromosomes with extranuclear cellular DNA. *Molecular Cell, 4*, 873–881.

Further Reading

Ashrafi, G., & Schwarz, T. L. (2013). The pathways of mitophagy for quality control and clearance of mitochondria. *Cell Death Differ, 20*(1), 31–42.

Balciuniene, J., & Balciunas, D. (2019). A Nuclear mtDNA Concatemer (Mega-NUMT) Could Mimic Paternal Inheritance of Mitochondrial Genome. *Front Genet, 10*, 518. https://doi.org/10.3389/fgene.2019.00518 PMID: 31244882; PMCID: PMC6563850.

Bolisetty, S., et al. (2013). Mitochondria and Reactive Oxygen Species: Physiology and Pathophysiology. *Int J Mol Sci, 14*(3), 6306–6344.

Caro, et al. (2010). Mitochondrial DNA sequences are present inside nuclear DNA in rat tissues and increase with age. *Mitochondrion, 10*(5), 479–486.

Chan, C. Y., & Kiechle, M. (2007). Ionizing radiation and restriction enzymes induce microhomology-mediated illegitimate recombination in Saccharomyces cerevisiae. *Nucleic Acids Res, 35*, 5051–5059.

Cheng, X., & Ivessa, A. S. (2010). The migration of mitochondrial DNA fragments to the nucleus affects the chronological aging process of Saccharomyces cerevisiae. *Aging Cell, 9*, 919–923.

Delsite, R., et al. (2002). Nuclear genes involved in mitochondria-to-nucleus communication in breast cancer cells. *Mol Cancer, 1*, 6.

Falkenberg, M. (2018). Mitochondrial DNA replication in mammalian cells: overview of the pathway. *Essays Biochem, 62*(3), 287–296.

Fields, P. D., Waneka, G., Naish, M., Schatz, M. C., Henderson, I. R., & Sloan, D. B. (2022). Complete Sequence of a 641-kb Insertion of Mitochondrial DNA in the Arabidopsis thaliana Nuclear Genome. *Genome Biol Evol, 14*(5), evac059. https://doi.org/10.1093/gbe/evac059.

Gaziev, A. I., & Shaikhaev, G. O. (2010). Nuclear Mitochondrial Pseudogenes. *Mol Biol Mol Biol, 44*(3), 358–368.

Gaziev, A. I., & Shaĭkhaev, G. O. (2007). Ionizing radiation can activate the insertion of mitochondrial DNA fragments in the nuclear genome. *Radiats Biol Radioecol, 47*, 673–683.

Golubenko, M. V., & Puzyrev, V. P. (2024). Liberties of the genome: insertions of mitochondrial DNA fragments into nuclear genome. *Vavilovskii Zhurnal Genet Selektsii, 28*(5), 467–475. https://doi.org/10.18699/vjgb-24-53.

Hamanaka, R. B., & Chandel, N. S. (2010). Mitochondrial reactive oxygen species regulate cellular signaling and dictate biological outcomes. *Trends Biochem Sci, 35*(9), 505–513.

Help me understand Genetics". Genetics Home Referrence. US National Library of medicine (2016).

Henz, K., et al. (2001). How Do Mitochondrial Genes Get into the Nucleus. *Trends Genet, 17*(7), 383–387.

Hisayuki, N., et al. (1985). Molecular Structures of Mitochondrial-DNA-Like Sequences in Human Nuclear DNA. *Nucleic Acids Res Nucleic AcidsResearch, 13*(5), 1649–1658.

Hoy, M. A. (2013). *Insect Molecular Genetics: An Introduction to Principles and Applications*"SanDiego. *Academic Press*, 613–620.

Ju, Y. S., et al. (2015). "Frequent somatic transfer of mitochondrial DNA into the nuclear genome of human cancer cells. *Genome Res, 25*(6), 814–824.

Kamimura, S., et al. (1989). Three separate mitochondrial DNA sequences are contiguous in human genome cDNA. *J Mol Biol, 210*, 703–707.

Michael, R. L. (2010). The Mechanism of Double-Strand DNA Break Repair by the Nonhomologous DNA End Joining Pathway. *Annu Rev Biochem, 79*, 181–211.

Mishmar, D., et al. (2004). Mitochondrial DNA-like sequences in the nucleus (NUMTs): Insights into our African origins and the mechanism of foreign DNA integration. *Hum Mutat, 23*(2), 125–133.

Nielsen, T., & Piatyszek, M. (1994). Autonomous replication activity of a human mitochondrial–DNA sequence inserted into genomic DNA. *Int J Oncol, 5*, 1003–1008.

Oeggerli, M., Tian, Y., et al. (2012). *PLoS One, 7*, e41664.

Pal, A., & Banerjee, S. (2018). Mitochondrial replacement therapy - a new remedy for defects in reproduction. *IndianJournalofAnimalSciences, 88*(6), 637–644.

Pal, A., Banerjee, S., Pathak, K., & Das, M. (2019). Whole mitochondrial genome sequencing Chotanagpuri sheep West Bengal, www.ncbi.nlm.nih.in. MN025257.

Pal, A., Pal, A., Banerjee, S., Batobyal, S., & Chatterjee, P. N. (2019). Mutation in Cytochrome B gene causes debility and adverse effects on health of sheep. *Mitochondrion, 46*, 393–404. https://doi.org/10.1016/j.mito.2018.10.003.

Pradhan, M., Pal, A., Samanta, A. K., Banerjee, S., & Samanta, R. (2019). Mutations in cytochrome B gene effects female reproduction of Ghungroo pig. *Theriogenology, 119*, 121–130.

Ricchetti, M. (2014). Continued colonization of the human genome by mitochondrial DNA. *PLoS Biol*, E273.

Richly, E. (2004). NUMTs in Sequenced Eukaryotic Genomes. *Mol Biologyand Evolution, 21*(6), 1081–1084.

Rogers, et al. (2012). Mitochondrial pseudogenes in the nuclear genomes of Drosophila. *PLoS ONE, 7*(3), e32593.

Sarah, F. P., et al. (2017). Regulation of Mammalian Mitochondrial Gene Expression: Recent Advances Trends. *Biochem Sci, 42*(8), 625–639.

Schwartz, M., & Vissing, J. (2002). Paternal inheritance of mitochondrial DNA. *N engl Jmed, 347*(8), 576–580.

Seok, J. Y. (2015). Abstract LB-161: Frequent Somatic Transfer of Mitochondrial DNA into the Nuclear Genome of Human Cancer Cells. *Cancer Res Cancer Res, 75*(15).

Shay, J. W. (1992). New evidence for the insertion of mitochondrial DNA into the human genome: significance for cancerand aging. *Mutat Res, 2752*, 227–235.

Shen, S., et al. (2016). Exceptionally High Cumulative Percentage of NUMTs Originating from Linear Mitochondrial DNA Molecules in the Hydra Magnipapillata Genome. *BMC Genomics, 14*(1), 447.

Thorsness, & Fox (2011). Nuclear Mitochondrial DNA Activates Replication in *Saccharomyces cerevisiae. PLoS One, 6*(3), e17235.

Thorsness, P. E., & Fox, T. D. (1990). Escape of DNA from Mitochondria to the Nucleus in Saccharomyces Cerevisiae. *Nature, 346*(6282), 376–379.

Thyagarajan, B., et al. (2012). Mitochondrial copy number is associated with colorectal cancer risk. *Cancer", Epidemiol Biomarkers Prev, 21*, 1574–1581.

Wai, T., & Langer, T. (2016). Mitochondrial dynamics and metabolic regulation. *TrendsEndocrinol Metab, 27*(2), 105–117.

Wang, D., et al. (2012). Environmental stress increases the entry of cytoplasmic organellar DNA into the nucleus in plants. *Proc Natl Acad Sci U S A, 109*, 2444–2448.

Wei, W., Pagnamenta, A. T., Gleadall, N., et al. (2020). Nuclear-mitochondrial DNA segments resemble paternally inherited mitochondrial DNA in humans. *Nat Commun, 11*, 1740. https://doi.org/10.1038/s41467-020-15336-3.

Weinberg, S. E., Sena, L. A., & Chandel, N. S. (2015). Mitochondria in the regulation of innate and adaptive immunity. *Immunity, 42*(3), 406–417 17.

Xianlong, L., et al. (2001). Relationship between integration of mtDNA fragments in the nuclei of gastric mucosal cells and Helicobacterpyiori infection. *Acta Acad Med Militaris Tertiae, 23*, 1043–1046.

CHAPTER

13

Mitochondrial replacement therapy: its prospects

Introduction

In addition to genomic DNA, the present emphasis is also on mitochondrial DNA (mtDNA). Mitochondria are crucial for supplying energy to cells via oxidative phosphorylation, including reproductive cells at any age; thus any failure in mitochondrial function results in diseases of the reproductive process. Heteroplasmy is a prevalent occurrence in mitochondrial mutations, with levels approaching or exceeding the threshold (>60%) posing a significant risk of transmission. Owing to the erratic epigenetic, genetic, genotypic and phenotypic interactions between mitochondria and nuclei, disorders resulting from mitochondrial defects are commonly referred to as 'mystondria' (mystery mitochondrial diseases). Mitochondrial illnesses are profoundly debilitating, frequently lethal and inherently complicated. Given the absence of potential for cure or treatment, the sole recourse is mitochondrial replacement therapy. The current topic addresses the impact of mtDNA on reproduction, the severe phenotypic effects resulting from mtDNA abnormalities, and potential strategies to mitigate these issues. Mitochondrial replacement treatment is an advanced topic to prevent the birth of offspring with mitochondrial defects, with high mortality due to defective heart, lungs, brain, kidney or any other vital organs of the body.

Fundamentals of mitochondria:

Although we have already discussed in detail in our earlier chapters, we try to mention the concept for better grasping of the topic. mtDNA comprises the genetic material found within the mitochondria of eukaryotic cells, commonly referred to as the 'powerhouse'. Humans possess 16,569 base pairs of mtDNA that encode a mere 37 genes. mtDNA was the crucial component of the human genome to be sequenced. Thirteen genes participate in oxidative phosphorylation, while the remaining genes are important for the synthesis of transfer RNA and ribosomal RNA, which facilitate the conversion of amino acids into proteins. A comparable trend is also evident for cattle species. mtDNA, with a distinct evolutionary origin from nuclear DNA, changes at a more rapid pace than nuclear genetic markers. mtDNA regulates cellular energy metabolism through oxidative phosphorylation, rendering it vulnerable to oxidative damage due to its proximity to the electron transport chain, where reactive oxygen species are produced, thus making it significantly more prone to mutations than nuclear DNA. The mutation-prone character of mtDNA is variably linked

Mitochondrial Genomics
DOI: https://doi.org/10.1016/B978-0-443-26655-3.00014-7

to numerous genetic diseases in mammals. mtDNA is predominantly passed matrilineally in most organisms, although paternal inheritance of mtDNA can occur in certain instances. Mammalian offspring inherit their mtDNA from the mother's oocyte, containing the cytoplasm through the process of fertilisation. Mitochondrial natural selection occurs exclusively in females, underscoring the phenomenon known as 'Mother's Curse', whereby males inherit detrimental mitochondrial genome changes from their mothers, but these mutations may be helpful, neutral, or less harmful to females (see Fig. 13.1).

Mitochondrial replacement therapy: Mitochondrial replacement therapy is a specialised type of in vitro fertilisation in which the mitochondria of the prospective child are sourced from a different individual. This procedure is conducted when the mother has faulty or damaged mitochondria. Numerous critical mitochondrial disorders arise from malfunctioning or altered mitochondria in offspring, as they are maternally inherited. Mitochondrial dysfunctions can result in various multiorgan abnormalities and failures. This chapter emphasises the impact of defective mtDNA on animal reproduction. Research on this issue is sparse, and there is a paucity of evidence concerning the impact of mitochondrial abnormalities on animal reproduction or reproductive organs. It has been effectively executed by certain researchers in primates. Our practical experience with farm animals indicates that

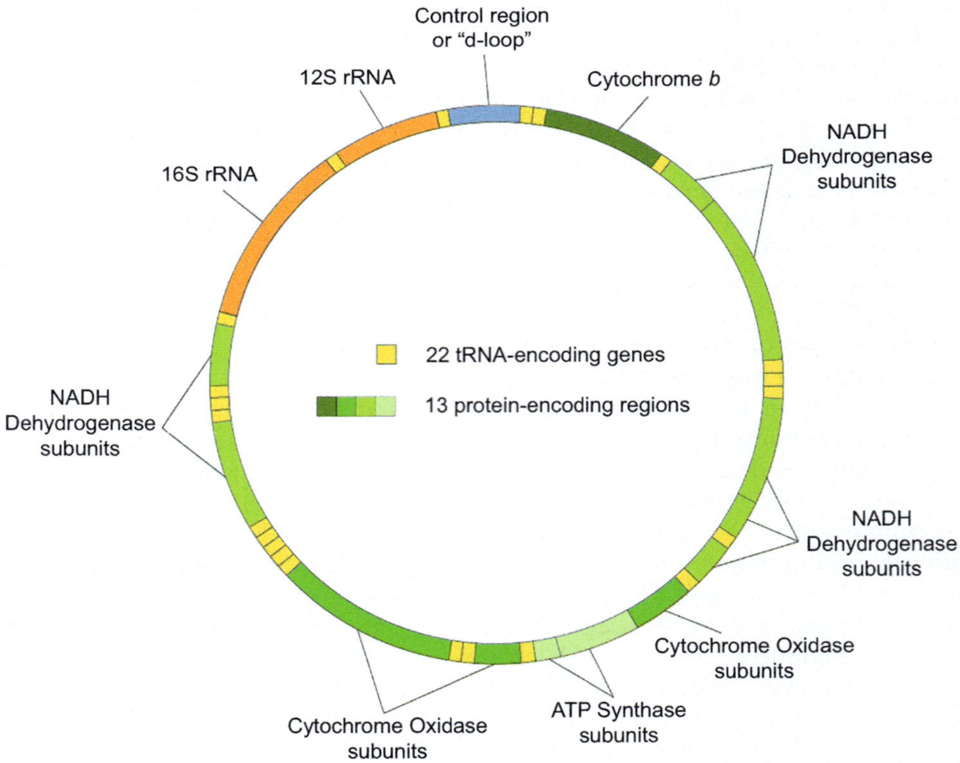

FIGURE 13.1 Mitochondrial DNA. Left: Genes for Mitochondria

30% of the overall population suffers from debility, which then results in reproductive failure and, progressively, multiorgan failure, finally leading to mortality in severe cases. **Concept of a three-parent baby:** Mitochondrial replacement treatment was first implemented in 2016 when a baby was born using this approach to avert the transmission of mitochondrial disorders from mother to kid. He was designated as the first 'three-parent baby' since the nuclear DNA was inherited from the parents, but the mtDNA was derived from an unrelated lady. Previously, offspring from this specific parent did not survive due to the mother's defective mitochondria. Since mitochondria are inherited from the maternal cytoplasm, the dysfunctional mitochondria were transmitted to the earlier progeny, resulting in their demise. The recent infant conceived by the three-parent method is devoid of any inherited genetic abnormalities. The procedure was conducted by in vitro fertilisation.

1. **Pronuclear transfer in mitochondrial replacement therapy:** The initial endeavour by scientists to prevent the transmission of defective mitochondria from mother to offspring is pronuclear transfer. The initial 'mitochondrial replacement' method, designed to prevent mitochondrial disorders, is referred to as pronuclear transfer. The initial occurrence took place in mouse embryos in 1983. Pronuclei are the nuclei derived from the egg and sperm within the fertilised egg, known as a zygote, that have not yet amalgamated into a singular nucleus (Fig. 13.2). This procedure is carried out by fertilising an egg from a mother with defective mitochondria and an egg from a donor with normal mitochondria. The fertilised

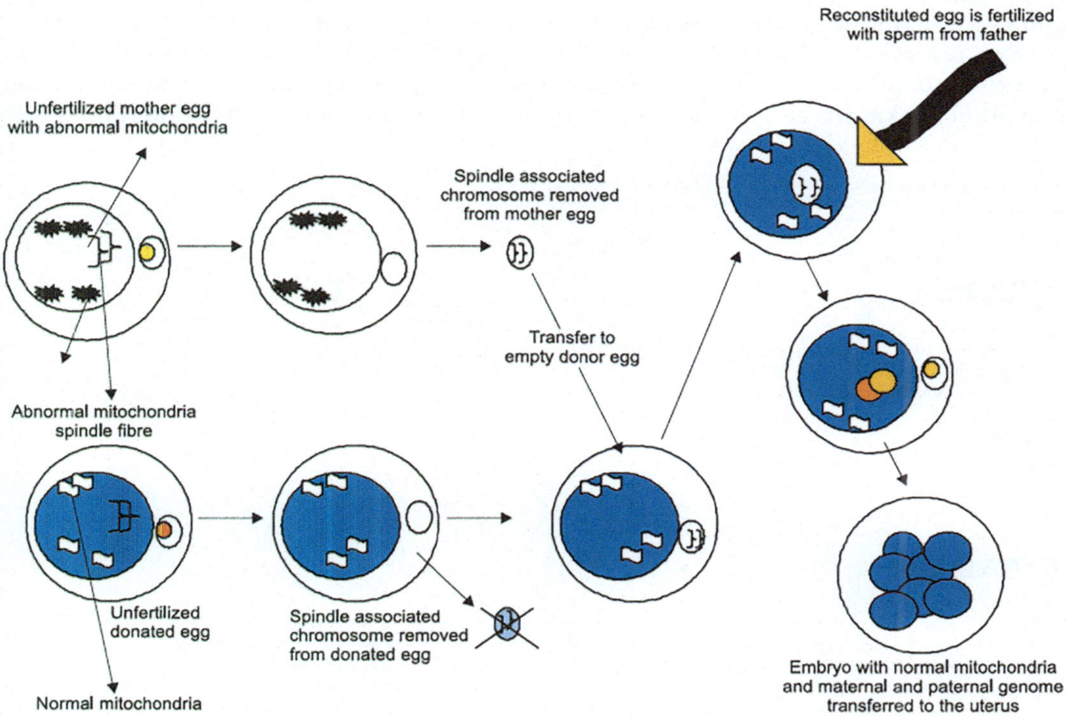

FIGURE 13.2 Spindle transfer for mitochondrial replacement therapy.

pronuclei from the donor's egg were extracted and discarded, after which the pronuclei from the mother's fertilised egg were transferred into the enucleated donor's egg containing normal mitochondria. Subsequently, the embryo, possessing normal mitochondria and both maternal and paternal genomes, was implanted into the uterus.

2. Pronuclear transfer of mitochondrial replacement therapy Fig. 13.3

Limitations for pronuclear transfer

The technique of pronuclear transfer appears to be unethical due to the destruction of two embryos. Secondly, from a technical perspective, the aberrant mitochondria may infiltrate the nucleus during the transfer process, resulting in an unacceptably elevated number of mitochondria from the maternal egg, including those carrying diseases, being incorporated into the donor's egg. Subsequently, scientists indicated that enhancements in the process resulted in embryos containing less than 2% of the mitochondria transferred from the maternal egg to the donor egg. An earlier study indicated that even a 1% carryover could be perilous, since mutated mitochondria may proliferate, ultimately dominating the cell and impairing its energy production. This procedure is regarded not as a curative treatment for infants with malfunctioning mitochondria, but rather as a form of infertility treatment, given that these infants may experience health complications associated with dysfunctional mitochondria in the future. **2. Spindle transfer for mitochondrial replacement therapy:**

DNA is located within the body's 46 chromosomes. Cell division leads to the division of 46 chromosomes into two sets of 23 chromosomes each. The chromosomes adhere to many protein fibres known as spindles. This approach involves the use of two unfertilised eggs,

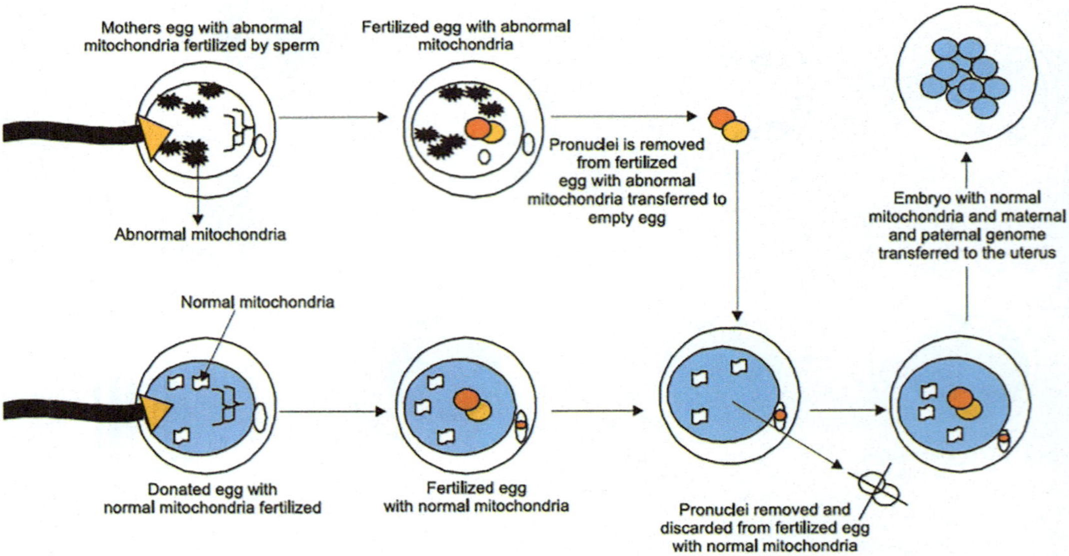

FIGURE 13.3 Pronuclear transfer of mitochondrial replacement therapy.

one from the mother and another from the donor. The nuclear membrane encasing the nucleus is compromised. At that stage, the spindle does not segregate into the chromosomes. The spindle and its associated chromosomes are extracted from the donor's egg and eliminated. The same process is applied to the maternal ovum, with the spindle fibres connected to the chromosomes retained. The spindle connected to the chromosome is injected into the donor's enucleated egg, and the father's sperm is utilised to fertilise the egg. Subsequently, the embryo possessing a normal paternal and maternal genome, together with normal mitochondria, is delivered to the mother's uterus (see Fig. 13.2).

The first report of the birth of a baby boy conceived through *the spindle transfer procedure* was in 2016. Mitalipov in Portland is the pioneer of spindle transfer. In 2009, he demonstrated that this procedure can result in the birth of healthy newborn monkeys. **Constraints of the spindle transfer technique:** The spindle fibre transfer technique also possesses several disadvantages. During the transfer of the spindle attached to the chromosome, chromosomes may detach, potentially resulting in the transfer of either a reduced number of chromosomes to the embryo or an excess if some remain in the donor's egg. In both instances, this may result in abnormalities. Zhang conducted spindle fibre procedures on five embryos, resulting in the successful birth of a newborn boy last year from one of them. It was discovered that he has barely 1% of his mother's mtDNA. At three months of age, he seemed to be in good health. However, there remains a concern about whether his health would be adversely impacted in the long term. His health may be compromised due to the incompatibility between the parents' nuclear DNA and the donor's mtDNA. In addition, to the possibility of trace levels of mitochondria proliferating, another study indicates that discrepancies between the parents' nuclear DNA and the donor mtDNA may influence ageing, as demonstrated in a study involving mice. There exist certain controversies over the procedures and the terminology. Cohen asserts that the term 'three parent baby' is inaccurate, as the kids do not inherit any characteristics from the donor, given that mitochondria do not play a role in transmitting fundamental traits from the parents; thus the donor cannot be appropriately classified as a parent. Andrew R. La Barbera concurs with this assertion, as mitochondria do not constitute genetic material, thus a child conceived using this approach possesses only two biological parents. A panel of specialists stated that it is ethical to employ this technology for the conception of male infants, but not for female infants. mtDNA is inherited exclusively from the mother, not the father. Consequently, the male infants conceived using this method will not transmit the donor's mtDNA.

A case study: three-parent baby

A very recent report in July 2025 indicates eight babies born in UK with DNA from three people through in vitro fertilization. It is regarded as a major breakthrough that could prevent genetic disorders. The procedure, known as mitochondrial donation treatment, is not a step towards 'designer babies' but a targeted medical intervention designed to stop incurable diseases from being passed from mother to child. The UK has legalised this technique in 2015 under strict vigilance.

Throughout its development, the UK has been leading, becoming the first country to officially legalise the method in 2015. Strict regulation and approval from the human fertilisation and embryology authority is the prime necessity (https://www.deccanherald.com/science/babies-born-in-uk-with-dna-from-3-people-medical-breakthrough-could-prevent-genetic-disorders-3636962). **Significance of tri-parental progeny in animal science:** The current study indicates that mitochondria are a crucial cellular organelle with significant roles in energy production. Any aberration in mtDNA results in illnesses and impacts reproduction. The approach to addressing the issue in animal science is twofold.

Marker-assisted selection:

Initially, identify animals with defective mitochondria, eliminate them, and select only those with healthy mitochondria. An essential concern is the heteroplasmy of dysfunctional mitochondria. A percentage of faulty mitochondria (containing mutant DNA) above 60% is deemed defective or harmful. Individuals with above 60% faulty mitochondria will be detected and eliminated throughout the selection phase. When the selection is predicated on all 37 mitochondrial genes, the procedure may be classified as genomic selection. This strategy is pertinent to livestock raised on farms. Consequently, a robust population of livestock will be established.

Production of elite high-producing animals with mitochondrial defects can be effectively treated with this procedure. Mitochondrial replacement therapy, also known as the three-parent infant concept, although seen as an expensive procedure, its extensive application can render it cost-effective. In animal science, elite high-producing animals with mitochondrial defects need not be killed, as they can be effectively treated with this procedure. mtDNA from a donor may be obtained from a female who has healthy mitochondria. Consequently, the distinctive characteristics of the animal's high productivity can be preserved with ease. This approach can also develop disease-resistant animals.

Generation of endangered creatures in the wild. This procedure can also be used to generate endangered creatures in the wild. There is no justification for culling an endangered species. However, when a faulty mitochondrial mutation occurs in an endangered species, it can be addressed with mitochondrial replacement treatment. Mitochondrial abnormalities result in severe debilitation, exercise intolerance and ultimately mortality owing to mitochondrial replacement therapy. To date, there is a deficiency of research on mitochondrial gene abnormalities in animals and their impact on debility and diseases. Our laboratory is investigating mutations in mitochondrial genes that result in debility, exercise intolerance and adverse effects on the essential functions of key organs, such as the liver, kidneys and heart, in sheep and pigs raised on our farm. We initially focused on the cytochrome B gene and are currently analysing all 37 mitochondrial genes using full mitochondrial genome sequencing using next-generation sequencing techniques. There remains significant potential for addressing mitochondrial abnormalities, as they contribute to considerable mortality and morbidity in animals.

Conclusion

Research on mitochondrial genes and their impact on reproduction and diseases in animal science is exceedingly restricted. We have commenced it in our laboratory. Despite the majority of instances remaining undiagnosed on a farm, this aligns with our practical experience. Approximately 30% of the animals exhibit mitochondrial defects, manifesting as severe debility culminating in multiorgan failure and subsequent mortality. The three-parent baby concept is a recent advancement that effectively addresses mitochondrial abnormalities, hence mitigating reproductive problems and other diseases while preserving indigenous germplasm. The absence of curative treatments and prenatal diagnosis has resulted in this concerning condition, which could be mitigated with mitochondrial replacement therapy.

Further Reading

Agarwal, A., Nallela, K. P., Allamaneni, S. S. R., & Said, T. M. (2004). Role of antioxidants in treatment of male infertility: an overview of the literature. *Reprod Biomed online, 8*, 616–627.

Aitken, R. J., & Clarkson, J. S. (1988). Significance of reactive oxygen species and antioxidants in defining the efficacy of sperm preparation techniques. *J Androl, 9*, 367–376.

Ajduk, A., Llozue, T., Windsor, S., et al. (2011). Rhythmic actomyosin-driven contractions induced by sperm entry predict mammalian embryo viability. *Nat Commun, 2*, 417.

Alexandre, H., Junghwa, A., Anne, R., Thanh, N. T., & Arnaud, C. (2013). Combining multiple autosomal introns for studying shallow phylogeny and taxonomy of Laurasiatherian mammals: Application to the tribe Bovini (Cetartiodactyla, Bovidae). *Mol Phylogenetics Evolution, 66*(3), 766–775.

Babayev, E., & Seli, E. (2015). Oocyte mitochondrial function and reproduction. *Curr Opin Obstet Gynecol PLoS Genet*e1005241.

Barbehenn, E. K., Wales, R. G., & Lowry, O. H. (1974). The explanation for the blockade of glycolysis in early mouse embryos. *Proc Natl Acad Sci USA, 71*, 1056–1060.

Barritt, J. A., Willadsen, S., Brenner, C., & Cohen, J. (2001). Cytoplasmic transfer in assisted reproduction. *Hum Reprod Update, 7*, 428–435.

Cardullo, R. A., & Baltz, J. M. (1991). Metabolic regulation in mammalian sperm: mitochondrial volume determines sperm length and flagellar beat frequency. *Cell Motil Cytoskeleton, 19*(3), 180–188.

Chappel, S. (2013). The Role of Mitochondria from Mature Oocyte to Viable Blastocyst. *Obstet Gynecol Int*.

Connel, M. O., McClure, N., & Lewis, S. E. (2002). A comparison of mitochondrial and nuclear DNA status in testicular sperm from fertile men and those with obstructive azoospermia. *Hum Rep, 17*, 1571–1577.

Fogleman, S., Santana, C., Bishop, C., Miller, A., & Capco, D. G. (2016). CRISPR/Cas9 and mitochondrial gene replacement therapy: promising techniques and ethical considerations. *Am J Stem Cell, 5*(2), 39–52 eCollection 2016.

Frank, S. A., & Hurst, L. D. (1996). Mitochondria and male disease. *Nature, 383*, 224.

Griveau, J. F., & Le Lannou, D. (1997). Reactive oxygen species and human spermatozoa: Physiology and pathology. *Int J Androl, 20*, 61–69.

Hamers, L. (2016). Scientists find clue to why mitochondrial DNA comes only from mom. *Sci N, 190*, 6.

Hyslop, L. A., Blakeley, P., Craven, L., Richardson, J., Fogarty, N. M., Fragouli, E., Lamb, M., Wamaitha, S. E., Prathalingam, N., Zhang, Q., O'Keefe, H., Takeda, Y., Arizzi, L., Alfarawati, S., Tuppen, H. A., Irving, L., Kalleas, D., Choudhary, M., Wells, D., ... Herbert, M. (2016). Towards clinical application of pronuclear transfer to prevent mitochondrial DNA disease. *Nature, 534*(7607), 383–386. https://doi.org/10.1038/nature18303 Epub 2016 Jun 8.

Kumar, D. P., & Sangeetha, N. (2009). Mitochondrial DNA mutations and male infertility. *Indian J Hum Genet, 15*(3), 93–97.

Kumar, R., Bamezai, R. N. K., Kumar, R., Gupta, N. P., & Dada, R. (2007). Increased lipid peroxidation, low antioxidant levels and nucleotide changes in mt Dna in idiopathic asthenozoopermic men of Indian origin. *Indian J Clin Biochem, 22*(suppl), 355.

Lal, R. Z., Biyani, S., & Lodha, R. (2016). Correlation of Thyroid Hormones with FSH, LH and Prolactin in Infertility in the Reproductive Age Group Women. *IAIM, 3*(5), 146–150.

Lee G., "Shoukhrat Mitalipov and Masahito Tachibana's Mitochondrial Gene Replacement in Primate Offspring and Embryonic Stem Cells (2009)". *Embryo Project Encyclopedia* (2018-01-10). ISSN: 1940-5030 http://embryo.asu.edu/handle/10776/13041.

Lee, S. K., Zhao, M. H., Kwon, J. W., et al. (2014). The association of mitochondrial potential and copy number with pig oocyte maturation and developmental potential. *J Reprod Dev, 60*, 128–135.

MacLeod, J. (1943). The role of oxygen in the metabolism and motilitity of human spermatozoa. *Am J Physiol, 138*, 512–518.

May-Panloup, P., Boucret, L., Chao de la Barca, J. M., Desquiret-Dumas, V., Ferré-L'Hotellier, V., Morinière, C., Descamps, P., Procaccio, V., & Reynier, P. (2016). Ovarian ageing: the role of mitochondria in oocytes and follicles. *Hum Reprod update, 22*(6), 725–743.

McDermott, A. (2016). Donor mitochondria could influence metabolism, aging. *Sci N, 190*, 8.

Onteru, S. K., Fan, B., Du, Z. Q., Garrick, D. J., Stalder, K. J., & Rothchild, M. F. (2011). A whole genome association study for pig reproductive traits. *Anim Genet, 43*, 18–26.

Pal, A. (2017a). Studies on Whole mitochondrial genome sequencing and its association with reproductive traits in Garole sheep. *PLoS One* communicated.

Pal, A., & Banerjee, S. (2018). Mitochondrial replacement therapy—a new remedy for defects in reproduction. *Indian J Anim Sci, 88*(6), 637–644.

Pal A., Banerjee S., Chatterjee P.N., Batobyal S. Cytochrome B gene affects the male and female reproduction in Sheep. In the Proceedings of ISSRF, Anand 2016.

Pal A., Banerjee S., Chatterjee P.N., Batobyal S., Biswas P. and Sharma, A. 2017. Next Generation Sequencing Confirms The Distinct Breed Identity of Birbhum Sheep – An Indigenous Breed of Sheep Identified From The Birbhum District of West Bengal. Proceedings of National Seminar on Biodyanmic Animal Farming for the management od Livestock Diversity under Changing climatic condition held at College of Veterinary Science, Mannuthy, Thrissur, Kerala from 8-10 February, 2017b.

Pal, A., & Chakravarty, A. (2006). Strategy for genetic selection of dairy bulls for higher libido and fertility. *J dairying, Food Home Sci, 22*(3&4), 170–175.

Pal, A., Chakravarty, A. K., & Chatterjee, P. N. (2014). Polymorphism of Growth hormone gene and its association with seminal and sexual behavioral traits in crossbred cattle. *Theriogenology, 81*, 474–480 201.

Pal, A., Chatterjee, P. N., Kumar, R., & Sharma, A. (2004). Growth hormone gene: molecular marker for improved animal production. *Rev Int J Cow Sci, 2*(2), 1–8.

Patel, M. R., Miriyala, G. K., Littleton, A. J., Yang, H., Trinh, K., Young, J. M., Kennedy, S. R., Yamashita, Y. M., Pallanck, L. J., & Malik, H. S. (2016). A mitochondrial DNA hypomorph of cytochrome oxidase specifically impairs male fertility in Drosophila melanogaster. *Elife*, 16923.

Pradhan M., Pal A., Samanta A.K., Banerjee S., Samanta R. Mutations in cytochrome B gene effects reproduction of Ghungroo pig. In the Proceedings of ISSRF, AIIMS, New Delhi, 2017.

Saey, T. H. (2016a). Risk identified in procedure for three parent babies. *Sci N Online, 2016*. https://www.sciencenews.org.

Saey, T. H. (2016b). First 'three-parent baby' born from nuclear transfer. *Sci N Online*. https://www.sciencenews.org.

Saey, T. H. (2016c). Refined 'three-parent-baby' procedure improves chances for healthy infant. *Sci N Online*.

Saey, T. H. (2016d). Risk identified in procedure for 'three-parent babies' *Sci N, 189*, 8.

Sampson, M. J., Decker, W. K., Beaudet, A. L., Ruitenbeek, W., Armstrong, D., & Hicks, M. J. (2001). Immotile sperm and infertility in mice lacking mitochondrial voltage dependent anion channel type 3. *J Biol Chem, 276*, 39206–39212.

Sanders, L. (2009). Mitochondrial DNA replacement successful in Rhesus monkeys. *Sci N, 176*, 8.

Schwartz, M., & Vissing, J. (2002). Paternal inheritance of mitochondrial DNA. *N engl Jmed, 347*(8), 576–580.

Shamsi, M. B., Kumar, R., Bhatt, A., Bamezai, R. N. K., Kumar, R., Gupta, N. P., Das, T. K., & Dada, R. (2008). Mitochondrial DNA Mutations in etiopathogenesis of male infertility. *Indian J Urol, 24*(2), 150–154.

Tibbitts, T. (2014). Third scientific review of the safety and efficacy of methods to avoid mitochondrial disease through assisted conception. *Hum Fertilisation Embryology Auth*.

Van, B. J., Davis, P. W., & Lee, J. (1995). ATP content of human oocytes and developmental potential and outcome after in-vitro fertilization and embryo transfer. *Hum Reprod, 10*, 415–424.

Venkatesh, S., Deecaraman, M., Kumar, R., Shamsi, M. B., & Dada, R. (1991). Role of reactive oxygen species in the pathogenesis of mitochondrial DNA (mtDNA) mutations in male infertility. *Indian J Med Res, 129*(2), 127–137.

Zhang, J., Liu, H., Luo, A., Chavez-Badiola, A., Liu, Z., Yang, M., & Munne, S. (2016). First live birth using human oocytes reconstituted by spindle nuclear transfer for mitochondrial DNA mutation causing Leigh syndrome. *Am Soc Reprod Medicine's Sci Congr, 106*, e375–e376.

Zhang, J., Zhuang, G., Zeng, Y., Grifo, J., Acosta, C., Shu, Y., & Liu, H. (2016). Pregnancy derived from human zygote pronuclear transfer in a patient who had arrested embryos after IVF. *Reprod Biomedicine Online, 33*, 529–533.

Zini, A., De Lamirande, E., & Gagnon, C. (1996). Low levels of nitric oxide promote human sperm capacitation *in vitro*. *J Androl, 16*, 424–431.

CHAPTER

14

Mitochondrial transfer and transplantation

Introduction

In the last chapter, we discussed mitochondrial replacement therapy (MRT). This markedly contrasts with mitochondrial transfer and transplantation.

Mitochondrial transfer and transplantation areuseful tools to rectify abnormalities due to mitochondrial DNA mutations. Healthy mitochondria, when exceeds diseased ones, in heteroplasmy, can function properly, ignoring mutant defective ones. This methodology markedly differs from MRT, an in vitro fertilisation method designed to rectify or reduce the heteroplasmy of deleterious mtDNA mutations in progeny, via microinjection to bypass the plasma membrane. In contrast to MRT, mitochondrial transplantation (MT) is a postnatal intervention, and it is not expected to induce genetic modifications in the germline. Mitochondria are not directly introduced into cells as in MRT. Rather, they are assimilated by mechanisms similar to contact-independent mitochondrial transport. MT can be utilised to treat many ailments (beyond hereditary conditions caused by mtDNA mutations), may be autologous or allogeneic, and can be conducted multiple times per patient, unlike MRT.

Mitochondrial replacement therapy and mitochondrial transfer: groundbreaking first findings

The preliminary discoveries about intercellular mitochondrial transport were based on fundamental principles of yeast protoplasts, which lack mtDNA, by fusing them with protoplasts from wild-type cells. The findings were subsequently broadened a few years later by employing isolated mitochondria or enucleated cells. In 1982, it was shown that isolated mitochondria from chloramphenicol and efrapeptin-resistant mouse tumour cells could be endocytosed by drug-sensitive tumour cells in culture, thereby imparting resistance, indicating that the uptake of extracellular mitochondria (ex-Mito) was able to genetically modify the drug-sensitive cells.

In 1977, it was found that mitochondrial genotypes could be obtained. The findings were subsequently reaffirmed using the microinjection of isolated mitochondria into human cancer cell lines. It was subsequently shown that injecting myoblast mitochondria with

Mitochondrial Genomics
DOI: https://doi.org/10.1016/B978-0-443-26655-3.00002-0

a pathogenic mtDNA tRNALys mutation into mtDNA-deficient cell lines led to the creation of cytoplasmic hybrid cells, known as 'cybrids', and that the donor mtDNA caused deficits in mitochondrial protein translation and cytochrome c oxidase activity, leading to the creation of cybrids with restored aerobic respiration, and subsequently, breast cancer. Subsequently, it was shown that platelets and sperm from healthy persons can transfer mtDNA to cybrids with the G10398A mtDNA mutation, which demonstrated resistance to apoptosis and enhanced metastatic potential. Although horizontal mtDNA gene transfer has been documented among cells in flowering plants, thereby broadening the spectrum of organisms known to exchange mitochondrial components, definitive evidence for the transfer of whole mitochondria between plant cells is lacking.

Recent enquiries regarding mitochondrial transfer

Tumours get mtDNA30 via intact mitochondria from host cells in mice, promoting carcinogenesis by increasing de novo pyrimidine synthesis. In 2016, it was shown that the maturation of mouse oocytes is partially affected by the transfer of mitochondria from sister cyst germ cells. In that year, retinal ganglion cells were shown to eject faulty mitochondria and move them to adjacent astrocytes for degradation, thereby establishing the concept of transmitophagy or licensed mitophagy, in which one cell type leverages the degradative function of another. Transmitophagy is now acknowledged to occur in multiple organ systems, including the heart, brown adipose tissue and retina, to sustain mitochondrial homoeostasis in donor cells. Furthermore, stimulated platelets were shown to release mitochondria, a mechanism recently linked to wound healing. It is now clear that various cell types participate in mitochondrial transport inside the central nervous system. r0 cells; nonetheless, it is recognised that in healthy cells, mouse mitochondria can fuse with human mitochondria but are not maintained for prolonged periods due to mitonuclear conflict. Multiple studies have demonstrated that mitochondrial transfer occurs in vivo inside complex animal systems. In 2011, it was revealed that canine transmissible venereal cancer cells assimilate mitochondria from host cells, and the subsequent year, it was shown that bone marrow stromal cells convey mitochondria to pulmonary alveoli to alleviate endotoxin-induced acute lung injury in mice. Years later, it was shown that r0 lung carcinoma cells in culture, rectifying inherent deficiencies in aerobic respiration and providing preliminary evidence that mitochondrial transfer could confer metabolic traits to recipient cells.

This was further substantiated by the xenogenic transfer of mouse mitochondria into humans. In 2006, it was documented that human bone marrow-derived mesenchymal stem cells or fibroblasts were able to transfer mitochondria to neighbouring central nervous system, peripheral nervous system, pulmonary system, cardiovascular system, white adipose tissue, brown adipose tissue, osseous tissue, haematopoietic system and the cancer microenvironment. In specific cases, respiration-competent mitochondria are disseminated into the circulation of mice and humans, serving as a conduit for mitochondrial transfer among cells of other organs (i.e. interorgan mitochondrial transfer). The transfer of mitochondria from adipocytes to the heart in obesity indicates that multiple cell types probably contribute to the reservoir of circulating ex-Mito in the bloodstream.

Mitochondrial transfer has been recognised as a regulator of numerous physiological processes, including the maintenance of adipose tissue homoeostasis, cardiovascular health, wound healing, angiogenesis, haematopoiesis and inflammation, which have been lately explored in greater detail elsewhere.

Progression of therapeutic techniques

Substantial attempts have been undertaken to employ mitochondrial transfer pathways for therapeutic purposes. These approaches incorporate aspects of mitochondrial transfer biology, such as cellular absorption pathways, although they are specifically tailored to transport mitochondria to a particular organ and/or cell type. The first studies employing isolated mitochondria for therapeutic purposes included the extraction and transplantation of cardiac mitochondria into the hearts of rabbits subjected to ischaemia and subsequent reperfusion, a method that reduced ischaemia-reperfusion injury and improved functional recovery. A similar outcome was later observed in pigs. In subsequent clinical trials, skeletal muscle mitochondria were retrieved from paediatric patients requiring extracorporeal membrane oxygenation (ECMO) for autologous transplantation, facilitating the rapid weaning of most children from ECMO. MT is now being investigated for several therapeutic uses, including ischaemic stroke, ischaemic limb injury, spinal cord trauma, neurodegeneration, cardiac resuscitation, refractory dermatomyositis or polymyositis and inherited mitochondrial diseases. Furthermore, isolated mitochondria can be modified or included into gel matrices to improve the efficacy of mitochondrial absorption or therapeutic benefits. Various groups have utilised isolated mitochondria to metabolically engineer cultured cells. Coculture of hematopoietic stem cells (HSCs) with isolated mitochondria has been shown to enhance engraftment efficiency in murine bone marrow transplantation models. This strategy was further applied in a clinical study, in which HSCs from patients with severe mitochondrial DNA deletion syndromes were cocultured in vitro with maternal mitochondria prior to autologous transplantation. Treated patients demonstrated reduced anaemia and improved quality of life. These promising findings suggest that medications employing mitochondrial transfer biology may effectively treat various disorders, requiring a comprehensive nomenclature to improve and unify academic and industry efforts focused on understanding and utilising this biology to promote human health.

Essential principles and nomenclature: transference of mitochondria

Mitochondria are transferred to the two daughter cells during cellular differentiation and division. This process is referred to as 'vertical inheritance of mitochondria', which involves the distribution of mitochondria during cell division, potentially demonstrating asymmetry in cases such as stem cell differentiation. Conversely, horizontal or intercellular 'mitochondria transfer' denotes a technique in which one cell donates a fraction of its mitochondria to another cell without the process of cell division. The 'donor cell' refers to the origin of the transferred mitochondria, whereas the recipient of these mitochondria is referred to as the 'acceptor cell'. The term 'acceptor cell' is used irrespective of whether the cell preserves, degrades, or re-exports the transferred mitochondria, and is intended to remain neutral concerning their fate or processing. Donor and acceptor cells are frequently, though not always, different in their embryonic origins. A 'mitochondria transfer axis' implies that when both donor and recipient cell types are established in vivo. Adipocytes move mitochondria to macrophages in both white and brown adipose tissues in mice, establishing an adipocyte-to-macrophage mitochondrial transfer axis. Adipocytes appear to convey a portion of their mitochondria to several other cell types, implying the concept of a 'mitochondria transfer network', which suggests that cells may participate in multiple concurrent or successive mitochondria transfer pathways. Furthermore, adipocytes possess the ability to release their mitochondria into the circulatory system, enabling their distribution to various organs, such as the heart. These results clarify the concept of 'interorgan mitochondria transfer', which is characterised as the long-distance transfer of mitochondria from a donor cell in one organ to an acceptor cell in another organ. In this scenario, blood, lymph, or cerebrospinal fluid may serve as a conduit for the transfer of 'extracellular mitochondria (ex-Mito)' from the donor cell to the recipient cell. The population is heterogeneous, consisting of various types from varied cellular origins, as discussed in greater detail below. Numerous research have utilised the phrases 'mitochondria transfer' and 'mitochondrial transfer' interchangeably, as they are equivalent. Therefore we assert that both phrases are permissible, with the choice depending on the investigator's preference.

Therapeutic approaches

Various innovative therapeutic approaches employ mitochondrial transfer channels or mechanisms. Mitochondria transplantation is a procedure that entails isolating mitochondria from a cellular or tissue source and then directly giving them to an animal to elicit a therapeutic response. Many studies have alternatively utilised the term 'mitochondrial transplant', which is also acceptable. In this article, we use the term 'mitochondria transplant' to conform to conventional transplant terminology (e.g. heart transplant). MT has been employed in preclinical studies using model organisms and in recent clinical trials that have produced promising preliminary results, including ongoing clinical trials that are yet to be published. Another therapy technique involves in vitro cell engineering using isolated mitochondria, where patient-derived cells are integrated with isolated mitochondria. Manufacturing process methodology has been utilised in a recent clinical trial, producing encouraging outcomes, as detailed below. It has also been employed to improve the vitality and efficacy of modified cells, including chimeric antigen receptor T-cells.

Methods for identifying mitochondrial transfer

Reporter proteins for mitochondria

These essential concepts offer a methodological examination of the practical definition of mitochondrial transfer. Diverse methodological approaches are available to identify mitochondrial transfer. The optimal strategy involves utilising fluorescent reporter proteins along with mitochondrial localisation signals, which are short polypeptide sequences that direct the fluorescent protein to the inner or outer mitochondrial membrane or the mitochondrial matrix. Various mitochondrial reporter protein constructs are available for use inside a stop-flox system, facilitating the identification of a donor cell based on the cell type specificity of a cre recombinase. A mitochondria transfer axis or network is explained through the identification of the labelled mitochondria across different cell types. Additionally, one may express mitochondrially specific proteins (e.g. halo) that enhance the detection of mitochondrial translocation. A major limitation of this approach is its ability to detect mitochondrial transfer solely when the reporter protein is incorporated within the transferred mitochondrial structure. This is not universally applicable. A subset of mitochondria-derived vesicles (MDVs) possesses selectively selected cargo and is developed as either single outer membrane vesicles or double membrane vesicles. As a result, a fluorescent reporter protein targeting the inner mitochondrial membrane (IMM) will not be able to detect all MDV transfer occurrences. MDVs are transported into multivesicular bodies and may be released as exosomes. Evidence indicates the selective sorting of mitochondrial components in MDVs that are released from cells. Brown adipocytes transfer MDVs to macrophages, which lack the distinctive brown adipocyte IMM protein uncoupling protein 1.

Fluorescent stains for mitochondria

A fluorescence-based alternative method utilised in several studies entails labelling donor cells with employ fluorescent mitochondrial dyes to observe the transfer of mitochondria to another cell type in vitro. While this method may first seem attractive due to the low effort and cost associated with evaluating a hypothesis related to mitochondrial transfer, these dyes are susceptible to leakage and are very prone to producing false-positive results, even after extensive washing. All mitochondrial dyes likely drain from the donor cell in situ during coculture, leading to dye transfer artefacts that either overstate mitochondrial transfer efficiency or suggest that mitochondrial transfer occurs when it does not. Mature red blood cells, without mitochondria, can transfer MitoTracker dyes to 293 T-cells. Therefore we strongly recommend against using mitochondrial dyes to identify mitochondrial transfer. When employing dyes, it is essential to establish multiple controls to address issues related to dye leakage and to titrate the dye to determine the least concentration required to reduce off-target staining of other organelles. Interpreting mitochondrial transfer results derived from dyes necessitates extreme caution, and all findings must be corroborated. Utilising a dye-independent methodology to avert the occurrence of false-positive results.

Alternative methodologies

Alternative approaches for defining mitochondrial transfer entail using donor cell types with unique mtDNA sequence variants, serving as a separate molecular barcode recognisable through DNA sequencing or PCR. However, the simple identification of transferred mtDNA is insufficient to confirm mitochondrial transfer, as mitochondria may attach to the cell surface without being internalised, and free mtDNA can also be transported between cells. Alternatively, one may recognise an acquired mitochondria-associated protein or trait conferred to the recipient cell, such as the reestablishment of respiratory competence in cells deficient in oxidative phosphorylation.

Methods for enhancing mitochondrial transport

Various efficient in vitro methods are available to examine certain phases of mitochondrial transport, including the mechanisms of mitochondrial uptake and release. Examples include the incorporation of isolated mitochondria into cultured cells or the detection of extracellularly released mitochondria in culture media; however, in the latter context, it is critical to account for cell death occurring under in vitro conditions. Certain cell types exhibit high efficiency in the uptake of ex-Mito, whereas others display limited uptake capacity.

Various methods are available to enhance the uptake efficiency by acceptor cells, such as the integration of a centrifugation phase (e.g. MitoCeption), the use of pressure (e.g. MitoPunch), or conjugation with a cell-penetrating peptide (e.g. phosphoenolpyruvate-1). Cell culture methodologies can utilise transwell systems to physically separate donor and acceptor cells, hence enabling the distinction between contact-dependent and contact-independent transfer mechanisms.

Mitochondrial fates upon cellular entry

Another essential consideration when analysing mitochondrial transfer axes is to evaluate the fate of the transferred mitochondria. Are the transferred mitochondria swiftly degraded, and if so, what is the mechanism behind this process? Do transplanted mitochondria penetrate the cytoplasm and integrate into the endogenous mitochondrial network of the recipient cell? Do the transferred mitochondria penetrate the nucleus and promote the integration of mtDNA pieces into the nuclear genome (i.e. numtogenesis)? Does the mtDNA in the transferred mitochondria induce stable genetic alterations in the receiving cell? Are the transplanted mitochondria reassembled and subsequently extruded from the cell for transfer to another cell type? The functional outcomes of internalised mitochondria are likely influenced by the characteristics of the donor and recipient cell types, as well as the prevailing physiological or pathological context. We caution against assuming that the effects of transplanted mitochondria in one cell type are the same as in another. We assert that determining the destiny and processing of transferred mitochondria is not essential to validate mitochondrial transfer; yet, investigating these elements may provide valuable insights into the roles of mitochondrial transfer within the studied context. In the realm of MT, understanding the biodistribution, fate and processing of the transplanted mitochondria may clarify the mechanisms underlying this therapeutic strategy.

Terminology focused on mechanisms

Recent findings demonstrate that cells utilise various distinct mechanisms for mitochondrial translocation. The transfer mechanisms can be categorised into two types: 'contact-dependent mitochondria transfer' and 'contact-independent mitochondria transfer'. The identification of one transfer mechanism does not exclude the possibility of others, as cells may utilise multiple mitochondrial transfer mechanisms simultaneously. This section is on the terminology used to describe the molecular mechanisms of mitochondrial transport pathways, as emphasised in various recent reviews.

Contact-dependent mitochondrial transfer: *tunnelling nanotubes*

Touch-dependent mitochondrial transfer occurs when the donor and acceptor are in direct physical touch for the transfer to occur. The main contact-dependent mechanism identified is via tunnelling nanotubes (TNTs), which are elongated, slender membranous structures that form between two cells with the aid of filamentous (F)-actin (Fig 3e). Mitochondria are transported along a microtubular channel through motor/adaptor complexes that include mitochondrial Rho GTPase 1 (Miro1). The development of TNT is a complex process, partially facilitated by growth-associated protein 43, utilised by neurones at the axonal growth cone to enhance axonal outgrowth, or alternatively, the M-Sec protein induces the reorganisation of plasma membrane components preceding TNT formation. Key characteristics of TNTs include their narrow diameter (generally 0.5–1.5 μm), the presence of filamentous actin (F-actin) and the establishment of direct cytoplasmic continuity between connected cells. However, not all contact-dependent mitochondrial transport occurs via TNTs.

Dendritic structures

Cells are capable of establishing intercellular connections through dendritic projections that enable mitochondrial transfer, a process termed 'dendritic structure–mediated mitochondrial transfer' (Fig. 3e). Three examples of this process include the transfer of mitochondria from osteocytes to other osteocytes, from osteocytes to endothelial cells and from astrocytes to neurones. These dendritic structures generally range from 1.5 to 3 μm in diameter and connect with other cells via end feet, which are enlarged termini of the dendritic structure. The terminal ends contact, but does not fuse with the receiving cell's plasma membrane. The models suggest that end-foot structures spatially define the sites of ex-Mito release in a manner dependent on CD38, mitofusin 2 and/or MIRO1, thereby facilitating transfer to the recipient cell, although alternative transfer mechanisms may be present, possibly involving the release of ex-Mito for delivery to nearby cells adjacent to the end feet. In practice, distinguishing between TNT- and dendrite-mediated mitochondrial transport is difficult. This can be partially resolved by illustrating the discontinuity between the cytoplasm of neighbouring cells, quantifying the diameter of the tubular structure emanating from the donor cell and detecting end feet at the junction of the dendritic structure and the receiving cell.

Adhesion via gap junction channels

A subsequent contact-dependent transfer mechanism involves connexin 43 (Cx43), which we define as 'adhesion-mediated mitochondria transfer' (Fig 3e). This gap junction protein contains a minute pore that is inadequate for the passage of mitochondria between various cell types. Cx43 promotes cell adhesion, enabling the recipient cell to obtain mitochondria from the donor cell through the formation and internalisation of annular gap junction vesicles. This mechanism is analogous to trogocytosis inside the immune system. In contrast, Cx43-mediated gap junctional channels require intimate proximity between two cells, possibly enabling mitochondrial transfer through different processes, such as the transfer of ex-Mito. When a donor cell releases ex-Mito, the expelled mitochondria are statistically more likely to be incorporated by neighbouring cells. Dendritic structures or TNTs may hypothetically form at the Cx43 contact site. These possibilities must be distinguished when adhesion-mediated mitochondrial transport is observed and will be clarified as scientific knowledge progresses.

Cell fusion is not synonymous with mitochondrial transfer

A prevalent question in the research is whether cell fusion that leads to the formation of multinucleated cells is a mechanism of mitochondrial transfer. Giant cells and osteoclasts are representative of multinucleated cells formed during cell fusion. We assert that this process does not constitute mitochondrial transfer, as all cellular components merge into a unified cellular entity. As a result, cell fusion does not possess a specific donor or acceptor cell. This procedure is more precisely characterised as a reverse modality of vertical mitochondrial inheritance.

Mitochondrial transfer occurring without physical touch, the extracellular release of mitochondria

Cells can perform mitochondrial transfer without necessitating direct physical contact. Donor cells can release their mitochondria to produce ex-Mito, which can be absorbed by another cell type. The mechanisms underlying contact-independent mitochondrial transfer can be elucidated by examining the structural features of the transferred mitochondria (see Fig. 14.3f).

We support a hierarchical nomenclature system. Ex-Mito is an all-encompassing term that includes all forms of mitochondria or structurally intact mitochondrial fragments (e.g. MDVs) found in extracellular fluid. On the contrary, 479 extracellular vesicles (EVs), which include free-floating, cytosolic mtDNA and are devoid of mitochondria or MDVs, are not ex-mito.

Extracellular vesicles that encapsulate mitochondria

Specific ex-mito are enclosed by EVs, which may demonstrate diversity, substantially concerning both dimensions and capability. EVs containing mitochondria (EV-Mito) can range in diameter from roughly 120 nm to 3–4 μm and may include MDVs or complete mitochondria, in addition to specific EVs. Mitochondrion with several mitochondria. These structures have considerable diversity, and cells can secrete several forms of EV-Mito. Upon considering the feasibility of employing mitochondrial cargo categories to delineate EV-Mito, we concluded that there is insufficient data to formulate cargo-based classifications at this time. Researchers are encouraged to characterise EV-Mito cargo as practicable.

Various release mechanisms are present, including exosomal release via multivesicular bodies and ectosomal release through vesicle synthesis from the plasma membrane. However, after EV-Mito is expelled from the cell, there is no reliable marker to distinguish between the exosomal and ectosomal release pathways; hence, these mechanisms are not currently included in EV-Mito nomenclature. In accordance with the recent guidelines from the International Society for EVs (MISEV2023), we recommend refraining from the use of the terms exosome and ectosome unless their subcellular origin is explicitly defined. Therefore we recommend that EVs carrying mitochondria be termed 'EV-Mito' and that their characterisation should largely focus on size, as this is a measurable physical property that can be evaluated using several approaches. EV-Mito must be characterised in accordance with the nomenclature guidelines set out by MISEV2023.

Mitochondria that are not bound

Additionally, cells can release ex-Mito that are not enclosed by an EV membrane. We propose that these ex-Mito be classified as 'free mitochondria'. While free mitochondria have been observed in tissues and circulation, the mechanisms by which cells release free mitochondria remain unclear and require further exploration. Free mitochondria can be classified as EVs because they are membrane-bound structures generated by cells; nevertheless, referring to free mitochondria as EVs is inaccurate and inadvisable. Electron microscopy and small-particle flow cytometry analyses demonstrate that free mitochondria generally exhibit diameters ranging from 0.5 to 1.1 μm. In cases where electron imaging is unfeasible, free mitochondria can be characterised using immunoprecipitation with antibodies that specifically target outer mitochondrial membrane proteins (e.g. antitranslocases of the outer mitochondrial membrane22) and lack EV markers (e.g. the tetraspanins CD9, CD63 and CD81). Nevertheless, small free mitochondria, roughly 100–200 nm in diameter, have been discovered in cerebrospinal fluid, which may represent either MDVs or small mitochondria generated during fission. Electron microscopy and the identification of monoamine oxidase-B, an outer mitochondrial membrane protein, on the surface of these structures, suggest that they are free mitochondria, including inner and outer mitochondrial membranes but lacking cristae. These entities are classified as tiny free mitochondria.

Circulating ex-mitochondria

Both EV-Mito and unbound mitochondria have been detected in the plasma of mice and humans. Multiple investigations have recorded the existence of cell-free mtDNA in plasma, predominantly located within exosomes. Evidence for this includes size-exclusion chromatography, which demonstrates that cell-free mtDNA can be removed from plasma using 0.45 μm filters. Additionally, centrifugation of plasma at 15,000–16,000 x g eliminates the mtDNA signal from the plasma and nullifies its respiratory capability. The detection of mtDNA in plasma may suggest the presence of circulating ex-Mito; nevertheless, the mere identification of mtDNA is insufficient for recognising this blood component, as mtDNA can circulate independently of ex-Mito. If the mtDNA is not encapsulated by ex-Mito, it will be susceptible to hydrolysis during a DNase I digestion assay. In contrast, mtDNA located in ex-Mito should be susceptible to DNase I degradation with the application of detergent.

Circulating ex-Mito can be defined by their dimensions, the ratios of EV-Mito to free mitochondria, the kind of donor cell, morphology and other relevant characteristics. Mitochondria can be transmitted across cells of different organs, as demonstrated by contacts between adipocytes and cardiac cells, and between platelets and arterial wall cells. Various supplementary donor cell types, including immune cells, are expected to release mitochondria into the circulatory system. This suggests that interorgan mitochondrial transfer may represent a widespread phenomenon involving multiple donor cell types and diverse target sites. A promising future direction for the study entails clarifying the diversity, interconnections and functions of interorgan mitochondrial transport networks.

Therapeutic approaches

The study of mitochondrial transport has facilitated the development of novel therapeutic techniques. A multitude of companies have recently been founded to promote medicines or technologies focused on mitochondrial transfer or transplantation. These firms and academic researchers are spearheading projects to develop and evaluate the safety and efficacy of therapeutic treatments, including MT and the utilisation of isolated mitochondria in cell engineering or manufacturing. A novel concept pertains to the creation of pharmaceuticals that either promote or obstruct mitochondrial transfer for therapeutic purposes (see Fig. 14.1).

Fig. 14.2.

Very recently, we have reports for clinical trials for the mitochondrial transfer by different groups of scientists against disease conditions, namely sepsis, cancer (malignancy of melanoma cells and gastric cancer (GC)) and neural condition as peripheral neuropathy. All these mitochondrial transfer methodologies deal with the correction of defective mitochondria with mutations in mitochondrial DNA. Scientists have also assessed mitochondrial function and protein composition in platelet-derived EVs. Experts from this field have also recommended for mitochondria transfer and transplantation nomenclature and characterisation.

Mitochondrial transplantation treats sepsis through immune cell metabolism: a metabolomics study—a case study

Sepsis is characterised by profound immune dysregulation and metabolic derangement largely attributable to mitochondrial dysfunction. MT has emerged as a potential therapeutic strategy to restore mitochondrial bioenergetic capacity; however, its effects on immune cell metabolism remain incompletely understood. In the present study, gas chromatography–time-of-flight mass spectrometry–based metabolomic profiling was employed to assess metabolic alterations in peripheral blood mononuclear cells and splenocytes derived from a rat model of polymicrobial sepsis following MT treatment. Multivariate analyses, including principal component analysis and partial least-squares discriminant analysis, demonstrated distinct metabolic clustering among sham, sepsis and MT-treated groups. Sepsis induced a pronounced reduction in metabolites associated with amino acid, carbohydrate and lipid

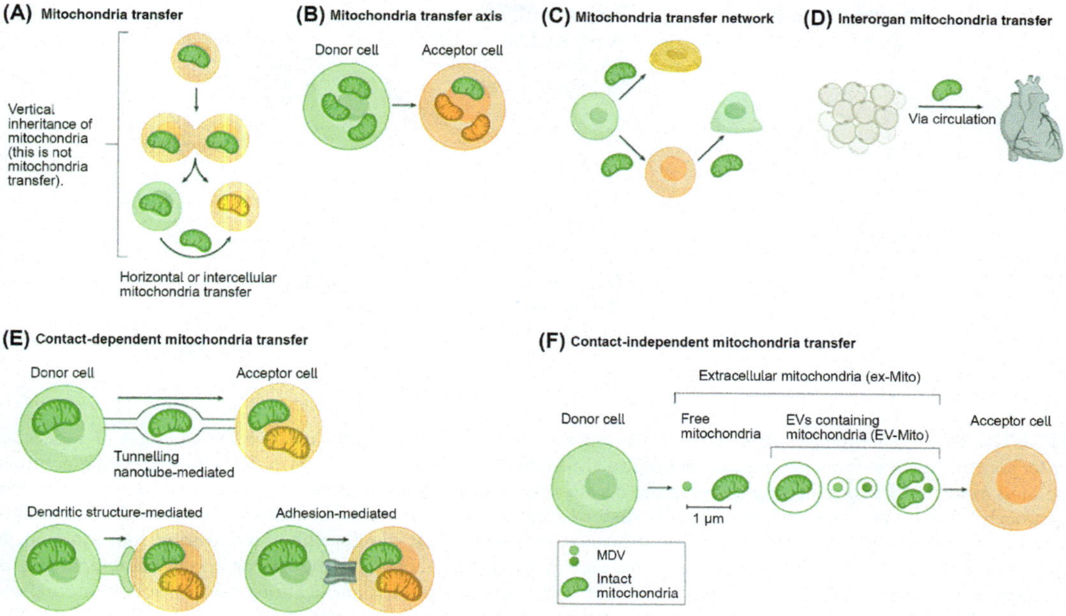

FIGURE 14.1 Terminology describing mitochondrial transfer. (A) Vertical inheritance of mitochondria takes place during cell division, enabling transmission of mitochondria to daughter cells. In contrast, horizontal or intercellular mitochondrial transfer takes place independently of cell division and may occur between developmentally or functionally distinct cell types. (B) The cell releasing mitochondria is referred to as the donor cell, whereas the cell acquiring the transferred mitochondria is termed the acceptor cell. When donor–acceptor relationships are defined in vivo, the interaction is described as a mitochondrial transfer axis. (C) A mitochondrial transfer network refers to the simultaneous or sequential occurrence of multiple mitochondrial transfer axes. (D) Interorgan mitochondrial transfer occurs when cells from one organ release mitochondria into circulating body fluids, such as blood, enabling delivery to cells in distant organs. (E) Contact-dependent mechanisms of mitochondrial transfer include tunnelling nanotube–mediated transfer, in which direct cytoplasmic connections form between two or more cells, allowing mitochondria to be transported directly from one cell to another. Dendritic structure–mediated transfer involves the extension of tubular protrusions from one cell that establish contact with an acceptor cell via end feet, without direct cytoplasmic continuity. Adhesion-mediated mitochondrial transfer occurs when donor and acceptor cells establish direct physical contact, sometimes involving gap junction proteins such as connexin 43 (Cx43). In this case, mitochondria are not transferred through the gap junction channel itself but are internalised by the acceptor cell through endocytic processes at the cell–cell interface. (F) Contact-independent mechanisms of mitochondrial transfer involve the release of mitochondria from donor cells into the extracellular environment, followed by uptake by acceptor cells. These mitochondria are collectively referred to as extracellular mitochondria (ex-Mito). Ex-Mito includes free mitochondria, which are not enclosed within extracellular vesicles (EVs), as well as EV-associated mitochondria (EV-Mito). EV-Mito populations vary in size, mitochondrial cargo and mechanisms of release, and may contain intact mitochondria or mitochondria-derived vesicles (MDVs). The molecular pathways governing the release, stability and uptake of ex-Mito remain an active area of investigation.

metabolism, including aspartic acid, glutamic acid, AMP and myo-inositol, indicative of mitochondrial metabolic impairment. MT partially restored the levels of these metabolites toward those observed in sham controls, accompanied by reactivation of tricarboxylic acid cycle–associated, nucleotide and lipid metabolic pathways. Pathway enrichment analysis

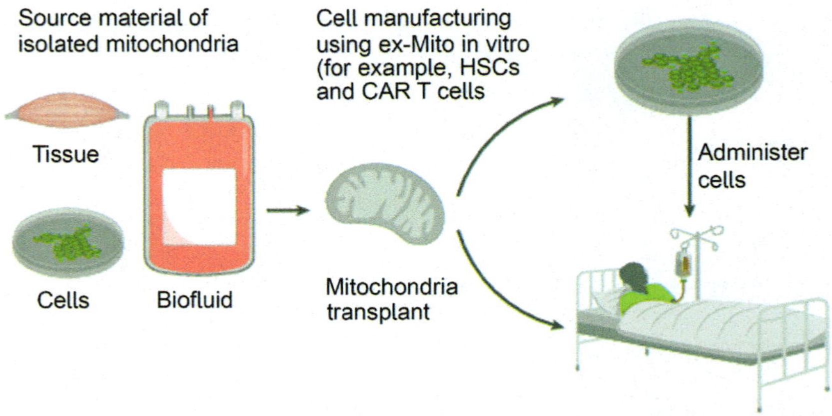

FIGURE 14.2 Therapeutic strategies employing extracellular mitochondria involve the isolation of functional mitochondria from cellular sources or biological fluids for clinical application. In mitochondrial transplantation, isolated mitochondria are directly delivered to a patient or experimental model to induce a therapeutic effect. This approach is mechanistically and procedurally distinct from mitochondrial replacement therapy, which is an in vitro fertilisation–based technique that alters the mitochondrial genome at the germline level during reproduction. In addition, isolated mitochondria may be introduced into cultured cells during ex vivo cell manufacturing or processing to enhance cellular function prior to transplantation. In such applications, excess or noninternalised mitochondria are removed before the engineered cells are infused into the recipient.

further confirmed that administration of exogenous mitochondria mitigated sepsis-induced metabolic suppression and promoted bioenergetic recovery in immune cells. Collectively, these findings provide metabolomic evidence that MT modulates immune cell metabolism and restores mitochondrial oxidative and biosynthetic functions during sepsis, supporting its therapeutic potential as a mitochondria-targeted metabolic intervention (see Fig. 14.3).

Investigation of the effects of interspecies mitochondrial transplantation on melanoma cell malignancy—a case study

The present study investigated the potential of allogenic and interspecies MT to modulate cancer malignancy. Melanoma, a highly metastatic tumour type, is critically dependent on mitochondrial function for energy production and invasive behaviour. Mitochondrial membrane potential (MMP) and ATP levels were assessed in human (A375) and murine (B16F10) melanoma cells, as well as in four potential mitochondrial donor cell types: human (HaCaT) and murine (MPEK-BL6) keratinocytes and human (HUVEC) and murine (MUVEC) endothelial cells. Successful mitochondrial transfer between human and mouse cells was confirmed. Among the donor sources tested, mitochondria derived from HUVECs uniquely suppressed the migratory capacity of B16F10 cells. HUVEC-derived mitochondria were efficiently internalised by B16F10 cells and colocalized with endogenous mitochondria, accompanied by a marked shift in mitochondrial morphology from a branched network to a more globular structure. This morphological remodelling was associated with reduced

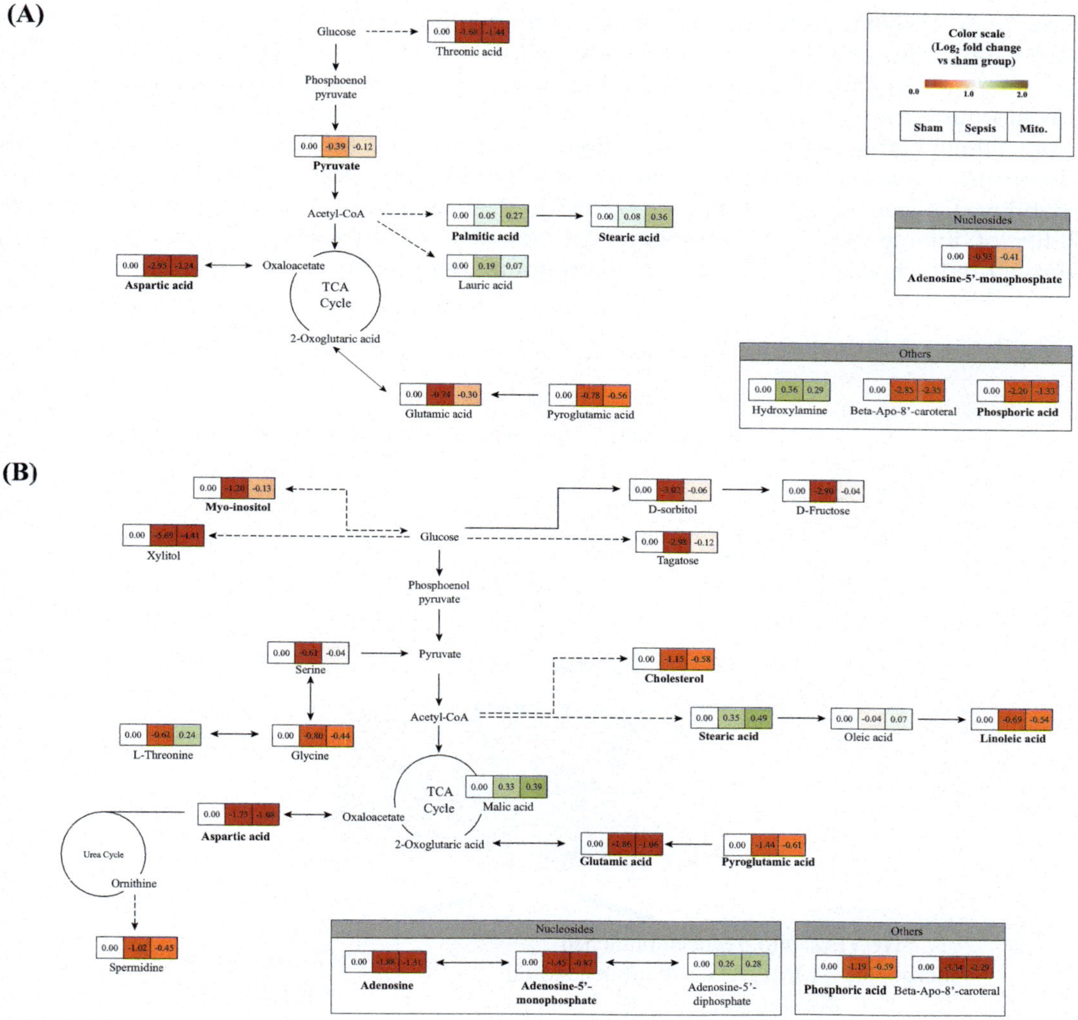

FIGURE 14.3 Metabolic pathway alterations associated with mitochondrial transplantation (MT) in sepsis. Pathway mapping of discriminant metabolites illustrates the metabolic perturbations induced by sepsis and the restorative effects of MT in (A) peripheral blood mononuclear cells and (B) splenocytes. Metabolites significantly downregulated during sepsis and subsequently restored following MT are highlighted, encompassing key intermediates involved in amino acid metabolism, the tricarboxylic acid (TCA) cycle, and carbohydrate metabolism. Collectively, these findings indicate that administration of exogenous mitochondria facilitates the reactivation of oxidative and biosynthetic metabolic pathways in immune cells.

expression of dynamin-related protein 1 and microtubule-associated protein 1 light chain 3-II (LC3-II), concomitant with decreases in MMP and ATP production. Furthermore, MT led to downregulation of key signalling and stemness-associated molecules, including transforming growth factor β, NANOG, SOX2, SMAD2/3, Ak strain transforming (AKT), extracellular signal-regulated kinase (ERK), N-cadherin and MMP-9, which correlated with diminished invasive potential, increased reactive oxygen species levels and reduced cell viability. Collectively, these findings provide preliminary evidence supporting the feasibility of interspecies MT and suggest its potential to attenuate melanoma malignancy through mitochondrial and metabolic reprogramming (see Fig. 14.4).

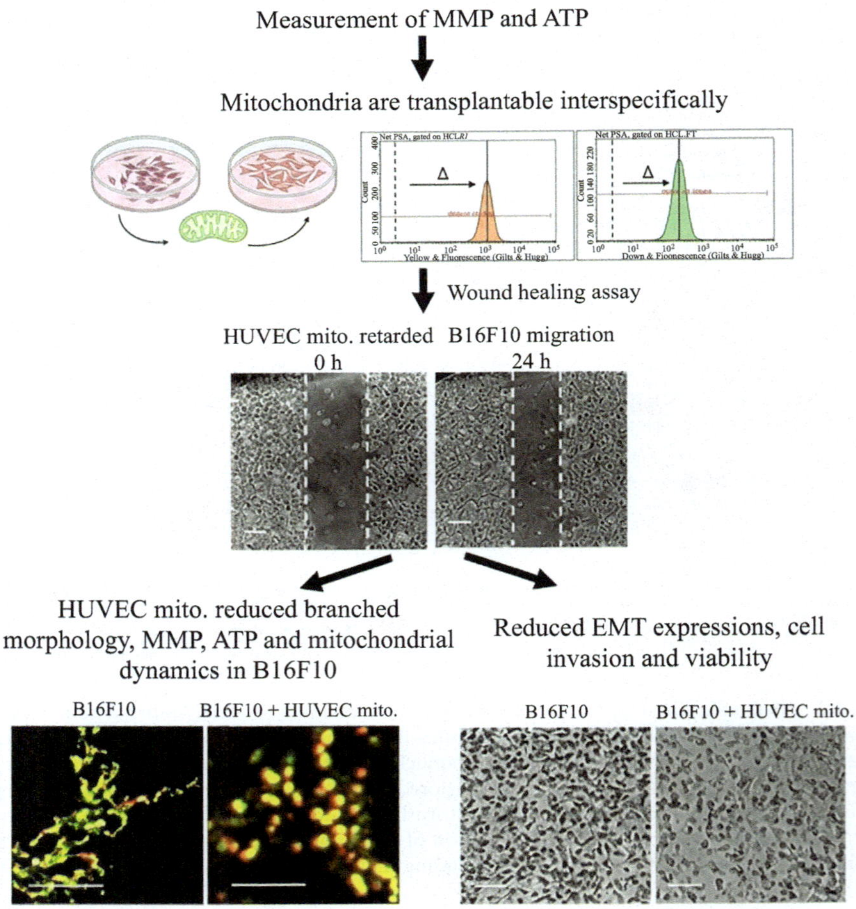

FIGURE 14.4 Mitochondria were successfully transferred between cells. Transplantation of HUVEC-derived mitochondria into mouse melanoma cells significantly attenuated cell migration, disrupted mitochondrial network morphology, impaired mitochondrial dynamics and reduced mitochondrial membrane potential (MMP) and ATP production. Additionally, epithelial–mesenchymal transition (EMT), invasive capacity and cell survival were markedly decreased following mitochondrial transplantation.

Mitochondrial transplantation derived from gastric epithelial cells suppresses malignant characteristics in gastric cancer—a case study

Mitochondria play a central role in energy production and metabolic homoeostasis in cancer cells. In this study, MT was explored as a strategy to attenuate the malignant phenotype of GC cells. The normal human gastric mucous epithelial cell line GES-1 exhibited a lower MMP compared with the immortalised human vascular endothelial cell line EA.hy926 and the human gastric adenocarcinoma cell line adeno gastric stomach (AGS). Successful transplantation of gastric epithelial cell line 1 (GES-1)–derived mitochondria into AGS cells was confirmed by confocal microscopy and flow cytometry.

Following mitochondrial transfer, AGS cells displayed significantly reduced migratory and invasive capacities without detectable effects on cell viability or apoptosis. Analysis of epithelial–mesenchymal transition–associated proteins revealed that GES-1 MT led to downregulation of mesenchymal markers, including α-smooth muscle actin, matrix metalloproteinase-9 , snail, vimentin and N-cadherin, whereas the expression of epithelial markers E-cadherin and claudin-1 remained unchanged. In addition, the cell cycle–related proteins cyclin B1 and cyclin D1 were significantly reduced.

In vivo, inoculation of mice with AGS cells containing transplanted GES-1 mitochondria resulted in the formation of smaller tumours compared with controls. Further investigation of mitochondrial homoeostasis demonstrated that the exogenous GES-1 mitochondria were more stably retained than endogenous AGS mitochondria. Transplanted GES-1 mitochondria exhibited reduced matrix metalloproteinase, ATP production and mitochondrial mass, accompanied by upregulation of mitophagy-associated proteins LC3-II and phosphatase and tensin homolog (PTEN)-induced putative kinase 1.

Collectively, these findings indicate that attenuation of AGS malignancy is mediated by the transplantation of exogenous GES-1 mitochondria. The results support the concept that transplantation of mitochondria with lower bioenergetic efficiency may represent a novel therapeutic approach for the treatment of cancer.

Mitochondrial transfer from glial cells to neurons confers protection against peripheral neuropathy—a case study

Primary sensory neurons within the dorsal root ganglia (DRG) possess long axons and exhibit a high demand for mitochondrial support. Mitochondrial dysfunction has been strongly implicated in the development of peripheral neuropathy associated with diabetes and chemotherapy. However, the mechanisms by which primary sensory neurons maintain mitochondrial homoeostasis remain incompletely understood. Satellite glial cells (SGCs), which ensheathe sensory neurons within the DRG, play key roles in regulating neuronal excitability and pain signalling.

In this study, we demonstrate that SGCs actively transfer mitochondria to DRG sensory neurons in vitro, ex vivo and in vivo through the formation of TNTs mediated by SGC-derived myosin 10 (MYO10). Scanning and transmission electron microscopy identified nanotube-like ultrastructures connecting SGCs and sensory neurons in both murine and

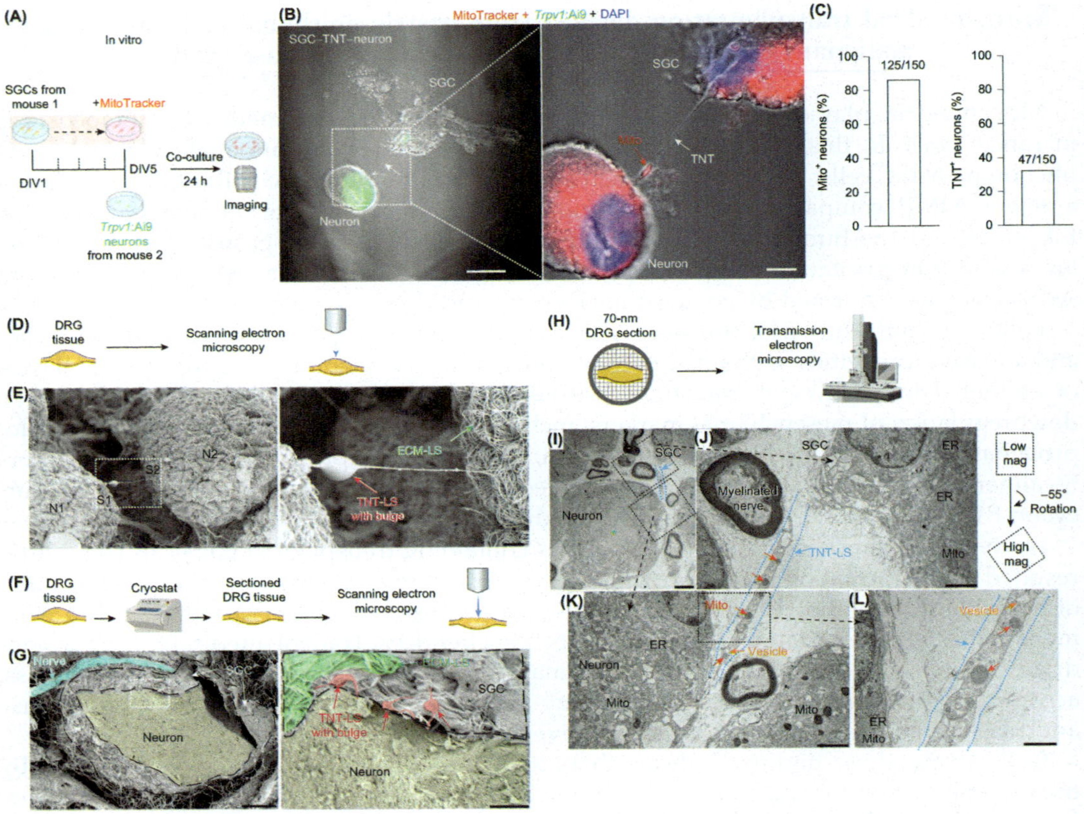

FIGURE 14.5 Mitochondrial transfer from satellite glial cells (SGCs) to neurons and tunnelling nanotube (TNT)–like structures in mouse dorsal root ganglia (DRG). (A) Schematic representation of SGC–neuron coculture systems derived from mouse DRG. SGCs were labelled with MitoTracker dye, and DRG neurons were obtained from Trpv1:Ai9 reporter mice. (B) Left, representative image of an SGC–neuron coculture showing interaction between a Trpv1-positive neuron and an adjacent SGC (scale bar, 20 μm). Right, magnified view of the boxed region illustrating a TNT; white arrow containing a mitochondrion (Mito; red arrow) (scale bar, 5 μm). (C) Quantification of the proportion of neurons positive for MitoTracker-labelled mitochondria (Mito$^+$) and TNT structures. A total of 109 neurons from six independent experiments were analysed. (D) Schematic illustrating the preparation of whole-mount DRG samples for scanning electron microscopy (SEM) without sectioning. (E) High-magnification SEM images of whole-mount mouse DRG revealing a TNT-like structure (TNT-LS) exhibiting a characteristic bulge (n = 4 biological replicates). Scale bars: 5 μm (left) and 1 μm (right). (F) Schematic of mitochondrial and endoplasmic reticulum SEM analysis of sectioned mouse DRG. (G) Representative SEM images of sectioned mouse DRG showing TNT-LS with bulges extending from SGCs to neurons (n = 4 biological replicates). Scale bars: 5 μm (left) and 1 μm (right). (H) Schematic depicting the preparation of mouse DRG for transmission electron microscopy (TEM). (I) Representative low-magnification TEM image showing a DRG neuron and surrounding SGCs (n = 4 biological replicates). The green asterisk denotes the neuronal nucleus. Scale bar, 5 μm. (J–L) Representative high-magnification TEM images illustrating a TNT-LS between an SGC and a neuron. Black 'Mito' labels indicate mitochondria within the cell (J–L), whereas red 'Mito' labels denote mitochondria located within the TNT-LS. (K) 'Vesicle' indicates vesicular structures within the TNT-LS (k, l), and the black asterisk marks the SGC nucleus. Panel (l) shows an enlarged view of panel (k), highlighting mitochondria (red arrows) and a vesicle within a smooth TNT-LS. ER, endoplasmic reticulum. Scale bars: 2 μm (J, K) and 800 nm. (l) Biological replicates: n = 3.

human DRG. Functional inhibition of mitochondrial transfer in naïve mice resulted in axonal degeneration and the development of neuropathic pain. Single-nucleus RNA sequencing and in situ hybridisation analyses revealed high MYO10 expression in human SGCs. Notably, SGCs isolated from individuals with diabetes exhibited reduced MYO10 expression and impaired mitochondrial transfer to sensory neurons. Furthermore, adoptive transfer of human SGCs into the mouse DRG conferred MYO10-dependent neuroprotection against peripheral neuropathy.

Collectively, these findings identify a previously unrecognised role for peripheral glial cells in supporting neuronal mitochondrial homoeostasis and provide mechanistic insights into the pathogenesis of diabetic small-fibre neuropathy. This work highlights mitochondrial transfer as a potential therapeutic target for the treatment of neuropathic pain (see Fig. 14.5.

Thus the most recent evidences indicate MT as one of the most efficient techniques to treat mitochondrial disorders affecting mutiorgans and multidiseases. Since heteroplasmy exists for mitochondria in the cell, if we introduce more healthy mitochondria in the cell, a healthy phenotype would be exhibited. We have already explained in earlier chapters that heteroplasmy exists for mitochondria, when both healthy and mutant mitochondria can coexists.

Further Reading

Brestoff, J. R., Singh, K. K., Aquilano, K., et al. (2025). Recommendations for mitochondria transfer and transplantation nomenclature and characterization. *Nat Metab, 7*, 53–67. https://doi.org/10.1038/s42255-024-01200-x.

Chung, T. N., Choi, S. R., Kim, S. H., Lee, C. H., & Kim, K. (2025). Mitochondrial Transplantation Restores Immune Cell Metabolism in Sepsis: A Metabolomics Study. *Int J Mol Sci; 27*(1), 332. https://doi.org/10.3390/ijms27010332.

Kuo, F. C., Chen, P. C., You, S. Y., Tsai, C. C., Tsai, H. Y., Liu, C. J., Wu, D. C., Lin, M. W., & Huang, B. (2024). Transplanting mitochondria from gastric epithelial cells reduces the malignancy of gastric cancer. *Mitochondrion, 78*, 101939. https://doi.org/10.1016/j.mito.2024.101939 2024 Sep.

Kuo, F. C., Cheng, B. L., Tsai, C. C., Chen, P. C., Sung, W. W., Tsai, K. J., Tsai, H. Y., Huang, Y. B., Liu, C. J., Wu, D. C., Lin, M. W., & Huang, B. (2026). Investigating interspecies mitochondrial transplantation on the malignancy of melanoma cells. *Biosci Biotechnol Biochem: zbag004*. https://doi.org/10.1093/bbb/zbag004.

Veilleux, V., Pichaud, N., Robichaud, G. A., & Boudreau, L. H. (2026). Assessing mitochondrial function and protein composition in platelet-derived extracellular vesicles. *Platelets, 37*(1), 2610264. https://doi.org/10.1080/09537104.2025.2610264.

Xu, J., Li, Y., Novak, C., Lee, M., Yan, Z., Bang, S., McGinnis, A., Chandra, S., Zhang, V., He, W., Lechler, T., Rodriguez Salazar, M. P., Eroglu, C., Becker, M. L., Velmeshev, D., Cheney, R. E., & Ji, R. R. (2026). Mitochondrial transfer from glia to neurons protects against peripheral neuropathy. *Nature*. https://doi.org/10.1038/s41586-025-09896-x.

CHAPTER

15

Whole mitochondrial genome sequencing and its implication in diseases in animal models

With the advances in sequencing technologies, referred to as next-generation sequencing, it is possible to obtain the complete DNA sequencing for the mitochondrial genome of about 16.5 bp, referred to as whole mitochondrial genome sequencing (WMGS). As discussed in the earlier chapters, mitochondria possess 37 genes involved in vital functions. We have also discussed how mutations in mitochondrial genes affect different physiological functions, including metabolic, viral, bacterial, parasitic, fungal diseases, neurodegenerative disorder, cancer and reproductive disorders. We employed a variety of techniques to detect mutagenesis, including Sanger sequencing. WMGS is a next-generation sequencing technique, where we can sequence all 37 genes at a time. This technique was widely used for molecular evolution studies. But in this current chapter, we will discuss solely how mutations in mitochondria affect diseases.

In this chapter, we will first discuss the basic methodology of WMGS, followed by certain case studies.

Methodology

Basic technique for whole mitochondrial genome sequencing with appropriate examples

We are describing here the steps followed in our lab with our own experiments.

Whole mitochondrial genome analysis was conducted for Bengal duck from West Bengal, India (genebank accession number MN011574). The mitochondrial map was generated. The steps for next-generation sequencing studies (WMGS) involve isolation of mtDNA, qualitative and quantitative analysis of g-DNA: PCR amplified with cytochrome oxidase 2 (COX2) (mt specific), glyceraldehyde 3-phosphate dehydrogenase (GAPDH), and beta-actin primers to validate. The next step involves preparation of the library: The paired-end sequencing library (NEBNext Ultra DNA Library Preparation Kit), quantity and quality check (QC) of the library on Bioanalyzer: Bioanalyzer 2100 (Agilent Technologies) using high sensitivity (HS) DNA chip. The final step is cluster generation and sequencing. The adaptors are designed so as to allow selective cleavage. (Fig. 15.1).

Mitochondrial Genomics
DOI: https://doi.org/10.1016/B978-0-443-26655-3.00005-6

Project details:
Service type Mitochondrial DNA sequencing
Platform Illumina
Read length 2 X 150 PE
Data ~3GB/Sample

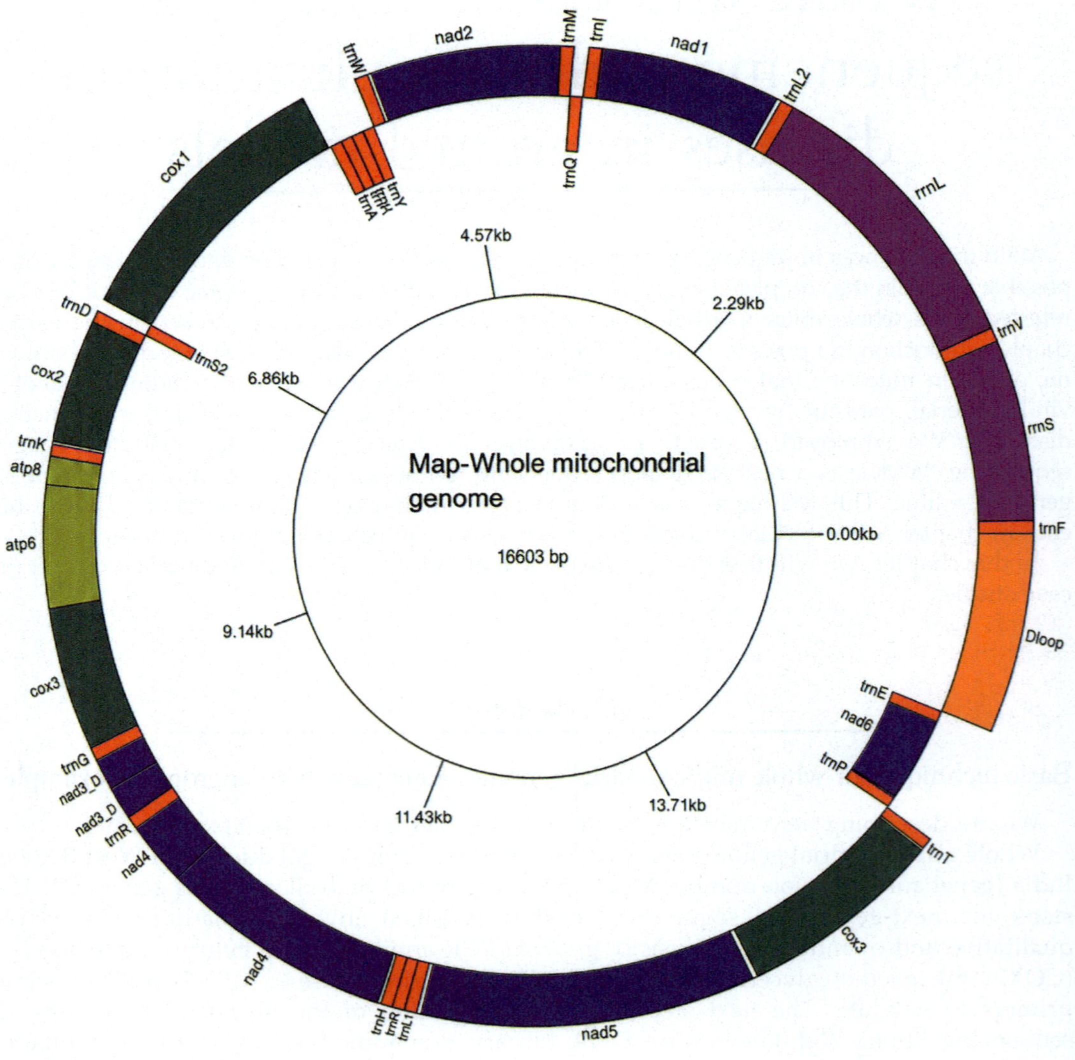

FIGURE 15.1 Mitochondrial map.

Methods

Isolation and quantitative analysis of DNA

DNA was isolated from the given samples using the Abcam Mitochondrial DNA Isolation Kit (ab65321). The volume of 1 µl of each sample was used for determining concentration using Qubit® 2.0 Fluorometer.

Preparation of the library

The paired-end sequencing library was prepared using TruSeq Nano DNA Library Prep Kit. The library preparation process was initiated with 50 ng input. DNA was mechanically sheared into smaller fragments by Covaris, followed by a continuous step of end-repair where an 'A' is added to the 3′ ends, making the DNA fragments ready for adaptor ligation. Following this step, platform-specific adaptors are ligated to both ends of the DNA fragments. These adaptors contain sequences essential for binding dual-barcoded libraries to a flow cell for sequencing, allowing for PCR amplification of adaptor-ligated fragments and binding standard Illumina sequencing primers. To ensure maximum yields from limited amounts of starting material, a high-fidelity amplification step was performed using a HiFi PCR master mix.

Quantity and quality check of library on bioanalyzer

The amplified library was analysed in Bioanalyzer 2100 (Agilent Technologies) using HS DNA chip as per the manufacturer's instructions.

Cluster generation and sequencing

After obtaining the Qubit concentration for the library and the mean peak size from Bioanalyzer profile, library will be loaded onto the Illumina platform for cluster generation and sequencing. Paired-end sequencing allows the template fragments to be sequenced in both the forward and reverse directions. The library molecules will bind to complementary adaptor oligos on a paired-end flow cell. The adaptors are designed to allow selective cleavage of the forward strands after re-synthesis of the reverse strand during sequencing. The copied reverse strand is then used to sequence from the opposite end of the fragment.

Quantification using Qubit fluorometer

It becomes essential to measure the concentration of isolated mitochondria.

Sr	Sample ID	Concentration	Volume	Yield (ng)
1	AP-1	35	100	3500
2	AP-2	2.16	100	216
3	AP-3	47.2	100	4720

4	HD-1	4.42	100	**442**
5	AA-1	1.45	100	**145**
6	AA-2	2.90	100	**290**
7	AA-3	0.94	100	**94**

Bioanalyzer profiles of the library loaded in the Agilent DNA High Sensitivity chip

Although bioanalyzer profile laboratories are available for all the samples, here we will represent a single one to describe. HD1 represents a healthy duck, while others were infected ones (see Fig. 15.2).

Wetlab inference

The libraries were prepared from the samples by *TruSeq Nano DNA Library Preparation Kit*.

The average size of libraries is between 415 bp and 545 bp.

The libraries were sequenced on the Illumina platform (2 x 150 bp chemistry) to generate ~3GB of data/Sample.

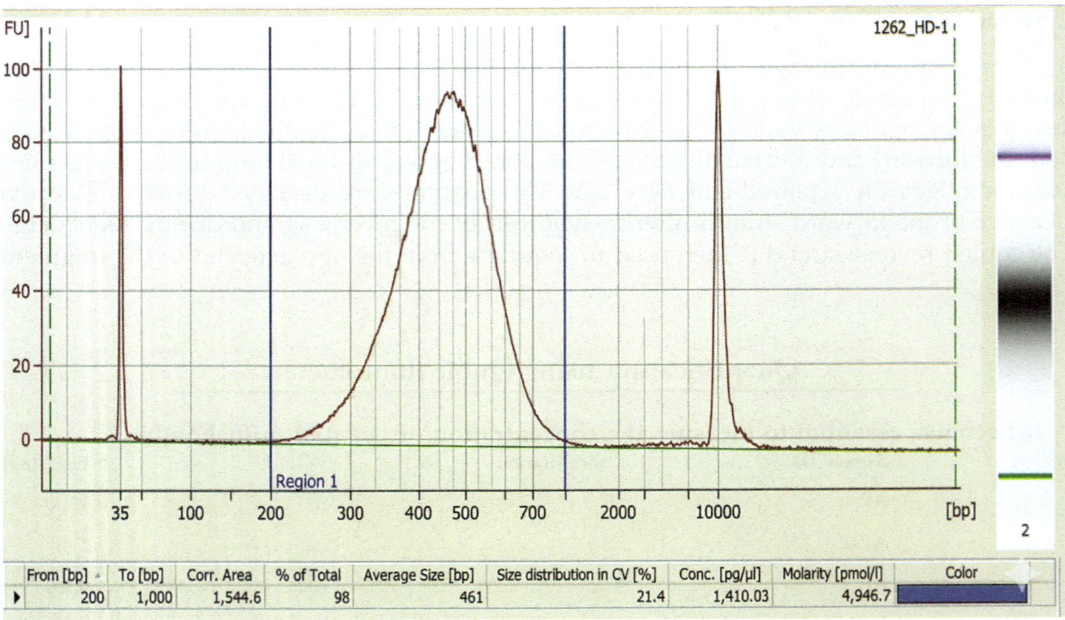

FIGURE 15.2 Bioanalyzer profile of HD-1 (healthy duck).

Bioinformatics analysis

Reads data statistics

The next-generation sequencing for all seven samples was performed on the Illumina platform. The statistics of the data generated are shown in the table given below (see FIg. 15.3 and Table 15.1, Table 15.2).

Mitochondrial genome analysis

De novo assembly

The mitochondrial genome assembly for all samples was carried out using NOVOPlasty v4.2.1 and CLC Genomics Workbench-6 with default parameters. NOVOPlasty is a seed-and-extend algorithm that assembles organelle genomes from whole genome sequencing data, starting from a related or distant single seed sequence. The sequences of the COX1 gene of *Anas platyrhynchos* (NC_009684.1) and *Gallus gallus* (NC_053523.1) were provided as seed for assembly. We obtained good assembly in the form of a single scaffold using NOVOPlasty for four samples, namely AP1, AP3, HD1 and AA2. Thus they were considered for annotation purposes.

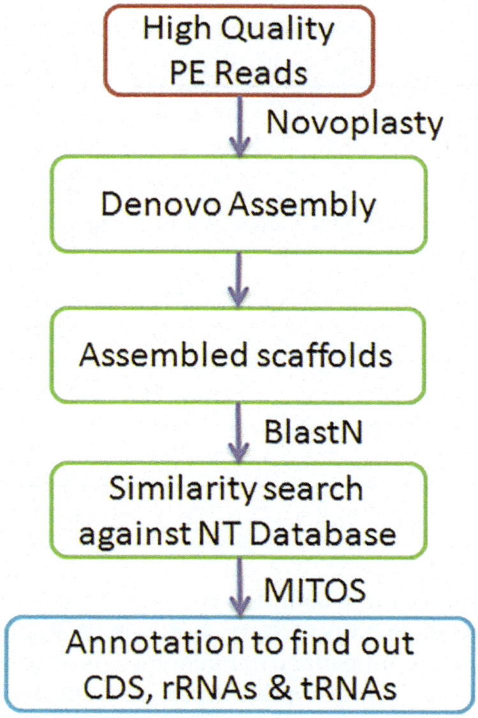

FIGURE 15.3 Bioinformatics analysis workflow.

TABLE 15.1 Next-generation sequencing reads statistics.

Sample ID	Total reads(R1+R2)	Total bases(R1+R2)	Total data(Gb)
AP1	102,273,614	15,443,315,714	15.44
AP2	35,516,558	5,363,000,258	5.36
AP3	24,713,758	3,731,777,458	3.73
HD1	45,191,908	6,823,978,108	6.82
AA1	3,259,038	492,114,738	0.49
AA2	8,702,674	1,314,103,774	1.31
AA3	5,064,208	764,695,408	0.76

TABLE 15.2 List of software used for bioinformatics analysis.

Software	*Version*	*Application*
NOVOPlasty	4.2.1	For de novo assembly of reads
CLC Genomics Workbench	6.1	For de novo assembly of reads
BLASTN	webserver	For similarity search against NT database and reference genome D-loop
MITOS webserver	-	For annotation of mitochondrial genome
GenomeVX webserver	-	For circular representation of mitochondrial genome

TABLE 15.3 NOVOPlasty de novo assembly statistics for four samples.

Description	AP1	AP3	HD1	AA2
# Scaffolds	1	1	1	1
Total scaffold size (bp)	16,603	16,603	16,606	16,783
Max scaffold size (bp)	16,603	16,603	16,606	16,783
Mean scaffold size (bp)	16,603	16,603	16,606	16,783
N50 value (bp)	16,603	16,603	16,606	16,783

However, for the remaining three samples, namely AP2, AA1 and AA3, we used CLC Genomics Workbench as well as NOVOPlasty, but were not able to assemble reads into a single scaffold representing the mitochondrial genome. Assembly data stats were obtained for all seven samples using an in-house per script (see Table 15.3):

TABLE 15.4 CLC de novo assembly statistics for four samples.

Description	AA1	AA3	AP2
# Scaffolds	1227	9948	16,570
Total scaffold size (bp)	322,272	3,790,792	5,345,434
Max scaffold size (bp)	3636	3571	2754
Mean scaffold size (bp)	263	381	323
N50 value (bp)	252	392	316

The de novo assembly thus generated is provided in the data deliverables file '*01_samplename_denovo_assembly.fasta*'.

Table 15.4

Identification of a scaffold representing the mitochondria genome

De novo assembled scaffolds of all samples were searched for similarity against NCBI's nucleotide (NT) database using the BLASTN algorithm. Summary of results for novoplasty assembled four sample scaffolds is as follows

Table 15.5

Summary for CLC assembled sample is not provided as there is a large number of scaffolds in all three samples. Result obtained after doing blastN for these samples is provided in data deliverables file '*02_Samplename_CLC_assembly_blastn.xls*'. Based on the blastn results of AA1, AA3 and AP2 samples, it was observed that there are small scaffolds representing mitochondrial genome, but **no single scaffolds representing complete mitochondrial genome**.

Further single scaffold obtained for AP1, AP3, HD1 and AA2 samples using NOVOPlasty was subjected to mitochondria genome annotation using MITOS webserver. Whereas no scaffold sequence for samples AA1, AA3 and AP2 were considered for annotation using MITOS and further analysis.

Table 15.6

Here, **AT skew** = $[A-T]/[A+T]$; **GC skew** = $[G-C]/[G+C]$.

The NT composition of all four samples mitochondrial genomes is $A+T$ rich. The AT skew is slightly positive and the GC skew is negative, indicating an obvious bias towards the use of As and Cs.

Annotation of the mitochondrial genome

MITOS was used for the annotation of **scaffolds representing mitochondria in AP1, AP3, HD1 and AA2 samples**. MITOS is a web server for the automatic annotation of

TABLE 15.5 BlastN summary for NOVOPlasty assembled samples.

Sample ID	Hitdescription	Hit accession	Query cover	E-value	Percentage Identity
AP1	Anas platyrhynchos mitochondrion, complete genome	MH744426.1	100%	0	99.98%
AP3	Anas platyrhynchos mitochondrion, complete genome	MH744426.1	100%	0	99.98%
HD1	Anas platyrhynchos mitochondrion, complete genome	MH744426.1	100%	0	99.95%
AA2	Gallus gallus isolate DC1-1 mitochondrion, complete genome	MK163564.1	88%	0	99.96%

TABLE 15.6 Nucleotide (NT) summary for NOVOPlasty assembled four samples.

Description	AP1		AP3		HD1		AA2	
NT	**Base pair**	**%**	**Base pair**	**%**	**Base pair**	**%**	**Base pair**	**%**
A	4848	29.20	4849	29.2056	4850	29.21	4474	26.66
T	3686	22.20	3686	22.2008	3686	22.20	3495	20.82
C	5448	32.81	5448	32.8133	5450	32.82	4836	28.81
G	2621	15.79	2620	15.7803	2620	15.78	1999	11.91
N	0	0.00	0	0	0	0.00	1979	11.79
Total length	16,603	100.00	16,603	100	16,606	100.00	16,783	100.00
A + T%	8534	51.40	8535	51.4064	8536	51.40	7969	47.48
G + C%	8069	48.60	8068	48.5936	8070	48.60	6835	40.73
AT skew	0.136161237		0.136262449		0.1362624		0.122851048	
GC skew	-0.350353204		-0.35052058		-0.350521		-0.415069495	

mitochondrial genomes. MITOS allows a reliable and consistent annotation of proteins and noncoding RNAs (tRNAs and rRNAs) using blastx against known annotated mitochondrial proteins and using a covariance model for tRNAs and rRNAs, respectively.

MITOS annotation resulted in the identification of 14 protein-coding genes (duplicate copy of nad3), 2 rRNAs and 22 tRNAs (except AA2, where three tRNAs were missing) taking total to 38 genes in all four samples thus the whole mitochondrial genome was covered in a single scaffold.

Protein coding genes identified by MITOS for all four samples are provided in the table below (Tables 15.7, 15.8, 15.9, 15.10) with their respective length. Whereas complete gene set identified in all four samples are listed in Tables 15.11, 15.12, 15.13, 15.14 with their respective product, start, end, length, strand information, anticodon information for tRNAs, and intergenic spacer.

From a total of 13 coding genes, 12 genes were on the H strand (+ve strand), whereas only 1 coding gene was located on the L strand (-ve strand). ***Coding gene NT and protein sequences have been provided as a part of data deliverables file*** *'Samplename_complete_gene_set.fasta'* ***and*** *'Samplename_coding_sequence_protein.faa'*.

In the AA2 sample, 3tRNAs that is trnI, trnM, trnV, were missing

D-Loop (Control region) Identification:

For identification of the D-loop, region which was not coding for any of the genes was extracted from AP1, AP3, HD1 and AA2 samples as follows:
AP1 probable D-loop regions: 15556 to 16603
AP3 probable D-loop regions: 15556 to 16603
HD1 probable D-loop regions: 15556 to 16606
AA2 probable D-loop regions: 15553 to 16783

TABLE 15.7 Size of genes in bp and their translated protein in amino acids of AP1 mitochondrial genome.

Name	Start	Stop	Strand	Size of nucleotide (bp)	Size of amino acid (bp)
nad1	2822	3775	+	954	318
nad2	4001	5026	+	1026	342
cox1	5410	6942	+	1533	511
cox2	7088	7762	+	675	225
atp8	7846	8007	+	162	54
atp6	8004	8684	+	681	227
cox3	8687	9469	+	783	261
nad3_a	9540	9713	+	174	58
nad3_b	9715	9888	+	174	58
nad4l	9963	10,256	+	294	98
nad4	10,253	11,620	+	1368	456
nad5	11,854	13,641	+	1788	596
cob	13,671	14,792	+	1122	374
nad6	14,966	15,484	-	519	173

TABLE 15.8 Size of genes in bp and their translated protein in amino acids of AP3 mitochondrial genome.

Name	Start	Stop	Strand	Size of nucleotide (bp)	Size of amino acid (bp)
nad1	2822	3775	+	954	318
nad2	4001	5026	+	1026	342
cox1	5410	6942	+	1533	511
cox2	7088	7762	+	675	225
atp8	7846	8007	+	162	54
atp6	8004	8684	+	681	227
cox3	8687	9469	+	783	261
nad3_a	9540	9713	+	174	58
nad3_b	9715	9888	+	174	58
nad4l	9963	10,256	+	294	98
nad4	10,253	11,620	+	1368	456
nad5	11,854	13,641	+	1788	596
cob	13,671	14,792	+	1122	374
nad6	14,966	15,484	-	519	173

TABLE 15.9 Size of genes in bp and their translated protein in amino acids of HD1 mitochondrial genome.

Name	Start	Stop	Strand	Size of nucleotide (bp)	Size of amino acid (bp)
nad1	2822	3775	+	954	318
nad2	4001	5026	+	1026	342
cox1	5410	6942	+	1533	511
cox2	7088	7762	+	675	225
atp8	7846	8007	+	162	54
atp6	8004	8684	+	681	227
cox3	8687	9469	+	783	261
nad3_a	9540	9713	+	174	58
nad3_b	9715	9888	+	174	58
nad4l	9963	10,256	+	294	98
nad4	10,253	11,620	+	1368	456
nad5	11,854	13,641	+	1788	596
cob	13,671	14,792	+	1122	374
nad6	14,966	15,484	-	519	173

TABLE 15.10 Size of genes in bp and their translated protein in amino acids of AA2 mitochondrial genome.

Name	*Start*	*Stop*	*Strand*	*Size of nucleotide (bp)*	*Size of amino acid (bp)*
nad1	2853	3791	+	939	313
nad2	4053	5048	+	996	332
cox1	5427	6959	+	1533	511
cox2	7107	7781	+	675	225
atp8	7861	8022	+	162	54
atp6	8016	8696	+	681	227
cox3	8699	9481	+	783	261
nad3_a	9552	9725	+	174	58
nad3_b	9727	9900	+	174	58
nad4l	9973	10,266	+	294	98
nad4	10,263	11,414	+	1152	384
nad5	11,950	13,053	+	1104	368
cob	13,682	14,806	+	1125	375
nad6	14,964	15,482	-	519	173

TABLE 15.11 Summary table for complete mitochondrial genome of AP1.

S. no	Name	Short name	Start	Stop	Strand	Length	Anticodon	Intergenic spacer
1.	tRNA-PHE	trnF(ttc)	1	70	+	70	ttc	-1
2.	s-rRNA	rrnS	70	1055	+	986		0
3.	tRNA-VAL	trnV(gta)	1056	1126	+	71	gta	0
4.	l-rRNA	rrnL	1127	2728	+	1602		0
5.	tRNA-LEU2	trnL2(tta)	2729	2802	+	74	tta	19
6.	NADH dehydrogenase subunit 1	nad1	2822	3775	+	954		7
7.	tRNA-ILE	trnI(atc)	3783	3854	+	72	atc	7
8.	tRNA-GLN	trnQ(caa)	3862	3932	-	71	caa	-1
9.	tRNA-MET	trnM(atg)	3932	4000	+	69	atg	0
10.	NADH dehydrogenase subunit 2	nad2	4001	5026	+	1026		13
11.	tRNA-TRP	trnW(tga)	5040	5115	+	76	tga	3
12.	tRNA-ALA	trnA(gca)	5119	5187	-	69	gca	2
13.	tRNA-ASN	trnN(aac)	5190	5262	-	73	aac	0
14.	tRNA-CYS	trnC(tgc)	5263	5328	-	66	tgc	0
15.	tRNA-TYR	trnY(tac)	5329	5399	-	71	tac	10
16.	cytochrome c oxidase subunit 1	cox1	5410	6942	+	1533		0
17.	tRNA-SER2	trnS2(tca)	6943	7015	-	73	tca	2
18.	tRNA-ASP	trnD(gac)	7018	7086	+	69	gac	1
19.	cytochrome c oxidase subunit 2	cox2	7088	7762	+	675		13
20.	tRNA-LYS	trnK(aaa)	7776	7844	+	69	aaa	1
21.	ATP synthase F0 subunit 8	atp8	7846	8007	+	162		-4
22.	ATP synthase F0 subunit 6	atp6	8004	8684	+	681		2
23.	cytochrome c oxidase subunit 3	cox3	8687	9469	+	783		1
24.	tRNA-GLY	trnG(gga)	9471	9539	+	69	gga	0

(*Continued*)

TABLE 15.11 (Continued)

S. no	Name	Short name	Start	Stop	Strand	Length	Anticodon	Intergenic spacer
25.	NADH dehydrogenase subunit 3	nad3_a	9540	9713	+	174		1
26.	NADH dehydrogenase subunit 3	nad3_b	9715	9888	+	174		4
27.	tRNA-ARG	trnR(cga)	9893	9962	+	70	cga	0
28.	NADH dehydrogenase subunit 4 L	nad4l	9963	10,256	+	294		-4
29.	**NADH dehydrogenase subunit 4**	**nad4**	**10,253**	**11,620**	**+**	**1368**		**10**
30.	tRNA-HIS	trnH(cac)	11,631	11,699	+	69	cac	0
31.	tRNA-SER1	trnS1(agc)	11,700	11,765	+	66	agc	-1
32.	tRNA-LEU1	trnL1(cta)	11,765	11,835	+	71	cta	18
33.	NADH dehydrogenase subunit 5	nad5	11,854	13,641	+	1788		29
34.	cytochrome b	cob	13,671	14,792	+	1122		11
35.	tRNA-THR	trnT(aca)	14,804	14,872	+	69	aca	10
36.	tRNA-PRO	trnP(cca)	14,883	14,952	-	70	cca	13
37.	NADH dehydrogenase subunit 6	nad6	14,966	15,484	-	519		0
38.	tRNA-GLU	trnE(gaa)	15,485	15,555	-	71	gaa	0
39.	Probable D-loop	D-loop	15,556	16,603		1048		0

Further, a probable D-loop was decided based on:
AT%
Homology search against reference genome D-loop/control region
As one of the major features of D-loop region in the mitochondrial genome is that it has high AT content, so D-loop region was first checked for AT content.
AT content of AP1 probable D-loop regions: 53.1
AT content of AP3 probable D-loop regions: 53.1
AT content of HD1 probable D-loop regions: 53.2
AT content of AA2 probable D-loop regions: 61.2

TABLE 15.12 Summary table for the complete mitochondrial genome of AP3.

S. no	Name	Short name	Start	Stop	Strand	Length	Anticodon	Intergenic spacer
1.	tRNA-PHE	trnF(ttc)	1	70	+	70	ttc	-1
2.	s-rRNA	rrnS	70	1055	+	986		0
3.	tRNA-VAL	trnV(gta)	1056	1126	+	71	gta	0
4.	l-rRNA	rrnL	1127	2728	+	1602		0
5.	tRNA-LEU2	trnL2(tta)	2729	2802	+	74	tta	19
6.	NADH dehydrogenase subunit 1	nad1	2822	3775	+	954		7
7.	tRNA-ILE	trnI(atc)	3783	3854	+	72	atc	7
8.	tRNA-GLN	trnQ(caa)	3862	3932	-	71	caa	-1
9.	tRNA-MET	trnM(atg)	3932	4000	+	69	atg	0
10.	NADH dehydrogenase subunit 2	nad2	4001	5026	+	1026		13
11.	tRNA-TRP	trnW(tga)	5040	5115	+	76	tga	3
12.	tRNA-ALA	trnA(gca)	5119	5187	-	69	gca	2
13.	tRNA-ASN	trnN(aac)	5190	5262	-	73	aac	0
14.	tRNA-CYS	trnC(tgc)	5263	5328	-	66	tgc	0
15.	tRNA-TYR	trnY(tac)	5329	5399	-	71	tac	10
16.	cytochrome c oxidase subunit 1	cox1	5410	6942	+	1533		0
17.	tRNA-SER2	trnS2(tca)	6943	7015	-	73	tca	2
18.	tRNA-ASP	trnD(gac)	7018	7086	+	69	gac	1
19.	cytochrome c oxidase subunit 2	cox2	7088	7762	+	675		13
20.	tRNA-LYS	trnK(aaa)	7776	7844	+	69	aaa	1
21.	ATP synthase F0 subunit 8	atp8	7846	8007	+	162		-4
22.	ATP synthase F0 subunit 6	atp6	8004	8684	+	681		2
23.	cytochrome c oxidase subunit 3	cox3	8687	9469	+	783		1
24.	tRNA-GLY	trnG(gga)	9471	9539	+	69	gga	0

(*Continued*)

TABLE 15.12 (Continued)

S. no	Name	Short name	Start	Stop	Strand	Length	Anticodon	Intergenic spacer
25.	NADH dehydrogenase subunit 3	nad3_a	9540	9713	+	174		1
26.	NADH dehydrogenase subunit 3	nad3_b	9715	9888	+	174		4
27.	tRNA-ARG	trnR(cga)	9893	9962	+	70	cga	0
28.	NADH dehydrogenase subunit 4 L	nad4l	9963	10,256	+	294		-4
29.	**NADH dehydrogenase subunit 4**	**nad4**	**10,253**	**11,620**	**+**	**1368**		**10**
30.	tRNA-HIS	trnH(cac)	11,631	11,699	+	69	cac	0
31.	tRNA-SER1	trnS1(agc)	11,700	11,765	+	66	agc	-1
32.	tRNA-LEU1	trnL1(cta)	11,765	11,835	+	71	cta	18
33.	NADH dehydrogenase subunit 5	nad5	11,854	13,641	+	1788		29
34.	cytochrome b	cob	13,671	14,792	+	1122		11
35.	tRNA-THR	trnT(aca)	14,804	14,872	+	69	aca	10
36.	tRNA-PRO	trnP(cca)	14,883	14,952	-	70	cca	13
37.	NADH dehydrogenase subunit 6	nad6	14,966	15,484	-	519		0
38.	tRNA-GLU	trnE(gaa)	15,485	15,555	-	71	gaa	0
39.	*Probable D-loop*	D-loop	15,556	16,603		1048		0

Further probable D-loop regions from three samples, namely AP1, AP3 and HD1, were searched for homology against D-loop of the duck mitochondrial reference genome (NC_009684.1), whereas the probable D-loop region of the AA2 sample was aligned against D-loop region of the chicken mitochondrial reference genome (NC_053523.1) using BLASTN. Here D-loop of reference genomes was taken as query and the probable D-loop of AP1, AP3, HD1 and AA2 were taken as subjects. Alignment screenshot for all four samples is shown below. (Fig. 15.4

From results of AT% and homology search, it was proved that the extra noncoding region at the end of both genomes were probable D-loop only.

TABLE 15.13 Summary table for the complete mitochondrial genome of HD1.

S. no	Name	Short name	Start	Stop	Strand	Length	Anticodon	Intergenic spacer
1.	tRNA-PHE	trnF(ttc)	1	70	+	70	-1	ttc
2.	s-rRNA	rrnS	70	1055	+	986	0	
3.	tRNA-VAL	trnV(gta)	1056	1126	+	71	0	gta
4.	l-rRNA	rrnL	1127	2728	+	1602	0	
5.	tRNA-LEU2	trnL2(tta)	2729	2802	+	74	19	tta
6.	NADH dehydrogenase subunit 1	nad1	2822	3775	+	954	7	
7.	tRNA-ILE	trnI(atc)	3783	3854	+	72	7	atc
8.	tRNA-GLN	trnQ(caa)	3862	3932	-	71	-1	caa
9.	tRNA-MET	trnM(atg)	3932	4000	+	69	0	atg
10.	NADH dehydrogenase subunit 2	nad2	4001	5026	+	1026	13	
11.	tRNA-TRP	trnW(tga)	5040	5115	+	76	3	tga
12.	tRNA-ALA	trnA(gca)	5119	5187	-	69	2	gca
13.	tRNA-ASN	trnN(aac)	5190	5262	-	73	0	aac
14.	tRNA-CYS	trnC(tgc)	5263	5328	-	66	0	tgc
15.	tRNA-TYR	trnY(tac)	5329	5399	-	71	10	tac
16.	cytochrome c oxidase subunit 1	cox1	5410	6942	+	1533	0	
17.	tRNA-SER2	trnS2(tca)	6943	7015	-	73	2	tca
18.	tRNA-ASP	trnD(gac)	7018	7086	+	69	1	gac
19.	cytochrome c oxidase subunit 2	cox2	7088	7762	+	675	13	
20.	tRNA-LYS	trnK(aaa)	7776	7844	+	69	1	aaa
21.	ATP synthase F0 subunit 8	atp8	7846	8007	+	162	-4	
22.	ATP synthase F0 subunit 6	atp6	8004	8684	+	681	2	
23.	cytochrome c oxidase subunit 3	cox3	8687	9469	+	783	1	

(Continued)

TABLE 15.13 (Continued)

S. no	Name	Short name	Start	Stop	Strand	Length	Anticodon	Intergenic spacer
24.	tRNA-GLY	trnG(gga)	9471	9539	+	69	0	gga
25.	NADH dehydrogenase subunit 3	nad3_a	9540	9713	+	174	1	
26.	NADH dehydrogenase subunit 3	nad3_b	9715	9888	+	174	4	
27.	tRNA-ARG	trnR(cga)	9893	9962	+	70	0	cga
28.	NADH dehydrogenase subunit 4 L	nad4l	9963	10,256	+	294	-4	
29.	NADH dehydrogenase subunit 4	nad4	10,253	11,620	+	1368	10	
30.	**tRNA-HIS**	**trnH(cac)**	**11,631**	**11,699**	**+**	**69**	**0**	**cac**
3.	tRNA-SER1	trnS1(agc)	11,700	11,765	+	66	-1	agc
32.	tRNA-LEU1	trnL1(cta)	11,765	11,835	+	71	18	cta
33.	NADH dehydrogenase subunit 5	nad5	11,854	13,641	+	1788	29	
34.	cytochrome b	cob	13,671	14,792	+	1122	11	
35.	tRNA-THR	trnT(aca)	14,804	14,872	+	69	10	aca
36.	tRNA-PRO	trnP(cca)	14,883	14,952	-	70	13	cca
37.	NADH dehydrogenase subunit 6	nad6	14,966	15,484	-	519	0	
38.	tRNA-GLU	trnE(gaa)	15,485	15,555	-	71	0	gaa
39.	*Probable D-loop*	D-loop	15,556	16,606		1051		0

Mitochondrial map

GenomeVX web server (http://wolfe.ucd.ie/GenomeVx/) was used to generate a circular visual overview of four mitochondrial genomes. All genes and control region representing their location and features are displayed clockwise. Outer circle of circular map represents genes in positive (+) strand and the inner part represents genes in negative (-) strand. (Fig. 15.5

TABLE 15.14 Summary table for the complete mitochondrial genome of AA2.

S. no	Name	Short name	Start	Stop	Strand	Length	Anticodon	Intergenic spacer
1.	tRNA-PHE	trnF(ttc)	1	70	+	70	ttc	-1
2.	s-rRNA	rrnS	70	1006	+	937		131
3.	l-rRNA	rrnL	1138	2738	+	1601		1
4.	tRNA-LEU2	trnL2(tta)	2740	2813	+	74	tta	39
5.	NADH dehydrogenase subunit 1	nad1	2853	3791	+	939		83
6.	tRNA-GLN	trnQ(caa)	3875	3945	-	71	caa	107
7.	NADH dehydrogenase subunit 2	nad2	4053	5048	+	996		4
8.	tRNA-TRP	trnW(tga)	5053	5128	+	76	tga	6
9.	tRNA-ALA	trnA(gca)	5135	5203	-	69	gca	3
10.	tRNA-ASN	trnN(aac)	5207	5279	-	73	aac	1
11.	tRNA-CYS	trnC(tgc)	5281	5346	-	66	tgc	0
12.	tRNA-TYR	trnY(tac)	5347	5416	-	70	tac	10
13.	cytochrome c oxidase subunit 1	cox1	5427	6959	+	1533		0
14.	tRNA-SER2	trnS2(tca)	6960	7034	-	75	tca	2
15.	tRNA-ASP	trnD(gac)	7037	7105	+	69	gac	1
16.	cytochrome c oxidase subunit 2	cox2	7107	7781	+	675		10
17.	tRNA-LYS	trnK(aaa)	7792	7859	+	68	aaa	1
18.	ATP synthase F0 subunit 8	atp8	7861	8022	+	162		-7
19.	ATP synthase F0 subunit 6	atp6	8016	8696	+	681		2
20.	cytochrome c oxidase subunit 3	cox3	8699	9481	+	783		1
21.	tRNA-GLY	trnG(gga)	9483	9551	+	69	gga	0
22.	NADH dehydrogenase subunit 3	nad3_a	9552	9725	+	174		1

(Continued)

TABLE 15.14 (Continued)

S. no	Name	Short name	Start	Stop	Strand	Length	Anticodon	Intergenic spacer
23.	NADH dehydrogenase subunit 3	nad3_b	9727	9900	+	174		4
24.	tRNA-ARG	trnR(cga)	9905	9972	+	68	cga	0
25.	NADH dehydrogenase subunit 4 L	nad4l	9973	10,266	+	294		-4
26.	NADH dehydrogenase subunit 4	nad4	10,263	11,414	+	1152		241
27.	tRNA-HIS	trnH(cac)	11,656	11,701	+	46	cac	8
28.	tRNA-SER1	trnS1(agc)	11,710	11,776	+	67	agc	0
29.	tRNA-LEU1	trnL1(cta)	11,777	11,847	+	71	cta	102
30.	NADH dehydrogenase subunit 5	nad5	11,950	13,053	+	1104		628
31.	cytochrome b	cob	13,682	14,806	+	1125		9
32.	tRNA-THR	trnT(aca)	14,816	14,884	+	69	aca	0
33.	tRNA-PRO	trnP(cca)	14,885	14,954	-	70	cca	9
34.	NADH dehydrogenase subunit 6	nad6	14,964	15,482	-	519		2
35.	tRNA-GLU	trnE(gaa)	15,485	15,552	-	68	gaa	0
36.	Probable D-loop	D-loop	15,553	16,783		1231		0

Analysis summary

NGS sequencing data was generated for samples on the Illumina platform using chemistry of 2 x 150 bp.

De novo-based analysis was carried out for all samples using CLC Workbench and NOVOPlasty. The result obtained from NOVOPlasty was better compared to CLC.

Single scaffold sequence representing mitochondrial genome obtained for AP1, AP3, HD1 and AA3 samples using NOVOPlasty were of size 16,603, 16,603, 16,606 and 16,783 bp, respectively.

These scaffolds were further subjected to annotation using MITOS to obtain 38 mitochondrial genes in case of AP1, AP3 and HD1 samples with a duplicate copy of nad3. Whereas sample AA2 had 35 genes.

```
Score             Expect   Identities          Gaps          Strand
1901 bits(1029)   0.0      1044/1051(99%)      2/1051(0%)    Plus/Plus

Query  1     AGCTAGAATAGCCTAATAATGCTCTCAGGAcccccccccccccccccccccAGGGGTTGC  60
             |||||||||||||||||||||||||||||||||||||||| | |||||||||||||||||
Sbjct  1     AGCTAGAATAGCCTAATAATGCTCTCAGGACCCCCCCCCCTTCCCCCCCCCAGGGGTTGC  60

Query  61    GGGGTTATTTGGTTATACATATCGTGGATACATTTATATTCCCCATATATTAACCTATGG  120
             ||||||||||||||||||||||||||||||||||||||||||||||||||||||||||||
Sbjct  61    GGGGTTATTTGGTTATACATATCGTGGATACATTTATATTCCCCATATATTAACCTATGG  160

Query  121   TCCCGGTAATAAACACTATTAACCAACTATCCTACATGCACGGACTAAACCCATCACATG  180
             ||||||||||||||||||||||||||||||||||||||||||||||||||||||||||||
Sbjct  121   TCCCGGTAATAAACACTATTAACCAACTATCCTACATGCACGGACTAAACCCATCACATA  180

Query  181   TCAAGGGACATACCCTACCTATCGGACTACCCTCCCAATGGACCCAGAATGAATGCTCTA  240
             |||||||||||||||||||||||||||||||||||||| |||||||||||||||||||||
Sbjct  241   TCAAGGGACATACCCTACCTATCGGACTACCCTCCCAACGGACCCAGAGTGAATGCTCTA  240

Query  241   ATACCCAACACCTCAACACCACATAACATGTCCCCAACCAGAACAAGGGCCCATAATGAT  360
             |||||||||||||||||||||||||||||| |||||||||||||||||||||||||||||
Sbjct  241   ATACCCAACACCTCAACACCACATAACATGCCCCCAACCAGAACAAGGGCCCATAATGAT  300

Query  301   GAATGCTTGACAGACATACCCTACCAACACTCCAAATTCCTCTGCACCGACCCATTACTC  360
             ||||||||||||||||||||||||||||||||||||||||||||||||||||||||||||
Sbjct  301   GAATGCTTGACAGACATACCCTACCAACACTCCAAATTCCTCTGCACCCACCCATTACTC  380

Query  361   ATGAAGCTGCGTACCAGATGGATTTATTAATCGTACACCTCAGGTGAAATCAGCAATCCT  420
             ||||||||||||||||||||||||||||||||||||||||||||||||||||||||||||
Sbjct  361   ATGAAGCTGCGTACCAGATGGATTTATTAATCGTACACCTCACGTGAAATCAGCAATCCT  420

Query  421   TGCACATAATGTCCGAGGTGACTAGCTTCAGGCCCATACGTTCCCCGTAAACCCCTCGCC  450
             ||||||||||||||||||||||||||||||||||||||||||||||||||||||||||||
Sbjct  421   TGCACATAATGTCCGAGGTGACTAGCTTCAGGCCCATACGTTCGCCCTAAACCCCTCGCC  480

Query  481   CTCCTCACATTTTTGCGCCTCTGGTTCCTCGGTCAGGGCCATCAATTGGGTTCACTCACC  540
             ||||||||||||||||||||||||||||||||||||||||||||||||||||||||||||
Sbjct  481   CTCCTCACATTTTTGCGCCTCTGGTTCCTCGGTCAGGGCCATCAATTGGGTTCACTCACC  540

Query  541   TCTCCTTGCCCTTCAAAGTGGCATCTGTGGAATACTTCCACCATCTCAATGCGTAATCGC  560
             ||||||||||||||||||||||||||||||||||||||||||||||||||||||||||||
Sbjct  541   TCTCCTTGCCCTTCAAAGTGGCATCTGTGGAATACTTCCACCATCTCAATGCGTAATCGC  600

Query  541   GGCATCCTCCAGCTTTTTGGCGCCTCTGGTTCGTTTTATTTTTTCCAGGGTTACCTCACA  660
             |||||| |||||||||||||||||||||||||||||||||||||||||||||||||||||
Sbjct  661   GGCATCTTCCAGCTTTTTGGCGCCTCTGGTTCCTTTTATTTTTTCCGGGGTTACCTCACA  660

Query  661   GCTGGCCCTTCCCAGTGACTTtGGGGGTCCCACAATCTAAGCCTGGACACACCTGGGTTA  720
             ||||||||||||||||||||||||||||||||||||||||||||||||||||||||||||
Sbjct  661   GCTGGCCCTTCCCAGTGACTTCGGGGGTCCCACAATCTAAGCCTGGACACACCTGGGTTA  780

Query  721   TCGCGCTATCCTATATCTCAGGGATTACTCAATGAGACGGTTGGCGTATATGGGGAATCA  780
             ||||||||||||||||||||||||||||||||||||||||||||||||||||||||||||
Sbjct  721   TCGCGCTATCCTATATCTCAGGGATTACTCAATGAGACGGTTGGCATATATGGGGAATCA  780

Query  781   CCTTGACACTGATGCACTTTGACCACATTCAGTTAATGCTCTCTCCACAGCTCTATATAA  840
             ||||||||||||||||||||||||||||||||||||||||||||||||||||||||||||
Sbjct  781   CCTTGACACTGATGCACTTTGACCACATTCAGTTAATGCTCTCTCCACAGCTCTATATAA  840

Query  841   TAGGGCTATTTAGTGAATGGTCGATGGACATACTTTAAAAACAAAACCACCCCAACCACA  900
             ||||||||||||||||||||||||||||||||||||||||||||||||||||||||||||
Sbjct  841   TAGGGCTATTTAGTGAATGGTCGATGGACATACTTTAAAAACAAAACCACCCCAACCACA  900

Query  901   ACCCCACA--ATATATATATACATACAAAACGAAATGCATAACATGACCTAAATTTATTA  958
             ||||||||  ||||||||||||||||||||||||||||||||||||||||||||||||||
Sbjct  901   ACCCCACAATATATATATATACATACAAAACGAAATGCATAACATGACCTAAATTTATTA  900

Query  959   GAGAAACTCCAGCACTAAAGACGATCCAAACCCGATGACAATCATTACTTTGACCTAACA  1018
             ||||||||||||||||||||||||||||||||||||||||||||||||||||||||||||
Sbjct  961   GAGAAACTCCAGCACTAAAGACGATCCAAACCCGATGACAATCATTACTTTGACCTAACA  1020

Query  1019  ACCATTACCCAATTAACCAGCCACCTGCCCC  1049
             |||||||||||||||||||||||||||||||
Sbjct  1021  AACATTACCCAATTAACCAGCCACCTGCCCC  1051
```

FIGURE 15.4 Alignment of HD1 probable D-loop region against D-loop region of *Anas platyrhynchos* mitochondrion (NC_009684.1).

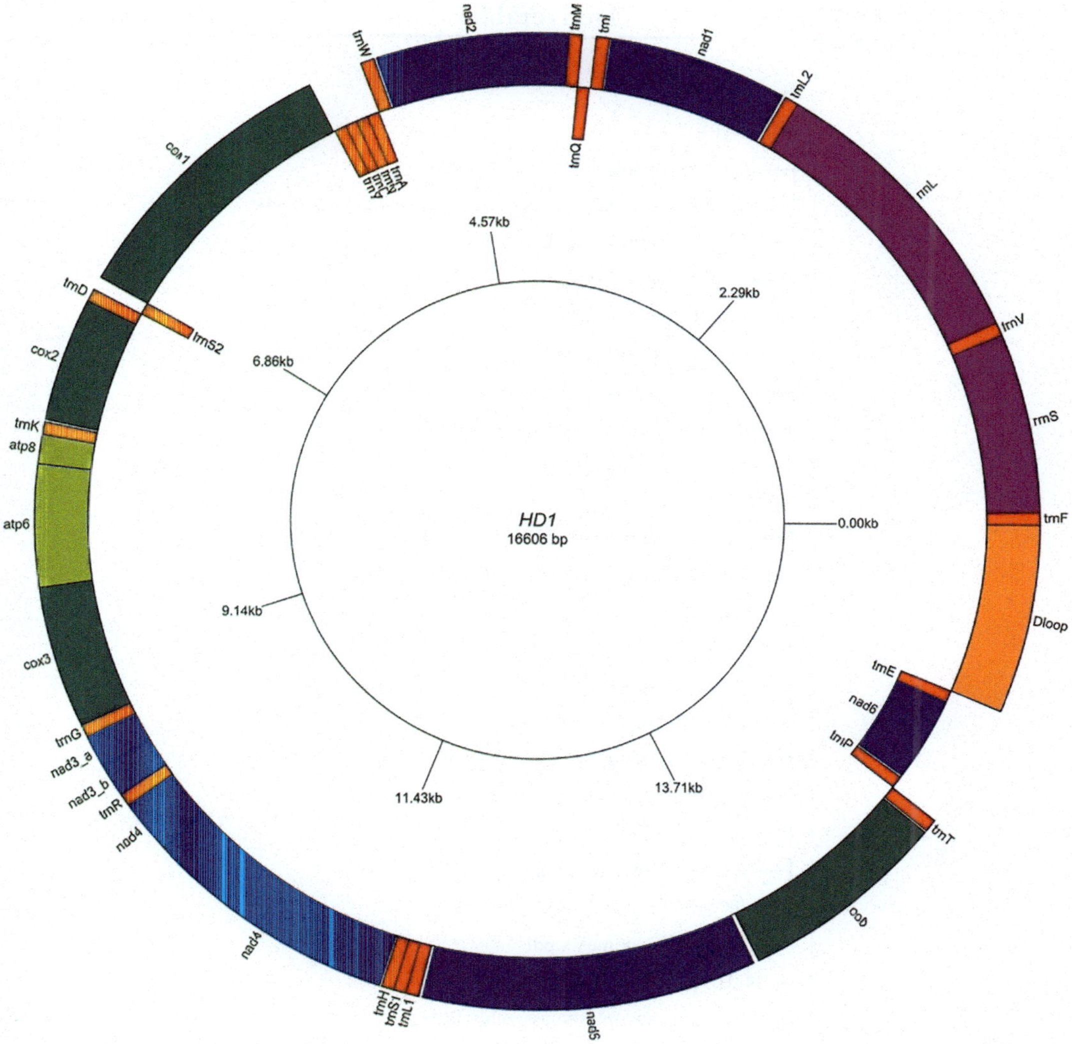

FIGURE 15.5 Circular map of HD1 mitochondrial genome drawn using ***GenomeVx***, representing 38 genes and D-loop, with a total genome size of 16,606 bp.

D-loop identification was carried out based on AT% and homology analysis against D-loop region of the reference mitochondrial genome sequence.

Nucleic acid content for each mitochondrial genome was as follows:

Deliverables

```
├── 01_AP1
│ ├── 01_AP1_novoplasty_assembly.fasta
│ ├── 02_AP1-mitochondria_genome_confirmation_blastN.txt
│ ├── 03_AP1_complete_gene_set.fasta
│ ├── 04_AP1_coding_sequence_protein.faa
│ ├── 05_AP1_rRNAs.fasta
│ ├── 06_AP1_tRNAs.fasta
│ ├── 07_AP1_probable_Dloop.fasta
│ ├── 08_AP1-probable_blast_alignment.txt
│ └── 09_AP1_GenomeVx.pdf
├── 02_AP2
│ ├── 01_AP2_CLC_assembly.fa
│ └── 02_AP2_CLC_assembly_blastn.xlsx
├── 03_AP3
│ ├── 01_AP3_novoplasty_assembly.fasta
│ ├── 02_AP3-mitochondria_genome_confirmation_blastN.txt
│ ├── 03_AP3_complete_gene_set.fasta
│ ├── 04_AP3_coding_sequence_protein.faa
│ ├── 05_AP3_rRNAs.fasta
│ ├── 06_AP3_tRNAs.fasta
│ ├── 07_AP3_probable_Dloop.fasta
│ ├── 08_AP3-probable_blast_alignment.txt
│ └── 09_AP3_GenomeVx.pdf
├── 04_HD1
│ ├── 01_HD1_novoplasty_assembly.fasta
│ ├── 02_HD1-mitochondria_genome_confirmation_blastN.txt
│ ├── 03_HD1_complete_gene_set.fasta
│ ├── 04_HD1_coding_sequence_protein.faa
│ ├── 05_HD1_rRNAs.fasta
│ ├── 06_HD1_tRNAs.fasta
│ ├── 07_HD1_probable_Dloop.fasta
│ ├── 08_HD1-probable_blast_alignment.txt
│ └── 09_HD1_GenomeVx.pdf
```

```
├── 05_AA1
│ ├── 01_AA1_CLC_assembly.fa
│ └── 02_AA1_CLC_assembly.xlsx
├── 06_AA2
│ ├── 01_AA2_novoplasty_assembly.fa
│ ├── 02_AA2-mitochondria_genome_confirmation_blastN.txt
│ ├── 03_AA2_complete_gene_set.fasta
│ ├── 04_AP1_coding_sequence_protein.faa
│ ├── 05_AA2_rRNAs.fasta
│ ├── 06_AA2_tRNAs.fasta
│ ├── 07_AA2_probable_Dloop.fasta
│ ├── 08_AA2-probable_blast_alignment.txt
│ └── 09_AA2_GenomeVx.pdf
├── 07_AA3
│ ├── 01_AA3_CLC_assembly.fa
│ └── 02_AA3_CLC_assembly.xlsx
├── NGS-1262_final_report_Mitogenome_30.07.21.pdf
└── Readme.txt
```

Whole mitochondrial genome sequencing for sheep Fig. 15.6

Uses of WMGS.

In contrast to traditional sequencing for mitochondrial genome, this next-generation sequencing technique has been proven to be useful for sequencing of all 37 genes at a time. Hence, it is a very useful technique for the identification of mutations existing in mitochondrial genes.

►WMGS is useful for molecular evolutionary studies.

►It has a great importance in the detection of mutations in mitochondrial genes in relation to a variety of diseases.

WMGS in disease: a brief outline of the work conducted in our lab.

WMGS in association with metabolic diseases

Studies on the role of mitochondrial genes against metabolic disease in the sheep model as assessed through WMGS—a case study

In our farm, we isolated some sheep with extreme debility and assessed the haematological and biochemical parameters. We observed reduced metabolic rates for such sheep. At the initial step, we assessed the mitogenomic composition for cytochrome B and detected certain deleterious mutations. We have already discussed in detail in Chapter 5.

We have sequenced the whole mitochondrial genome sequence for both healthy and debilitated sheep and observed novel D-loop mutation in addition to cytochrome B mutations discussed in Chapter 5.

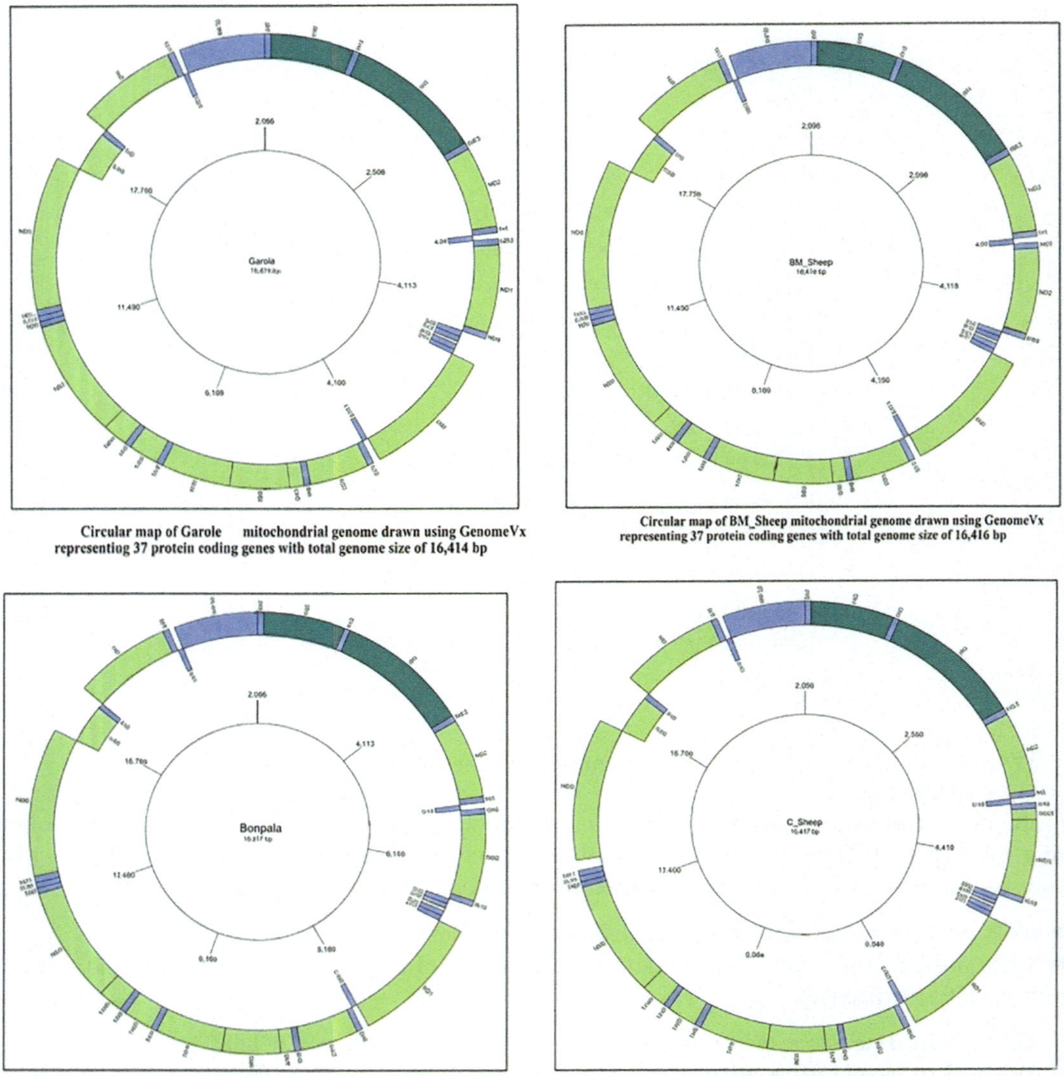

Circular map of Garole mitochondrial genome drawn using GenomeVx representing 37 protein coding genes with total genome size of 16,414 bp

Circular map of BM_Sheep mitochondrial genome drawn using GenomeVx representing 37 protein coding genes with total genome size of 16,416 bp

Circular map of Bonpala mitochondrial genome drawn using GenomeVx representing 37 protein coding genes with total genome size of 16,417 bp

Circular map of C_Sheep mitochondrial genome drawn using GenomeVx representing 37 protein coding genes with total genome size of 16,417 bp

FIGURE 15.6 Whole mitochondrial genome sequencing for different breeds of sheep.

The gene bank accession numbers for the complete mitochondrial sequence as available in the public domain (NCBI) are mentioned below:

Diseased sheep1: D-loop mutation:

MN038345.1 Ovis aries breed Garole mitochondrion, complete genome

Sheep healthy:

MN025259.1 Ovis aries breed Garole mitochondrion, complete genome

In this current chapter, we discuss the screening for the entire mitochondrial genome of 16.5 kb and identify mutations in the D-loop, the controlling region, also through next-generation sequencing technology (WMGS).

Methodology

Characterisation of the WMGS in Garole sheep. Mitochondrial DNA isolation, Illumina platform, 2 x 150 PE, Data: 3GB per sample, pair-end sequencing library preparation, quantity and QC of library on Bioanalyzer, cluster generation and sequencing, quantification using Qubit Fluorometer.

Bioanalyzer profiles of the library loaded in the Agilent DNA HS chip, bioinformatics analysis.

3D structural analysis, model quality assessment, important domain identification, I-mutant, Provean, protein-protein interaction network depiction-Cyt B.

Copy number variation of the mitochondrial genome through differential mRNA expression profiling of host immune response genes with respect to healthy and debilitated sheep.

Statistical analysis

Parasitic disease

Studies on role of mitochondrial genes against parasitic resistance in the sheep model as assessed through WMGS—a case study

In a similar way, we have sequenced the complete genome sequence with respect to healthy sheep and parasitic (*Haemonchus contortus*) infected sheep exposed to natural challenge and identified single nucleotide polymorphisms (SNPs) in cytochrome B and D-loop.

Bacterial disease

Studies on the role of mitochondrial genes against duck Pasteurellosis in the duck model as assessed through WMGS. We identified D-loop mutations in mitochondria as T15773C, T15825C, A15830G and C16161T.

Viral disease

Studies on the role of mitochondrial genes against avian influenza resistance in duck and chicken models as assessed through WMGS.

Reproductive disorder

WMGS reveal certain SNPs in the mitochondrial genome with respect to reproductive disorders in sheep and pig models.

Further Reading

Chakraborty, A., Pal, A., Batabyal, S., Chatterjee, P.N. 2025. Mitochondrial mutations causes susceptibility to bacterial infection (Duck Pasteurellosis) in duck model- a study revealed through *whole mitochondrial genome sequencing (NGS) study*, Proceedings on National conference on *"Advancing Animal Health and production through Biochemical and Biotechnological innovations"* University of Animal and Fishery Sciences, 37, K.B. Sarani, Kolkata-37. West Bengal. India. Pin-700037.

Deb, D., Pal, A., Rahman, S., Samanta, R. 2025. Whole mitochondrial genome sequencing reveal certain mutations responsible for reproductive disorders in sheep model. In Proceedings on National symposium on "Reinforcing BioSecurity: Paving the way for a safer One Health Environment". Jan, 2025. venue: West Bengal University of animal and fishery sciences, 37, K.B. Sarani, Kolkata-37. West Bengal, India.

Debnath, M., Chakraborty, A., Pal, A. 2025a. The role of mitochondrial genes affecting host resistance against avian influenza in Haringhata Black chicken has been revealed through Whole mitochondrial genome sequencing. In Proceedings on National symposium on "Reinforcing BioSecurity: Paving the way for a safer One Health Environment". Jan, 2025. venue: West Bengal University of animal and fishery sciences, 37, K.B. Sarani, Kolkata-37. West Bengal, India.

Debnath, M., Chakraborty, A., Pal, A. 2025b. Whole mitochondrial genome sequencing reveals the role mitochondrial genes affecting host resistance against avian influenza in Bengal duck. In Proceedings on National symposium on "Reinforcing BioSecurity: Paving the way for a safer One Health Environment". Jan, 2025. venue: West Bengal University of animal and fishery sciences, 37, K.B. Sarani, Kolkata-37. West Bengal, India.

Pal, A., Samiddha Banerjee, S. Batabyal, P.N. Chatterjee. 2025. Whole mitochondrial genome sequencing reveal certain mutations in cytochrome B and D loop adversely affecting the health of sheep. In the Proceedings of International Conference on *"Precision Animal Breeding through Genomics, artificial intelligence and machine learning"*, 13-14th November, 2025. ISAGB conference at West Bengal University of Animal and Fishery Sciences, Kolkata, West Bengal, India.

Rahman, S., Pal, A., Deb, D., Samanta, R. 2025. Mutations in Mitochondrial genes have role in reproductive disorders in pig model- a study through Whole mitochondrial genome sequencing. In Proceedings on National symposium on "Reinforcing BioSecurity: Paving the way for a safer One Health Environment". Jan, 2025. venue: West Bengal University of animal and fishery sciences, 37, K.B. Sarani, Kolkata-37. West Bengal, India.

CHAPTER

16

Implication of gene editing in mitochondrial repair

Introduction

In our earlier chapters, we have studied in detail the mitochondrial genome. We have also studied the role of the mitochondrial genome in different disease conditions, including infective and noninfective conditions.

Mitochondrial defects are naturally corrected by our body through **mitochondrial fission** followed by **mitophagy**. Third, as a therapeutic measure, there is the concept of **mitochondrial replacement** from a healthy human. Fourth, we have mentioned the concept of **three-parent baby concept. Heteroplasmy** is a common condition of coexistence of both defective and healthy mitochondria. Disease phenotypes appear when the proportion of mutant mitochondria exceeds to normal ones. In mitochondrial replacement, we are inserting the healthy mitochondria with the hope of having a better proportion of healthy mitochondria compared to mutant ones. Mitochondrial fusion is another condition of the fusion of healthy and mutant mitochondria. In that case, there is a natural increase in the number of mitochondrial DNA (mtDNA) coding for a healthy phenotype.

Now the question arises if there is any way to correct the defect at the genomic level of mitochondria. In a previous chapter, we discussed mitochondrial replacement therapy for transplantation of mitochondria from a healthy donor. In this chapter, we will describe the correction of mitochondrial mutations through gene editing approaches.

Gene editing

Gene editing is a genetic engineering technique, wherein a particular fragment of DNA is:

Inserted—gene knock in
Replaced—knock out coupled with knock in
Gene knock out—removed from a genome
with an artificially engineered nuclease, referred to as 'Molecular scissors'. It is also known as genome editing with engineered nucleases.

Mitochondrial Genomics
DOI: https://doi.org/10.1016/B978-0-443-26655-3.00004-4

Clustered regularly interspaced short palindromic repeats a (CRISPRa)
CRISPRi

There are three available tools for gene editing:

1. **Zinc finger technology**
2. **Transcription activator-like effector nuclease (TALEN) technology**
3. **CRISPR-Cas technology**

Among these available techniques, **CRISPR-Cas technology is** claimed to be the most effective.

Let us discuss in brief about these techniques

1. Zinc finger technology: Zinc finger technology is a genome editing technique that involves zinc finger nucleases (ZFNs). These are synthetic restriction enzymes developed through the combination of a zinc finger DNA-binding domain with a DNA-cleavage domain. Consequently, specialised software facilitates the construction of zinc finger domains that target specific DNA sequences, enabling zinc-finger nucleases to identify unique sequences within complicated genomes. Endogenous DNA repair mechanisms are effective for accurately modifying the genomes of higher species. Similar to Cas9 and TALEN proteins, ZFN is also gaining significance as an effective method in genome editing. The figure illustrates a schematic representation of zinc finger technology. A ZNF is a highly site-specific endonuclease. Specific software designs ZNF to facilitate the binding and cleavage of DNA at designated locations. It comprises two protein domains: the initial domain is the DNA-binding domain. It contains eukaryotic transcription factors and zinc finger motifs. The second domain is the nuclease domain. It possesses the FokI restriction enzyme, which facilitates the catalytic breakage of DNA.

Domain that binds to DNA: The primary constituents of DNA-binding domains in each ZFN are three to six distinct zinc finger repeats. These repeats may distinguish between 9 and 18 base pairs, respectively. In case the zinc finger domains are specific to their designated target site, a pair of three-finger ZFNs can recognise a total of eighteen base pairs. The modular assembly of individual zinc-finger domains with predefined DNA-binding specificities represents a highly effective strategy for engineering novel zinc-finger protein arrays. The predominant modular assembly approach integrates three distinct zinc fingers. These zinc fingers identify a three-base pair DNA sequence, forming a three-finger array that recognises a nine-base pair target location.

Domain for DNA breakage: FokI is frequently used as the cleavage domain in ZFNs. FolkI is a cleavage domain of a type II restriction endonuclease. The cleavage domain undergoes dimerisation to facilitate DNA-cleavage. Two ZFNs are required to target nonpalindromic DNA sequences. The C-terminus of each zinc finger domain is fused with the cleavage domain of conventional zinc fingers. The two distinct zinc fingers interact with opposite strands of DNA positioned at a specific distance from the C-terminus. This facilitates the dimerisation of two cleavage domains and the cleavage of DNA. FokI nuclease functions as the DNA-cleavage domain and interacts with DNA through designed Cys2His2 zinc fingers. Particular zinc fingers facilitate the identification of various nucleotide triplets and the dimerisation of the FokI nuclease, resulting in the induction of double-strand breaks (DSB).

1. TALEN technology TALEN systems originate from a fusion of transcription activator-like effectors (TALEs), derived from *Xanthomonas* spp., and the restriction endonuclease FokI. By modifying the amino acid repeats in TALEs, researchers may tailor TALEN systems to selectively bind to target DNA and produce DNA breaks.

2. CRISPR-Cas technology

RNA-guided endonucleases utilise a short guide RNA (gRNA) to identify DNA, associate with an endonuclease, resulting in site-specific cleavage. Originating from the CRISPR in bacteria, which function to recognise and eliminate foreign DNA (see Fig. 16.1

CRISPR is a better, cost-effective technique compared to TALEN technology. The genome editing technique of CRISPR is cheaper, but there are certain limitations. CRISPR technology may lead to off-target effects, which means it may cause the unintentional modification of nontargeted genes.

Can clustered regularly interspaced short palindromic repeats edit mitochondria?

A case study

Studies indicate that **mtDNA editing is possible using CRISPR/Cas9**. Mitochondria-targeted Cas9 (**mitoCas9)** with specific localisation to the mitochondria was observed to be applicable for mitochondrial genome editing. A significant barrier in the advancement of CRISPR-Cas9-mediated gene editing of the mitochondrial genome is the inadequacy of efficient techniques for delivering gRNA across the mitochondrial membrane. This **limitation exist** since the import of mRNAs into the mitochondria is largely dependent on tRNA.

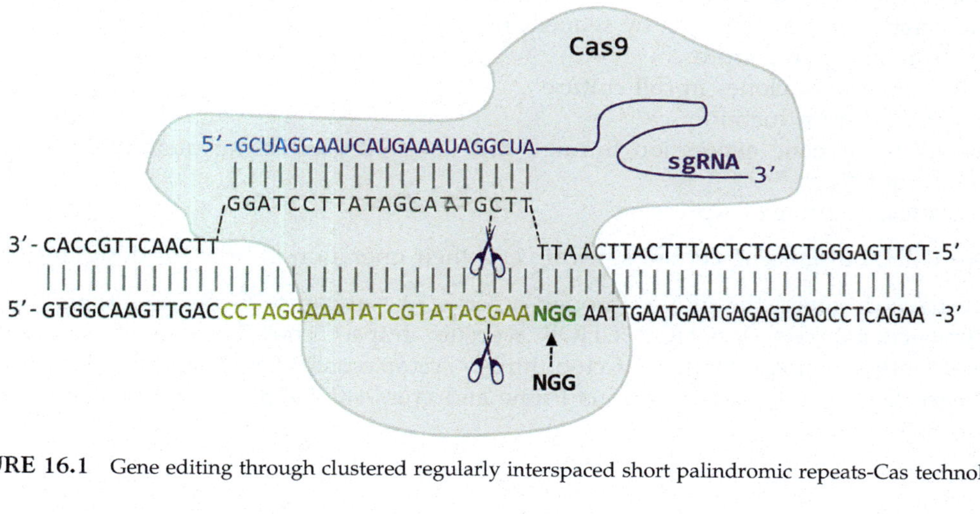

FIGURE 16.1 Gene editing through clustered regularly interspaced short palindromic repeats-Cas technology.

Can mitochondrial DNA be modified?

mtDNA are a significant contributor to various disorders and pose a hazard to human health. Consequently, a viable method for the editing of mtDNA remains in its nascent stages. There exist certain open questions:

1. The Mitochondrial genome exists in multiple copies in a single cell, in contrast to the nuclear genome.
2. There exists heteroplasmy, that is, there is the presence of both wild-type and mutant types of mitochondria in a cell. Whenever the mutant types exceed those of the wild-type, phenotypic defects are observed.

Whenever we target gene editing for the mitochondrial genome, does it exclusively target only the mutant one or both wild and mutant types?

Is there the existence of protospacer-adjacent motif (PAM) site within the mitochondrial genome, to be targeted?

The steps for gene editing may be enumerated as follows:

1. Designing of gRNA: it is one of the most basic and important steps for any CRISPR-based gene editing. In case of CRISPRa and CRISPRi, we target the promoter region, whereas in case of CRISPR knock out and CRISPR knock in, we target the particular region to insert our designed gene to be expressed.
2. Vector/plasmid isolation from glycerol stock
3. Plasmid digestion/linearization of vector
4. PCR clean up: cleaning up of linear plasmid
5. Primer/gRNA
   -dilution
   -annealing and phosphorylation of single guide RNA (sgRNA)
6. Final ligation: plasmid, sgRNA
7. Transformation in bacteria
8. Bacterial culture
9. Screening of recombinant bacteria: through a selective marker, PCR-based screening
10. Plasmid isolation followed by sequencing
11. Mammalian cell culture
12. Introduction of clones in cell culture
13. Lentiviral transduction
14. Analysis of gene expression through mRNA expression using quantitative real-time PCR (qPCR)
15. Analysis of protein expression

We will be discussing certain software and their operation in brief for designing gRNA.

1 We have designed the gRNA through a series of softwares—Geneious Prime: (https://www.geneious.com/), CHOP CHOP website: https://chopchop.cbu.uib.no/, CRISPR direct: https://crispr.dbcls.jp/, Cctop: https://cctop.cos.uni-heidelberg.de/. We designed our primers mainly with Geneious Prime and cross-validated with others (see Fig. 16.2, Fig. 16.3

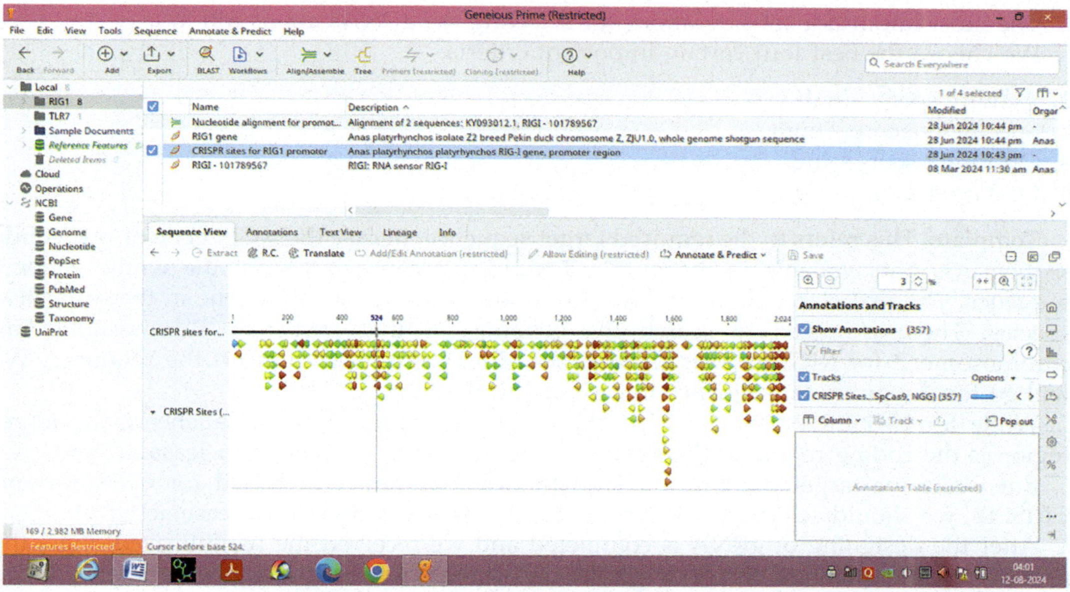

FIGURE 16.2 An example for the designing of guide RNA for CRISPR sites with Geneious Prime software in sequence view.

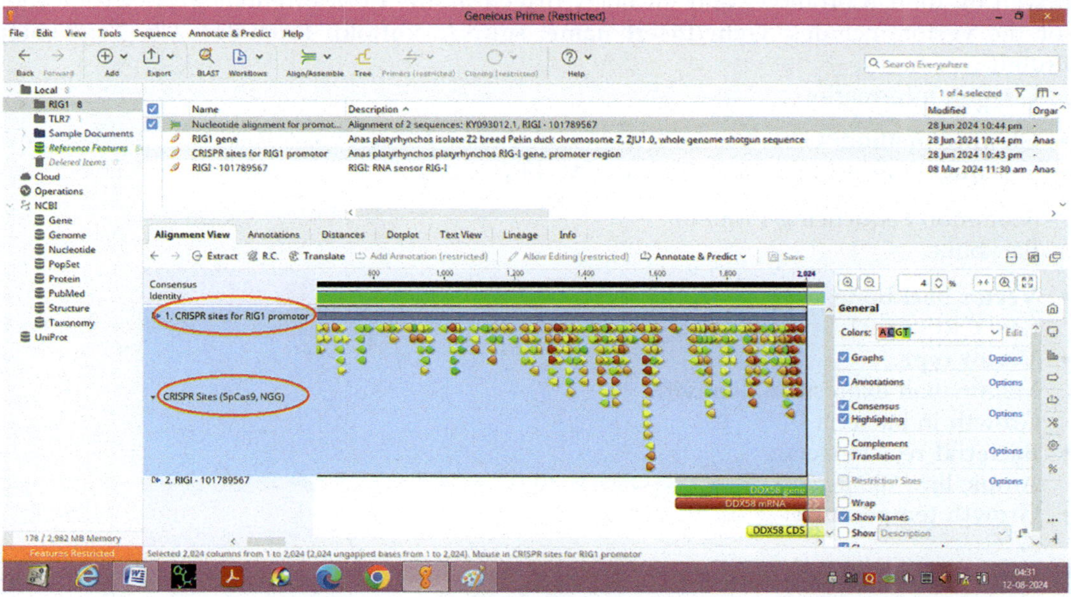

FIGURE 16.3 Designing of guide RNA for CRISPR sites with Geneious Prime software in alignment view.

The most important considerations for selecting the gRNAs are as follows:
We choose the best four certain important criteria:

1. Activity score—high
2. Specificity score—high
3. Off-target score—low
4. On-target score—high

Template: This refers to the required target sequence. In case the NCBI gene or nucleotide sequence is not available in the public domain, we can select from the whole genome sequence. The first step is to locate the chromosomal location of the gene in the respective species. Then, we need to search the genome for that particular chromosome. Say chromosome Z for RIGI in *Anas platyrynchos* (duck). We need to insert the sequence. We will go for the promoter sequence in case of CRISPRa and CRISPRi.

While for CRISPR knock out, CRISPR knock in, and CRISPR mutagenesis, the target region is the coding region of the gene. Cas9 is the choice with endonuclease activity.

But while we target the CRISPR sites for RIGI promoter designed for CRISPRa and CRISPRi, we should select dCas9 (which only binds without endonuclease activity).

After the designing of gRNA is completed and we received the lyophilisized oligo, the next step is dilution of guide RNAs (gRNA and sgRNA). We have designed separate gRNAs pertaining to separate experiments for CRISPRa, CRISPRi, CRISPR knock out and CRISPR knock in. Then, we followed annealing and phosphorylation of sgRNA.

We used the expression vector as a lentiviral vector. The vector or the plasmid is a circular double-stranded DNA molecule (see Fig. 16.4

Lentivirus compatible *Streptococcus pyogenes* Cas9 (SpCas9)/dCas9 sgRNA scaffold driven by the U6 promoter and mCherry driven by the EF1a promoter. The specification for the vector includes: **Gene/Insert name:** sgRNA scaffold, **Promoter:** U6, **Tag/Fusion Protein:** mCherry.

Cloning information:

- **Cloning method:** Gibson cloning
- **5′ sequencing primer:** TTTCTTGGGTAGTTTGCAGTTTT

(Common Sequencing Primers)
Backbone

- **Vector backbone:**
  LentiCRISPR v2
- **Vector type:**
  Mammalian expression, lentiviral
  Growth in bacteria
- **Bacterial resistance(s):**
  Ampicillin, 100 μg/mL
- **Growth temperature:**
  37°C
- **Growth strain(s):**

  NEB stable

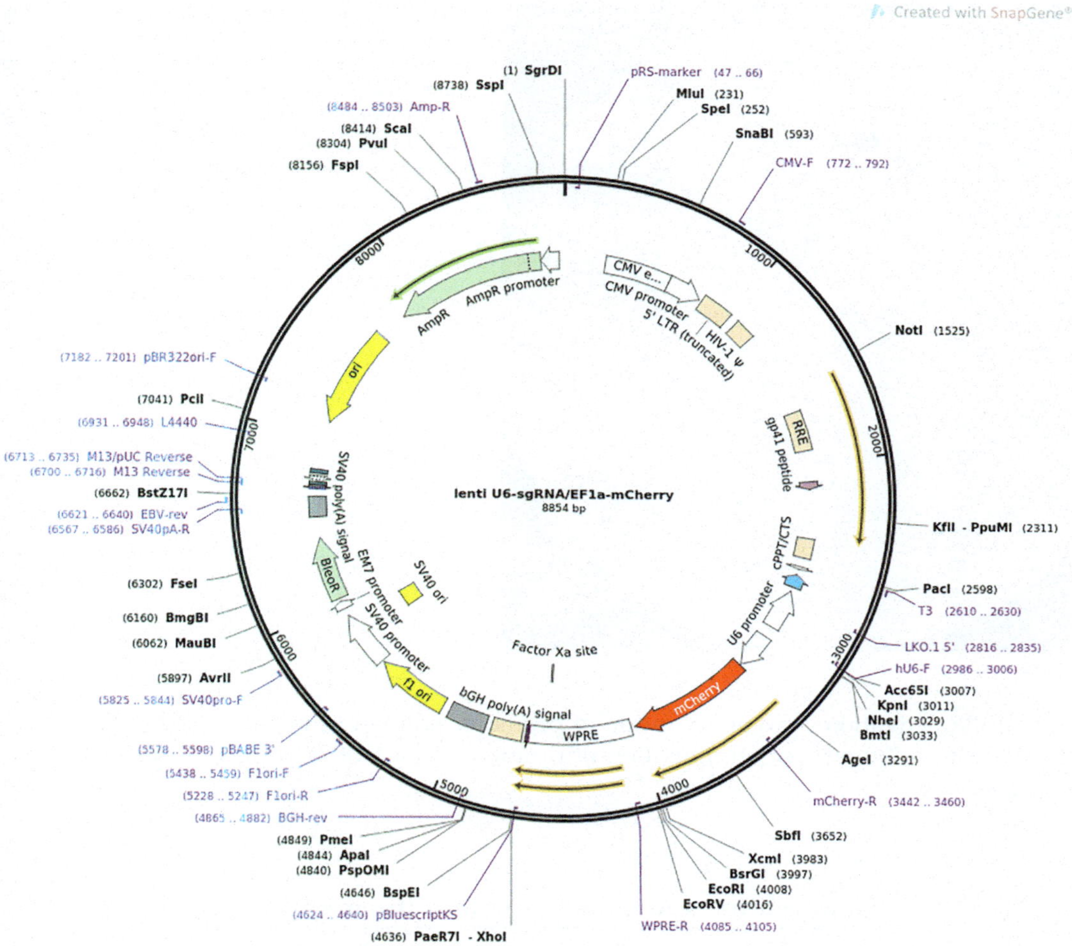

FIGURE 16.4 Lentiviral mammalian expression vector used in our study. Lenti U6-sgRNA/EF1a mCherry.

In our next step, we cut the vector with restriction enzyme BbSI to linearise it and purified it with purification kit (QiAQuick PCR purification kit). In the next step, we employed **In-Fusion Snap Assembly**, gRNA (pairs fused): 200 ng, linearized vector: 200 ng, 5x fusion HD enzyme premix: 2 µL, nuclease free water upto 10 µL. Incubate for 15 minutes at 50°C then place on ice.

In the next step, we transformed the stellar competent cells (*E. coli* HST08 strain) and selected five colonies growing in LB agar plates with ampicillin. Screened them finally through sequencing (see Fig. 16.5).

CRISPR-based genome editing may be of following four types:

CRISPR knock out

CRISPR knock in

CRISPR mutagenesis

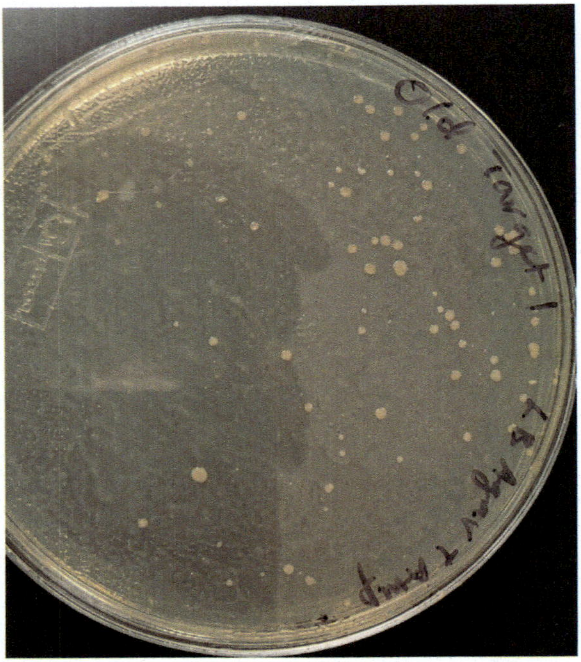

FIGURE 16.5 Positive colonies for transformed stellar competent cells (*E. coli* HST08 strain) with recombinant lenti U6-sgRNA/EF1a mCherry with guide RNA insert, cultured in LB agar with ampicillin. Colonies picked up for subculturing, picked up for LB broth culture, followed by plasmid isolation, sent for sequencing.

CRISPR activation—target activation of promoter region of the gene—SPAH
CRISPR inhibition—target inhibition of promoter region of target gene—KRAB
Table 16.1

Gene editing system delivery into the mitochondria

In current days, gene editing is well documented in nuclear genes, but gene editing directly within the mitochondria is rather complicate due to certain factors- multiple number of mitochondria, copy number variation for mitochondrial DNA within a mitochondria, heteroplasmy and many other factors, already discussed in earlier chapters.

Fig. 16.6

DNA components might be introduced via liposome transfection or alternative techniques. The blue mitochondria represent those carrying altered mtDNA (the target site of the gene editing system), whereas the green mitochondria signify wild-type mitochondria. Restriction endonuclease; ZFN; TALEN; CRISPR; prokaryotic argonaute proteins (pAgo) (see Fig. 16.7).

Fig. 16.8

Table: Comparison of several gene editing methods.

In this current section, we attempt to compare different methodologies for gene editing. However, in our lab, we are currently working with CRISPR- Cas9, and we find it easy to apply.

Table 16.2, Table 16.3

TABLE 16.1 Comparison of mitochondrial intervention.

Mitochondrial intervention	Explanation	Specific	Autologous
Mitochondrial replacement therapy			
PNT	Transfer of the patients' pronuclei	-	-
PBT	Transfer of the patients' polar body	-	-
MST	Transfer of the patients' MII spindle complex	-	-
Mitochondrial gene transfer			
Viral vectors	Adenovirus and its associated vectors	-	-
Nonviral strategy	Gene overexpression, gene knockdown and chemical agents	-	+
Mitochondrial gene editing			
Mito-RE	Specifically targets and cuts DNA at RE-binding site	+	+
Mito-CRISPR	Specifically cuts DNA at CRISPR/Cas9-binding site	+	+
Mito-ZFN	Specifically targets and cuts DNA at ZFN-binding site	+	+
Mito-TALEN	Specifically targets and cuts DNA at TALEN-binding site	+	+
DdCBE	mtDNA base editor sequence via fused TALE	+	+
In vitro oogenesis	Oocyte derived from selected patient iPSC lines	+	+
Medication	Drug therapy for the enhancement of mitochondrial quality	+	+

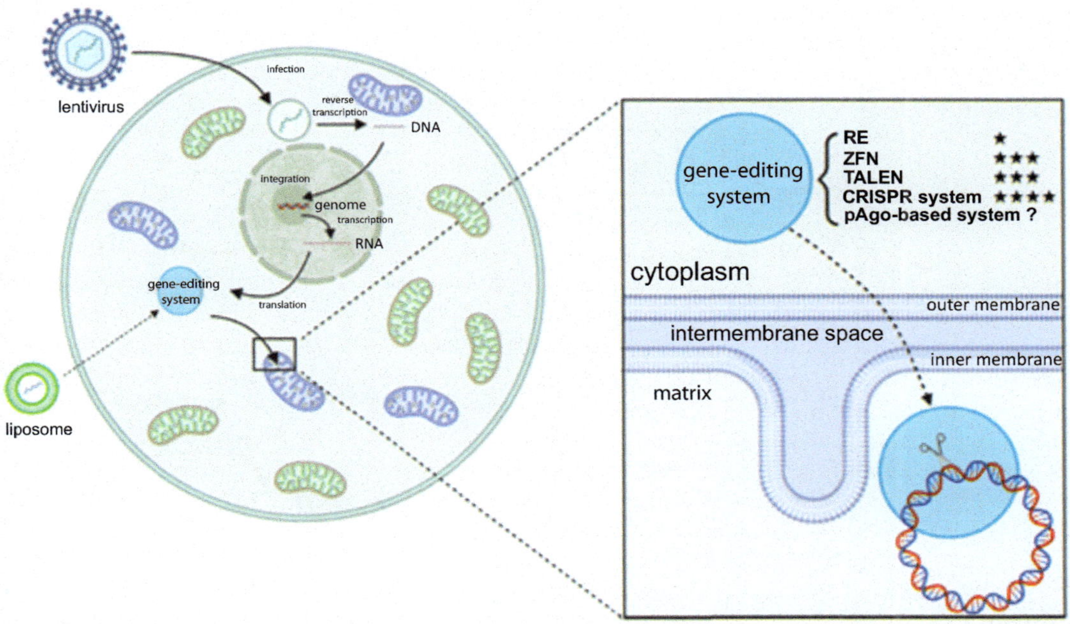

FIGURE 16.6 Graphical representation of the transport of gene editing systems into the mitochondria. The protein or RNA component can be introduced through lentivirus transfection and expressed in the nuclear genome, altered by mitochondrial targeting sequence or RP-loop.

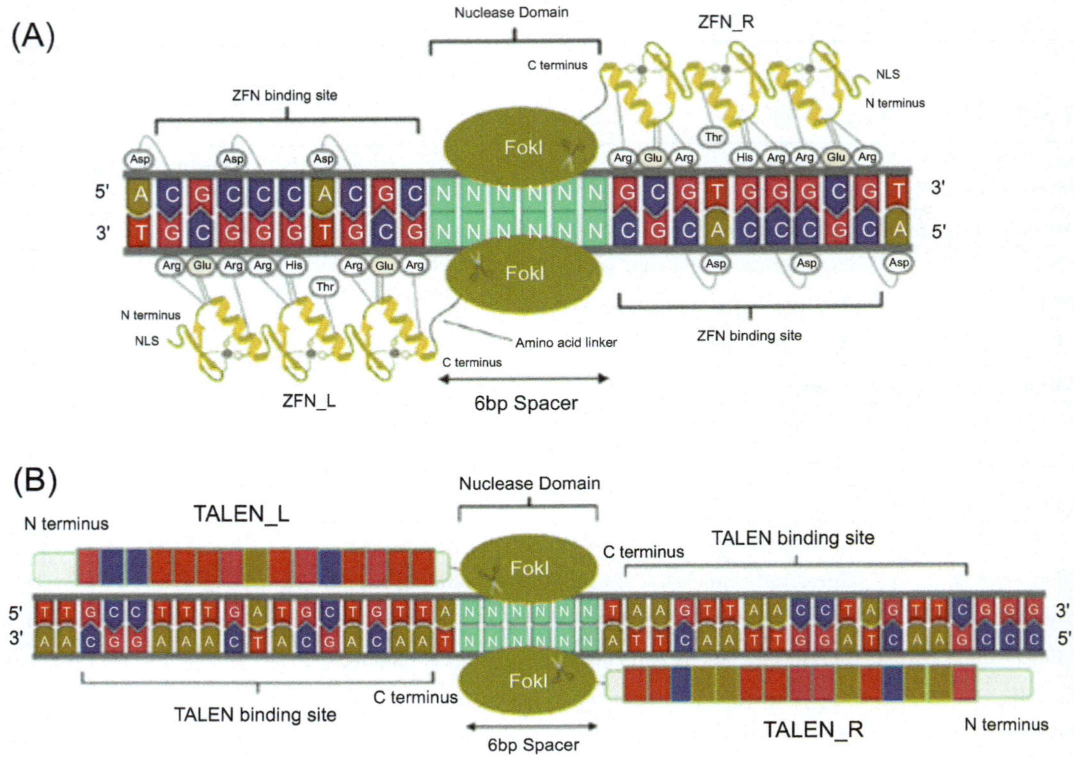

FIGURE 16.7 Several methods of gene editing. (A) Graphical abstract of the mechanism of zinc finger nuclease (ZFN). (B) Graphical abstract of the mechanism of transcription activator-like effector nuclease (TALEN).

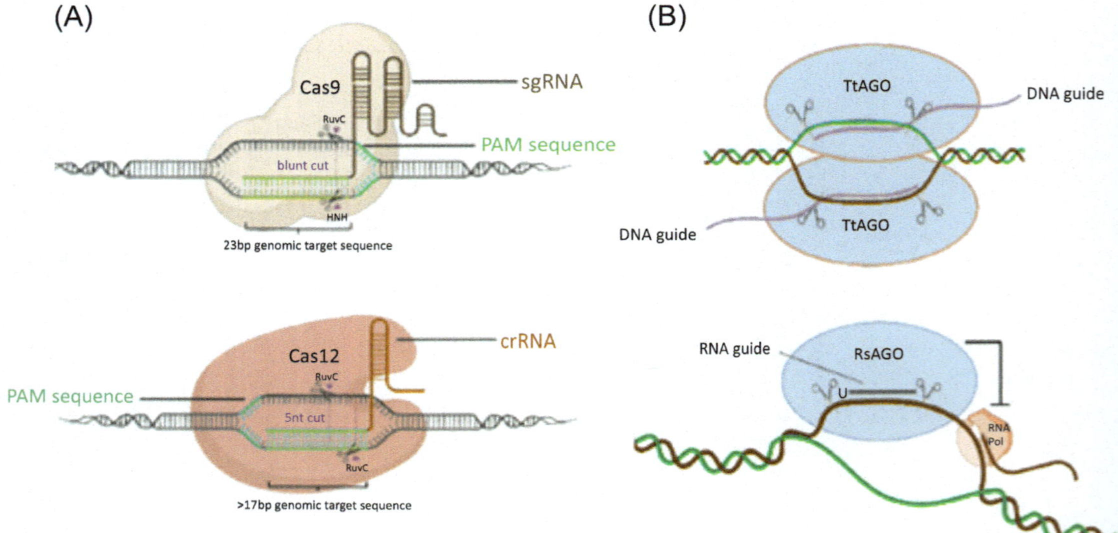

FIGURE 16.8 Several methods of gene editing. (A) Graphical abstract of the mechanism of CRISPR/Cas9 and CRISPR/Cas12 system. (B) Graphic abstract of the mechanism of the pAgo-based system, taking *Thermus thermophilus*(DNA-guide) and *Rhodobacter sphaeroides*(RNA-guide) as examples.

TABLE 16.2 Comparison of several gene editing methods.

Characteristics/types	Restriction enzyme	Zinc finger	TALEN	CRISPR-Cas	CRISPR-Cpf
Recognition method	RE self-recognition	ZF array protein recognition	TALE array protein recognition	Guide RNA recognition	Mature crRNA recognition
Cleavage method	RE	FokI dimer	FokI dimer	Cas protein monomer	Cpf1 protein
Recognition cleavage features	Depends on restriction enzyme types	Recognising 9–18 bp from two sides of double chin respectively with 3 bp as a unit	Recognising 14–20 bp from two sides of double chain respectively, recognising one to one with T on the '0'th base	Recognising 20 bp single chain, 3′ terminal sequence is NGC	Recognising multiple sites, the 3′ end is rich in t sequence, forming sticky ends
Application scope	DNA and RNA	Only DNA	Only DNA	DNA and RNA	DNA and RNA
Advantage	High recognition accuracy and cleavage efficiency	Low immunogenicity and easy transportation	High specificity, easy transportation and operation	High specificity, easy operation and short cycle	High specificity, easy operation and wide target
Disadvantage	Limited target positions, no artificial design	Poor specificity and miss target	Cumbersome operation and miss target	Miss target, PAM dependent	Miss target, PAM dependant and low activity

TABLE 16.3 Cases of different mitochondrial gene editing technologies.

Gene editing approaches	Cell type	Principle	Target	Outcome
RE	Enucleated skin fibroblasts Simian virus 40 (SV40) immortalised NZB/BALB fibroblasts	Restriction nuclease SmaI modified by MTS Restriction enzyme ApaLI modified by 5′ and 3′UTR sequences of ATP5b	m.8993T>G mutation BALB/NZB mtDNA	Decreasing the ratio of mutated mtDNA preventing the transmission of mitochondrial genomes
ZFN	COS-7143B(TK-) wild-type cells and 143B(TK-) NARP cybrid cells; cell lines 143B(TK-) and 143B (TK-) NARP cybribds; Flp-In TREx HEK-293 cell-line; H39 HOS; HOS 143B; human osteosarcoma 143B	ZFP modified by MTS and NES Single-chain ZFNs conjugating two FokI nuclease domains, together with an N-terminal mitochondrial targeting sequence Replace homodimer FokI with heterodimer in mtZFNs Precisely control the expression of mtZFNs by using interative editing strategy and hammerhead ribozyme	m.8993T>G mutation m.8993T>G mutation m.8993T>G mutation; common deletion(CD) m.8993T>G mutation	Cleavage of target site ZFNs are efficiently transported into mitochondria and cleavage dsDNA at predicted sites adjacent to the mutation, increasing the proportion of wt mtDNA molecules in the cell reduction in mutant mtDNA haplotype load, and subsequent repopulation of wild-type mtDNA restored mitochondrial respiratory function in a CD cybrid cell model realizing the nearly complete transformation of mtDNA heterogeneity to wild-type biochemical phenotypes
TALEN	Human osteosarcoma cells	SOD2 mitochondria localisation signal (MLS) modified TALENs	m.8993T>G mutation;common deletion (CD)	Permanent reductions in deletion or point mutant mtDNA in patient-derived cells
	MELAS-iPSCs	Platinum Gate TALEN construction system	m.13513G>A mutation	The m.13513G>A heteroplasmy level in MELAS-iPSCs was decreased in the short term by transduction of G13513A-mpTALEN

Fusions of an osteosarcoma cell line and dermal fibroblasts	Hybrid molecules mito-Tev-TALEs from T4 phage with MTS	m.8344A>G mutation	Shifting the mtDNA ratio toward the wild-type; improvement of oxidative phosphorylation function
Fusions of an osteosarcoma cell line and dermal fibroblasts	MLS modified TALENs	m.8344A>G t RNA Lys; m.13513G>A ND5 mutation	Reduction of the levels of the targeted pathogenic mtDNAs; recovery of respiratory capacity and oxidative phosphorylation enzymes activity
MELAS-iPSCs; porcine oocytes	MLS modified TALENs	m.3243A>G mutation	Elimination of the m.3243A>G mutation in MELAS-iPSCs; reduction in the human m.3243A>G mtDNA mutation in porcine oocytes
NZB/BALB oocytes; artificial mammalian oocytes	TALENs modified by MTS	NZB mtDNA; m.14459G>A mutation	Preventing the transmission of mitochondrial genomes; specific reduction of human mutated mitochondrial genomes responsible for mitochondrial diseases in mammalian oocytes
BTA calli; SW18 which was originally created by asymmetric cell fusion between a rapeseed 'Westar' as the recipient and a CMS radish as the donor E6-E7 gene from the human papilloma virus immortalised MEFs derived from the heteroplasmic tRNA ^{Ala} m.5024C>T	MLS modified TALENs MLS modified TALENs; intramuscular, intravenous and intraperitoneal injections of AAV9-mitoTALE	CMS-associated genes (orf79 and orf125) m.5024C>T mutation	Knock out of CMS-associated genes (orf79 and orf125), strongly suggesting that these genes are causes of CMS Muscle and heart were efficiently transduced showed a robust reduction in mutant mtDNA, Which was stable over time; the molecular defect, namely a decrease in transfer RNA Ala levels, was restored by the treatment

(*Continued*)

TABLE 16.3 (Continued)

Gene editing approaches	Cell type	Principle	Target	Outcome
CRISPR	HEK-293 T Zebrafish embroyo of the AB strain, HEK-293 T	Using MLS to import Cas9 together with gRNA into mitochondria Modifing Cas9 protein with MTS in zebrafish to insert DNA fragments by HR after generating DSB on target site	Cox1; Cox3 human mitochondrial genes ND1 and ND4 and two sites of zebrafish mitochondrial gene Dloop	Building a mtDNA mutation cell model Reduction mtDNA copy number in both human cells and zebrafish; an exogenous single-stranded DNA arm was knocked into the targeting loci accurately, and could be steadily transmitted to F1 generation of zebrafish
	Clostidium reinhardtii wild-type line(CC-125); yeast strains; MCC109p0, MCC125, CUY563, NB80	Directly introduced the 'editing plasmid' into the yeast and chloroplast of *Arabidopsis thaliana* by microinjection, then inserted DNA fragments by HR after generating DSB on target site	psaA	Confirming donor DNA insertion at the target sites facilitated by homologous recombination only in the presence of Cas9/gRNA activity in yeast mitochondria and chlamydomonas chloroplasts

Some studies have been conducted in mice and humans for gene editing of mitochondria. Although we have mentioned three gene editing technologies, we will be confined to CRISPR-based technology for mitochondrial gene editing.

A case study

We will be confined to methodology and results in detail.

Procedure for gene editing in mitochondrial DNA utilising Clustered regularly interspaced short palindromic repeats-Cas technology

Targeted mitogens: Cox1, Cox2, Cox3, ATP6/8.

Materials and methods: reagents, plasmids and antibodies

The following are the reagents for primary antibodies: 1. mouse antibody targeting poly (ADP-ribose) polymerase 1 (PARP1) (cat# 556494, dilution 1:1000, BD Bioscience), 2. mitochondrial marker antibody sampler kit (cat# 8674, dilution 1:3000, Cell Signalling), 3. rabbit antibody to CoxIV, cytochrome C (CytC), heat shock protein 60 (HSP60), prohibitin 1 (PHB1), pyruvate dehydrogenase (PDH), succinate dehydrogenase complex subunit A (SDHA) and mouse antibody to superoxide dismutase 1 (SOD1), 4. mouse antibody against glyceraldehyde 3-phosphate dehydrogenase (GAPDH) (GT239, dilution 1:5000, GeneTex), 5. mouse antibody against FLAG (clone M2, Sigma), 6. rabbit antibody against PTEN-induced putative kinase 1 (PINK1) (cat# 100–494, dilution 1:2000, Novus Biologicals) and 7. mouse antibody against UQCRC2 (cat# ab14745, dilution 1:3000, Abcam).

The reagents for secondary antibodies are enumerated below

1. Horseradish peroxidase (HRP) conjugated mouse antibody targeting FLAG (catalogue number A8592, dilution 1:5000, Sigma-Aldrich), HRP-conjugated antibody targeting hemagglutinin (HA) (1:3000, Roche), HRP-conjugated mouse antibody targeting β-actin (AC15, dilution 1:10000, Sigma-Aldrich), HRP-conjugated sheep antibody targeting mouse IgG (cat# RPN4301, dilution 1:5000, GE Healthcare), HRP-conjugated donkey antibody specific to rabbit IgG (cat# RPN4101, dilution 1:5000, GE Healthcare), 6. Alexa Fluor 488-conjugated donkey antibody for mouse IgG (H+L) (catalogue number A21202, dilution 1:1000, Invitrogen) and 7. Alexa Fluor 568-conjugated donkey antibody for rabbit IgG (catalogue number A10042, dilution 1:1000, Invitrogen). LentiCRISPR and lentiCRISPR-

sgRNA-enhanced green fluorescent protein (eGFP)#2 constructs were acquired from Addgene. Oligonucleotides were procured and synthesised by Cosmo Genetech.

Development of mitochondria-targeting Cas9 and guide RNAs for DNA expression

gRNAs were designed using appropriate software as necessary. sgRNA targeting mtDNA was included in the lentiCRISPR construct according to the manufacturer's instructions (provided by Dr Feng Zhang's laboratory). The oligonucleotides utilised for the synthesis of each sgRNA are enumerated in Table 16.2. The NCBI and Ensembl genome browser databases were employed to retrieve the mitochondrial genome sequence. The integrity of the sequence was confirmed via sequencing. The lentiCRISPR-EGFP sgRNA 2 construct was engineered to express mitoCas9 by replacing the N-terminal FLAG epitope and nuclear localisation signal (NLS) of Cas9 with sequences encoding a mitochondrial targeting sequence (MTS) and an HA epitope tag: ATGT CCGTCCTGACGCCGCTGCTGCTGCGGGGCTTGACAGGCTCGGCCCGGCGGCTCCCAGTG CCGCGCGCCAAGATCCATTCGTTG that encodes an MTS derived from subunit VIII of human CytC oxidase and peptides MSVLTPLLLRGLTGSARRLPVPRAKIHSL and TACCCATACGAC GTCCCAGACTACGCT, that encode the HA peptide YPYDVPDYA. To acquire DNA that expresses gRNA targeting the Cox1 area without the expression of FLAG-NLS-Cas9, the U6 promoter, gRNA targeting sequence, scaffold and termination signal were PCR-amplified utilising the specified primers and lentiCRISPR-sgRNA-Cox1 constructs as PCR templates. LentiCRISPR-sgRNA-eGFP#2 acted as the template for the control gRNA PCR. The forward primer sequence is: CAGAGAGGGCCTATTTCCCATG. The reverse primer sequence is: CTAGAATTCAAAAAAGCACCGACTC.PCR results were analysed by agarose gel electrophoresis and subsequently purified utilising the Qiagen gel extraction kit. Purified PCR products were used for transient transfection in conjunction with a mitoCas9-expressing plasmid.

Cell cultivation and transfection: Human embryonic kidney 293T (HEK-293T) cells were maintained in Dulbecco's modified Eagle medium supplemented with 10% (v/v) foetal bovine serum and antibiotic-antimycotic solution, under standard culture conditions (37°C, 5% CO_2, 95% air, humidified atmosphere). Transient transfection was performed using the X-tremeGENE transfection reagent according to the manufacturer's protocol, utilising the indicated plasmid constructs. For assessment of cellular proliferation, HEK-293T cells were transfected with constructs expressing mitoCas9 and gRNAs targeting Cox1 and Cox3, or an eGFP construct as a control. Five days posttransfection, cells were reseeded into 24-well plates and cultured in complete medium lacking pyruvate. Cell proliferation was quantified at defined time points using trypan blue exclusion and automated cell counting with the EVE automatic cell counter (NanoEnTeK).

Subcellular fractionation: HEK-293T cells were separated into subcellular components—cytosol, mitochondria and nucleus, utilising the Qproteome Mitochondria Isolation Kit (Qiagen) in accordance with the manual's recommendations. The cytosolic (Cyt) components were further concentrated using acetone precipitation. The purity of each fraction was confirmed using Western blots employing antibodies targeting Cyt marker proteins (GAPDH, 2% of BioMed Research International 3).

Table 16.2: Oligonucleotides used for gRNA construction.

mtDNA	Forward	Reverse
Cox1	CACCGGGCCCAGCTCGGCTCGAATA	AAACTATTCGAGCCGAGCTGGGCCC
Cox2	CACCGTATGAGGGCGTGATCATGAA	AAACTTCATGATCACGCCCTCATAC
Cox3	CACCGGCCTAGTATGAGGAGCGTTA	AAACTAACGCTCCTCATACTAGGCC
ATP8/6	CACCGTCGTCCTTTAGTGTTGTGTA	AAACTACACAACACTAAAGGACGAC

Table 16.3: Real-time PCR primers.

Gene	Forward	Reverse
Cox1	CGCCGACCGTTGACTATTCT	GGGGGCACCGATTATTAGGG
Cox2	TTCATGATCACGCCCTCATA	TAAAGGATGCGTAGGGATGG
Cox3	CAGCCCATGACCCCTAACAG	TGTGGTGGCCTTGGTATGTG
ND1	TCTCACCATCGCTCTTCTAC	TTGGTCTCTGCTAGTGTGGA
GAPDH	CATGTGCAAGGCCGGCTTCG	CTGGGTCATCTTCTCGCGGT
GAPDH for cDNA	AAACCCATCACCATCTTCCAG	AGGGGCCATCCACAGTCTTCT

Forward and reverse primers are engineered to anneal to the upstream and downstream regions around gRNA target locations for each mtDNA locus or individual genes. The entire Cyt fraction was used, together with 20% of the total mitochondrial fraction (SDHA), and 5% of the total nuclear fraction (PARP1).

Real-time quantitative polymerase chain reaction

Genomic DNA and mtDNA were extracted utilising DirectPCR tail lysis buffer (Viagen) augmented with proteinase K (Roche), following techniques that were significantly modified from the manufacturer's guidelines. Two or five days posttransient transfection, HEK-293T cells were rapidly rinsed with PBS and lysed in 100 µL of DirectPCR lysis solution, augmented with 3 µL of proteinase K per well (12-well culture). Proteinase K was inactivated by incubating samples at 80°C for 1 hour after incubation at 55°C for 2 hours. The DNA sample was diluted 100-fold and utilised directly for real-time quantitative PCR (Rotor-Gene Q, Qiagen). For cDNA synthesis, total RNA was collected 5 days posttransient transfection of mitoCas9 and the matching gRNA constructs, utilising the Qiagen RNeasy Mini Kit in accordance with the manufacturer's instructions. cDNA was synthesised from total RNA using the iScript cDNA synthesis kit (Bio-Rad) for further real-time PCR applications. Utilising the primers specified in Table 16.3, mtDNA areas encompassing each gene or mitochondrially-encoded genes of interest were amplified. Primers for Cox1 and Cox3 were annealed to mtDNA sequences adjacent to the cleavage sites targeted by sgRNA-Cox1 and sgRNA-Cox3, respectively, to assess specific cleavage of mtDNA. GAPDH served as an internal loading control for the amplification of genomic DNA. RT-qPCR was conducted utilising SYBR green master mix (QuantiFast SYBR Green PCR Kit, Qiagen) in accordance with the manufacturer's guidelines.

Western blot analysis

Total protein lysates were generated by directly adding 2x sample buffer containing β-mercaptoethanol (Bio-Rad) to HEK-293T cells that were briefly rinsed with ice-cold PBS. Following a 5 minute boiling of the samples, they were subjected to SDS-PAGE separation and subsequently transferred to a nitrocellulose membrane for immunoblotting investigations. The nitrocellulose membrane, previously blotted, was stained with Ponceau (Sigma) to visualise uniform protein transfer. Immunoblotting was conducted using a specific antibody, followed by chemiluminescence visualisation (Pierce). Densitometric studies of the bands were conducted utilising ImageJ (NIH, http://rsb.info.nih.gov/ij/).

Immunofluorescence

HEK-293T cells, 24 hours posttransient transfection with the specified constructs, were fixed in 4% paraformaldehyde/PBS (pH 7.4), blocked with 10% donkey serum (Sigma-Aldrich)/PBS containing 0.3% Triton X-100, and subsequently incubated with antibodies targeting FLAG or HA and CoxIV.Cells were treated with Mitotracker Red CMXRos (200 nM, 10 minute, Molecular Probes, Invitrogen) for the labelling of functioning mitochondria prior to PFA fixation. Following brief washes with PBS, cells on glass plates were treated with appropriate secondary antibodies conjugated to fluorescent dyes. Following DAPI nuclear staining, samples were affixed with a mounting solution (Aqua-Mount, Thermoscientific). Images were acquired via a fluorescent microscope (Zeiss Axiovert 200 M).

Statistical analysis

Quantitative data are represented as *mean ± standard error of the mean* (*s. e. m.*). Statistical significance was evaluated using an unpaired two-tailed Student's *t*-test for comparisons between two groups (control and test) or an ANOVA test followed by Student–Newman–Keuls post hoc analysis for comparisons among multiple groups exceeding three, as specified in each figure legend. *P* values below 0.05 were deemed indicative of a significant difference among groups.

Results

Distribution of clustered regularly interspaced short palindromic repeats/Cas9 to mitochondria

The CRISPR defence mechanism in bacteria has been effectively utilised to modify specific genomic loci across many species. CRISPR-associated (Cas) nucleases, guided by a sgRNA, enable sequence-specific recognition and cleavage of target DNA, thereby facilitating precise genome editing and modification. The presently utilised CRISPR/Cas9 approach employs nuclear localisation sequence-tagged SpCas9 to enable the nuclear translocation of SpCas9, where genomic editing is intended to take place. Nonetheless, the exact subcellular localisation has not been investigated. To determine the subcellular localisation of overexpressed FLAG-tagged SpCas9, HEK-293T cells were transiently transfected with the lentiCRISPR-sgRNA-eGFP#2 construct, which encodes FLAG-tagged Cas9 and a sgRNA targeting *eGFP*. Subcellular fractionation was performed to separate cytoplasmic, mitochondrial, and nuclear compartments. As anticipated, immunoblot analysis revealed the presence of FLAG-tagged Cas9 predominantly within the nuclear fraction. In accordance with the findings from subcellular fractionation, immunofluorescence for overexpressed FLAG-Cas9 reveals its colocalization with mitochondria labelled by MitoTracker Red and the nucleus stained with DAPI.

CRISPR/Cas9-facilitated mitochondrial DNA cleavage

Given the observed distribution of FLAG-Cas9 to mitochondria, we postulated that the CRISPR/Cas9 system may be utilised to alter mtDNA when paired with sgRNA directed at specific mtDNA loci. To evaluate this hypothesis, we designed several sgRNAs targeting mtDNA areas that encode peptides (Cox1 and Cox3) involved in oxidative phosphorylation complexes. Oligonucleotides targeting specific regions of mtDNA were cloned into the lentiCRISPR vector, which drives expression of the sgRNA under the U6 promoter and SpCas9 under the EFS promoter. To evaluate whether the CRISPR/Cas9 system can mediate site-specific cleavage of mtDNA in a manner analogous to its activity on nuclear genomic DNA, HEK-293T cells were transfected with lentiCRISPR constructs encoding sgRNAs targeting Cox1 (lentiCRISPR-sgRNA-Cox1) and Cox3 (lentiCRISPR-sgRNA-Cox3). These structures should break mtDNA and produce mtDNA fragments. LentiCRISPR-sgRNA-eGFP#2 served as the control. The specific cleavage of mtDNA was assessed using primers surrounding the cleavage sites (primers for Cox1 and Cox3).

qPCR analysis revealed an approximately 90% reduction in intact mtDNA at the Cox1 locus in HEK-293T cells cotransfected with lentiCRISPR-sgRNA-Cox1 and lentiCRISPR-sgRNA-Cox3 constructs, compared to control cells transfected with the lentiCRISPR-sgRNA-eGFP#2 plasmid (Fig. 16.1D). Furthermore, an ~80% reduction in intact mtDNA at the Cox3 locus was observed under the same conditions (Fig. 16.1D), indicating efficient CRISPR/Cas9-mediated cleavage. To assess specificity, the *ND1* locus—a nontargeted region—was analysed using locus-specific primers (Table 16.3), confirming the absence of off-target cleavage at this site.

The mtDNA copy number, as assessed by ND1 site RT-qPCR, rose by roughly 50%. The results demonstrate that FLAG-Cas9 can penetrate the mitochondrial matrix and, when paired with sgRNA targeting mtDNA, it cleaves mtDNA in a site-specific manner.

Modification of mitochondrial proteins resulting from CRISPR/Cas9-induced mitochondrial DNA cleavage

Subsequently, we investigated changes in mitochondria-associated proteins as a marker of mitochondrial disruption potentially produced by CRISPR/Cas9-mediated mtDNA cleavage. Truncation of mtDNA at the Cox1 and Cox3 loci via CRISPR/Cas9 may impact the transcription of mtDNA heavy strands and consequently the expression of downstream genes from the Cox1 locus (Fig. 16.1B). Upon monitoring various mitochondrial marker proteins, we observed modifications in numerous proteins: the levels of SDHA, HSP60, and PHB1 diminished, while the levels of PDH and SOD1 elevated. Nonetheless, no substantial alterations were seen in CoxIV or CytC. The distribution of tRNA and peptide-encoding regions, along with the bidirectional transcription of mtDNA, suggests that breakage at various loci of mtDNA may provide distinct consequences. In this context, we also investigated whether the cleavage of certain mtDNA single loci resulted in differential regulation of mitochondria-associated proteins. Transfection was carried out for HEK-293T cells with sgCox1, sgCox2, sgCox3, or sgATPase8/6. Each sgRNA transfection produced varying effects on mitochondria-associated proteins. Cleavage of the Cox1, Cox3, or ATP8/6 domains resulted in a little decrease in ubiquinol-CytC reductase core protein 2 (UQCRC2) and PHB1, while the amount of PDH exhibited a slight rise.

Cleavage of the Cox1 or Cox3 loci by the CRISPR/Cas9 system resulted in a reduction in the expression of PINK1, a mitochondrial protein implicated in Parkinson's disease pathogenesis. In contrast, targeted cleavage at the *Cox2* locus led to an upregulation of PINK1 levels. Additionally, **expression of Mitofusin-2 (Mfn2), a key regulator of mitochondrial fusion, was decreased following CRISPR/Cas9-mediated cleavage at the Cox1, Cox3, or ATP8/6 loci. Collectively, these results indicate that site-specific cleavage of mtDNA by CRISPR/Cas9 perturbs mitochondrial proteostasis**. Moreover, the nature and extent of mitochondrial protein dysregulation in the early cellular response to mtDNA damage appear to be locus-dependent. Alterations in Mfn2 or PINK1 indicate a possible compensatory mechanism for mitochondrial quality control in cells after mtDNA damage.

mitoCas9. While we established that FLAG-Cas9 with NLS peptide can efficiently target mitochondria for functional mtDNA editing, the particular localisation of Cas9 to mitochondria is essential for the safe execution of mtDNA editing without impacting genomic DNA. Consequently, we sought to develop a novel variant of Cas9 capable of selectively targeting mtDNA. mitoCas9 was generated by modifying the lentiCRISPR-sgRNA-eGFP#2 construct through the removal of the N-terminal FLAG epitope and NLS, and the insertion of a MTS and HA epitope tag at the N-terminus of SpCas9. The MTS derived from CytC, a mitochondrial matrix-localised protein, was employed to direct the import of mitoCas9 into the mitochondrial matrix, thereby facilitating its interaction with mtDNA. Expression of mitoCas9 was confirmed via immunodetection using an HA-specific antibody (BioMed Research International).

FLAG-NLS-Cas9 is localised to the mitochondria. Subcellular localisation of FLAG-Cas9 was evaluated in the Cyt, mitochondrial (Mit) and nuclear (Nu) fractions of HEK-293T cells transfected with lentiCRISPR-sgRNA-eGFP#2 and analysed via Western blot. GAPDH functioned as a Cyt marker, PARP1 acted as a nuclear marker, and SDHA served as a mitochondrial marker. The Ponceau stain of the blotted nitrocellulose membrane is displayed in the bottom lane to illustrate the relative protein loading quantities. Immunofluorescence photos of FLAG-NLS-Cas9 illustrating its distribution in the nucleus, cytoplasm and mitochondria. HEK-293T cells were subjected to immunostaining using a mouse antibody against FLAG (green) 24 hours post-transient transfection with the FLAG-NLS-Cas9 construct. Mitochondria were stained with MitoTracker Red, whereas the nucleus was stained with DAPI.

Depiction of human mtDNA. A simplified representation of mtDNA is shown to illustrate the areas that encode peptides or tRNA for the specified amino acids. Filled triangles denote locations targeted by gRNAs directed against Cox1 or Cox3. Arrows denote the primer annealing sites utilised for real-time PCR of the Cox1 or Cox3 regions.

Quantitative analysis of mtDNA copy number at the Cox1, Cox3, and ND1 loci was performed in HEK-293T cells transiently transfected with lentiCRISPR-sgRNA-Cox1 and lentiCRISPR-sgRNA-Cox3 constructs, or with lentiCRISPR-sgRNA-eGFP#2 as a control, and was assessed by real-time quantitative PCR utilising the primers. GAPDH acted as an internal loading control or housekeeping gene (n=3 per group). Quantitative data are shown as mean ± s. e. m., P<.001, utilising an unpaired two-tailed Student's t-test. Changes in mitochondria-associated proteins subsequent to CRISPR/Cas9-mediated mtDNA modification. Mitochondrial proteins in HEK-293T cells following CRISPR/Cas9-induced cleavage of mtDNA at the Cox1 and Cox3 loci, as assessed by Western blot analysis with specified antibodies. β-actin served as a loading control. Quantification of mitochondrial proteins in HEK-293T cells transfected with lentiCRISPR-sgRNA-Cox1 + Cox3 or lentiCRISPR-sgRNA-eGFP#2 control, as depicted in panel (a), normalised to β-actin. Depiction of human mtDNA. Specific loci targeted by lentiCRISPR-sgRNAs (Cox1, Cox2, Cox3 and ATP8/6) are represented by filled triangles. Representative Western blots demonstrating the differential modification of mitochondrial proteins subsequent to the cleavage of certain mtDNA loci facilitated by the specified sgRNAs in HEK-293T cells.

Development of mitochondria-targeted MTS-HA-Cas9. Schematic representation of mitoCas9. Information regarding the MTS and HA tag is provided. Expression of mitoCas9 in HEK-293T cells, transiently transfected with the MTS-HA-Cas9 construct, was assessed via Western blot analysis employing an HA antibody. β-actin served as a loading control. The subcellular localisation of MTS-HA-Cas9 was evaluated in the Cyt, mitochondrial (Mit) and nuclear (Nu) fractions of HEK-293T cells transfected with lentiCRISPR-sgRNA-eGFP#2 and analysed via Western blot. GAPDH acted as a Cyt marker, PARP1 acted as a nuclear marker, while SDHA acted as a mitochondrial marker. An immunofluorescence microscopy image represented the subcellular locations of MTS-HA-Cas9 (HA, green) and CoxIV (red) in HEK-293T cells transfected with the MTS-HA-Cas9 construct. The combined image (yellow, right panel) illustrates the colocalization of mitoCas9 and CoxIV. Representative gel images of the PCR products of hU6-sgRMA-eGFP and hU6-sgRNA-Cox1, which were purified via gel extraction (bottom panel). Primers and lentiCRISPR-sgRNA templates were used to amplify the U6 promoter and corresponding sgRNA components for transfection (upper panel).

Quantification of copy counts for the Cox1, Cox3 and ND1 regions of mtDNA taken from HEK-293T cells transiently transfected with specified constructs (mitoCas9 is a plasmid, while U6-sgRNAs are PCR products), estimated through real-time quantitative PCR utilising primers. GAPDH served as an internal loading control (n=3 per group). Quantified data (b) are shown as mean ± s. e. m., P<.001, analysed using analysis of variance (ANOVA) followed by Student–Newman–Keuls post hoc analysis. The anticipated molecular weight, along with subcellular fractionation, indicated that mitoCas9 predominantly localised to the mitochondria. Furthermore, immunofluorescence validated the colocalization of HA-tagged mitoCas9 with the mitochondrial marker protein CoxIV.

Mitochondrial DNA modification by mitochondrial-targeted Cas9

To ascertain the functionality of this novel Cas9 variant for mtDNA editing, we transfected HEK-293T cells with both mitoCas9 and DNA encoding sgRNA for Cox1. For this application, we performed PCR amplification of the hU6 promoter and sgRNA-Cox1 regions from the lentiCRISPR-sgRNA-Cox1 plasmid. U6-sgRNA-eGFP was additionally PCR-amplified for use as a control. mitoCas9 and hU6-sgRNA-Cox1 effectively cleaved the mtDNA target region, as confirmed by real-time quantitative PCR exclusively at the Cox1 locus. Collectively, our findings indicate that mitochondria-targeting mitoCas9 can be utilised to modify mtDNA alongside custom-designed sgRNAs.

Mitochondrial dysfunction induced by mitochondrial-targeted Cas9-mediated mitochondrial DNA damage

The targeting of mitochondria by specific restriction enzymes, such as EcoRI or PstI, has been utilised to degrade mtDNA, alongside chemical inducement for mtDNA removal. To ascertain if mitoCas9-mediated site-specific mtDNA cleavage results in functional impairment of mitochondria, we initially assessed mtDNA copy quantity and the expression of various mitochondrial genes. Five days posttransient transfection of HEK-293T cells with mitoCas9 and sgRNA targeting the Cox1 and Cox3 areas, real-time PCR measurement of mtDNA copy number was conducted utilising ND1 primers. Cell division over a period of 5 days, accompanied by site-specific mitoCas9 cleavage at the Cox1 and Cox3 loci, results in an approximate 30% decrease in mtDNA copy numbers. In accordance with the observation that mtDNA was cleaved and the steady-state mtDNA copy number diminished due to the CRISPR/Cas9 system, the levels of messenger RNAs encoding mitochondrial ND1, Cox1 and Cox2 genes were approximately reduced by 50% relative to those in sgRNA control HEK-293T cells. Research indicates that mitochondrial dysfunction caused by mtDNA damage leads to changes in mitochondrial membrane potential and cellular growth capacity. To evaluate the impact of mitoCas9-mediated mtDNA breakage and reduction on mitochondrial and cellular functioning, we assessed mitochondrial membrane potential using MitoTracker Red, which selectively stains functional mitochondria in a membrane potential-sensitive manner. In accordance with the premise that mitochondrial gene expression and proteomic equilibrium are disrupted by mitoCas9-induced mtDNA damage, the mitochondrial membrane potential was diminished in HEK-293T cells transfected with mitoCas9 and sgRNA targeting Cox1 and Cox3, relative to control cells. The growth rates of cells exhibiting mitochondrial dysfunction produced by mitoCas9 in HEK-293T were also analysed. After four consecutive days of observation postseeding, HEK-293T cells transfected with sgRNA targeting Cox1 and Cox3 demonstrate a reduced growth rate in comparison to the control cell line. Collectively, our findings indicate that mitoCas9-induced mtDNA damage results in mitochondrial dysfunction and ultimately diminishes cellular proliferative potential.

Adapting Clustered regularly interspaced short palindromic repeats/Cas9 system for targeting mitochondrial genome

Another case study: Gene editing of the mitochondrial genome using the CRISPR-Cas9 system is highly challenging, mainly due to subefficient delivery of gRNA and Cas9 enzyme complexes into the mitochondria. In this study, we achieved targeted editing of mtDNA by appending a gRNA specific to *NADH-ubiquinone oxidoreductase chain 4* (*ND4*) with an RNA transport-derived stem-loop element (RP-loop), and coexpressing Cas9 fused to an N-terminal MTS. Confocal imaging revealed colocalization of the RP-loop–modified gRNA with mitochondria, and a significant reduction in ND4 transcript levels was observed in cells harbouring the pathogenic 11205G variant within the ND4 gene. This reduction was also accompanied by a concomitant decrease in overall mtDNA levels. These findings provide proof-of-concept that RP-loop–tagged gRNAs can be imported into mitochondria and functionally engage with mitochondrially localised Cas9 to mediate sequence-specific cleavage of mtDNA. This approach offers a promising strategy for the development of CRISPR-Cas9-based therapeutics targeting mitochondrial disorders (see Fig. 16.9).

1. Mitochondrial RP loop and gRNA targeting ND4 is in the downstream of U6 promoter, and chicken beta actin controls the expression of mSpCas9 flanked by mitochondrial localisation sequence 1 (MLS1) and MLS2 sequences.
2. Mfold analysis of chimeric gRNA with RP loop.

Fig. 16.10

Immunofluorescence analysis conducted 24 hours posttransfection demonstrated that Cas9 constructs fused with a mitochondrial localisation sequence (MLS-Cas9) predominantly localised to the cytoplasm, whereas Cas9 constructs containing a NLS-Cas9 were confined to the nucleus. Additionally, mouse embryonic fibroblast (MEF) cells were transfected with Alexa Fluor 488-labelled ND4-specific CRISPR RNAs (ND4-crRNA and RP-loop–modified ND4-crRNA). Mitochondria were subsequently visualised using MitoTracker Red CMXRos (Deep Red FM) staining 18 hours posttransfection to assess colocalization of the RNA constructs with the mitochondrial network.

Cells were affixed in Vectashield anti-fade mounting medium containing DAPI (Vector Laboratories), and images were captured at 100x magnification using a Nikon Ti2-E inverted microscope; scale bar: 10 mm. Cells exhibiting colocalised mitochondrial gRNA (solid arrows in B) were quantified at 20x magnification, revealing an increased percentage of RP-ND4-crRNA fused with mitochondria (n=3, Student's t-test, P=.01). The RP loop elevates the concentrations of sgRNA seen in the mitochondrial fractions. qPCR was performed, and sgRNA levels were normalised to mitochondrial ND4 and ND1 as internal controls, represented as fold change 211CT (n=3, Student's t-test, *P<.05) (see Fig. 16.11

Fig. 16.12

A case study: FLAG-nuclear localisation signal (NLS)-Cas9 is localised to the mitochondria. We describe gene editing approach for mitochondria.

Jo et al. (2015).

Fig. 16.13

Fig. 16.14

Fig. 16.15

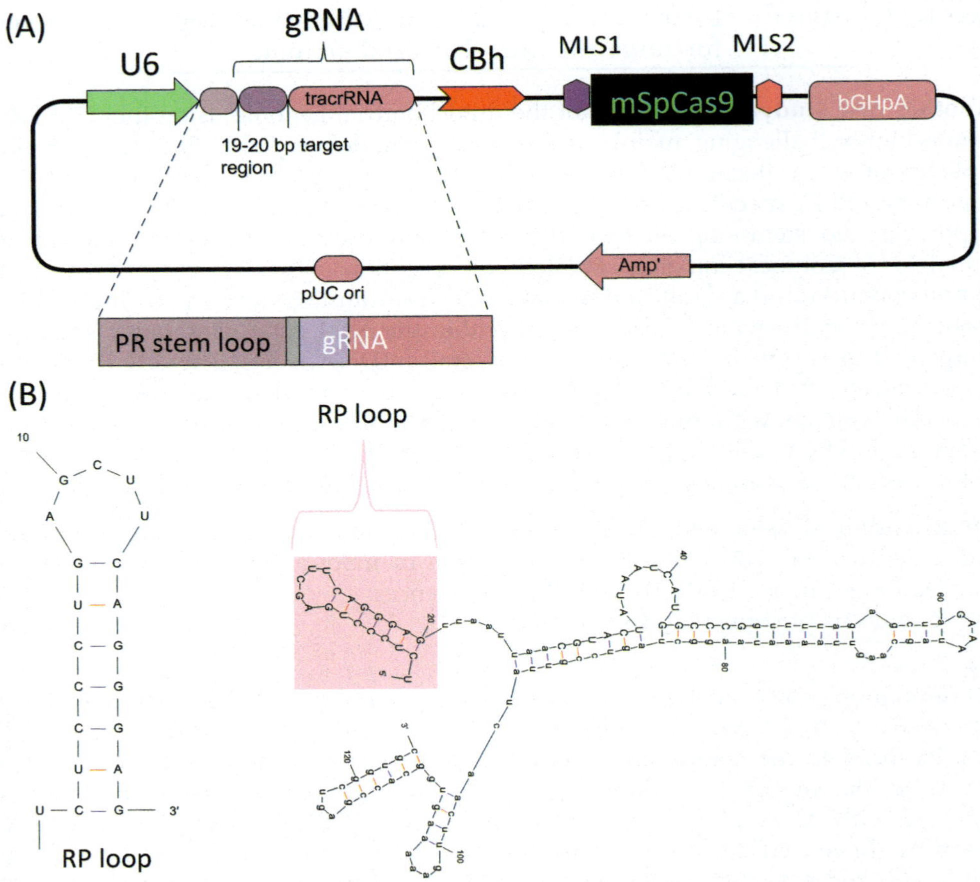

FIGURE 16.9 Generation and structure of mitochondria targeting the CRISPR, CRISPR-associated protein 9 nuclease (Cas9 complex).

GAPDH functioned as a Cyt marker, PARP1 acted as a nuclear marker, and SDHA served as a mitochondrial marker.

(d) Immunofluorescence microscopy image represents the subcellular locations of MTS-HA-Cas9 (HA, green) and CoxIV (red) in HEK-293T cells transfected with the MTS-HA-Cas9 construct. The combined image (yellow, right panel) illustrates the colocalization of mitoCas9 and CoxIV. Gel images of the PCR products hU6-sgRMA-eGFP and hU6-sgRNA-Cox1, which were purified via gel extraction (bottom panel). Primers and lentiCRISPR-sgRNA templates used to amplify the U6 promoter and corresponding sgRNA components for transfection (upper panel).

(f) Quantification of copy counts for the Cox1, Cox3 and ND1 regions of mtDNA taken from HEK-293T cells transiently transfected with specified constructs (mitoCas9 is a

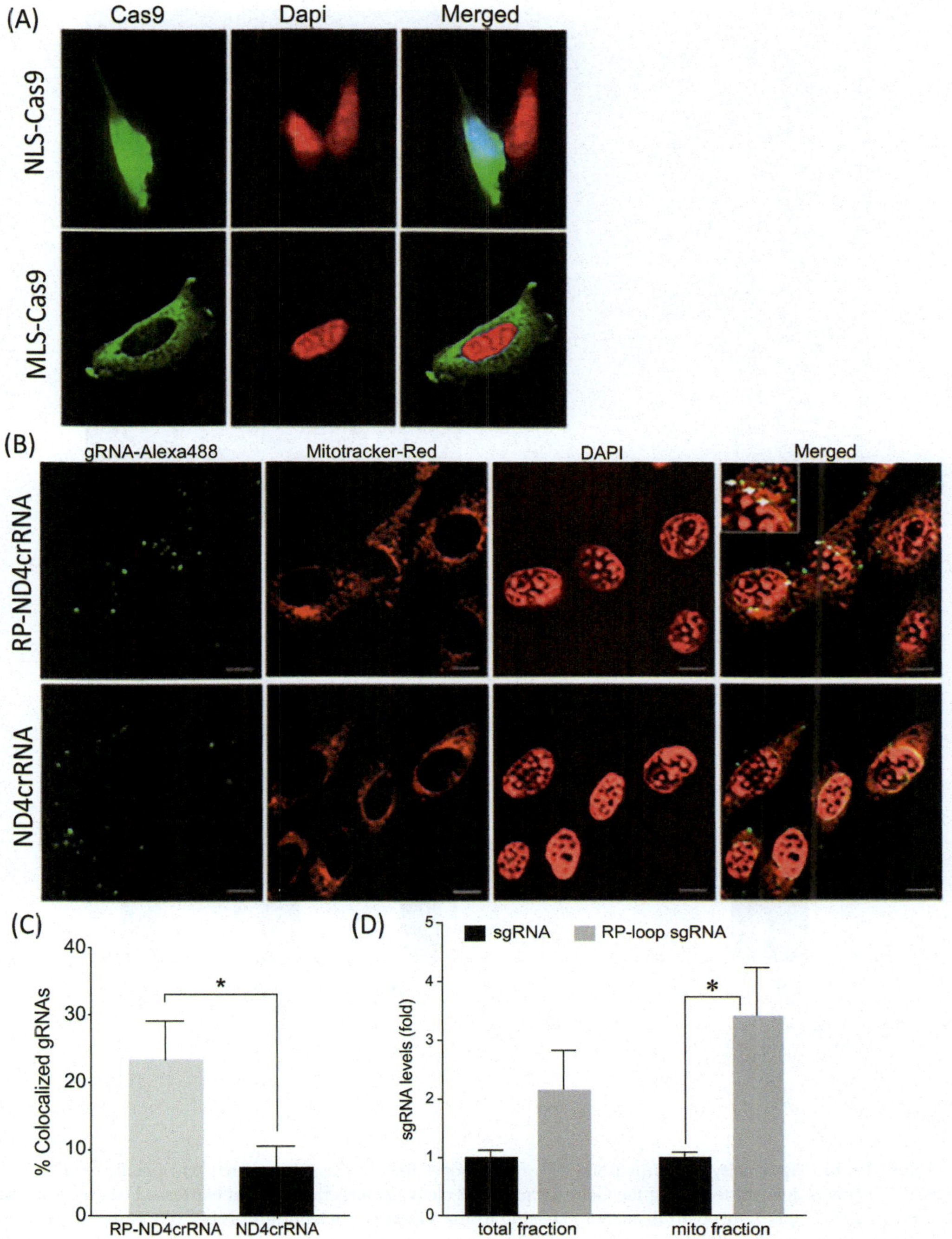

FIGURE 16.10 Introduction of CRISPR–Cas9 complex into the mitochondria.

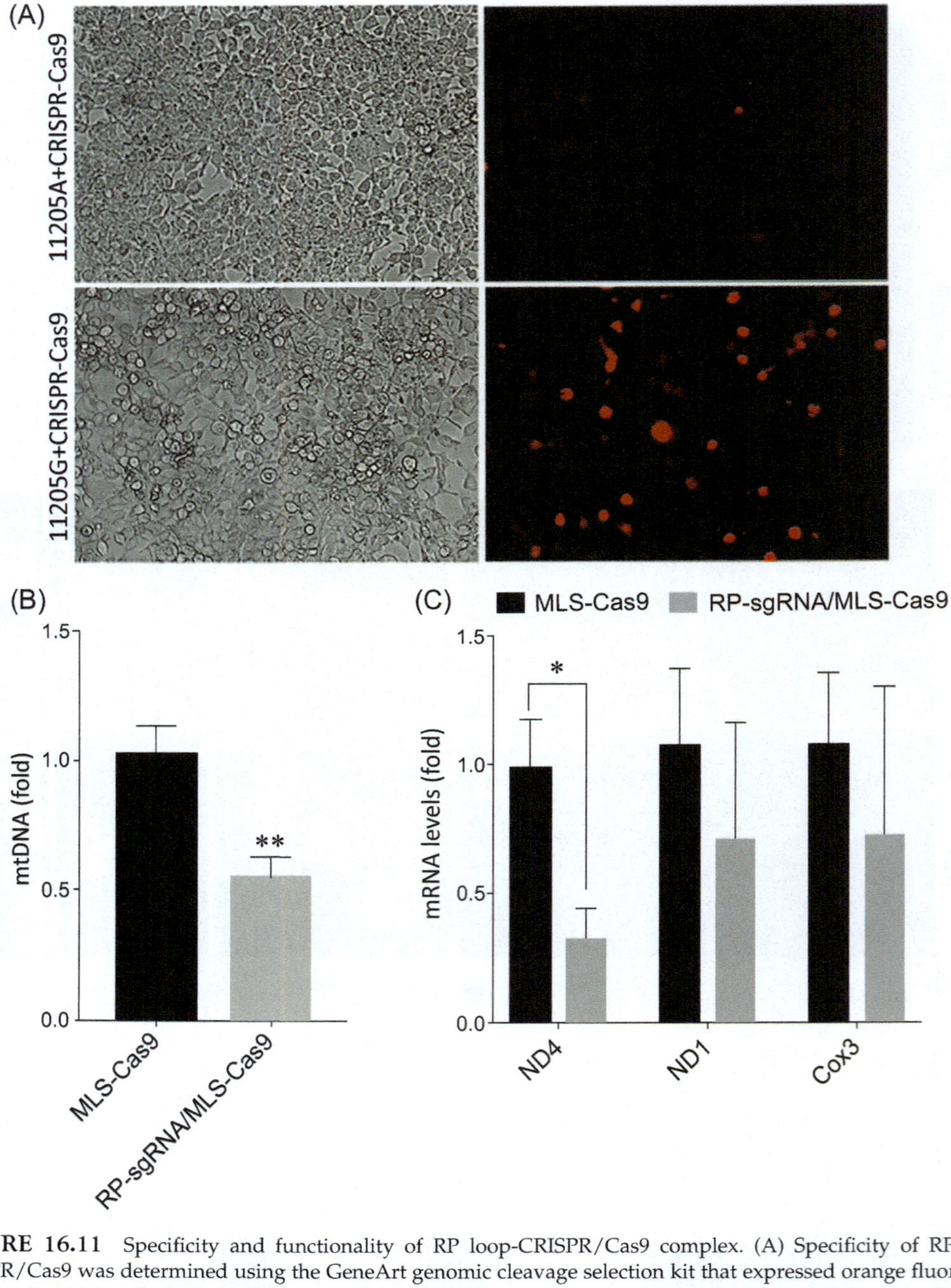

FIGURE 16.11 Specificity and functionality of RP loop-CRISPR/Cas9 complex. (A) Specificity of RP loop-CRISPR/Cas9 was determined using the GeneArt genomic cleavage selection kit that expressed orange fluorescent reporter. In 293 K cells 72 h posttransfection, specific to the 11206 G nucleotide of ND4; qPCR for mitochondrial DNA (mtDNA) levels. (B) qRT-PCR for ND4 mRNA expression. (C) Was significantly lower in RP loop sgRNA-CRISPR/Cas9 transfected MEF cells, at 24 h posttransfection, compared with the mitochondrial localisation sequence (MLS)-Cas9 controls (n = 5, Student's t test, *P=.016, **P=.005).

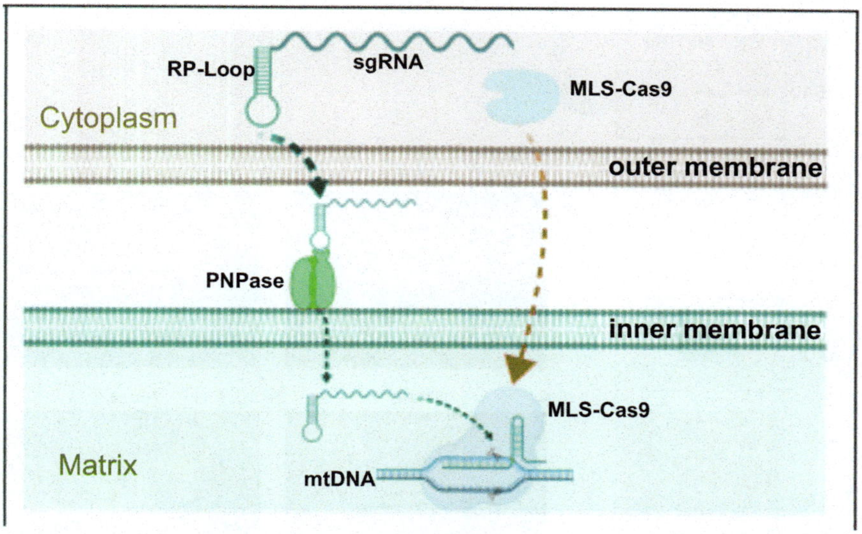

FIGURE 16.12 Graphical summary of CRISPR/Cas9 delivery into the mitochondria. Guide RNA with RP loop attached upon transfection is imported into the mitochondria by attaching to polynucleotide phosphorylase (PNPase) in the inner mitochondrial membrane (IM). Cas9 protein with N-terminus mitochondria localisation sequences (MLS) targets Cas9 to the mitochondria, where it forms a complex with the RP loop sgRNA to cause sequence-specific mitochondrial DNA (mtDNA) cleavage.

plasmid, while U6-sgRNAs are PCR products), quantified through real-time quantitative PCR utilising primers detailed in Table 16.3.

GAPDH acted as an internal loading control (n=3 per group). Quantified data (b) are represented as mean ± s. e. m., P<.001, analysed with ANOVA followed by Student–Newman–Keuls post hoc analysis (see Fig. 16.16).

Quantification of copy numbers for ND1 region of mtDNA extracted from HEK-293T cells transiently transfected with indicated constructs (gRNA control, mitoCas9 and U6-sgRNA to eGFP, gRNA Cox1, Cox3, mitoCas9 and U6-sgRNA to Cox1 and Cox3), determined by real-time qPCR using primers listed in Table 16.3 at the indicated time points (2 days and 5 days after transfection). GAPDH was used as an internal loading control.

Quantification of messenger RNAs for ND1, Cox1 and Cox2 genes in HEK-293T cells 5 days following transient transfection with lentiCRISPR-sgRNA-Cox1 + Cox3 or lentiCRISPR-sgRNA -eGFP2 control determined by real-time PCR and normalised to GADPH.

Representative images of MitoTracker Red staining for functional mitochondria in HEK-293T cells transfected with the indicated constructs (upper panel). Quantification of relative MitoTrackerr Red staining intensities in two groups was shown in the bottom panel. (n=50 mitochondria from three independent cultures per group)

Cell counts demonstrating proliferation rate of the HEK-293 cells following cleavage of specific mtDNA loci (n=6 individual measurements per group). Days indicate the time intervals proceeded after cell seeding following 5 days of transient transfection. Quantified data are expressed as mean±sem, *P≤.05, **P≤.001, unpaired two-tailed Student's t test.

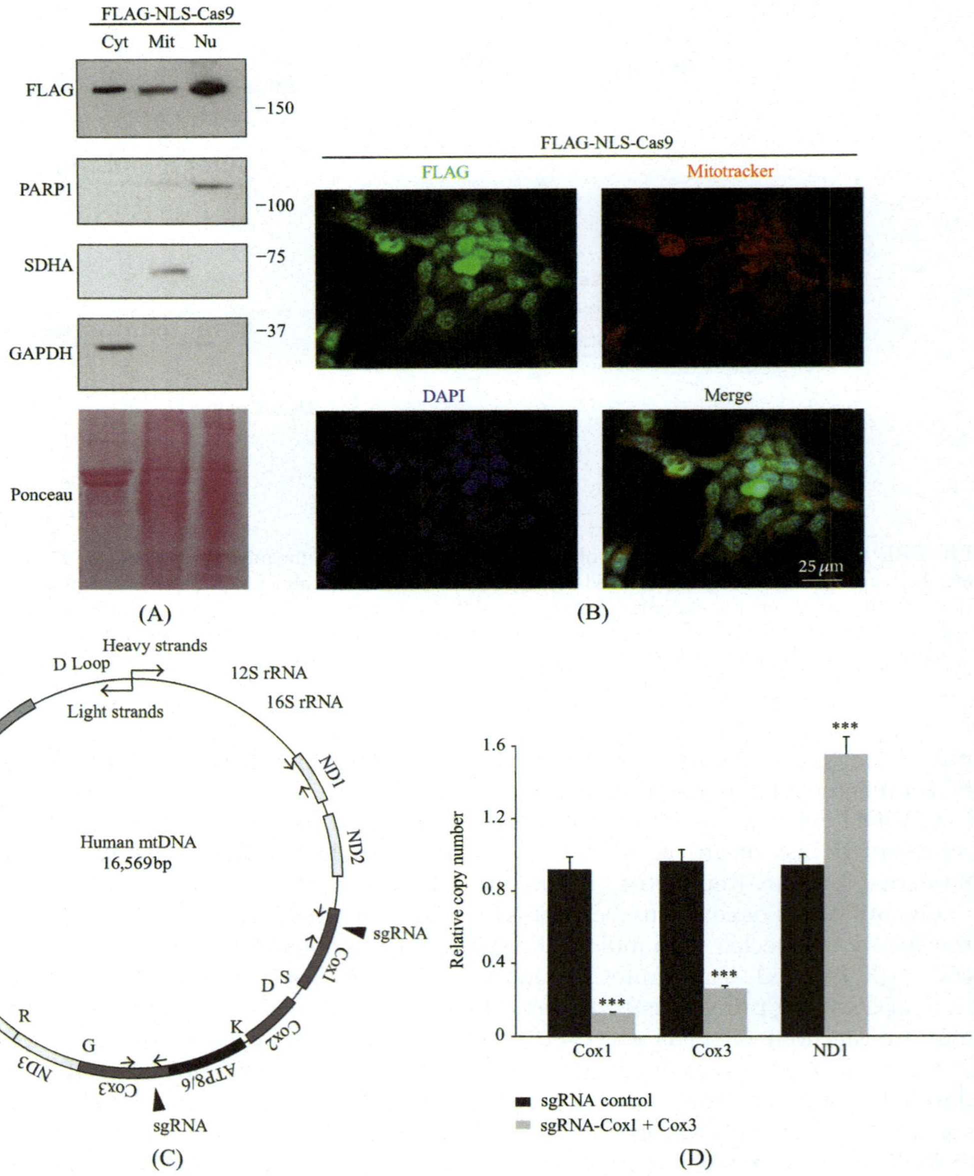

FIGURE 16.13 FLAG-nuclear localisation signal (NLS)-Cas9 is localised to the mitochondria. (A) The subcellular localisation of FLAG-Cas9 was evaluated in the cytosolic (Cyt), mitochondrial (Mit) and nuclear (Nu) fractions of HEK-293T cells transfected with lentiCRISPR-sgRNA-eGFP#2, and analysed via Western blot. Glyceraldehyde 3-phosphate dehydrogenase (GAPDH) acted as a Cyt marker, poly(ADP-ribose) polymerase 1 functioned as a nuclear marker, and succinate dehydrogenase complex subunit A acted as a mitochondrial marker. Ponceau staining of the blotted nitrocellulose membrane was displayed in the lower lane to illustrate relative

(continued)

Clustered regularly interspaced short palindromic repeats technology

The CRISPR and CRISPR-associated (Cas) protein system was initially identified in bacteria and archaea. It is an immunological defence mechanism developed through prolonged evolution to combat invading viruses and other foreign DNA intrusions. CRISPR consists of several short, conserved repeat sequence sections (Repeats) and spacer regions (Spacers). Repeats encompass palindrome sequences capable of forming hairpin structures, whereas spacers, which are foreign DNA sequences assimilated by bacteria, are distinctive. The upstream leader is regarded as the promoter of the CRISPR sequence. A polymorphic gene family, Cas, is located upstream of the sequence. Upon the invasion of foreign DNA, the Cas protein functions to cleave it into short sequences, which are then integrated as spacers between conserved repeat sequences that include palindromic sequences capable of forming. Hairpin configurations. Upon subsequent invasion by the identical foreign DNA, the spacer will serve as a template for the transcription of CRISPR RNA (crRNA), which associates with transactivating crRNA (tracrRNA) to direct Cas in cleaving the invading foreign DNA. It is important to recognise that pinpointing a specific region and facilitating cleavage must also comply with the existence of a PAM after the homologous sequence, so constraining the possible editing sites. The CRISPR/Cas system, as a gene editing tool, solely induces DNA DSB, while the subsequent cellular DNA repair mechanisms, including nonhomologous end-joining (NHEJ) and homology-directed repair (HDR), finalise gene editing at designated loci. NHEJ introduces or removes several bases at the DSB site during repair, resulting in frameshift mutations that effectively knock out the gene, whereas HDR facilitates error-free DNA repair when a homologous donor DNA serves as a template for the repair and insertion of foreign fragments. The CRISPR/Cas system is categorised into three types: I, II and III, based on the sequence and structure of the Cas protein. The most prevalent is the type II CRISPR/Cas system, which necessitates just one Cas protein, Cas9, because to its capacity to colocalise RNA and DNA, along with its significant transformative potential. In practice, gRNA, which encompasses the essential structural elements of crRNA and tracrRNA, is more practical and effective. The CRISPR/Cas system has demonstrated significant potential across various applications, including genome editing, transcription regulation and viral detection. The CRISPR/Cas9 technology has demonstrated efficacy in editing the genomes of humans, zebrafish, mice, fruit flies and

protein loading quantities. (B) Immunofluorescence photos of FLAG-NLS-Cas9 illustrate its localisation in the nucleus, cytoplasm and mitochondria. HEK-293T cells were immunostained with a mouse antibody against FLAG (green) 24 hours posttransient transfection with the FLAG-NLS-Cas9 construct. Mito-tracker Red was employed for staining of Mitochondria, whereas DAPI was used to stain the nucleus. (C) Depiction of human mtDNA. A simplified representation of mtDNA is presented to illustrate areas encoding peptides or tRNA for the specified amino acids. Filled triangles denote locations targeted by gRNAs directed against Cox1 or Cox3. Arrows denote primer annealing locations utilised for real-time PCR of the Cox1 or Cox3 regions. (D) mtDNA copy number at the Cox1, Cox3 and ND1 loci was quantified in HEK-293T cells transiently transfected with lentiCRISPR-sgRNA-Cox1 and lentiCRISPR-sgRNA-Cox3 constructs, or with lentiCRISPR-sgRNA-eGFP#2 as a control. Quantification was performed using real-time quantitative PCR (qPCR) with locus-specific primers listed in Table 16.3. *GAPDH* was used as an internal reference gene for normalisation. Data represent mean values from three biological replicates (n=3 per group). Quantitative data (d) are shown as mean ± standard error of the mean (s. e. m.), P<.001, utilising an unpaired two-tailed Student's t-test.

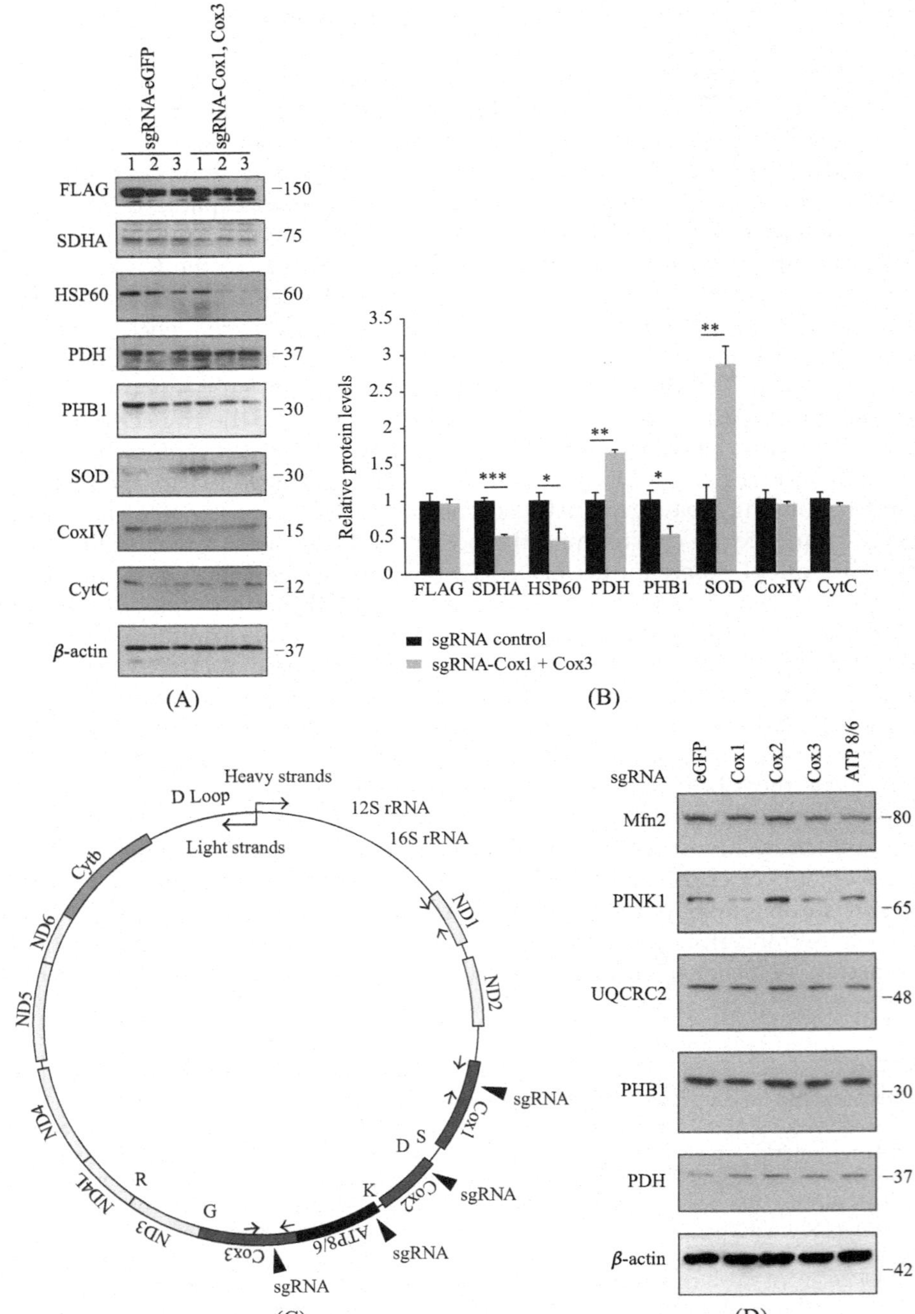
sgRNA-eGFP
sgRNA-Cox1, Cox3
1 2 3 1 2 3
FLAG –150
SDHA –75
HSP60 –60
PDH –37
PHB1 –30
SOD –30
CoxIV –15
CytC –12
β-actin –37
(A)
Relative protein levels
3.5
3
2.5
2
1.5
1
0.5
0
***
*
**
*
**
FLAG SDHA HSP60 PDH PHB1 SOD CoxIV CytC
sgRNA control
sgRNA-Cox1 + Cox3
(B)
Heavy strands
D Loop
12S rRNA
16S rRNA
Light strands
Cytb
ND6
ND5
ND4
ND4L
ND3
ND1
ND2
Cox1
Cox2
ATP8/6
Cox3
sgRNA
sgRNA
sgRNA
sgRNA
D S
K
G
R
(C)
sgRNA eGFP Cox1 Cox2 Cox3 ATP 8/6
Mfn2 –80
PINK1 –65
UQCRC2 –48
PHB1 –30
PDH –37
β-actin –42
(D)

various other organisms. Furthermore, catalytically inactive Cas9 (dCas9) can associate with a protein domain that modulates transcription, enabling precise and stable transcriptional regulation of target genes under the direction of gRNA. Moreover, the CRISPR/Cas13-based SHERLOCK technology can efficiently identify the virus and is also applicable for the rapid detection of the novel coronavirus from 2020. Significant advancements have been achieved in nuclear gene editing via the CRISPR method. Nonetheless, for comparatively smaller operational targets like mtDNA, utilising the CRISPR–Cas9 technology is challenging. Furthermore, mitochondria are incapable of importing RNA, so inhibiting the gRNA-Cas9 complex from identifying the target sequence. What methods can be employed to introduce RNA into mitochondria, and how can one guarantee the correct direction towards the altered gene following the creation of gRNA? All these issues constrain the viability and efficacy. The utilisation of the CRISPR–Cas9 system in mitochondrial gene editing. TALEN is generally preferable to ZFN because it does not require recognition of DNA sequences via base triplets, and its unique recognition provides enhanced specificity and flexibility in selecting target sequences. Furthermore, TALEN offers the benefits of more straightforward procedure and a reduced timeframe. The crucial aspect of gene editing is DNA-binding. ZFN and TALEN employ proteins as guides for targeted editing, necessitating intricate and time-consuming protein engineering, whereas the CRISPR-Cas system is directed by gRNA, which is comparatively inexpensive and straightforward to design and generate. In mitochondrial gene editing, it is essential to evaluate the capacity of RNA and proteins to traverse the mitochondrial membrane, the volume of the occupied space, and the feasibility of transport into the mitochondrial membrane. A novel CRISPR system, known as CRISPR-Cpf1, has just been identified and classified as a second-generation gene editing tool. It substitutes the Cas9 protein with the Cpf1 protein, which exhibits comparable genome editing effectiveness to Cas9, while also offering other advantages. The Cpf1 system requires only one RNA for gene editing, capable of autonomously processing precursor CRISPR-derived RNA (pre-crRNA) and then targeting and cleaving DNA precisely with the processed crRNA. The Cpf1 enzyme is smaller than the conventional Cas9, facilitating its transmission to cells and tissues. The Cpf1 system generates sticky ends rather than blunt ends when cleaving DNA, resulting in enhanced precision and reduced off-target effects, although some off-target effects persist in this system. Moreover, the Cpf1 incision occurs distantly from the recognition site, indicating that it can still exert an effect even if the target gene is altered at the cleavage site, so offering many opportunities for successful editing.

FIGURE 16.14 Modifications in mitochondria-associated proteins subsequent to CRISPR/Cas9-mediated mitochondrial DNA (mtDNA) editing. (A) Mitochondrial proteins in HEK-293T cells following CRISPR/Cas9-induced breakage of mtDNA at the Cox1 and Cox3 loci, as assessed by Western blot analysis utilising specified antibodies. β-actin served as a loading control. (B) Quantification of mitochondrial proteins in HEK-293T cells transfected with lentiCRISPR-sgRNA-Cox1 + Cox3 or lentiCRISPR-sgRNA-eGFP#2 control, as depicted in panel (a), normalised to β-actin. (C) Illustration of human mtDNA. Specific loci targeted by lentiCRISPR-sgRNAs (Cox1, Cox2, Cox3 and ATP8/6) are represented by filled triangles. (D) Representative Western blots illustrating the differential modification of mitochondrial proteins subsequent to the cleavage of certain mtDNA loci facilitated by the specified sgRNAs in HEK-293T cells. Quantified data (b) are presented as mean ± s. e. m., with significance levels indicated as * $P<.05$, ** $P<.01$ and *** $P<.001$, determined by an unpaired two-tailed Student's t-test.

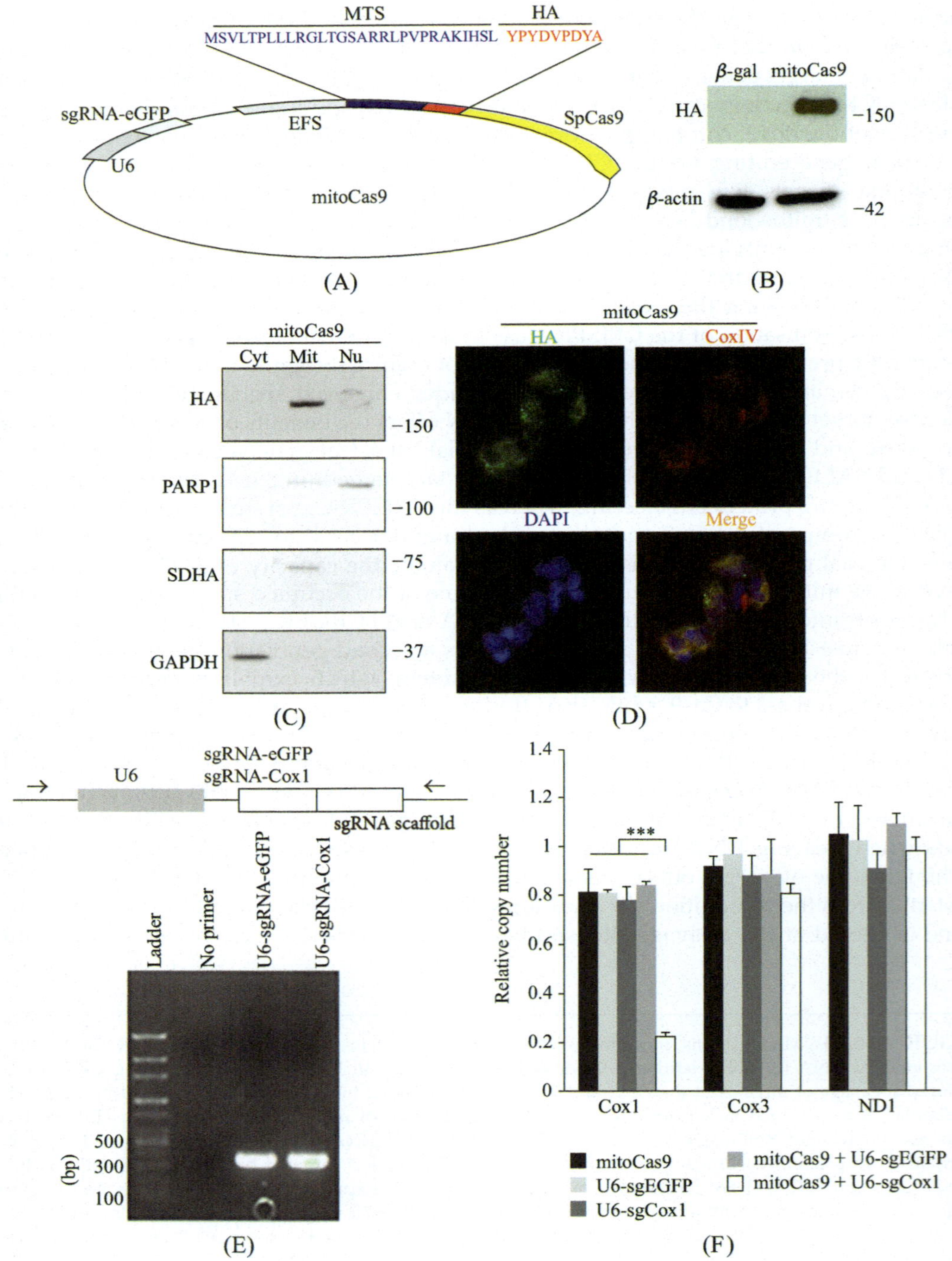
MTS
HA
MSVLTPLLLRGLTGSARRLPVPRAKIHSL
YPYDVPDYA
sgRNA-eGFP
EFS
SpCas9
U6
mitoCas9
(A)
β-gal
mitoCas9
HA
−150
β-actin
−42
(B)
mitoCas9
Cyt
Mit
Nu
HA
−150
PARP1
−100
SDHA
−75
GAPDH
−37
(C)
mitoCas9
HA
CoxIV
DAPI
Merge
(D)
U6
sgRNA-eGFP
sgRNA-Cox1
sgRNA scaffold
Ladder
No primer
U6-sgRNA-eGFP
U6-sgRNA-Cox1
(bp)
500
300
100
(E)
Relative copy number
1.4
1.2
1
0.8
0.6
0.4
0.2
0
***
Cox1
Cox3
ND1
mitoCas9
U6-sgEGFP
U6-sgCox1
mitoCas9 + U6-sgEGFP
mitoCas9 + U6-sgCox1
(F)

Cpf1 can direct RNAs to simultaneously edit many genes or various places within a single gene, enhancing the efficacy of gene knock out, hence facilitating the concurrent screening of numerous proteins. It broadens the selection range of gene editing target locations by employing a straightforward transport system to direct CRISPR effector proteins and gRNAs, and it is widely regarded as having significant potential in gene therapy applications. Despite Cpf1's superiority over Cas9, it exhibits reduced activity, and its target selection is constrained because of the relative fixing of the PAM sequence. The exploration of methods to enhance activity and broaden the target range for a more extensive gene editing effect is still necessary. As a novel gene editing method, the functionality of the CRISPR-Cpf1 technique is more straightforward and precise, enabling more effective editing of tiny mitochondrial genes; nonetheless, its application in mitochondrial gene editing has yet to be documented. In the future, the CRISPR-Cpf1 system may emerge as an efficient method for mitochondrial gene editing.

Mitochondrial gene editing utilising the CRISPR-Cas9 system

The successful application of the CRISPR/Cas9 system to mitochondrial gene editing was first described in 2015. However, the conclusions were controversial and were also questioned by another group. The process by which gRNA is introduced into mitochondria is yet unclear. The 20 ribonucleotide stem-ring series derived from H1 RNA (the RNA component of the RNase P enzyme), which is linked to the injected cytoplasmic RNA, can effectively target RNA to mitochondria. This approach is efficient for both noncoding RNA, such as tRNA and mRNA that encodes proteins, elucidating the importation of gRNA into mitochondria within the CRISPR/Cas9 system. Gammage et al. synthesised the prevailing RNA import idea and its rebuttal. They refuted the presence of RNA components in mtRNase P and posited that RNase P and RNase MRP primarily function in the nucleus. They also examined the function of mitochondrial ribosomes in facilitating rRNA and polynucleotide phosphorylase (PNPase) in cytoplasmic RNA transport. Reports indicate that two domains of yeast cytoplasmic tRNALys (CUU), specifically the F-arm and D-hairpin, along with certain domains of 5S RNA, may facilitate RNA entry into mitochondria. Given the absence of the NHEJ repair pathway and the presence of the homologous recombination (HR) repair pathway in mitochondria, attempts to insert DNA sequences into mtDNA were successful in human and zebrafish mitochondria, as well as in yeast mitochondria and Arabidopsis chloroplasts, employing the CRISPR-Cas9 system to generate

FIGURE 16.15 Development of mitochondria-targeting sequence (MTS)-HA-Cas9. (A) Schematic representation of mitochondria-targeting Cas9 (mitoCas9). Information regarding the MTS and HA tag is provided. (B) Expression of mitoCas9 in HEK-293T cells, transiently transfected with the MTS-HA-Cas9 construct, was assessed via Western blot analysis employing a HA antibody. β-actin served as a loading control. (C)The subcellular distribution of MTS-HA-Cas9 was assessed in cytosolic (Cyt), mitochondrial (Mt) and nuclear (Nu) fractions of HEK-293T cells transfected with the lentiCRISPR-sgRNA-eGFP#2 construct. Western blot analysis was performed to determine the compartment-specific localisation of the expressed MTS-HA-Cas9 protein.

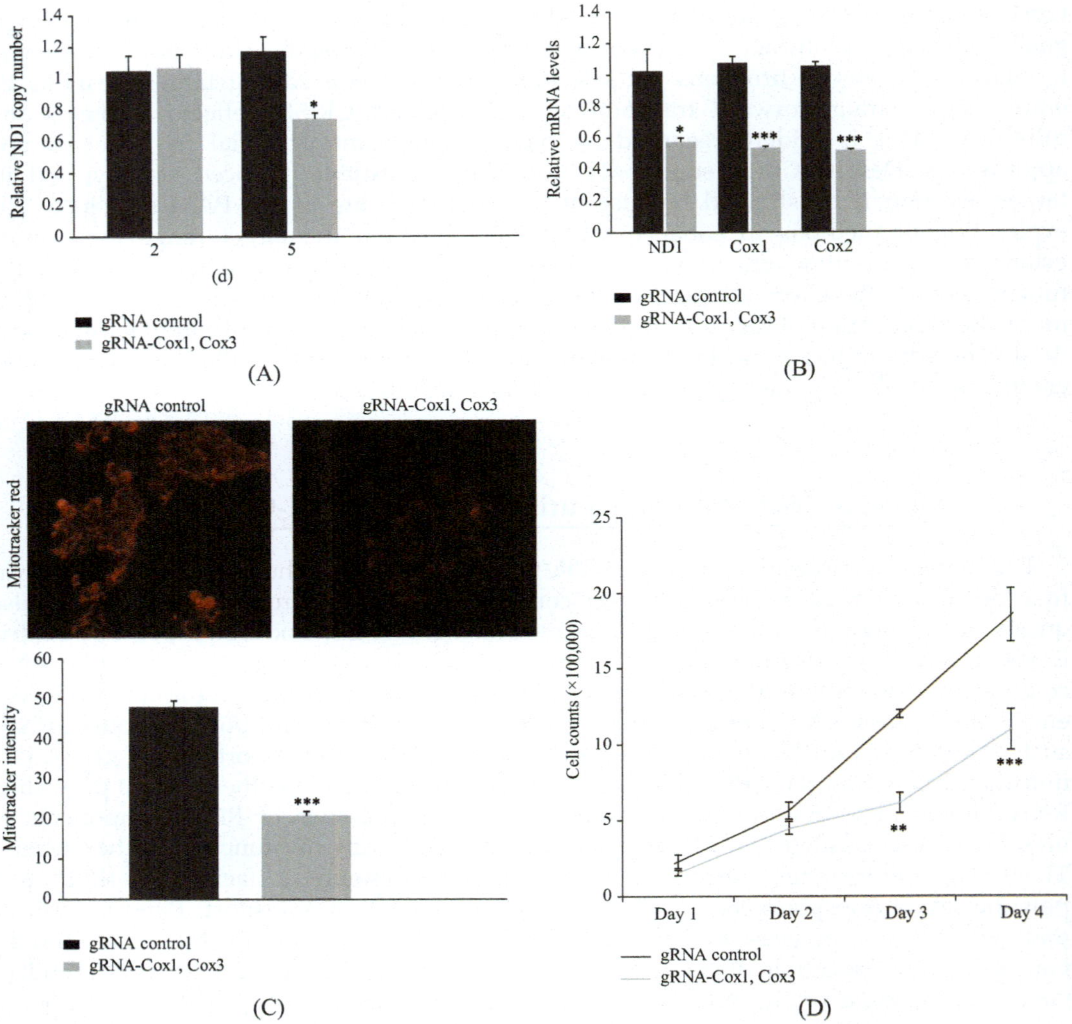

FIGURE 16.16 Mitochondrial-targeted Cas9 (mitoCas9) induced mitochondrial DNA (mtDNA) damage, leading to mitochondrial dysfunction. Quantification of copy numbers for the ND1 region of mtDNA extracted from HEK-293T cells transiently transfected with indicated constructs (gRNA control, mitoCas9 and U6-sgRNA to eGFP, gRNA Cox1, Cox3, mitoCas9 and U6-sgRNA to Cox1 and Cox3), determined by real-time quantitative PCR using primers listed in Table 16.3 at the indicated time points (2 days and 5 days after transfection). GAPDH was used as an internal loading control. Quantification of messenger RNAs for ND1, Cox1 and Cox2 genes in HEK-293T cells 5 days following transient transfection with lentiCRISPR-sgRNA-Cox1 + Cox3 or lentiCRISPR-sgRNA -eGFP2 control determined by real-time PCR and normalised to GADPH. Representative images of MitoTracker Red staining for functional mitochondria in HEK-293T cells transfected with the indicated constructs (upper panel). Quantification of relative MitoTracker Red staining intensities in two groups was shown in the bottom panel. (*n*=50 mitochondria from three independent cultures per group) Cell counts demonstrating the proliferation rate of the HEK-293 cells following cleavage of specific mtDNA loci (*n*=6 individual measurements per group). Days indicate the time intervals proceeded after cell seeding following 5 days of transient transfection. Quantified data are expressed as mean±sem, **P*≤.05, ***P*≤.001, unpaired two-tailed Student's t test.

DSB and exogenous DNA to facilitate homologous fragments. Historically, the entry of DNA into mitochondria encountered significant challenges due to membrane potential, pH levels and the phagocytosis of bacteriophages. Foreign DNA has been effectively delivered into yeast mitochondria using biolistic bombardment with DNA-coated tungsten particles, electroporation, DQAsome complexes and TAT. A separate work used MTS to alter the adeno-associated viral capsid VP2, successfully facilitating the transport of the ND4 gene into mitochondria, so introducing a novel approach for mitochondrial DNA sequence delivery. Alternative mitochondrial delivery systems for pharmaceuticals or proteins are engineered using metal-organic frameworks triphenylphosphonium and cell-penetrating poly(disulphide)s modified biodegradable silica nanoparticles (BS-NPs), which show promise for the mitochondrial delivery of exogenous DNA or RNA components. In the research conducted by Bian et al., they altered Cas9 with MTS and performed in vitro transcription, subsequently microinjecting gRNA and the in vitro transcription product into the cytoplasm. They also introduced an exogenous DNA vector carrying a homologous arm via ViaFect transfection. Exogenous ssDNA with a short homologous arm was accurately integrated into the target loci and this mutagenesis was reliably inherited by the F1 generation of zebrafish. This suggests that ssDNA and gRNA can be delivered into mitochondria without modification, corroborating recent findings published by others. The mitoCas9 protein can selectively target mtDNA and decrease mtDNA copy quantity in both human cells and zebrafish. A separate cohort of researchers presents 'Edit Plasmids', which encompass the Cas9 gene, gRNA expression sequence and homologous sequence, introduced into yeast mitochondria and Chlamydomonas chloroplasts via the microprojectile transformation technique. The findings indicate donor DNA insertion at the target loci enabled by HR, analogous to the research conducted in zebrafish. The presence of an endogenous pathway for nucleic acid import into mammalian mitochondria, essential for mitochondrial CRISPR/Cas9 gene editing, remains contentious.

pAgo-dependent system: argonaute proteins were initially discovered in eukaryotes; however, analogous pAgos have also been identified in archaea and bacteria. Argonaute proteins represent a varied family of nucleic acid-guided proteins with potential applications in genome editing. Eukaryotic Argonaute proteins facilitate RNA-guided RNA interference, but prokaryotic Argonautes predominantly employ DNA guides to target complementary DNA sequences, thereby safeguarding their hosts against invasive DNA. The cyanobacterium *Synechococcus elongatus* is a DNA-guided nuclease that mostly targets single-stranded DNA, while it has nonspecific, guide-independent action on double-stranded substrates. *Clostridium butyricum* (CbAgo) targets multicopy genetic elements, generates DNA interference among homologous sequences, and initiates DNA degradation. The loading of CbAgo with tiny DNA guides relies on its inherent endonuclease activity and the DSB repair mechanisms. Both *Methanocaldococcus jannaschii* and *Thermus thermophilus* exhibit two modes of action: the canonical guide-dependent endonuclease activity and a nonguided DNA endonuclease activity. Moreover, CbAgo and *Limnothrix rosea* function as DNA-guided DNA nucleases at physiological temperatures, hence broadening their applicability in genomic endeavours. Specifically, certain pAgos, such as *Rhodobacter sphaeroides*, obtain guides from plasmid-derived mRNA to target plasmid DNA, thereby diminishing the transcription of reporter genes and the overall plasmid content. Recently, multiple pAgos have been recognised for their role in augmenting the positive selection process of recombinase A (RecA)-mediated

DNA strand exchange via the interaction between their PIWI domain and RecA, facilitating the HR of target sequences. Given that mtDNA undergoes renewal and replication, it is logical to introduce pAgos homologous arms to target the mutated sequence in mitochondria. The homologous arm must be constructed as a standard mtDNA sequence of specific length to facilitate the conversion of the altered DNA sequence to the normal DNA during a subsequent replication cycle. If the mutant DNA is converted into normal DNA, the impaired mitochondria may be salvaged and restored. Nonetheless, the utilisation of the pAgo-based approach for editing mtDNA necessitates more investigation. The aggregated results from the papers examined above outline the primary techniques for mitochondrial gene editing. Although standard gene editing techniques such as RE, ZFN and TALEN have achieved some advancements in mitochondrial gene editing, a comprehensive, precise and systematic methodology for mitochondrial gene editing has yet to be developed. The updated CRISPR/Cas9 method has partially addressed the issue of gene orientation; nonetheless, significant shortcomings remain. To date, there is insufficient and credible experimental evidence to substantiate the conclusion. Nevertheless, informed by these investigations, researchers may endeavour to enhance CRISPR/Cas9 and other gene editing methodologies, such as the pAgo-based system, to target specific mitochondrial genes. Regardless of whether the defective mitochondria are eliminated following the customised CRISPR/Cas9 system's cleavage of mtDNA, or whether the mutant mtDNA fragments are directly repaired using pAgo-based gene editing approaches to restore mitochondrial function, mitochondrial dysfunction will be effectively rectified. If the issues of ineffective delivery of gRNA and Cas9 enzyme complexes into the mitochondria, along with low editing specificity, can be resolved. The modified CRISPR/Cas9 system and the pAgo-based gene editing system may emerge as the most promising methods for mitochondrial gene editing. These will undoubtedly yield insights and methodologies for the diagnosis and treatment of hereditary mitochondrial disorders in the future.

Mitochondrial DNA editing in mice with DddA-TALE fusion deaminases—a case study

In this section, we describe a case study depicting mitochondrial DNA editing in mice model. Fig. 16.17

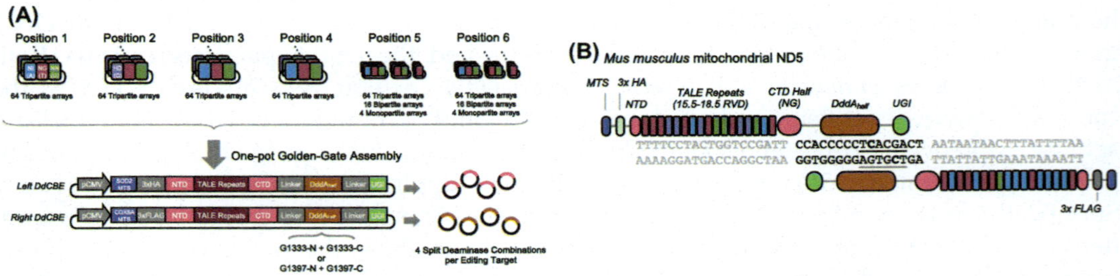

FIGURE 16.17 Schematic illustration for assembling DdCBE and its mitochondrial DNA editing.

Scheme of one-pot Golden Gate assembly for efficient DdCBE construction. A total of 424 arrays (64 v tripartitite arrays × 6 + 16 bipartite arrays × 2 + 4 monopartite arrays × 2) and an expression vector were mixed to generate left and right modules for the final plasmid constructs.

Illustration of DdCBE interacting with mouse mtDNA target ND5. TALE-binding regions are shown in grey and base editing windows are depicted in black. Different repeat variable diresidue modules are shown in orange and blue. Green and yellow, which represent 'NI' for adenine, NG for Thymine, 'NN' for guanine and 'HD' for cytosine recognition respectively (see Fig. 16.18).

Fig. 16.19

1 To observe the germline transmission of mtDNA mutations, the female F0 (sipup-3) mouse was crossed with a wild-type C57BL6/J male to obtain F1 pups, after which

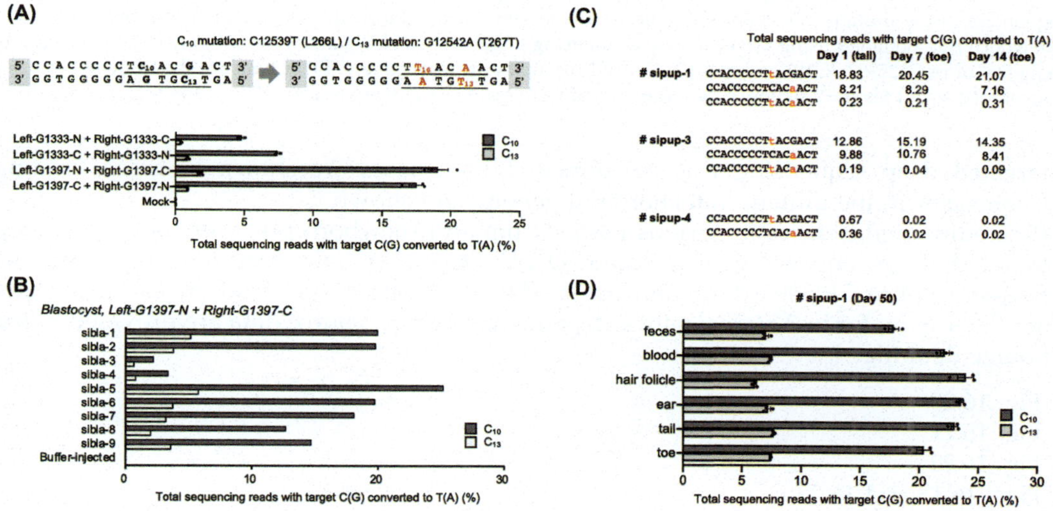

FIGURE 16.18 Illustrates the point mutation in the mouse mitochondrial ND5 gene produced by base editing derived from DdCBE. Target and efficiency of DdCBE deaminase-mediated cytosine-to-thymine base editing in NIH3T3 cells. In the target sequence, translation codons are emphasised with underlining, and potential editing locations are shown in red. Transfected combinations of DdCBE are designated as left or right, -G1333 or -G1397 and -N or -C. The *P* values for the combinations left-G1333-N+right-G1333-C, left-G1333-C+right-G1333-N, left-G1397-N+right-G1397-C, and left-G1397-C+right-G1397-N for the C10 mutation are 0.0012, 0.0003, 0.0014 and 0.0009, respectively. For the C13 mutation, the *P* values are 0.0116, 0.0076, 0.0030, and 0.0003, respectively (*P<.05 and **P<.01 using Student's two-tailed t test). Corresponding base editing efficacy in murine blastocysts. The sequencing data were acquired from blastocysts that formed following the microinjection of zygotes with mRNA encoding the left-G1397-N and right-G1397-C DdCBE. (C) Alignments of mutated sequences from neonatal pups. Targeted deep sequencing was conducted with genomic DNA extracted from the tails of neonates immediately postbirth, as well as from the toes at 7 and 14 days postbirth. Revised bases are indicated in red. The editing frequencies in the mutant mitochondrial genome are presented. Editing efficiencies in diverse tissues of an adult F0 mouse (sipup-1). The sequencing data were acquired from each tissue 50 days postnatally. In all graphs, the dark and light grey bars denote the frequency of m.C12539T (C10) and m.G12542A (C13) mutations, respectively. Error bars represent the standard error of the mean (s. e. m.) for n=3 physiologically independent samples. *The source data is contained in the source data file.*

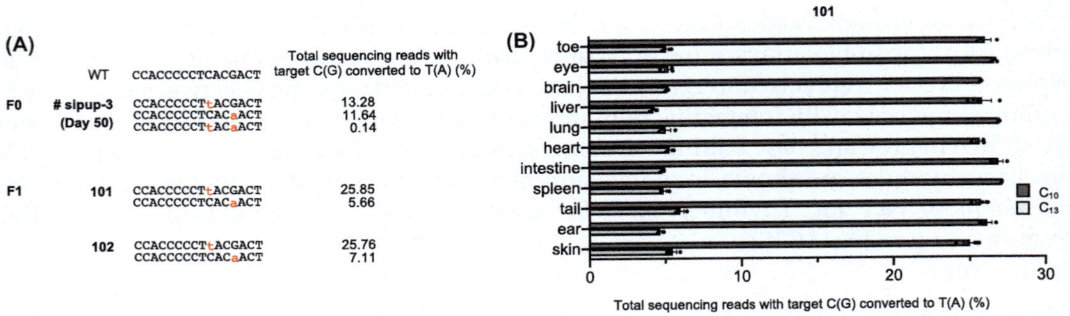

FIGURE 16.19 Germline transmission of mutant mitochondrial DNA (mtDNA). To observe the germline transmission of mtDNA mutations, the female F0 (sipup-3) mouse was crossed with a wild-type C57BL6/J male to obtain F1 pups, after which targeted deep sequencing was performed. Edited bases are shown in green. Editing frequencies in the mutant mitochondrial genome are shown. Base editing efficiencies in various tissues from an F1newborn pup obtained using targeted deep sequencing of genomic DNA. Dark and light grey bars represent the frequency of m.C12539T (C10) and m.G12542A (C13) mutations, respectively. Error bars are the standard error of the mean (s. e. m.) for $n = 3$ biologically independent samples. Source data are provided in the Source data file.

targeted deep sequencing was performed. Edited bases are shown in green. Editing frequencies in the mutant mitochondrial genome are shown.

2 Base editing efficiencies in various tissues from an F1newborn pup were obtained using targeted deep sequencing of genomic DNA. Dark and light grey bars represent the frequency of m.C12539T (C10) and m.G12542A (C13) mutations, respectively. Error bars are the s. e. m. for n=3 biologically independent samples. Source data are provided in the Source data file.

Fig. 16.20
Fig. 16.21
Fig. 16.22
Figs. 16.13 and 16.23

Editing of the mitochondrial genome

The advancement of gene editing technology has enabled the creation of precise methods to synthesise nonnative mtDNA sequences or to repair existing mtDNA sequences, hence altering the mtDNA heteroplasmy ratio. As a result, mitochondrial genome editing (MGE) has emerged as a new focal point for researchers. Moreover, the significance of mitochondrial gene function and the threshold effect of mitochondrial illnesses indicate that any genetic therapy must likely rectify the majority of cells in the body to render germline correction feasible. In 2015, it was initially claimed that targeted reduction of mitochondrial genomes in the germline to prevent the transmission of mitochondrial disorders might be accomplished using mitochondria-targeted nucleases. Mitochondrial genome editing involves the alteration of human zygotes or oocytes susceptible to mtDNA diseases through the intracytoplasmic microinjection of mitochondria-targeted nucleases to prevent the germline transmission of

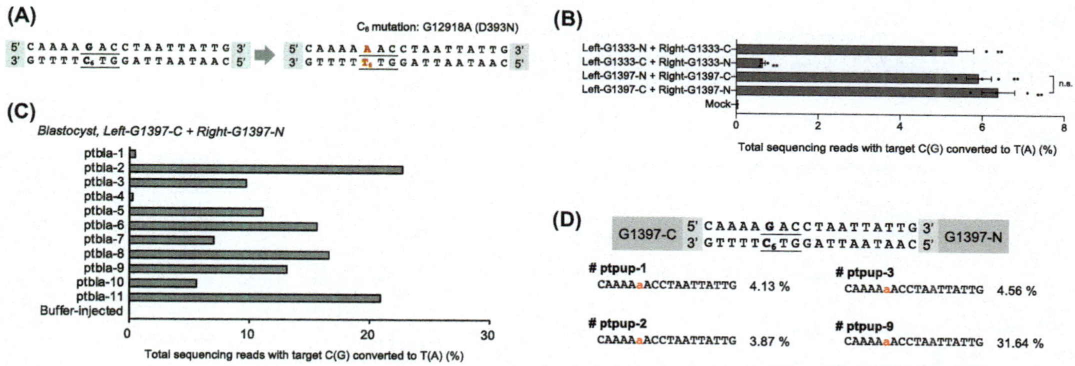

FIGURE 16.20 Illustrates the induction of the G12918A mutation in mouse mitochondrial ND5 by DdCBE. The DdCBE objective is to induce the m.G12918A point mutation, resulting in a D393N alteration in the ND5 protein. The target codon is emphasised, and the potential editing locus is shown in red. The efficacy of cytosine-to-thymine base editing utilising DdCBE in NIH3T3 cells. The annotations denote the combination of cotransfected DdCBE pairs. Error bars represent standard error of the mean for three biologically independent samples (n. s. indicates not significant, *P<.05, and **P<.01 as determined by Student's two-tailed t test). The P values for the C6 mutation are as follows: left-G1333-N + right-G1333-C: 0.0052, left-G1333-C + right-G1333-N: 0.0099, left-G1397-N + right-G1397-C: 0.0027 and left-G1397-C + right-G1397-N: 0.0040. The P value for nonsignificance is 0.4971. The efficiency of base editing for the c m.G12918A point mutation in mouse blastocysts. The sequencing results were acquired from cultured blastocysts that developed following the microinjection of one-cell stage embryos with mRNA encoding the left-G1397-C and right-G1397-N DdCBE. F0 mice possessing an ND5 point mutation. F0 pups possessing an ND5 point mutation, resulting from the microinjection of DdCBE mRNAs. Alignment of mutant sequences from neonatal pups. Edited nucleotides are highlighted in red, while the adjacent column denotes the editing frequencies in the mutant mitochondrial genome. The source data is contained in the source data file.

mutant mtDNA haplotypes. Targetable nucleases, including endonucleases, ZFN, TALENs, CRISPR and CRISPR-associated protein 9 (Cas9)—CRISPR/Cas9—facilitate direct and precise gene editing of the mitochondrial genome. MGE presents considerable potential as an independent approach for preventing mtDNA disorders, serving as an embryo-sparing donor option for the treatment of mitochondria-related infertility.

Mitochondria-targeted restriction endonucleases (mitoREs) are endonucleases fused with N-terminal polypeptides—MLS—which facilitate the protein's translocation into the mitochondria. The mitoREs, expressed in the nuclear-Cyt compartments, identify certain sequences in the mtDNA and induce DSBs upon their importation into the mitochondria. Originally designed for editing the nuclear genome, ZFNs and TALENs comprise the MLS, the DNA-binding domain, and the nuclease catalytic domain integrated into a single protein. DSBs are generated by a nuclease domain derived from FokI nuclease. The zinc finger DNA-binding domain or bacterial TALE protein enhances the binding specificity of designed nucleases to pathogenic mtDNA molecules. CRISPR/Cas9, a unique nuclear genome editing method, utilises a single-guide RNA (sgRNA) to modify particular DNA target regions. The specificity of DNA recognition is enhanced due to its reliance on DNA–RNA interaction, surpassing other approaches. Furthermore, it is considered to be more efficacious in genome editing and more adaptable owing to the straightforward process of sgRNA customisation. The manipulation of mtDNA using MGE can be

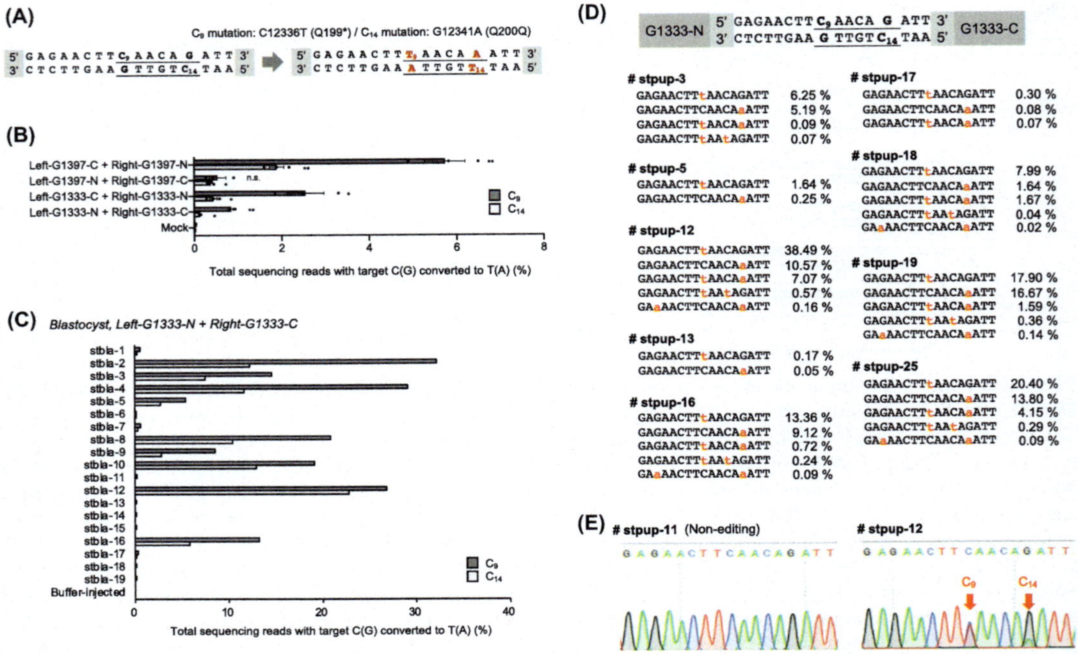

FIGURE 16.21 Illustrates the generation of a nonsense mutation in the mouse mitochondrial ND5 gene using cytidine deaminase-mediated base editing. a The DdCBE objective for producing the m.C12336T nonsense mutation and the m.G12341A silent mutation. The m.C12336T (C9) mutation results in a Q199stop mutation in the ND5 protein, while m.G12341A (C14) induces a quiet Q200Q mutation. Translation triplets are highlighted, and potential modification locations are indicated in red. b The efficacy of cytosine-to-thymine base editing resulting in a nonsense mutation in NIH3T3 cells. The annotations denote the combination of DdCBE pairs that were cotransfected into the cells. The dark and light grey bars denote the frequency of the m.C12336T (C9) and m.G12341A (C14) mutations, respectively. Error bars indicate standard error of the mean for $n=3$ physiologically independent samples (n. s. denotes not significant, *$P<.05$ and **$P<.01$ as determined by Student's two-tailed t test). P values for left-G1333-N combined with right-G1333-C, left-G1333-C combined with right-G1333-N, left-G1397-N combined with right-G1397-C and left-G1397-C combined with right-G1397-N The values for the C9 mutation are 0.0065, 0.1143, 0.0266 and 0.0037, whereas the values for the C14 mutation are 0.0077, 0.0144, 0.0406 and 0.0214, respectively. Efficiency of base editing in murine blastocysts. The sequencing data were acquired from blastocysts that formed following the microinjection of zygotes with mRNA encoding left-G1333-N and right-G1333-C DdCBE. The dark and light grey bars denote the frequency of the C9 and C14 mutations, respectively. Alignment of mutant sequences from neonatal pups. Edited bases are highlighted in red, whereas the column on the right displays the editing frequencies in the mutant mitochondrial genome. Sanger sequencing chromatograms from unedited and modified mice. *The red arrows represent the substituted nucleotides. The source data is contained in the source data file.*

categorised into the alteration of mtDNA copy number and the repair or removal of harmful mutant mtDNA. Cleavage and degradation of mtDNA have been accomplished utilising mitochondria-targeted TALENs (mito-TALENs) and mitochondria-targeted ZFNs (mito-ZFNs), which could diminish the mutant mitochondrial genome. Conversely, the possibly harmful mutant mtDNA can be modified to diminish mtDNA heteroplasmy to the normal wild type utilising mito-REs, mito-TALENs, or mito-ZFNs.

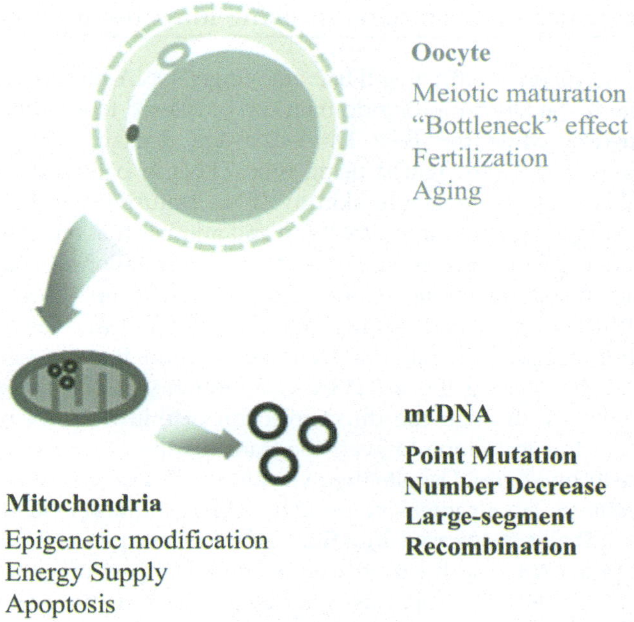

FIGURE 16.22 Summary of mitochondria and mitochondrial DNA (mtDNA) function in oocyte.

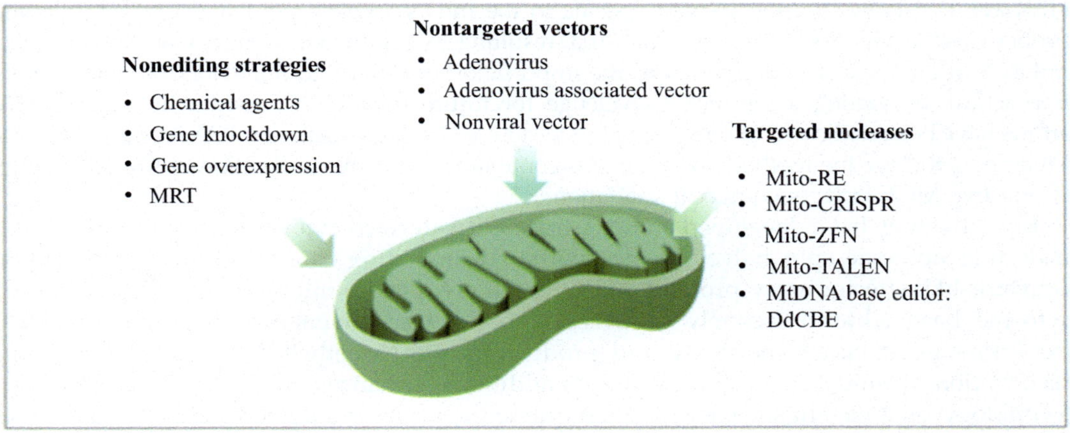

FIGURE 16.23 Possible manipulation strategies targeting mitochondrial DNA (mtDNA), MRT, mitochondrial replacement therapy, mito RE, mitochondria targeted restriction endonuclease, mito-CRISPR, mitochondria targeted clustered regularly interspaced short palindromic repeats. Mito-ZFN, mitochondria targeted zinc finger nuclease, and mito-TALEN, mitochondria-targetted transcription activator like effecter nuclease.

Possible therapeutic interventions utilising mitochondrial genome editing

ZFNs are dimers consisting of the Cys2His2 zinc-finger DNA-binding domain and the FokI DNA-cleavage domain. The DNA-binding domain can be altered to identify a rather lengthy and specific target sequence, while the FokI DNA-cleavage domain will generate DSBs upon dimerisation. Processes have been created using mito-ZFNs in patient-derived cybrid cells and mice models. TALENs possess a structure akin to ZFNs, featuring two DNA-binding domains: one that exclusively targets the mutant mtDNA site and another that binds to both wild-type and mutant mtDNA. Mito-TALENs have been utilised to target mutations associated with mitochondrial myopathy, encephalopathy, lactic acidosis and stroke-like episodes (MELAS) in induced pluripotent stem cells (iPSCs;). A study shown that modified mito-TALENs directed against the m.13513G>A mutation reduced mutant mtDNA levels upon transduction. A separate study indicated that an iPSC harbouring the m.3243A>G mutation was effectively eradicated, both in iPSCs and porcine oocytes, through the direct injection of mito-TALEN mRNA, implying the potential application of this technique to avert the transmission of mutant mtDNA in advanced assisted reproductive technologies. Despite their versatility, the creation of new nucleases for each genomic target necessitates labour-intensive protein engineering and assembly procedures using ZFNs and TALENs. Initially showcased in 2012, CRISPR/Cas9 genome editing technology has garnered significant public interest due to its transformative impact on gene editing, culminating in the Nobel Prise in Chemistry in 2020. Cas9 may be readily programmed by a gRNA to target almost any gRNA-complementary sequence followed by a protospacer-adjacent motif, offering answers to several challenging human genetic disorders. CRISPR-associated protein 9 (Cas9) exhibits versatility and efficacy in genome editing owing to the straightforward process of single-gRNA customisation. A report indicates that CRISPR/Cas9 can effectively cleave mtDNA in HEK-293 T cells by utilising gRNAs specific to the mtDNA genes Cox1 and Cox3, facilitating proper Cas9 localisation to the mitochondria, resulting in a reduction of nearly 80–90% in intact mtDNA in HEK-293 T cells. However, the importation of RNA into mitochondria is inherently impractical, presenting a significant challenge for future research. Consequently, considerable efforts have been dedicated to optimising this system to eradicate mutant mtDNA and modify the MGE; nonetheless, the methodology remains contentious, and the application of this technology to effective MGE lacks unequivocal validation.

Recently, they have developed an RNA-free DddA-derived cytosine base editor (DdCBE), including nontoxic split halves of the cytidine-deaminating catalyst DddA, transcription activator-like effector array proteins, and a uracil glycosylase inhibitor. The DddA-derived cytosine base editor effectively catalyses C•G-to-T•A conversions in human mtDNA, exhibiting great target specificity and product purity. Nonetheless, its utilisation in the rectification of unidentified illnesses needs additional enhancement and refinement of this technology, as it has thus far been limited due to the pronounced predilection for a 5'-TC-3' editing sequence. While it presently resembles a theoretical approach to addressing human infertility using MGE, the swift advancement of MGE technology will undoubtedly provide viable approaches in the near future. Table 16.2 provides a summarised comparison of several mitochondrial interventions.

Present condition of mitochondrial genome editing

The swift advancement of genetic technology has significantly enhanced our understanding of the mitochondrial genome, its associated pathologies, and possible therapeutic methods. While the application of MGE technology in treating infertile people remains largely experimental, it demonstrates significant potential for additional investigation and research into mitochondrial illnesses. Nonetheless, obstacles persist, including the difficulty of targeting the mitochondrial genome with gene therapies due to the distinctive inheritance patterns of mitochondria, the swift manifestation of phenotypes postfertilisation, the abundant quantity of mtDNAs within each cell, and the physical and chemical barriers encasing them. The compact nature of mitochondrial genes and the double-membrane architecture render the nontargeted insertion of new genetic material into the mitochondrial genome unfeasible. Targetable nucleases possess both advantages and disadvantages. mitoREs are recombinant restriction endonucleases synthesised in the nuclear-Cyt compartments and transported to the organelle via MLSs, which are N-terminal polypeptides that facilitate the protein's entry into the mitochondria. Upon importation, mitoREs identify particular sequences in mtDNA and induce DSBs. This technique has been employed to selectively cleave a certain type of mtDNA in a heteroplasmic cell or mouse line, resulting in alterations in heteroplasmy. Despite the method's demonstrated efficacy and safety in cleaving mutant mtDNA and altering heteroplasmy in cellular and murine models, it is constrained by the scarcity of distinctive recognition sites generated by pathogenic mutations.

The precise and efficient editing of genomes utilising meganucleases, including ZFNs, TALENs and Cas proteins, has garnered significant interest in recent years. Scientists have revealed advancements in their research on the application of ZFN to target and cleave specific loci in mtDNA. They developed a mitochondrially targeted ZFN including two cleavage domains attached to the same protein, which specifically eradicates harmful mtDNA and an mtDNA region commonly linked to illnesses.

It encompasses many transfer RNAs and structural genes of the oxidative phosphorylation machinery. Concurrently, mito-TALENs are employed to cleave particular sequences in mtDNA to eradicate mitochondria harbouring harmful point mutations. A recent study indicated that this technique targets two clinically significant mtDNA point mutations linked to mitochondrial illnesses, specifically myoclonus epilepsy with ragged red fibres (MERRF) and MELAS/Leigh syndrome. The mutation burden was effectively diminished in vitro, evidenced by reduced biochemical OXPHOS deficiencies. The positioning of the mutations and the dimensions of the mito-TALEN continue to pose significant obstacles in the application of the mito-TALEN methodology within a therapeutic context. A separate study demonstrated that a mito-TALEN targeting the mtDNA region containing the m.5024 C>T mutation was inserted into an adeno-associated virus and phage vector to evaluate its function in modulating mtDNA heteroplasmy in vivo in a murine model. A notable reduction in the mutant/wild-type ratio in skeletal and cardiac muscle, relative to nontargeted tissue, was seen following the systemic administration of AAV9–mito-TALEN.

Nontargeted vectors for nuclease administration can be succinctly classified as adenoviruses, adeno-associated viruses and nonviral vectors. These vectors are essential vehicles for the implementation of MGE in vivo, as they are consistently utilised in conjunction with mitochondria-targeted nucleases or endonucleases for distribution and optimal functionality within cells. Research indicates that, in certain instances, mitochondrial genes can be assimilated into the nuclear genome through the application of nontargeted viral vectors, a mechanism that emulates the evolutionary translocation of mitochondrial genes to the nucleus.

Recombinant adenovirus type 5 (rAd5) was utilised in vivo in NZB/BALB mice to transport mito-ApaLI to the mitochondria. rAd5-expressing mitoREs effectively eradicated BALB mtDNA in the heart. Nonetheless, owing to the limited expression duration and the induction of a robust immune response, rAd5 is infrequently utilised for in vivo applications. Adeno-associated virus (AAV) is the predominant viral vector utilised in gene therapy, attributed to its prolonged transgene expression and a proven record of clinical safety. Various serotypes of AAV vectors can be produced to transduce a diverse array of tissues and cells. The episomal expression characteristic of AAV ensures prolonged presence in tissues, whereas the infrequent integration enhances its safety profile. Leber's hereditary optic neuropathy, a mitochondrial disorder characterised by acute vision loss in young adulthood due to mutations in NADH-ubiquinone oxidoreductase subunits (including ND4), has also been demonstrated through the application of AAV. The nuclear insertion of corrected ND4 gene copies led to vision enhancements in certain human patients, while the viral injection of MTS-modified AAVs with corrected ND4 copies into the eyes of affected mice also produced vision improvements. Nonetheless, the limitations of AAV encompass a packaging size restricted to under 5 kb, thereby constraining the dimensions of the transgene that can be transported and administered. Nonetheless, there have been documented immune responses in patients exposed to AAV, necessitating significant caution in its clinical use. Despite the availability of numerous AAV serotypes, many have broad tropisms that can transduce nontarget tissues. Furthermore, modification of AAV capsid proteins facilitates targeted and accurate tissue transduction while reducing off-target effects.

Third, nonviral vectors bypass the significant issue of immunological responses elicited by viral vectors. Nanoparticles utilised for nonviral nucleic acid delivery typically manifest as lipoplexes (DNA/cationic lipids), polyplexes (DNA/cationic polymers) and lipopolyplexes (DNA/cationic polymers/cationic lipids) for the transfer of substantial plasmid DNA and diminutive DNA molecules, including oligodeoxynucleotides and RNAs. Inorganic nanoparticles may consist of silica or gold, while organic particles might be composed of lipid emulsions, lipid nanoparticles and other organic polymers. The delivery of these molecules encompasses physical techniques, including direct injection, electroporation, sonoporation and magnetofection, to facilitate gene transfer. The utilisation of nanoparticles for MGE requires additional investigation.

Opportunities and challenges of MGE: Currently, the difficulties in delivering gRNA into the mitochondria remain the primary obstacle to utilising CRISPR/Cas editing technology for mtDNA. It was predicted that the integration of modified gRNA with MTS-mediated shuttling of CRISPR/Cas9 proteins could enable precise editing of the mitochondrial genome. As demonstrated in a 2010 study, PNPASE regulates the import of nuclear-encoded RNAs into the mitochondrial matrix via a 20-nucleotide stem-loop mitochondrial targeting signal (RP loop), which enables the relocation of nonmitochondrial transcripts into mitochondria in a PNPASE-dependent manner. Practically, this RP signal has been used to import wild-type tRNAs efficiently into human mitochondria. A recent study showed that the mtDNA could be targeted with engineered CRISPR/Cas9 to assess the mutagenesis outcome of nuclease-generated DSBs in human cells. Site-specific DSBs were created via Staphylococcus aureus Cas9 fused with the MTS of COX8A in mtDNA, presenting novel potential for the therapy of mitochondrial disorders with

MGE. Moreover, attaching alternative signal sequences to gRNAs has recently been explored, offering the first proof that this is actually a viable technique to produce mito-CRISPRs. Consequently, it remains unexpectedly promising to observe additional advancements in the application of mtDNA editing technologies for human infertility treatments.

Furthermore, the creation of animal models is a crucial and essential method for investigating the mitochondrial genome. Currently, mouse models of mtDNA disease have been created either by introducing pathogenic mtDNA mutations into mitochondria of mouse embryos and subsequently incorporating them into embryonic stem cells for implantation into embryos, or by generating and selecting mtDNA mutations in mtDNA mutator mice. Innovative MGE technologies present the potential for a more accurate and effective approach to developing animal models with specific mtDNA abnormalities, hence providing optimism for the treatment of mitochondrial disorders. It has been demonstrated that the targeting of functional nucleases, such as restriction enzymes, ZFNs, TALENs, or CRISPR, into animal mitochondria is feasible, but with questionable efficacy. Nonetheless, the advancement of DdCBE has demonstrated an ideal editing effectiveness ranging from 40% to 49% for the intended base in human cell lines. Comparable methodologies have been utilised to facilitate ZFN-mediated methylation of mtDNA and to guide the paired FokI domains with mitochondrially targeted TALE or ZFN nucleases, thereby confining nuclease activity to designated mtDNA target sequences. Furthermore, the availability of intracellular mtDNA mutation generation, when combined with other model development systems, warrants investigation in the study of infertility diseases. It has been proposed that iPSCs or organoid systems could be integrated with gene editing technology, thereby obviating the need to acquire cells from patients with rare diseases and enabling the examination of mutant and control cells or organoids under uniform nuclear genetic backgrounds.

In addition, targeted gene editing of mtDNA can bring fresh insights for understanding the novel activities of mtDNA and for generating new therapeutics for human maternally heritable mitochondrial infertility or other infertility illnesses. However, it cannot be ignored that mtDNA editing is still only a promising option, and further studies and investigations are still required prior to implementing it in different cell types, organelles, and diseases as a new clinical therapy. Although due to the ethical reasons, current studies on the human embryo are rather limited, we still believe that the development as well as the amelioration of the above-mentioned MGE technology will lead to the improvement of oocyte quality and the better prevention or treatment of certain mitochondrial improcreance and finally bring about the overall enhancement of the human fertility. Mitochondria are not only responsible for ATP production but also are related to a wide range of essential biological processes, including biosynthesis, calcium storage, cellular signalling and immunological response. They are the sole semiautonomous organelles possessing a distinct mitochondrial genome that collaborates with the nuclear genome to regulate organelle function. In contrast to nuclear DNA, circular mtDNA is densely organised and strongly compacted, exhibiting a comparatively elevated mutation rate. The inheritance of mtDNA is exclusively maternal, resulting in a distinct 'bottleneck' effect that typically 'purifies' cellular heteroplasmy into homoplasmy.

Mitochondria are essential for gametogenesis and fertilisation; thus, mitochondrial dysfunction and harmful mtDNA can lead to detrimental outcomes or effects on sperm or oocyte quality, ultimately resulting in infertility. The swift advancement of gene editing technology has led to heightened interest in MGE and significant breakthroughs; thus, targeting mtDNA presents a novel avenue for addressing mito-related infertility. Nevertheless, owing to obstacles related to accessibility, efficacy, controllability and safety, the advancement of this field and the clinical implementation of this strategy remain significantly underdeveloped.

Reference

Jo, A., Ham, S., Lee, G. H., Lee, Y. I., Kim, S., Lee, Y. S., Shin, J. H., & Lee, Y. (2015). Efficient Mitochondrial Genome Editing by CRISPR/Cas9. *Biomed Res Int, 2015*, 305716. Epub 2015 Sep 10 https://doi.org/10.1155/2015/305716, PMID:26448933, PMCID:PMC4581504.

Further Reading

Alexeyev, M., Shokolenko, I., Wilson, G., & Ledoux, S. (2013). The maintenance of mitochondrial DNA integrity–critical analysis and update. *Cold Spring Harb Persp Biol, 5*, a012641.

Allen, J. F., Raven, J. A., Andersson, G. E., Karlberg, O., Canbäck, B., & Kurland, C. G. (2003). On the origin of mitochondria: a genomics perspective. *Philos Trans R Soc Lond, 358*(1429), 165–179.

Anderson, S., Bankier, A. T., Barrell, B. G., de Bruijn, M. H. L., Coulson, A. R., Drouin, J., et al. (1981). Sequence and organization of the human mitochondrial genome. *Nature, 290*(5806), 457–465.

Anzalone, A. V., Koblan, L. W., & Liu, D. R. (2020). Genome editing with CRISPR–Cas nucleases, base editors, transposases and prime editors. *Nat Biotechnol, 38*(7), 824–844. https://doi.org/10.1038/s41587-020-0561-9.

Araiso, Y., Tsutsumi, A., Qiu, J., Imai, K., Shiota, T., Song, J., et al. (2019). Structure of the mitochondrial import gate reveals distinct preprotein paths. *Nature, 575*(7782), 395–401.

Bacman, S. R., Kauppila, J. H. K., Pereira, C. V., Nissanka, N., Miranda, M., Pinto, M., et al. (2018). MitoTALEN reduces mutant mtDNA load and restores tRNAAla levels in a mouse model of heteroplasmic mtDNA mutation. *Nat Med, 24*(11), 1696–1700.

Bacman, S. R., Williams, S. L., Pinto, M., Peralta, S., & Moraes, C. T. (2013). Specific elimination of mutant mitochondrial genomes in patient-derived cells by mitoTALENs. *Nat Med, 19*(9), 1111–1113. https://doi.org/10.1038/nm.3261.

Bacman, S. R., Williams, S. L., Pinto, M., Peralta, S., & Moraes, C. T. (2013). Specific elimination of mutant mitochondrial genomes in patient-derived cells by mitoTALENs. *Nat Med* 13.

Barchiesi, A., & Vascotto, C. (2019). Transcription, processing, and decay of mitochondrial RNA in health and disease. *Int J Mol Sci, 20*(9), 2221.

Barshad, G., Marom, S., Cohen, T., & Mishmar, D. (2018). Mitochondrial DNA transcription and its regulation: an evolutionary perspective. *Trends Genet, 34*(9), 682–692.

Becker, T., Song, J., & Pfanner, N. (2019). Versatility of preprotein transfer from the cytosol to mitochondria. *Trends Cell Biol, 29*(7), 534–548.

Bian, W.-P., Chen, Y.-L., Luo, J.-J., Wang, C., Xie, S.-L., & Pei, D.-S. (2019). Knock-in strategy for editing human and zebrafish mitochondrial DNA using mito-CRISPR/Cas9 system. *ACS Synth Biol, 8*(4), 621–632.

Callegari, S., Cruz-Zaragoza, L. D., & Rehling, P. (2020). From TOM to the TIM23 complex– handing over of a precursor. *Biol Chem, 401*, 709–721. https://doi.org/10.1515/hsz-2020-0101.

Carelli, V., Carbonelli, M., Coo, I. F. D., Kawasaki, A., & Barboni, P. (2017). International consensus statement on the clinical and therapeutic management of leber hereditary optic neuropathy. *J Neuro-Ophthalmol, 37*, 371.

Chinnery, P. F., & Turnbull, D. M. (1999). Mitochondrial DNA and disease. *Lancet, 354*(SI), S17–S21.

DiMauro, S., & Davidzon, G. (2005). Mitochondrial DNA and disease. *Ann Med, 37*(3), 222–232.

DiMauro, S., & Schon, E. A. (2003). Mitochondrial respiratory-chain diseases. *N Engl J Med, 348*(26), 2656–2668.

Fu, L., Luo, Y.-X., Liu, Y., Liu, H., H-z, Li, & Yu, Y. (2021). Potential of MitochondrialGenomeEditing for Human Fertility Health. *Front Genet, 12*, 673951. https://doi.org/10.3389/fgene.2021.673951.

Gaj, T., Gersbach, C. A., & Barbas, C. F. (2013). ZFN, TALEN, and CRISPR/Cas-based methods for genome engineering. *Trends Biotechnol, 31*(7), 397–405.

Gammage P.A., Moraes C.T., Minczuk M. Mitochondrial genome engineering: the revolution may not Be CRISPR-Ized. (2018).

Gammage, P. A., Rorbach, J., Vincent, A. I., Rebar, E. J., & Minczuk, M. (2014). Mitochondrially targeted ZFNs for selective degradation of pathogenic mitochondrial genomes bearing large-scale deletions or point mutations. *EMBO Mol Med, 6*, 458–466.

Gammage, P. A., Viscomi, C., Simard, M.-L., Costa, A. S. H., Gaude, E., Powell, C. A., et al. (2018). Genome editing in mitochondria corrects a pathogenic mtDNA mutation in vivo. *Nat Med, 24*(11), 1691–1695.

Gammage P.A. et al. Near-complete elimination of mutant mtDNA by iterative or dynamic dose-controlled treatment with mtZFNs. (2016).

Ghelli, A., Zanna, C., Porcelli, A. M., Schapira, A. H. V., Martinuzzi, A., Carelli, V., et al. (2003). Leber's hereditary optic neuropathy (LHON) pathogenic mutations induce mitochondrial-dependent apoptotic death in transmitochondrial cells incubated with galactose medium. *J Biol Chem, 278*(6), 4145–4150.

Greaves, L. C., Reeve, A. K., Taylor, R. W., & Turnbull, D. M. (2012). Mitochondrial DNA and disease. *J Pathol, 226*, 274–286.

Haddad, S., Abánades Lázaro, I., Fantham, M., Mishra, A., Silvestre-Albero, J., Osterrieth, J. W. M., et al. (2020). Design of a functionalized metal-organic framework system for enhanced targeted delivery to mitochondria. *J Am Chem Soc, 142*(14), 6661–6674.

Haihua, et al. (2018). SaCas9 requires 5′-NNGRRT-3′ PAM for sufficient cleavage and possesses higher cleavage activity than SpCas9 or FnCpf1 in human cells. *Biotechnol J.*

Hansen, K. G., & Herrmann, J. M. (2019). Transport of proteins into mitochondria. *Protein J, 38*(3), 330–342.

Hashimoto, M., Bacman, S. R., Peralta, S., Falk, M. J., Chomyn, A., Chan, D. C., et al. (2015). MitoTALEN: a general approach to reduce mutant mtDNA loads and restore oxidative phosphorylation function in mitochondrial diseases. *Mol Ther J Am Soc Gene Ther, 23*(10), 1592–1599.

Hussain, S. A., Yalvac, M. E., Khoo, B., Eckardt, S., & McLaughlin, K. J. (2021). Adapting CRISPR/Cas9 System for Targeting Mitochondrial Genome. *Front Genet, 12*, 627050. https://doi.org/10.3389/fgene.2021.627050.

Kazama, T., Okuno, M., Watari, Y., Yanase, S., Koizuka, C., Tsuruta Yu, et al. (2019). Curing cytoplasmic male sterility via TALEN-mediated mitochondrial genome editing. *Nat Plants, 5*(7), 722–730.

Kellner, M. J., Koob, J. G., Gootenberg, J. S., Abudayyeh, O. O., & Zhang, F. (2019). SHERLOCK: nucleic acid detection with CRISPR nucleases. *Nat Protoc, 14*(10), 2986–3012.

Kytvuori, L., et al. (2016). A novel mutation m.8561C>G in MT-ATP6/8 causing a mitochondrial syndrome with ataxia, peripheral neuropathy, diabetes mellitus, and hypergonadotropic hypogonadism. *J Neurol, 263*, 2188–2195.

Lee, H., Lee, S., Baek, G., et al. (2021). Mitochondrial DNA editing in mice with DddA-TALE fusion deaminases. *Nat Commun, 12*, 1190. https://doi.org/10.1038/s41467-021-21464-1.

Lei, C., et al. (2017). The CCTL (Cpf1-assisted Cutting and Taq DNA ligase-assisted Ligation) method for efficient editing of large DNA constructs in vitro. *Nucleic Acids Res*, e74.

Leslie, M. (2018). ,'Old' genome editors might treat mitochondrial diseases. *Science, 361*, 1302. 10.1126/science.361.6409.1302.

Loutre, R., Heckel, A. M., Smirnova, A., Entelis, N., & Tarassov, I. (2018). Can mitochondrial DNA be CRISPRized: pro and contra: can mitochondrial DNA be crisprized. *Int Union Biochem Mol Biol Life, 70.*

McCully, J. D., Cowan, D. B., Emani, S. M., & del Nido, P. J. (2017). Mitochondrial transplantation: From animal models to clinical use in humans. *Mitochondrion, 34*, 127–134.

Minczuk M., Papworth M.A., Kolasinska P., Murphy M.P., Klug, A. Sequence specific modification of mitochondrial DNA using a chimeric zinc finger methylase.

Mok, B. Y., de Moraes, M. H., Zeng, J., Bosch, D. E., Kotrys, A. V., Raguram, A., et al. (2020). A bacterial cytidine deaminase toxin enables CRISPR-free mitochondrial base editing. *Nature, 583*(7817), 631–637. https://doi.org/10.1038/s41586-020-2477-4.

Mordel, P., et al. (2017). A 2 bp deletion in the mitochondrial ATP 6 gene responsible for the NARP (neuropathy, ataxia, and retinitis pigmentosa) syndrome. *Biochem Biophys Res Commun*S0006291X17320442.

Neupert, W., & Herrmann, J. M. (2007). Translocation of proteins into mitochondria. *Annu Rev Biochem, 76*(1), 723–749.

Pelnena et al. Complete mtDNA sequencing reveals mutations m.9185T > C and m.13513G > A in three patients with Leigh syndrome.

Pereira, C. V., et al. (2018). mitoTev-TALE: a monomeric DNA editing enzyme to reduce mutant mitochondrial DNA levels. *EMBO Mol Med, 10*, e8084.

Reddy, P. et al. Selective Elimination of Mitochondrial Mutations in the Germline by Genome Editing. (2015).

Schapira, A. H. V. (2012). Mitochondrial diseases. *Lancet, 379*(9828), 1825–1834.

Schmidt, S., Strub, A., & Voos, W. (2001). Protein translocation into mitochondria. *Biol Signals Receptors, 10*(1-2), 14–25.

Smirnov, A., Tarassov, I., Mager-Heckel, A.-M., Letzelter, M., Martin, R. P., Krasheninnikov, I. A., et al. (2008). Two distinct structural elements of 5S rRNA are needed for its import into human mitochondria. *RNA (N Y), 14*(4), 749–759.

Sofou, K., de Coo, I. F. M., Ostergaard, E., Isohanni, P., Naess, K., De Meirleir, L., et al. (2018). Phenotype-genotype correlations in leigh syndrome: New insights from a multicentre study of 96 patients. *J Med Genet, 55*(1), 21–27.

Suomalainen, A., & Battersby, B. J. (2018). Mitochondrial diseases: the contribution of organelle stress responses to pathology. *Nat Rev Mol Cell Biol, 19*(2), 77–92.

Tanaka, M., et al. (2002). Gene therapy for mitochondrial disease by delivering restriction endonuclease SmaI into mitochondria. *J Biomed Sci, 9*, 534–541. https://doi.org/10.1159/000064726.

Taylor, R. W., & Turnbull, D. M. (2005). Mitochondrial DNA mutations in human disease. *Nat Rev Genet, 6*(5), 389–402.

Tonin, Y., Heckel, A.-M., Vysokikh, M., Dovydenko, I., Meschaninova, M., Rötig, A., et al. (2014). Modeling of antigenomic therapy of mitochondrial diseases by mitochondrially addressed RNA targeting a pathogenic point mutation in mitochondrial DNA. *J Biol Chem, 289*(19), 13323–13334.

Torchilin, V. P. (2006). Recent approaches to intracellular delivery of drugs and DNA and organelle targeting. *Annu Rev Biomed Eng, 8*(1), 343–375.

Tucker, K., & Park, E. (2019). Cryo-EM structure of the mitochondrial protein-import channel TOM complex at near-atomic resolution. *Nat Struct Mol Biol*, 1–9.

Wallace, D., Singh, G., Lott, M., Hodge, J., Schurr, T., Lezza, A., et al. (1988). Mitochondrial DNA mutation associated with Leber's hereditary optic neuropathy. *Science, 242*(4884), 1427–1430.

Wallace, D. C. (1999). Mitochondrial diseases in man and mouse. *Science*.

Wang, G., Chen, H.-W., Oktay, Y., Zhang, J., Allen, E. L., Smith, G. M., et al. (2010). PNPASE regulates RNA import into mitochondria. *Cell, 142*(3), 456–467.

Wang, G., Shimada, E., Zhang, J., Hong, J. S., Smith, G. M., Teitell, M. A., Koehler, C. M., et al. (2012). Correcting human mitochondrial mutations with targeted RNA import. *Proc Natl Acad Ences, 109*.

Yahata, N., Matsumoto, Y., Omi, M., Yamamoto, N., & Hata, R. (2017). TALEN-mediated shift of mitochondrial DNA heteroplasmy in MELAS-iPSCs with m.13513G>A mutation. *Sci Rep, 7*, 15557. https://doi.org/10.1038/s41598-017-15871-y.

Yamada, Y., Ito, M., Arai, M., Hibino, M., Tsujioka, T., & Harashima, H. (2020). Challenges in promoting mitochondrial transplantation therapy. *Int J Mol Sci, 21*(17), 6365. https://doi.org/10.3390/ijms21176365.

Yanping, et al. (2018). Mitochondrial DNA mutations in late-onset Leigh syndrome. *J Neurol*.

Yi, Y., Wu, H., Kang, X., Liang, Y., Lan, T., Li, T., et al. (2018). Targeted elimination of mutant mitochondrial DNA in MELAS-iPSCs by mitoTALENs. *Protein Cell, 9*(3), 283–297.

Yoo, B. C., Yadav, N. S., Orozco, E. M., & Sakai, H. (2020). Cas9/gRNA-mediated genome editing of yeast mitochondria and Chlamydomonas chloroplasts. *PeerJ, 8*.

Yu, H., Koilkonda, R. D., Chou, T.-H., Porciatti, V., Ozdemir, S. S., Chiodo, V., et al. (2012). Gene delivery to mitochondria by targeting modified adenoassociated virus suppresses Leber's hereditary optic neuropathy in a mouse model. *Proc Natl Acad Sci U S A, 109*(20), E1238–E1247.

Yuan, P., Mao, X., Wu, X., Liew, S. S., Li, L., & Yao, S. Q. (2019). Mitochondria-targeting, intracellular delivery of native proteins using biodegradable silica nanoparticles. *Angew Chem Int Ed, 131*(23), 7739–7743.

Zetsche, B., Gootenberg, J., Abudayyeh, O., Slaymaker, I., Makarova, K., Essletzbichler, P., et al. (2015). Cpf1 Is a Single RNA-Guided Endonuclease of a Class 2 CRISPR-Cas System. *Cell, 163*(3), 759–771.

Zetsche, B., Heidenreich, M., Mohanraju, P., Fedorova, I., Kneppers, J., DeGennaro, E. M., et al. (2017). Multiplex gene editing by CRISPR–Cpf1 using a single crRNA array. *Nat Biotechnol, 35*(1), 31–34.

Zhong, G., Wang, H., Li, Y., Tran, M. H., & Farzan, M. (2017). Cpf1 proteins excise CRISPR RNAs from mRNA transcripts in mammalian cells. *Nat Chem Biol, 13*(8), 839–841.

Index

Note: Page numbers followed by "*f*" and "*t*" refer to figures and tables, respectively.

D

E

H

I

Q

R

S

T

U

Printed and bound by CPI Group (UK) Ltd, Croydon, CR0 4YY
14/09/2026
02216410-0015